VALORIZATION OF MICROALGAL BIOMASS AND WASTEWATER TREATMENT

VALORIZATION OF MICROALGAL BIOMASS AND WASTEWATER TREATMENT

Edited by

SUHAIB A. BANDH

Assistant Professor, Environmental Science, Higher Education Department, Government of Jammu and Kashmir, Srinagar, India

FAYAZ A. MALLA

Assistant Professor, Department of Environmental Science, Govt. Degree College Tral, Jammu and Kashmir, India

ELSEVIER

Elsevier
Radarweg 29, PO Box 211, 1000 AE Amsterdam, Netherlands
The Boulevard, Langford Lane, Kidlington, Oxford OX5 1GB, United Kingdom
50 Hampshire Street, 5th Floor, Cambridge, MA 02139, United States

ISBN: 978-0-323-91869-5

For Information on all Elsevier publications
visit our website at https://www.elsevier.com/books-and-journals

Publisher: Candice G. Janco
Acquisitions Editor: Maria Elekidou
Editorial Project Manager: Sara Valentino
Production Project Manager: Kumar Anbazhagan
Cover Designer: Christian J. Bilbow

Typeset by MPS Limited, Chennai, India

Contents

8. Life cycle assessment of wastewater treatment by microalgae 137

Christy B.K. Sangma and Rokozeno Chalie-u

9. Life cycle assessment of microalgal biomass for valorization 179

Maria Lúcia Calijuri, Iara Barbosa Magalhães, Jessica Ferreira, Jackeline de Siqueira Castro and Bianca Barros Marangon

10. Biorefinery and bioremediation potential of microalgae 197

Eleni Koutra, Sameh Samir Ali, Myrsini Sakarika and Michael Kornaros

11. Recent developments and challenges: a prospectus of microalgal biomass valorization 219

Maria Lúcia Calijuri, Paula Assemany, Eduardo Couto, Adriana Paulo de Sousa Oliveira, Juliana F. Lorentz and Letícia Rodrigues de Assis

12. Nonconventional treatments of agro-industrial wastes and wastewaters by heterotrophic/mixotrophic cultivations of microalgae and Cyanobacteria 239

Rihab Hachicha, Fatma Elleuch, Hajer Ben Hlima, Pascal Dubessay, Helene de Baynast, Cedric Delattre, Guillaume Pierre, Ridha Hachicha, Slim Abdelkafi, Imen Fendri and Philippe Michaud

13. Ecological and environmental services of microalgae 261

Archita Sharma and Shailendra Kumar Arya

14. Valorization of microalgae for biogas methane enhancement 317

Fayaz A. Malla, Nazir Ahmad Sofi, Navindu Gupta and Suhaib A. Bandh

List of contributors

Slim Abdelkafi Laboratoire de Génie Enzymatique et Microbiologie, Equipe de Biotechnologie des Algues, Ecole Nationale d'Ingénieurs de Sfax, Université de Sfax, Sfax, Tunisia

Sameh Samir Ali Biofuels Institute, School of the Environment and Safety Engineering, Jiangsu University, Zhenjiang, P.R. China; Botany Department, Faculty of Science, Tanta University, Tanta, Egypt

Suhaib Al-Maawali Department of Petroleum and Chemical Engineering, College of Engineering, Sultan Qaboos University, Muscat, Oman

Ala'a H. Al-Muhtaseb Department of Petroleum and Chemical Engineering, College of Engineering, Sultan Qaboos University, Muscat, Oman

Aashia Altaf Sri Pratap College Campus, Cluster University Srinagar, Srinagar, Jammu and Kashmir, India

Shailendra Kumar Arya Department of Biotechnology, University Institute of Engineering and Technology (UIET), Panjab University (PU), Chandigarh, India

Paula Assemany Department of Environmental Engineering, Campus Universitário, Federal University of Lavras (Universidade Federal de Lavras/UFLA), Lavras, Minas Gerais, Brazil

Letícia Rodrigues de Assis Post-Graduate Program in Civil Engineering, Department of Civil Engineering, Campus Universitário, Federal University of Viçosa (Universidade Federal de Viçosa/UFV), Viçosa, Minas Gerais, Brazil

Suhaib A. Bandh Assistant Professor, Environmental Science, Higher Education Department, Government of Jammu and Kashmir, Srinagar, India

Suhail Bashir Al Noor Environment Consultants, Sharjah, United Arab Emirates

Zahid Bashir Sri Pratap College Campus, Cluster University Srinagar, Srinagar, Jammu and Kashmir, India

Hajer Ben Hlima Laboratoire de Génie Enzymatique et Microbiologie, Equipe de Biotechnologie des Algues, Ecole Nationale d'Ingénieurs de Sfax, Université de Sfax, Sfax, Tunisia

Maria Lúcia Calijuri Post-Graduate Program in Civil Engineering, Department of Civil Engineering, Campus Universitário, Federal University of Viçosa (Universidade Federal de Viçosa/UFV), Viçosa, Minas Gerais, Brazil

Jackeline de Siqueira Castro Post-Graduate Program in Civil Engineering, Department of Civil Engineering, Campus Universitário, Federal University of Viçosa (Universidade Federal de Viçosa/UFV), Viçosa, Minas Gerais, Brazil

Rokozeno Chalie-u ICAR Research Complex for NEH Region, Nagaland Centre, Dimapur, Nagaland, India

Eduardo Couto Institute of Applied and Pure Sciences, Rua Irmã Ivone Drumond, Federal University of Itajubá, Campus Itabira (Universidade Federal de Itajubá, Campus Itabira/Unifei), Itabira, Minas Gerais, Brazil

Achintya Das Department of Physics, Mahadevananda Mahavidyalaya, Barrackpore, West Bengal, India

Helene de Baynast Université Clermont Auvergne, CNRS, SIGMA Clermont, Institut Pascal, Clermont-Ferrand, France

Cedric Delattre Université Clermont Auvergne, CNRS, SIGMA Clermont, Institut Pascal, Clermont-Ferrand, France; Institut Universitaire de France (IUF), 1 rue Descartes Paris, France

Pascal Dubessay Université Clermont Auvergne, CNRS, SIGMA Clermont, Institut Pascal, Clermont-Ferrand, France

Fatma Elleuch Laboratoire de Génie Enzymatique et Microbiologie, Equipe de Biotechnologie des Algues, Ecole Nationale d'Ingénieurs de Sfax, Université de Sfax, Sfax, Tunisia

Imen Fendri Laboratoroire de Biotechnologies Végétales Appliquées à l'Amélioration des Cultures, Faculté des Sciences de Sfax, Université de Sfax, Sfax, Tunisia

Jessica Ferreira Post-Graduate Program in Civil Engineering, Department of Civil Engineering, Campus Universitário, Federal University of Viçosa (Universidade Federal de Viçosa/UFV), Viçosa, Minas Gerais, Brazil

Nédia de Castilhos Ghisi Department of Bioprocess Engineering and Biotechnology, Federal University of Technology of Paraná - Dois Vizinhos, Paraná, Brazil

Navindu Gupta Center for Environment science and climate-resilient agriculture (CESRA), Indian Agricultural Research Institute, Delhi, New Delhi, India

Mariliz Gutterres Laboratory for Leather and Environmental Studies (LACOURO), Porto Alegre, Brazil

Ridha Hachicha Laboratoire de Génie Enzymatique et Microbiologie, Equipe de Biotechnologie des Algues, Ecole Nationale d'Ingénieurs de Sfax, Université de Sfax, Sfax, Tunisia

Rihab Hachicha Laboratoroire de Biotechnologies Végétales Appliquées à l'Amélioration des Cultures, Faculté des Sciences de Sfax, Université de Sfax, Sfax, Tunisia; Université Clermont Auvergne, CNRS, SIGMA Clermont, Institut Pascal, Clermont-Ferrand, France

Farrukh Jamil Department of Chemical Engineering, COMSATS University Islamabad (CUI), Lahore, Pakistan

Umarin Jomnonkhaow Department of Biotechnology, Faculty of Technology, Khon Kaen University, Khon Kaen, Thailand; Research Group for Development of Microbial Hydrogen Production Process from Biomass, Khon Kaen University, Khon Kaen, Thailand

Michael Kornaros Laboratory of Biochemical Engineering and Environmental Technology (LBEET), Department of Chemical Engineering, University of Patras, Patras, Greece

Eleni Koutra Laboratory of Biochemical Engineering and Environmental Technology (LBEET), Department of Chemical Engineering, University of Patras, Patras, Greece

Juliana F. Lorentz Post-Graduate Program in Civil Engineering, Department of Civil Engineering, Campus Universitário, Federal University of Viçosa (Universidade Federal de Viçosa/UFV), Viçosa, Minas Gerais, Brazil

Iara Barbosa Magalhães Post-Graduate Program in Civil Engineering, Department of Civil Engineering, Campus Universitário, Federal University of Viçosa (Universidade Federal de Viçosa/UFV), Viçosa, Minas Gerais, Brazil

Fayaz A. Malla Assistant Professor, Department of Environmental Science, Govt. Degree College Tral, Jammu and Kashmir, India

Bianca Barros Marangon Post-Graduate Program in Civil Engineering, Department of Civil Engineering, Campus Universitário, Federal University of Viçosa (Universidade Federal de Viçosa/UFV), Viçosa, Minas Gerais, Brazil

Philippe Michaud Université Clermont Auvergne, CNRS, SIGMA Clermont, Institut Pascal, Clermont-Ferrand, France

Adriana Paulo de Sousa Oliveira Post-Graduate Program in Civil Engineering, Department of Civil Engineering, Campus Universitário, Federal University of Viçosa (Universidade Federal de Viçosa/UFV), Viçosa, Minas Gerais, Brazil

Aline de C.C. Pena Laboratory for Leather and Environmental Studies (LACOURO), Porto Alegre, Brazil; Group of Intensification, Modeling, Simulation, Control, and Optimization of Process (GIMSCOP), Porto Alegre, Brazil

Guillaume Pierre Université Clermont Auvergne, CNRS, SIGMA Clermont, Institut Pascal, Clermont-Ferrand, France

Irteza Qayoom Sri Pratap College Campus, Cluster University Srinagar, Srinagar, Jammu and Kashmir, India

Alissara Reungsang Department of Biotechnology, Faculty of Technology, Khon Kaen University, Khon Kaen, Thailand; Research Group for Development of Microbial Hydrogen Production Process from Biomass, Khon Kaen University, Khon Kaen, Thailand; Academy of Science, Royal Society of Thailand, Bangkok, Thailand

Ananya Roy Chowdhury Department of Botany, Chakdaha College, Chakdaha, Nadia, India

Myrsini Sakarika Laboratory of Biochemical Engineering and Environmental Technology (LBEET), Department of Chemical Engineering, University of Patras, Patras, Greece; Center for Microbial Ecology and Technology (CMET), Ghent University, Gent, Belgium

Christy B.K. Sangma ICAR Research Complex for NEH Region, Nagaland Centre, Dimapur, Nagaland, India

Asma Sarwer Department of Chemical Engineering, COMSATS University Islamabad (CUI), Lahore, Pakistan

Archita Sharma Department of Biotechnology, University Institute of Engineering and Technology (UIET), Panjab University (PU), Chandigarh, India

Ingrid Fernanda Silvano Pacheco Correa Furtado Department of Bioprocess Engineering and Biotechnology, Federal University of Technology of Paraná - Ponta Grossa, Paraná, Brazil

Sureewan Sittijunda Faculty of Environment and Resource Studies, Mahidol University, Nakhon Pathom, Thailand

Nazir Ahmad Sofi Department of Agriculture Research Information System, Sher-e-Kashmir University of Agricultural Sciences and Technology, Shalimar Campus, Srinagar, Jammu and Kashmir, India

Rhaianny Malucelli Stahlschmidt Department of Bioprocess Engineering and Biotechnology, Federal University of Technology of Paraná - Ponta Grossa, Paraná, Brazil

Alessandra Cristine Novak Sydney Department of Bioprocess Engineering and Biotechnology, Federal University of Technology of Paraná - Ponta Grossa, Paraná, Brazil

Eduardo Bittencourt Sydney Department of Bioprocess Engineering and Biotechnology, Federal University of Technology of Paraná - Ponta Grossa, Paraná, Brazil

Luciane F. Trierweiler Group of Intensification, Modeling, Simulation, Control, and Optimization of Process (GIMSCOP), Porto Alegre, Brazil

Konstantina Tsigkou Laboratory of Biochemical Engineering and Environmental Technology (LBEET), Department of Chemical Engineering, University of Patras, Patras, Greece

Marina Wust Vasconcelos Department of Bioprocess Engineering and Biotechnology, Federal University of Technology of Paraná - Dois Vizinhos, Paraná, Brazil

Dimitris P. Zagklis Laboratory of Biochemical Engineering and Environmental Technology (LBEET), Department of Chemical Engineering, University of Patras, Patras, Greece

About the editors

Dr. Suhaib A. Bandh is an assistant professor in the Department of Higher Education, Government of Jammu and Kashmir. Dr. Bandh is the president and founder of Academy of EcoScience besides being a life member of the Academy of Plant Sciences India and National Environmental Science Academy, India. Dr. Bandh, a recipient of many awards, has several scientific publications in some highly reputed and impacted journals to his credit, which attest to his scientific insight, fine experimental skills, and outstanding writing skills. Dr. Bandh has edited and authored many books with some leading scholarly publishing houses including Springer Nature, Elsevier Inc. USA, Callisto References, and AAP/CRC, A Taylor & Francis Group. Dr. Bandh, the managing editor of Micro Environer (https://microenvironer.com/editorial-board/), is an academic editor of the *Journal Advances in Agriculture* and *International Journal of Clinical Practices* published by Hindawi.

Dr. Fayaz A. Malla finished his PhD in environmental science on "Biogas methane enrichment using selective chemical scavengers and microalgae" from the Indian Agricultural Research Institute, New Delhi. He has a strong interest in waste management, pollution, and renewable energy as his PhD was focused on the same issue. Moreover, in the university, he has also dealt with water conservation and management, water auditing, and water use efficiency in agriculture. He has published many research articles in reputed, referred national, and international journals. He has also presented many research papers at national and international conferences. He has worked as a research associate with the Water Resources Division in The Energy and Research Institute (TERI) and is currently working as an assistant professor in the Higher Education Department, Government of Jammu and Kashmir.

Scientometric analysis of microalgae wastewater treatment

Ingrid Fernanda Silvano Pacheco Correa Furtado[1],,
Marina Wust Vasconcelos[2],*,
Rhaianny Malucelli Stahlschmidt[1],*,
Alessandra Cristine Novak Sydney[1],
Nédia de Castilhos Ghisi[2] and Eduardo Bittencourt Sydney[1]*

[1]Department of Bioprocess Engineering and Biotechnology, Federal University of Technology of Paraná - Ponta Grossa, Paraná, Brazil [2]Department of Bioprocess Engineering and Biotechnology, Federal University of Technology of Paraná - Dois Vizinhos, Paraná, Brazil

1.1 Introduction

Freshwater is an essential element for all living beings, and it accounts for the proper functioning of the entire terrestrial ecosystem (Arora et al., 2021). This resource is essential for the economic and social development experienced over the years. However, unlike what was once assumed, it is a finite source. According to estimates, there will be an increase by 25% in water consumption within the next 9 years due to the exacerbated and nonsustainable use of this input by industries, as well as due to groundwater contamination issues, a fact that can likely lead to the water crisis in the future (Sun et al., 2016; UNESCO, 2021).

For many decades, man believed that nature was an infinite resource, the reason why the most technological and industrial advances led to increasing amounts of wastewater being improperly disposed of in water bodies and causing environmental impacts such as eutrophication and animal deaths (Al-Jabri et al., 2021; Mohsenpour et al., 2021; (Wen et al., 2017). Environmental concern is a recent issue, thus great advances achieved so far mainly resulted from policies to meet the need to treat waste. Methodologies

---

* Authors have equally contributed to the current study.

Valorisation of Microalgal Biomass and Wastewater Treatment
DOI: https://doi.org/10.1016/B978-0-323-91869-5.00010-7

conventionally used for wastewater treatment are based on physical, chemical, and biological techniques. However, these traditional technologies cannot remove recalcitrant materials such as pharmaceutical compounds and heavy metals. Moreover, they are mostly incapable of removing persistent nutrients such as nitrogen and phosphorus and even generating undesirable gaseous components that end up reaching the atmosphere (Al-Jabri et al., 2021; Priya, 2014; Xiong et al., 2021). For many years, wastewater treatment was seen as a legal requirement for industrial operations. However, it demands investments, space, time, and labor, consequently decreasing the short-term profit. The development of technologies capable of promoting rational wastewater reuse to produce (new) bioproducts has gained global attention due to advances in understanding benefits deriving from sustainable methodologies.

Microalgae are photosynthetic microorganisms that can be used to reduce biochemical oxygen demand (BOD) and chemical oxygen demand (COD) and remove recalcitrant and inorganic materials such as nitrates and phosphates. Thus they help improve the physico-chemical features of the final wastes such as turbidity, color, oxygenation content, among others (Al-Jabri et al., 2021; Arora et al., 2021). The fact that microalgae grow in the absence of organic carbon makes their cultivation easily exploitable as an additional step to treat some wastewater types. Special attention must also be given to microalgae with mixotrophic metabolism due to their ability to use organic and inorganic carbon for heterotrophic and autotrophic growth, respectively.

Microalgal biomass produced during nutrient removal from wastewater has biotechnological interest due to its composition rich in proteins, lipids, carbohydrates, pigments, and active biomolecules, favoring the circular economy concept and leading to organic waste valorization (Al-Jabri et al., 2021). Significant scientific knowledge is produced and published weekly, making it hard to keep up with all the available knowledge. Thus scientometric tools can help to compile and better visualize trends and gaps in the literature about a given topic to understand the state-of-the-art of a given field better and anticipate future perspectives to guide efforts on technological development. The current chapter presents the scientometric analysis of the use of microalgae to treat effluents in the last 5 years.

1.2 Methodology

The Web of Science's (WoS) core collection database from Clarivate Analytics was used to develop the current scientometric review. WoS is considered the most robust and complete database among the available ones since it covers more than 34,000 journals comprising the main studies developed worldwide (Birkle et al., 2020). The scientific literature was searched by limiting the publication period to the last 5 years (2017−21) to highlight the latest trend of using microalgae in treating wastewater.

The search was carried out on March 31, 2021. Boolean operators such as AND, OR, and NOT were used as a strategy to search for the keywords. They were used to limit the search to terms based on the use of microalgae to treat wastewater and remove nutrients and other pollutants from it. Operator "NOT" was used to exclude treatments based on using bacteria, consortium, or more than one microalgal genera, whereas operators

"AND" and "OR" were used to add terms or to search for other keywords, respectively. Thus each of the following words was explored separately to carry out the research: treatment AND microalgae, microalgae AND removal NOT algae-bacteria, microalgae AND removal NOT consortium, microalgae AND sewage, microalgae AND nutrients removal, microalgae AND waste OR microalgae AND wastewater, microalgae AND bioremediation OR microalgae AND remediation, and microalgae AND microalgae AND valorization OR microalgae AND valorization.

These terms were searched in titles, abstracts, keywords, and keywords plus in all documents. Studies that only used microalgae belonging to a single genus and species to treat wastewater, such as domestic or industrial sewage and water contaminated with toxic pollutants or with any other material of the same kind, were the selected ones. Furthermore, the articles' inclusion process comprised two different stages: the first stage focused on selecting studies based on reading titles and abstracts; this stage enabled finding 1579 manuscripts in total. Subsequently, the second stage comprised the full reading of the entire selected collection (1579 documents) to ensure that all manuscripts addressed the objective mentioned above: 891 articles were selected, in total, 863 of them were derived from the primary WoS collection. Fig. 1.1 briefly depicts the search process mentioned above.

The final refinement resulted in 863 articles from WoS database, which were exported to CiteSpace software since it was one of the predefined search criteria. Some WoS data were exported to Excel and Statistic software (version 10) to help better exploring the scientific trends. CiteSpace is a free Java application aimed at illustrating trends and changes

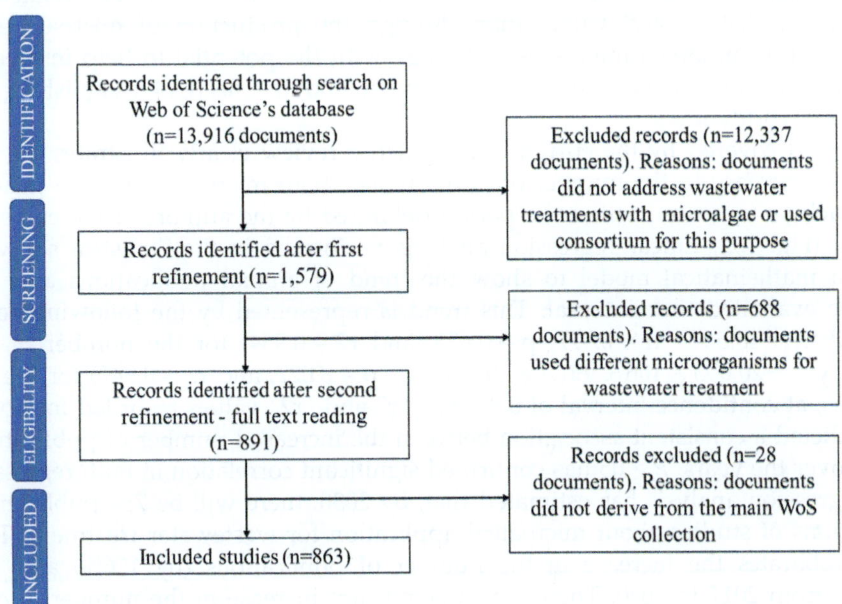

FIGURE 1.1 Summary of data search steps. Summary of data search steps conducted in WoS to find documents included in the scientometric analysis.

in progress in scientifically based subjects (Chen, 2015). Infographics showing the main countries publishing studies about the investigated topic was generated. It shows the top 10 manuscripts, the prominent journals, and their respective categories, as well as the main terms cited for microalgae used to treat wastewater.

1.3 Results and discussion

1.3.1 Analysis about the evolution of scientific production concerning microalgal use to treat wastewater

The final set comprised 863 studies published in WoS database from 2017 to 2021; it is considered significant to investigate features and characteristics of recent research on microalgae quantitatively. Among them, 99% were written in English, and it showed the prevalence of this language in studies conducted by the scientific community, regardless of the publication country (Cheng et al., 2020). Publications were classified into eight document types: articles, proceedings paper, reviews, early access, meeting abstracts, correction, editorial material, and reprint articles, which accounted for 809 (94%) of them.

The whole dataset received 7485 citations, 8.67 citations per item, on average. H-index equal to 33 was observed. It evidenced good scientific performance, mainly considering the investigated short-time interval since the H-index provides estimates on the quantitative influence of research's cumulative contributions . The importance given by the scientific community to wastewater treatments with microalgae is closely related to the understanding that wastes' valorization through the production of microalgae-derived products with commercial interest is a strategy with the potential to help in meeting the increasing demands for sustainable development and the need of establishing circular bioeconomy worldwide.

The period covered by the current scientometric review clearly illustrates the trend of consecutive increase in the number of publications about microalgae for the treatment of wastes; such an increase justifies the period delimited by the authors of the current study (2017−21) (Fig. 1.2). Linear regression analysis performed in the Statistica software was used as a mathematical model to show the trend of annual publications and citations within the evaluated time interval. This trend is represented by the following equations: $y = -97{,}904.1 + 48.6 \times x$ ($r = 0.977$, $p = 0.023$, and $r2 = 0.954$) for the number of publications and $y = -20{,}005 \times 106 + 991.7 \times$ ($r = 0.973$, $p = 0.027$, and $r2 = 0.947$) for the number of citations, at confidence interval of 0.95. The "r" and "r2" values recorded for both equations predicted a consistent association between the increased number of publications and citations over the years. $P > .05$ has confirmed significant correlation in both regressions.

The regression analysis has estimated that, by 2030, there will be 754 publications and 8151 citations of studies about microalgal application for wastewater treatment. This estimate corroborates the increase in the number of publications (by 106%) and citations (by 401%) from 2017 to 2020. There was a significant increase in the number of citations, which led to increased interest in the topic. Publications and citations observed up to March 2021 were not taken into consideration in the regression analyses.

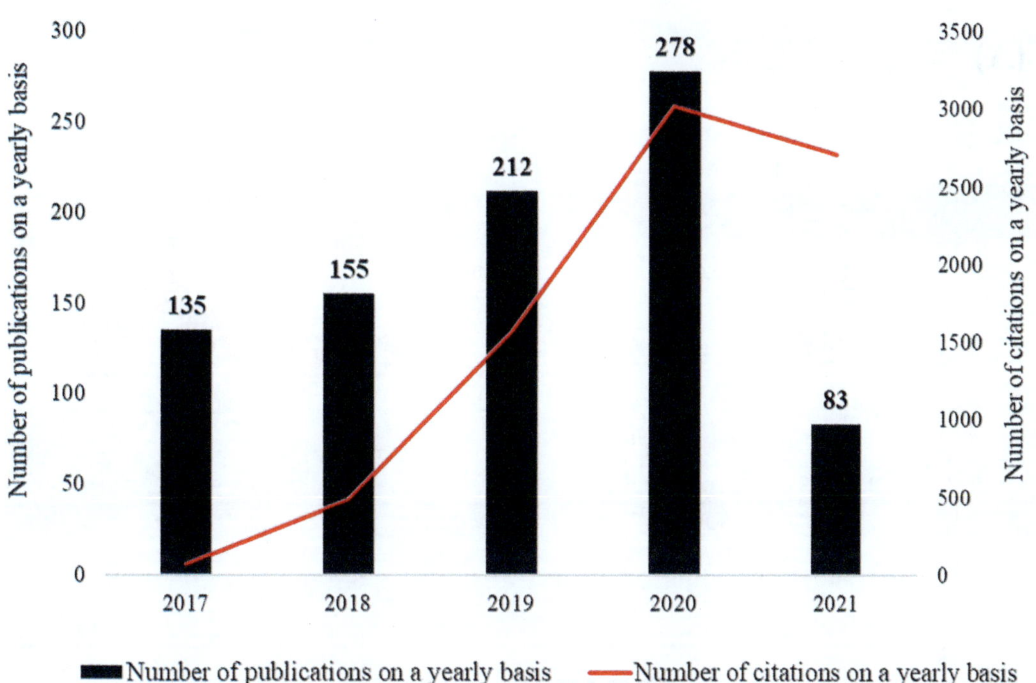

Number of publications on a yearly basis — Number of citations on a yearly basis

FIGURE 1.2 Comparison between the number of publications and annual citation. Comparison between the number of publications and annual citations, and consecutive increase in the number of publications about wastewater treatment with microalgae.

1.3.2 Analysis of the geographic distribution and contribution of studies about microalgal use to treat wastewater

The geographic distribution of publications about microalgal use to treat wastewater is an extremely important factor. It enables visualizing the countries standing out in research and technological development about this topic (Fig. 1.3). Fig. 1.3A refers to the geolocation of published studies deriving from countries presenting the highest rate of publications about the investigated topic. China was the most prominent country since it accounted for 28.74% of publications, followed by India (11.01%), the United States (8.57%), Brazil (7.53%), Malaysia (6.60%), and South Korea (5.68%). It is noteworthy that the first three countries are the most populous globally (United States Census, 2021). Consequently, they are most concerned about and interested in finding efficient methodologies to treat the generated wastes.

The use of microalgae for wastewater treatment does not generate waste products, unlike conventional methodologies, such as the ones based on activated sludge. These methodologies are not capable of removing recalcitrant materials, and they generate large amounts of dry sludge that, in turn, may present heavy metals and other toxic components likely to contaminate the soil it is deposited in (Al-Jabri et al., 2021; Mohsenpour et al., 2021; Xiong et al., 2021).

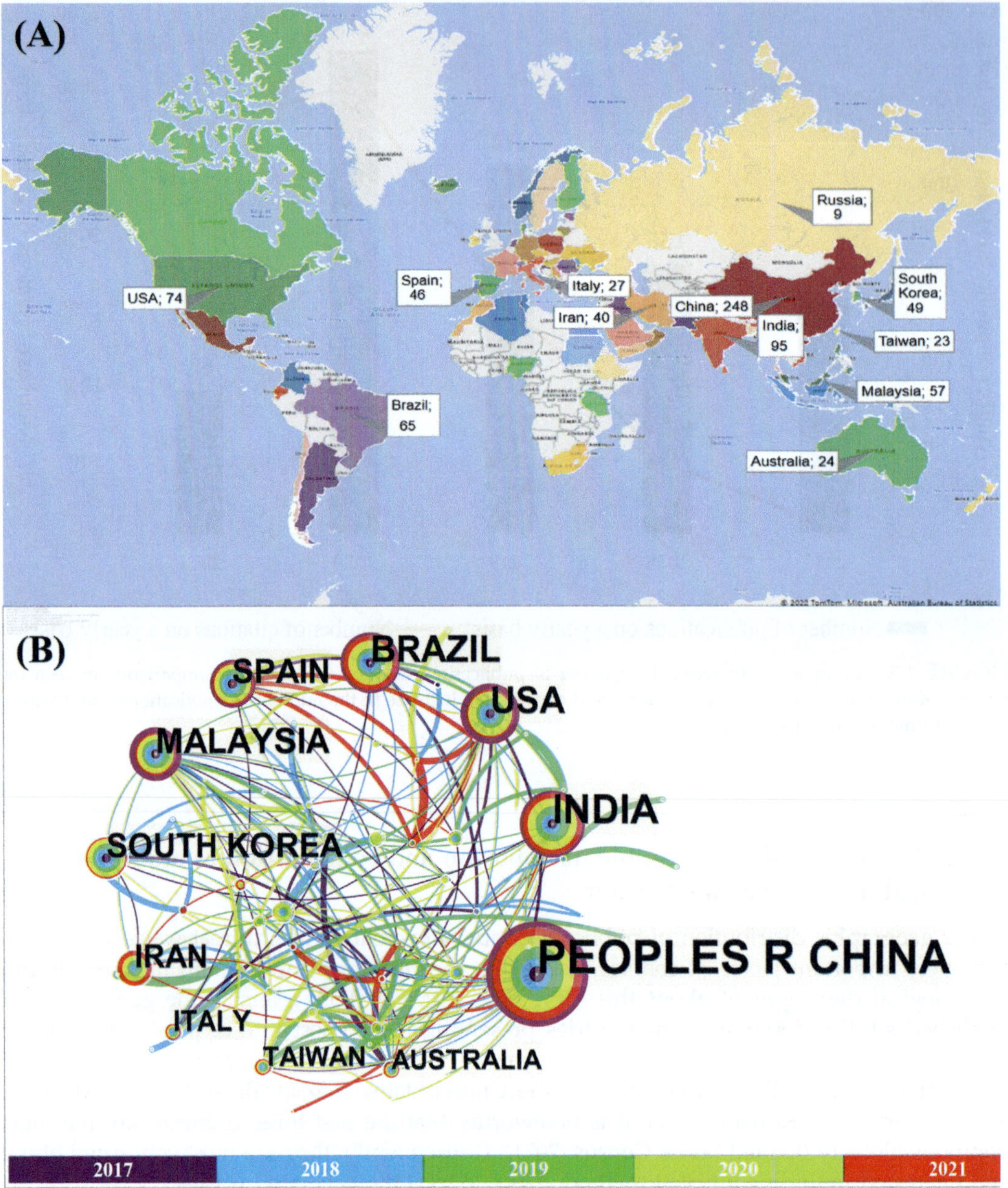

FIGURE 1.3   Publications about microalgae using for wastewater. (A) Geographic distribution of publications about microalgae using for wastewater treatment. (B) Cooperation network among countries publishing studies about microalgae using for wastewater treatment.

According to estimates, public properties in the United States generate approximately 13.8 million tons of dry sludge yearly (Coleman et al., 2017) —this value is considerably high, since dry sludge is an emerging environmental issue. This amount of waste has encouraged the implementation of research focused on using eco-efficient and sustainable methodologies to treat the generated effluents, such as treatments based on microalgae.

Furthermore, Fig. 1.3B shows the cooperation network between the main countries that have published studies about the use of microalgae for wastewater treatment. The size of each node in Fig. 1.3B expresses its frequency, that is, the country's visibility on the addressed topic. In contrast, the purple color in the outer ring of the node indicates its centrality. Thus the wider the purple ring, the higher the country's influence (Liu et al., 2020). Fig. 1.3B also shows the links formed, which are directly associated with the period when the studies were cited; purple links refer to the studies published in 2017; blue links to the studies published in 2018; green links to the studies published in 2019; yellow links to the studies published in 2020; and red links to the studies published in 2021 (Chen, 2020).

The thickness of the links corresponds to the strength of cooperation among authors of the analyzed countries. Based on the links formed in Fig. 1.3B, it is evident that researchers from China, who have great visibility and influence, have weak cooperation with researchers from other countries, indicating the production of individual structures. On the other hand, the association observed among researchers from the United States, Brazil and, Spain, mainly in 2021, has shown strong scientific cooperation among these countries, clearly visible through their links' thickness. Moreover, countries such as China, India, United States, Malaysia, and Brazil stood out for their centrality (Fig. 1.4). It is noteworthy that although the United States, India, and Brazil account for the highest rate of published manuscripts, 8.57%; 11%, and 7.53%, respectively, Malaysia had a stronger influence (Fig. 1.4). This finding indicates higher representativeness of manuscripts published by Malaysian researchers about using microalgae for waste treatment and biofuel production as a secondary aim (Low et al., 2021).

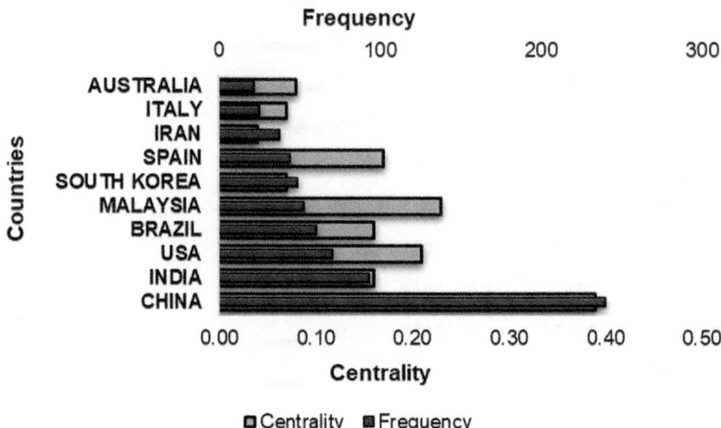

FIGURE 1.4 Frequency and centrality of the top 10 countries. Infographic of the frequency and centrality of the top 10 countries in wastewater treatment with microalgae.

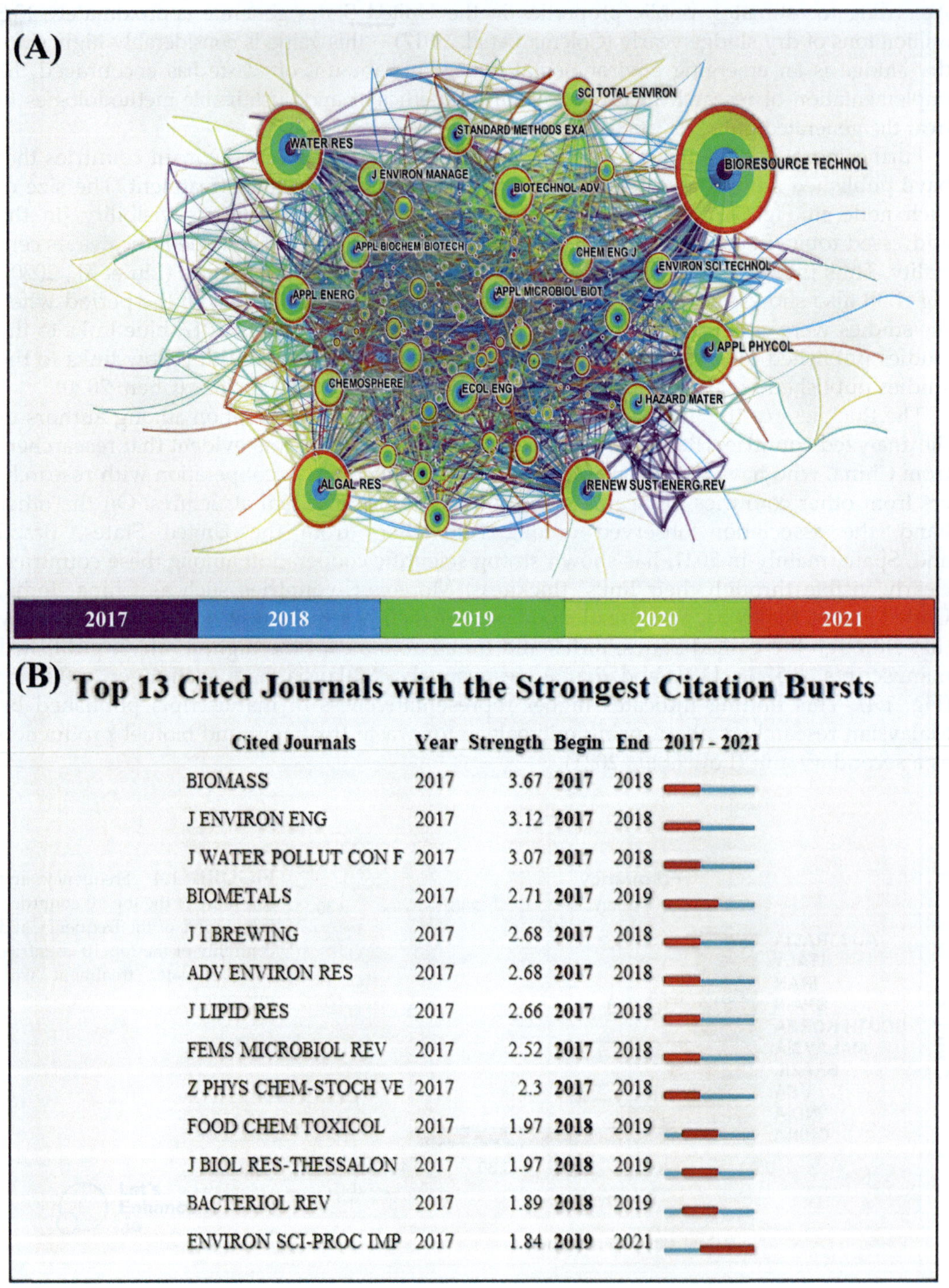

(B) Top 13 Cited Journals with the Strongest Citation Bursts

Cited Journals	Year	Strength	Begin	End	2017 – 2021
BIOMASS	2017	3.67	**2017**	2018	
J ENVIRON ENG	2017	3.12	**2017**	2018	
J WATER POLLUT CON F	2017	3.07	**2017**	2018	
BIOMETALS	2017	2.71	**2017**	2019	
J I BREWING	2017	2.68	**2017**	2018	
ADV ENVIRON RES	2017	2.68	**2017**	2018	
J LIPID RES	2017	2.66	**2017**	2018	
FEMS MICROBIOL REV	2017	2.52	**2017**	2018	
Z PHYS CHEM-STOCH VE	2017	2.3	**2017**	2018	
FOOD CHEM TOXICOL	2017	1.97	**2018**	2019	
J BIOL RES-THESSALON	2017	1.97	**2018**	2019	
BACTERIOL REV	2017	1.89	**2018**	2019	
ENVIRON SCI-PROC IMP	2017	1.84	**2019**	2021	

FIGURE 1.5 Citation burst of the cited journals. (A) Network organized with the most cited journals. (B) Citation burst of the most cited journals.

1.3.3 Analysis of the 10 main sources of publications, journals, and research fields

The network organized with the most cited journals is represented in Fig. 1.5A. The most cited journals are listed below, along with their respective Journal Impact Factors 2019–20: Bioresource Technology (7.539), Water Research (9.130), Algal Research (4.008), Renewable & Sustainable Energy Reviews (12.110), and Journal Applied Phycology (3.016). There may be significant scientific interest in this topic since it has been published in high-impact journals. According to the figure mentioned above, each node represents a given journal, and node size represents the journal's respective citation frequency, which shows the journals that mainly publish studies on the topic in question. The lines between the nodes mean the contribution between journals, and they are associated with publication year. These journals make connections with many others that present lower citation frequency, and it enables the formation of an extensive network of citations.

The analyzed journals did not differ based on the centrality criterion; this outcome implies that no journal significantly influenced this topic. Fig. 1.5B shows the top 13 journals presenting the citation burst. The journal Biomass presented the strongest citation burst between 2017 and 2018, whereas Environmental Science: Processes & Impacts currently has the strongest citation burst. These journals are considered references in microalgae application for wastewater treatment and the best journals for scientists to disclose their research results and scientific findings on this topic.

The highlighted journals present thematic axes consistent with the use of microalgae to treat wastewater. Bioresource Technology and Water Research focuses on wastewater treatment, biological waste treatment, environmental restoration, and others. On the other hand, Algal Research and Journal of Applied Phycology focus on research associated with algal biology and biotechnology, as well as with commercially useful microalgae and with their bioproducts. The journal Biomass focused on studies about the valorization of biomass and biologically based materials to be applied in biorefineries. In contrast, biometals have a multidisciplinary scope and focus on discussions about metals and stand out in studies about microalgae's ability to remove such compounds. Finally, Environmental Science: Processes & Impacts is a diversified journal encompassing all environmental science processes, such as anthropogenic and natural contaminants.

The most cited knowledge fields about waste treatment with microalgae are shown in Fig. 1.6A and B. Fields such as "Engineering," "Biotechnology," "Environment and Ecology," "Water Resources", "Energy and Fuels," and "Biotechnology and Applied Microbiology" appear at a higher frequency, highlighting the visibility of these fields in the investigated topic. The Engineering category also presents higher centrality since it is the most representative field focused on microalgal application to treat waste. It is a notorious field because it enables technology to convert microalgal biomass into energy sources and biofuels (Bahadar and Bilal Khan, 2013). Together with biotechnology and applied microbiology, engineering aims to develop and implement molecular tools and techniques to find improved microalgal strains based on genetic engineering and promote increased biomass yield (Kumar et al., 2020; Lu et al., 2021). Furthermore, microalgae are promising sources used for the bioremediation of wastewater and pollutants, based on the sustainable applications of their bioproducts,

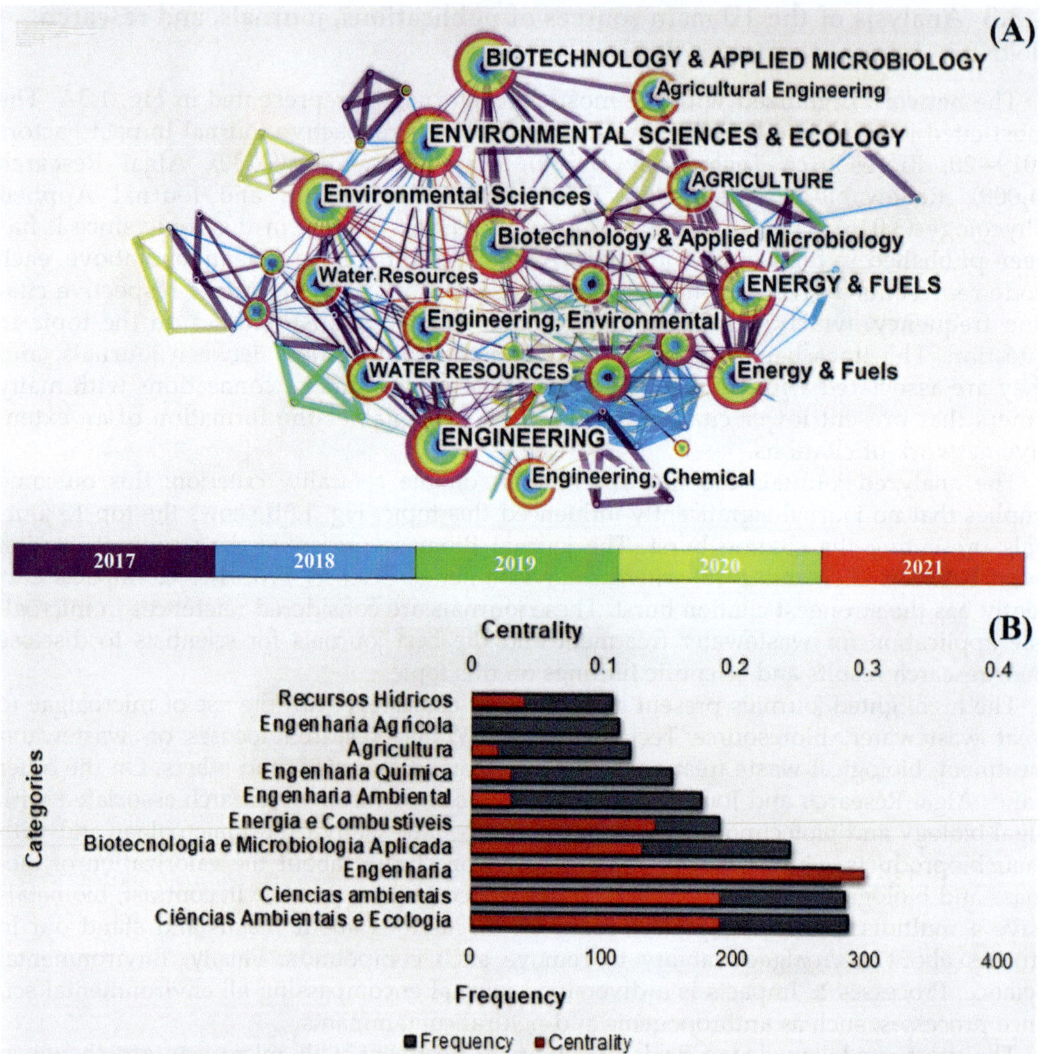

FIGURE 1.6　Studies about waste treatments with microalgae. (A) Network showing the main categories mentioned in studies about waste treatments with microalgae. (B) Graph showing the frequency and centrality of knowledge fields.

a fact that triggers the interest of fields such as environment, ecology, and water resources (Hussain, 2019).

Research fields focused on Energy and Fuels are associated with the topic since microalgal biomass enables large amounts of biodiesel, biogas, and other renewable energy sources (Al-Jabri et al., 2021; Hussian, 2018). Some challenges still need to be overcome, such as developing biomass conversion routes capable of generating economically and environmentally sustainable bioproducts and improving biotechnological methods capable

TABLE 1.1 Top 10 articles about wastewater treatment with microalgae.

No.	Title	Authors and year	Source title	Total no. of citations	Average per year
1	The promising future of microalgae: current status, challenges, and optimization of a sustainable and renewable industry for biofuels, feed, and other products	Khan et al. (2018)	Microbial cell factories	368	92.00
2	Recent progress in microalgal biomass production coupled with wastewater treatment for biofuel generation	Salama et al. (2017)	Renewable & sustainable energy reviews	172	34.40
3	Treatment of real wastewater using coculture of immobilized *Chlorella vulgaris* and suspended activated sludge	Mujtaba and Lee (2017)	Water research	78	15.60
4	Biosorption for removal of hexavalent chromium using microalgae *Scenedesmus* sp.	Pradhan et al. (2019)	Journal of cleaner production	70	23.33
5	Heterotrophic cultivation of microalgae using aquaculture wastewater: a biorefinery concept for biomass production and nutrient remediation	Guldhe et al. (2017)	Ecological engineering	69	13.80
6	Responses of microalgae *Coelastrella* sp. to stress of cupric ions in treatment of anaerobically digested swine wastewater	Li et al. (2018)	Bioresource technology	67	16.75
7	Microalgae from wastewater treatment to biochar—feedstock preparation and conversion technologies	Yu et al. (2017)	Energy conversion and management	65	13.00
8	Performance of a microalgal photobioreactor treating toilet wastewater: pharmaceutically active compound removal and biomass harvesting	Hom-Diaz et al. (2017)	Science of the total environment	59	11.80
9	Centrate wastewater treatment with *C. vulgaris*: simultaneous enhancement of nutrient removal, biomass, and lipid production	Ge et al. (2018)	Chemical engineering journal	54	13.50
10	High-efficiency nutrients reclamation from landfill leachate by microalgae *C. vulgaris* in membrane photobioreactor for biolipid production	Chang et al. (2018)	Bioresource technology	53	13.25

of making positive contributions to this field (Mohsenpour et al., 2021). Thus it is essential to encourage research on the subject and contributions among authors from different regions to help spread new technologies, bioproducts, and scientific information globally.

Table 1.1 shows the top 10 articles with the most significant number of citations between 2017 and 2021. The study by Kim (2018), titled "The promising future of microalgae: current status, challenges and optimization of a sustainable and renewable industry for biofuels, feed and other products," presented the largest number of citations—92 citations a year, on average.This study focuses on investigating large-scale microalgae cultivation to produce biomass, as well as the generation of high-value products for the nutraceutical, pharmaceutical, and bioenergy industry based on the application of microalgae for waste treatment and carbon dioxide (CO_2) consumption. It also explains extensive microalgal applications, their respective challenges and limitations, and what can be done to overcome them to make microalgae suitable to the market.

The study by Salama et al. (2017), titled "Recent progress in microalgal biomass production coupled with wastewater treatment for biofuel generation," published in Renewable & Sustainable Energy Reviews, had 172 citations. This review discusses how wastewater can be used as a potential source of nutrients to produce biomass and biofuels from microalgae. It suggests a new approach involving mixing wastewater to enable an ideal nitrogen/phosphorus ratio to optimize biomass production for sustainable and economically viable production.

Both articles mentioned above refer to what was discussed above about the applications and challenges involved in using microalgae. The journals that appear in sequence had less than 100 citations each; they covered topics such as the use of microalgal species *Chlorella vulgaris*, *Scenedesmus* sp., and *Coelastrella* sp., and of other technologies for waste treatment purposes.

1.3.4 Keyword analysis focusing on the use of microalgae for wastewater treatment

The keyword analysis, called hotspot analysis, is the most informative method used in scientometric reviews since it shows the trends and lacks in the state of the art of a given field. Fig. 1.7A shows the cocitation network. The term "microalgae" is the keyword presenting the highest citation frequency and centrality, therefore it is the most effective and most visible term among the searched meshes. Links represent cocitations between words that are strongly correlated to each other, reinforcing how they stand out in this field. Colors are associated with the number of times the keywords were mentioned. In contrast, node size refers to the frequency of each cited word, a fact that indicated its visibility in research about wastewater treatment with microalgae.

Keywords are descriptive and differentiated words used to compile and understand the concepts and contents of articles. In addition, they make it much easier to monitor changes in the research field. All words shown in Fig. 1.6A are closely related to the use of microalgae for wastewater treatment. The prominent word "microalgae" corresponds to recent scientific efforts to combine the use of these microorganisms in environmental wastewater remediation processes and as raw material for the next bioenergy generation . This term is closely related to growth, cultivation, nutrient removal, removal, wastewater, nitrogen,

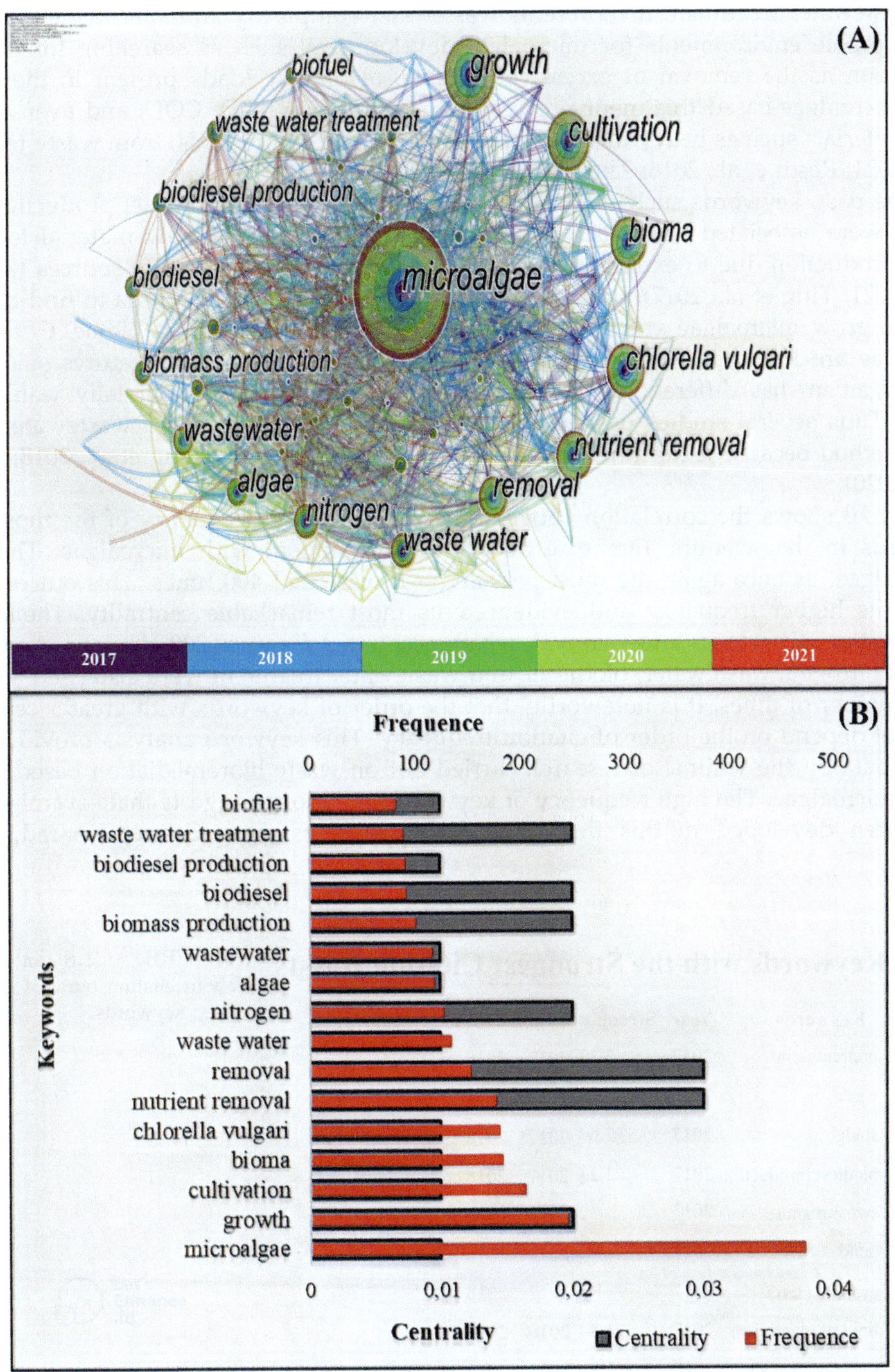

FIGURE 1.7 Scientific studies about waste treatment with microalgae. (A) Network of keywords ranked based on citation frequency. (B) Graph of correlation between the centrality and frequency of the most cited keywords in scientific studies about waste treatment with microalgae.

and wastewater treatment. It represents wastewater complexity and specificities to establish favorable environments for microalgal development, such as searching for biomass production as the removal of excessive amounts of organic loads present in the waste. Thus microalgae-based treatment processes aim to remove BOD, COD, and even recalcitrant materials such as heavy metals and pharmaceutical compounds, from waste (Al-Jabri et al., 2021; Resdi et al., 2016; Zhu, 2016).

In the past, keywords such as biomass production, biodiesel, biodiesel production, and biofuel were associated with microalgae's potential to be used as raw material for bioenergy production due to extreme urgency in finding sustainable energy sources (Al-Jabri et al., 2021; Ting et al., 2017). It is also necessary to make scientific efforts to find cheaper ways to grow microalgae at a large scale (Zhu, 2016). Words such as biome, *C. vulgaris*, and algae are linked to the diversity of species and their respective features since each microorganism has different abilities. Some species are more commercially viable than others. Thus several studies use the microalga species *C. vulgaris* as a wastewater treatment method because it can produce large amounts of biomass (Chang et al., 2018; Cheng et al., 2020).

Fig. 1.7B shows the correlation between the centrality and frequency of the most cited keywords in the scientific literature about waste treatment with microalgae. The term "microalgae" is once again the most prevalent—it was cited 400 times. This outcome has shown its higher frequency and evidenced its most remarkable centrality. Then, there is cultivation, *C. vulgaris* and removal, which accounted for over 200 citations. Keywords such as nitrogen, wastewater, biodiesel, and wastewater treatment were also cited a significant number of times. It is noteworthy that the order of keywords with greater centrality does not depend on the order of citation frequency. This keyword analysis provided consistent data on the volume of research carried out on waste bioremediation based on the use of microalgae. The high frequency of keyword repetitions suggests that several studies have been developed in this field, but the opposite is true (Castro; Konrad, 2020).

Top 9 Keywords with the Strongest Citation Bursts

Keywords	Year	Strength	Begin	End	2017 - 2021
pretreatment	2017	3.13	2017	2018	
waste	2017	3	2017	2018	
sludge	2017	2.67	2017	2018	
biodieselproduction	2017	2.23	2017	2018	
swine manure	2017	2	2017	2018	
pond	2017	2	2017	2018	
cyanobacteria	2017	1.9	2017	2018	
lipid productivity	2017	1.84	2018	2019	
scenedesmus sp	2017	1.81	2018	2019	

FIGURE 1.8 Information about citation burst of the most cited keywords.

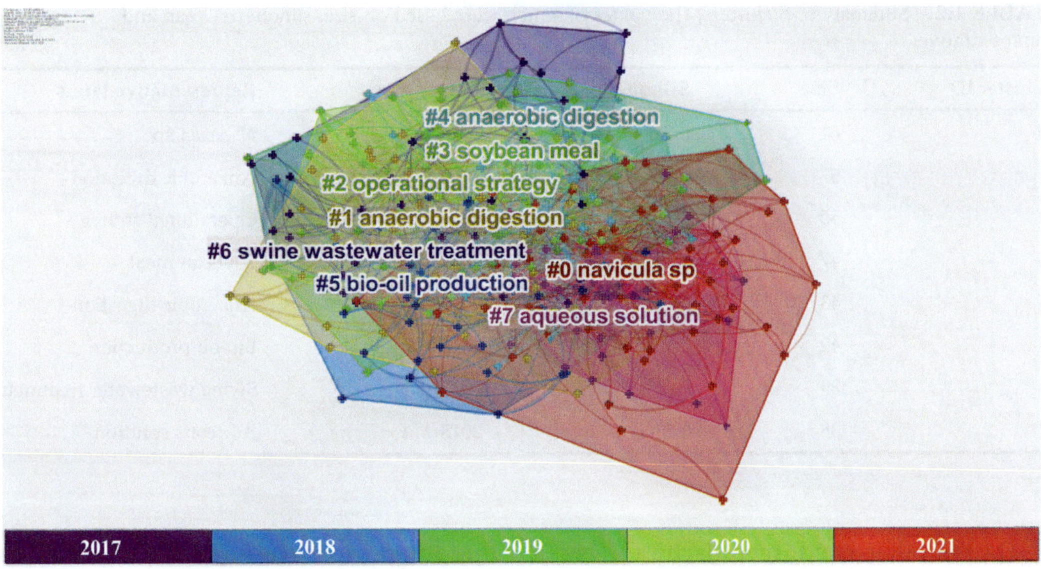

FIGURE 1.9 Keyword clusters.

In addition, keywords' representativeness can also express research gaps, trends, and hotspots concerning this topic.

The keyword citation burst (Fig. 1.8) enables observing the term-use duration and the attention given by researchers to a specific keyword in each analyzed period. Pretreatment and waste were the most searched terms between 2017 and 2018, whereas lipid productivity and *Scenedesmus* sp. presented citation burst in this very same period; it indicated indices with high search trends. On the other hand, 2020 and 2021 did not record citation bursts for specific terms; this may reveal a change of trend and hotspots. The interdisciplinary profile of this topic and the large number of knowledge fields associated with it corroborate the diversity of words in the keyword network (Fig. 1.7), bursts (Fig. 1.8), and clusters (Fig. 1.9); it is noticeable that the words barely overlap and represent different topics.

Cluster analysis is an exciting tool used in systematic reviews to investigate and organize integral terms and contexts and visualize research patterns and connections. It is possible to organize a large number of data and classifying them based on grouping strength. Clusters group the most similar keywords and name them with the most representative terms for each group. Cluster formation processes consider criteria such as Silhouette and modularity Q. Silhouette expresses clusters' homogeneity through values ranging from -1 to 1—the closer to 1, the more similar the group is. On the other hand, modularity Q represents how well-disposed the division of these groups is in values ranging from 0 to 1—values higher than 0.50 are considered as good separation (Chen, 2020). The grouping shown in Fig. 1.9 recorded Silhouette equal to 0.70; this value has evidenced clusters' uniformity.

TABLE 1.2 Summary of 8 clusters, the values of which were based on size, silhouette, year, and representative.

Cluster ID	Size	Silhouette	Mean (year)	Representative terms
0	61	0.717	2018	*Navicula* sp.
1	57	0.653	2017	Anaerobic digestion
2	50	0.686	2018	Operational strategy
3	47	0.647	2018	Soybean meal
4	43	0.781	2017	Anaerobic digestion
5	42	0.685	2018	Bio-oil production
6	39	0.649	2018	Swine wastewater treatment
7	38	0.836	2018	Aqueous solution

On the other hand, modularity Q was 0.39 due to an excessive number of cocitations among the words represented in the collaboration network seen in Fig. 1.7A. The color of the links is associated with publication year, and clusters' font colors are associated with the grouping. The featuring of these groups can indicate the main research directions.

Table 1.2 presents the featuring of each cluster and enables seeing that all clusters expressed Silhouette higher than 0.64, which indicates good cluster homogeneity based on its size and the diversity of keywords. Research fronts addressing the use of microalgae for waste treatment can be identified in Fig. 1.9 and Table 1.2. Suppose one considers that the representative term of the first cluster (0) was *Navicula* sp., as well as the incidence of genus *Scenedesmus* sp. in citation bursts and the fact that *C. vulgaris* was the fifth most cited keyword. In that case, it is possible inferring that most research focused on investigating the efficiency of different microalgal species and genera in organic wastewater compound bioremediation processes and biomass used for bioenergy production. *Navicula* sp. is often used to remove pharmaceutically active compounds from wastewater. According to Ding et al. (2020), it can remove up to 90% of the materials mentioned above from wastewater. *Scenedesmus* sp. and *C. vulgaris* were reported to efficiently remove N and P from wastewater (Salama et al., 2017; Khan et al., 2018; Mohsenpour et al., 2021).

The second and fifth clusters were represented by the term "anaerobic digestion," which is widely used as an extra or combined step to improve microalgae resource recovery and to increase biogas yield (Solé-Bundó et al., 2019). The sixth cluster (5) corresponds to the use of microalgal biomass as a source of bio-oil production. Microalgal biomass is a sustainable and eco-friendly alternative to replace large monocultures, such as soybean, that, besides other attributions, is widely used for oil production and is represented in cluster 3. This alternative has emerged among methodologies traditionally based on oilseeds, such as soybean monocultures because the oil generated by microalgae is of high quality and rich in proteins. However, it is necessary to conduct further studies to make the process feasible at an industrial scale, based on different ways to get these bio-oils (Xue et al., 2020).

Clusters 2 and 7 were named operational strategy and aqueous solution, respectively; they refer to strategies and solutions focused on implementing microalgae for wastewater treatment and as bioenergy source, as it is a complexity of compounds, processes, and assignments. Cluster 6 represents a waste type rich in organic material and nitrogen compounds. In its untreated form, this waste can be a polluting source that limits the development of aerobic organisms due to eutrophication. Thus, using microalgae, in this case, is an excellent biotechnological alternative since they can remove 99% of N and P from swine wastewater, which is why they are the object of high interest within the scientific community (Li et al., 2018). The distinction among clusters highlights the interdisciplinary profile of this topic and how promising and beneficial the use of microalgae for wastewater treatment is, given the sustainable alternatives provided by this methodology.

Microalgae applicability potential goes far beyond its use for waste treatment since it is just one of the benefits provided by these microorganisms. They are capable of producing large amounts of biomass with high added value, as previously mentioned. Thus microalgae have opened a new horizon to wastewater, which is now considered a rich and promising carbon source to produce value-added products.

Several practical and economic challenges have been faced in using microalgae for effluent treatment, therefore many studies have been carried out to find a way to implement this technology at an industrial scale. Al Ketifea, et al. (2019) carried out a technical and economic feasibility analysis that evaluated implementing a large-scale system to replace microalgae with activated sludge lagoons in the Persian Gulf Channel. Results have shown a selling price of $0.544 per kg of biomass, which is equivalent to $0.9 L^{-1} of the extracted biocrude—this value can cover the system's operating expenses. According to Fernández, et al. (2018), well-designed microalgal-use strategies enable obtaining 200 tons of high-value biomass per year and remove approximately 90% of nutrients found in wastewater. Using microalgae in wastewater systems enables removing nutrients from them in a sustainable manner, mitigating greenhouse gas emissions and reducing the total costs with wastewater treatment by half, in comparison to traditional methods.

1.4 Conclusion

This scientometric review provided a global analysis of research and information and enabled visualizing scientific gaps in and trends about wastewater treatment with microalgae. Thus data were compiled to rank in the network of collaborations and cocitations per country, journal, category, and keyword. The final dataset comprised 863 publications that were cited 7485 times and presented H-index equal to 33. Countries that stood out for publications about the use of microalgae for waste treatment included China (28.74% of publications), India (11.01%), the United States (8.57%), Brazil (7.53%), Malaysia (6.60%), and South Korea (5.68%).

The keyword "network" has shown that most studies focus on associating waste bioremediation and the production of biomass, presenting high biotechnological value for other applications by mainly using *Navicula* sp., *Scenedesmus* sp., and *C. vulgaris*, which are associated with the removal of pharmaceutics compounds, recalcitrant materials, as well as nitrogen and phosphorus from different wastewater types, such as swine effluent, among other pollutants. Due to the potential of biomass for microalgae, studies focused on replacing the oil produced by oilseeds, such as soybeans, with microalgae oil which is currently in progress. Because, in addition to having good lipid features, microalgae oil is also a rich source of proteins. Another applicability in the use of microalgae biomass lies in the production of biofuels such as biodiesel.

The information available in this chapter has shown that research on microalgae is an influential and relevant topic within the scientific community since they are seen as sustainable and eco-friendly solutions. The global panorama can provide directions for future research since there are clear gaps yet to be addressed, such as the implementation of efficient microalgae-based effluent bioremediation processes that are economically and environmentally sustainable and that can be applied at a large industrial scale. Current studies have shown the potential to produce 200 tons of biomass a year and sell at $0.544 per kg of biomass. This outcome has evidenced how promising this process is and how it tends to be in the future.

References

Al-Jabri, H., et al., 2021. Treatment of wastewaters by microalgae and the potential applications of the produced biomass—a review. Water (Switzerland) 13 (1). Available from: https://doi.org/10.3390/w13010027. Qatar: MDPI AG.

Al Ketifea, A.M.D., Almomani, F., EL-Naas, M., Judd, S., 2019. A technoeconomic assessment of microalgal culture technology implementation for combined wastewater treatment and CO_2 mitigation in the Arabian Gulf—A technoeconomic assessment of microalgal culture technology implementation for combined wastewater treatment and CO_2 mitigation in the Arabian Gulf. Process Safety and Environmental Protection. Elsevier.

Arora, K., et al., 2021. Valorization of wastewater resources into biofuel and value-added products using microalgal system. Frontiers in Energy Research 9, 1–25.

Bahadar, A., Bilal Khan, M., 2013. Progress in energy from microalgae: a review, Renewable and Sustainable Energy Reviews, 27. Elsevier Ltd., Pakistan, pp. 128–148. Available from: http://doi.org/10.1016/j.rser.2013.06.029.

Birkle, C., et al., 2020. Web of Science as a data source for research on scientific and scholarly activity. Quantitative Science Studies 1 (1), 363–376. Available from: https://doi.org/10.1162/qss_a_00018. MIT Press - Journals.

Castro, A., Konrad, O., 2020. Global overview of scientific production on microalgae in wastewater treatment. Valore, Volta Redonda.

Chang, H., et al., 2018. High-efficiency nutrients reclamation from landfill leachate by microalgae *Chlorella vulgaris* in membrane photobioreactor for bio-lipid production, Bioresource Technology, 266. Elsevier Ltd., China, pp. 374–381. Available from: http://doi.org/10.1016/j.biortech.2018.06.077.

Chen, C., 2015. How to Use CiteSpace (c) 2015–2020.

Chen, C. 2020. How to Use CiteSpace (c) 2015-2020. p. 2015–2020. < https://leanpub.com/howtousecitespace >.

Cheng, Z., et al., 2020. A bibliometric-based analysis of the high-value application of Chlorella. 3. Biotech 10 (106), 1–14.

Coleman, M., Skaggs, R.L., Seiple, T.E., 2017. Municipal wastewater sludge as a sustainable bioresource in the United States. Journal of Environmental Management 197, 673–680.

Ding, T., et al., 2020. Biological removal of pharmaceuticals by *Navicula* sp. and biotransformation of bezafibrate. Chemosphere. Elsevier Ltd., China, p. 240. Available from: http://doi.org/10.1016/j.chemosphere.2019.124949.

Fernández, F.G.A., Gómez-Serrano, C., Fernández-Sevilla, J.M., 2018. Recovery of Nutrients from Wastewaters Using Microalgae. Frontiers in Sustainable Food Systems 2, 1–13.

Ge, S., et al., 2018. Centrate wastewater treatment with *Chlorella vulgaris*: simultaneous enhancement of nutrient removal, biomass and lipid production, Chemical Engineering Journal, 342. Elsevier B.V, China, pp. 310–320. Available from: http://doi.org/10.1016/j.cej.2018.02.058.

Guldhe, A., et al., 2017. Heterotrophic cultivation of microalgae using aquaculture wastewater: a biorefinery concept for biomass production and nutrient remediation, Ecological Engineering, 99. Elsevier B.V, South Africa, pp. 47–53. Available from: http://doi.org/10.1016/j.ecoleng.2016.11.013.

Hom-Diaz, A., et al., 2017. Performance of a microalgal photobioreactor treating toilet wastewater: Pharmaceutically active compound removal and biomass harvesting, Science of the Total Environment, 592. Elsevier B.V, Spain, pp. 1–11. Available from: http://doi.org/10.1016/j.scitotenv.2017.02.224.

Hussian, A.E., 2018. The role of microalgae in renewable energy production: challenges and opportunities. Marine Ecology - Biotic and Abiotic Interactions. IntechOpen.

Hussain, F., 2019. Microalgae an ecofriendly and sustainable wastewater treatment option: Biomass application in biofuel and bio-fertiliser production. A review. Renewable and Sustainable Energy Reviews .

Khan, M.I., Shin, J.H., Kim, J.D., 2018. The promising future of microalgae: current status, challenges, and optimization of a sustainable and renewable industry for biofuels, feed, and other products. Microbial Cell Factories 17 (1), 1–21.

Kumar, G., et al., 2020. Bioengineering of microalgae: recent advances, perspectives, and regulatory challenges for industrial application, Frontiers in Bioengineering and Biotechnology, 8. Frontiers Media S.A., India. Available from: http://doi.org/10.3389/fbioe.2020.00914.

Li, X., et al., 2018. Responses of microalgae *Coelastrella* sp. to stress of cupric ions in treatment of anaerobically digested swine wastewater. Bioresource Technology. Elsevier BV, pp. 274–279. Available from: http://doi.org/10.1016/j.biortech.2017.12.058.

Liu, C., et al., 2020. A scientometric analysis and visualization of research on Parkinson's disease associated with pesticide exposure. Front Public Health 8, 1–14.

Low, S.S., et al., 2021. Microalgae cultivation in palm oil mill effluent (Pome) treatment and biofuel production. Sustainability (Switzerland) 13 (6). Available from: https://doi.org/10.3390/su13063247. Malaysia: MDPI AG.

Lu, Y., et al., 2021. Engineering microalgae: transition from empirical design to programmable cells. Critical Reviews in Biotechnology. Taylor and Francis Ltd, China. Available from: http://doi.org/10.1080/07388551.2021.1917507.

Mohsenpour, S.F., et al., 2021. Integrating micro-algae into wastewater treatment: a review, Science of The Total Environment, 752. Elsevier BV, p. 142168. Available from: http://doi.org/10.1016/j.scitotenv.2020.142168.

Mujtaba, G., Lee, K., 2017. Treatment of real wastewater using co-culture of immobilized *Chlorella vulgaris* and suspended activated sludge, Water Research, 120. Elsevier Ltd., South Korea, pp. 174–184. Available from: http://doi.org/10.1016/j.watres.2017.04.078.

Pradhan, D., et al., 2019. Biosorption for removal of hexavalent chromium using microalgae *Scenedesmus* sp, Journal of Cleaner Production, 209. Elsevier Ltd., India, pp. 617–629. Available from: http://doi.org/10.1016/j.jclepro.2018.10.288.

Priya, M., Gurung, N., 2014. 23 Microalgae in removal of heavy metal and organic pollutants from soil. Microbial Biodegradation and Bioremediation. Elsevier, pp. 521–540.

Resdi, R., et al., 2016. Review of microalgae growth in palm oil mill effluent for lipid production. Clean Technologies and Environmental Policy 18 (8), 2347–2361. Available from: https://doi.org/10.1007/s10098-016-1204-1. Malaysia: Springer Verlag.

Salama, E.S., et al., 2017. Recent progress in microalgal biomass production coupled with wastewater treatment for biofuel generation, Renewable and Sustainable Energy Reviews, 79. Elsevier Ltd., South Korea, pp. 1189–1211. Available from: http://doi.org/10.1016/j.rser.2017.05.091.

Solé-Bundó, M., et al., 2019. Co-digestion strategies to enhance microalgae anaerobic digestion: a review, Renewable and Sustainable Energy Reviews, 112. Elsevier Ltd., Spain, pp. 471–482. Available from: http://doi.org/10.1016/j.rser.2019.05.036.

Sun, Y., et al., 2016. Characteristics of water quality of municipal wastewater treatment plants in China: implications for resources utilization and management, Journal of Cleaner Production, 131. Elsevier Ltd., China, pp. 1–9. Available from: http://doi.org/10.1016/j.jclepro.2016.05.068.

UNESCO, 2021. The United Nations World Water Development Report 2021.U.S. Census Bureau Current Population. Disponível em: < https://www.census.gov/popclock/print.php?component=counter >.

Wen, Y., Schoups, G., Van de Giesen, N., 2017. Organic pollution of rivers: combined threats of urbanization, livestock farming and global climate change. Scientific Reports 7, 1–9.

Xiong, Q., et al., 2021. Microalgae-based technology for antibiotics removal: from mechanisms to application of innovational hybrid systems, Environment International, 155. Elsevier BV, p. 106594. Available from: http://doi.org/10.1016/j.envint.2021.106594.

Xue, Z., et al., 2020. Development prospect and preparation technology of edible oil from microalgae, Frontiers in Marine Science, 7. Frontiers Media S.A., China. Available from: http://doi.org/10.3389/fmars.2020.00402.

Yu, K.L., et al., 2017. Microalgae from wastewater treatment to biochar — feedstock preparation and conversion technologies, Energy Conversion and Management, 150. Elsevier Ltd., Malaysia, pp. 1–13. Available from: http://doi.org/10.1016/j.enconman.2017.07.060.

Zhu, J., 2016. Cultivation of microalgae in wastewater: a review, pp. 305–310.

Scientometric analysis of consortium-based wastewater treatment

Aline de C.C. Pena[1,2], *Luciane F. Trierweiler*[2] *and Mariliz Gutterres*[1]

[1]Laboratory for Leather and Environmental Studies (LACOURO), Porto Alegre, Brazil
[2]Group of Intensification, Modeling, Simulation, Control, and Optimization of Process (GIMSCOP), Porto Alegre, Brazil

2.1 Introduction

Microalgae have been the subject of several studies. Over than 50,000 documents have been found in the literature since 1990 highlighting various microalgae applications, such as wastewater treatment (Fontoura et al., 2017), biofuel production (Aketo et al., 2020), drug development (Mehariya et al., 2021), food (de Carvalho et al., 2020), among others.

The wastewater treatment process carried out with microalgae represents a versatile possibility as these microorganisms have a high capacity for fixing carbon dioxide from the air and removing nutrients dissolved in the water (Shahid et al., 2020). In addition, at the end of the wastewater treatment process with microalgae, the generated biomass can be transformed into third-generation biofuels, giving rise to biodiesel, ethanol, biokerosene, biohydrogen, biogas (methane), and chemical intermediates for the petrochemical sector (Dasan et al., 2019).

Another advantage is that these microorganisms easily adapt changes in the environment, whether in temperature, pH, salinity, nutrient availability, and establish symbiotic relationships with other microorganisms such as fungi, bacteria or even with different species of microalgae, making it possible to carry out the treatment of different types of wastewater (Pérez et al., 2016; Wang et al., 2021).

The application of consortium of microalgae (bacteria, fungi or different species of microalgae) for the treatment of wastewater is reported in the literature due to the ease of maintaining the culture, efficiency in the pollutant removal process and superior biomass production, when compared to monocultures (Talapatra et al., 2021). The

Valorisation of Microalgal Biomass and Wastewater Treatment
DOI: https://doi.org/10.1016/B978-0-323-91869-5.00001-6

application of consortia for different types of wastewater is described, such as municipal (Solís-Salinas et al., 2021), leather industry (Pena et al., 2020), pharmaceutical industry (Murshid and Dhakshinamoorthy, 2021), agro-industrial (Chia et al., 2021), and dairy farm (Hena et al., 2015).

Thorough understanding of the process and the current research progress in this area could facilitate the establishment of effluent treatments with microalgae consortia and their applications. In this context, scientometric methods have been performed to quantitatively analyze information from the literature to report the current researches and predict future perspectives. Several scientometric analyzes were carried out on research in the area of microalgae, such as food (Konur, 2020a), bioenergy (Konur, 2020b), and bioremediation (Konur, 2020c). However, a scientometric study on the treatment of wastewater with microalgae consortium has not yet been found.

Assumed the gap presented, this work performed a scientometric analysis on the use of microalgae consortia in wastewater treatment. In this study, the scientometric characteristics and the interconnection among documents in the period between 1945 and 2021 were analyzed. Keywords were selected to carry out the search on the Web of Science (WoS) platform. Changes over the years, new research trends, and future prospects for research were identified.

2.2 Materials and methods

The WoS database was used to perform a scientometric analysis of consortium-based wastewater treatment. The survey was conducted in the period from January 1945 to September 2021 using four WoS databases: Science Citation Index-Expanded, Social Sciences Citation Index (SSCI), Arts & Humanities Citation Index (A&HCI) and Emerging Sources Citation Index. A set of relevant keywords was used to search the abstracts for keywords, titles and author of the articles searched.

To perform the search on the WoS platform, three groups of words were used to perform the search: the first referring to microalgae, the second to wastewater treatment, and the third to the consortium of microalgae, including fungi, bacteria, and different species of microalgae. The set of keywords used to perform the search is provided in Appendix A. The articles, reviews, proceedings paper, meeting abstracts, editorial material, notes, and reprints were selected.

The extracted documents were analyzed using scientometric parameters: (1) document, (2) publishing authors, (3) publishing countries, (4) publishing institutions, (5) publishing year, (6) publishing journals, and (7) citation analysis.

2.3 Results

2.3.1 Documents

Through the search performed, 2743 documents were found, including articles, reviews, meeting abstracts, editorial material, and notes, which are shown in Table 2.1.

TABLE 2.1   Distribution of documents found by type.

Document type	Number	%
Articles	2448	89.25
Reviews	287	10.46
Editorial material	4	0.14
Notes	3	0.11
Meeting abstracts	1	0.04
	2743	100

The total number of documents related to wastewater treatment with consortium found (2.743) shows that the area has been gaining space in research. This is due to the growing need to find clean, sustainable, and efficient ways to treat wastewater (Khan et al., 2022).

Records of the documents found were articles with 89.25%, followed by 10.46% of reviews. The others totaled 0.29% of the sample. However, there is still room for research on wastewater treatment with consortium, especially the discoveries of social impact, since 0.87% of the papers were indexed in the SSCI and A&HCI.

Based on the sample, the majority (2717) of the documents were written in the English language, followed by nine Spanish, six French, three Polish, three Russian, two Portuguese, one Korean, one Japanese, and one Turkish.

2.3.2 Publishing authors

Through the selection performed, the authors who contributed to the documents were analyzed. A total of 9298 authors were found, of whom 20 authors with the highest number of publications and the number of articles that had more than 100 citations on wastewater treatment with consortium are presented in Table 2.2. The 20 most prolific authors hold 15.38% documents, shaping research in this area.

Raul Munoz is the author who published the largest number of papers (65 papers), most of which deal with wastewater treatment with the consortium of microalgae and bacteria. The author has 11 articles with more than 100 citations, presenting a high impact on the subject.

2.3.3 Publishing countries

In this section, the countries that contributed to the selected documents were analyzed, totalizing 102 countries that contributed with publications on studies on wastewater treatment with consortium. Table 2.3 shows the 20 countries that provided the most published documents and the number of articles with more than 100 citations. The 20 most active countries had 107% of the articles from the total of 142.27%, since some articles have more than one country of authors.

TABLE 2.2　The 20 authors with the highest number of publications.

	Authors	Paper number	Paper number %	P100
1	Raul Muñoz	65	2.37	11
2	Yoav Bashan	34	1.24	11
3	Luz Bashan	31	1.13	10
4	Roger Ruan	21	0.77	1
5	Hee-Mock Oh	20	0.73	2
6	Yongjun Zhao	20	0.73	0
7	Eduardo Jacob-lopes	19	0.69	0
8	Elena Ficara	18	0.66	0
9	Pedro Garcia	18	0.66	0
10	Yang Liu	18	0.66	0
11	Joan Garcia	17	0.61	1
12	Jean Steyer	17	0.61	1
13	Yuhuan Liu	16	0.58	2
14	Juan Liu	16	0.58	0
15	Mallavarapu Megharaj	16	0.58	2
16	Benoit Guieysse	16	0.58	3
17	Esther Posadas	15	0.55	2
18	Saul Blanco	15	0.55	2
19	Qian Lu	15	0.55	0
20	Navid Moheimani	15	0.55	0

P100 = number of papers with at least 100 citations.

China is the country with the highest number of publications with 20.39%, strongly impacting research on consortium for wastewater treatment. This is noted as China has the greatest economic development in the world, and has invested heavily in recent years for scientific and technological progress (Kang and Liu, 2021). Then United States with 14.83% of publications and 30 articles with more than 100 citations, showing its pioneering spirit and the incentive offered for research in this area. An analysis of the countries that are members of the European Union found that they published 716 papers equivalent to 26.10% of these documents, surpassing China and the United States. On the other hand, South America represented by Brazil and Mexico, with 10.2% of publications in this area.

TABLE 2.3 The 20 countries with the highest number of publications.

	Country	Paper number	Paper number %	P100
1	China	561	20.39	17
2	United States	408	14.83	30
3	India	258	9.38	9
4	Spain	242	8.80	16
5	South Korea	147	5.34	9
6	Brazil	145	5.27	1
7	Australia	139	5.05	8
8	Mexico	136	4.94	13
9	Italy	116	4.22	3
10	France	95	3.45	4
11	Canada	93	3.38	5
12	Malaysia	81	2.94	3
13	Portugal	77	2.80	2
14	Japan	75	2.73	0
15	England	74	2.69	7
16	Germany	74	2.69	4
17	Egypt	61	2.22	1
18	The Netherlands	60	2.18	1
19	Taiwan	57	2.07	1
20	Belgium	52	1.89	3

P100 = number of papers with at least 100 citations.

2.3.4 Publishing institutions

A total of 2498 institutions contributed with the documents found. Research in this area is disseminated globally, with a diversity of research institutions and higher education. An analysis of the institutions that most collaborate with the studies of wastewater treatment with consortia was also carried out. Table 2.4 shows the 20 institutions with the highest number of documents published and the country of origin.

Once again, China stands out as it has 4 institutions with the highest number of documents. "Chinese Academy of Sciences" is the institution with the highest number of articles published worldwide, representing 3.93%. China and the United States together represent 11.68% of institutions with publications in this area, referring to a greater

TABLE 2.4 The 20 institutions with the highest number of publications.

	Institutions	Country	Paper number	Paper number %	P100
1	Chinese Academy of Sciences	China	108	3.93	5
2	Indian Institute of Technology System	India	68	2.47	1
3	Universidad de Valladolid	Spain	61	2.22	9
4	Centre National de La Recherche Scientifique	France	52	1.89	2
5	Harbin Institute of Technology	China	49	1.78	1
6	Universidad Nacional Autonoma De Mexico	Mexico	41	1.49	1
7	Egyptian Knowledge Bank	Egypt	39	1.42	1
8	University of Chinese Academy of Sciences	China	39	1.42	1
9	University of Minnesota System	United States	34	1.24	1
10	United States Department of Energy	United States	31	1.13	0
11	Nanchang University	China	30	1.09	1
12	University of Minnesota Twin Cities	United States	30	1.09	1
13	Universidade do Porto	Portugal	29	1.05	1
14	National Institute of Technology	India	28	1.02	1
15	Consejo Superior de Investigaciones Cientificas	Spain	27	0.98	2
16	Institut National de la Recherche Agronomique	France	26	0.94	1
17	Council of Scientific Industrial Research	India	25	0.91	1
18	Ghent University	Belgium	25	0.91	2
19	Korea Research Inst. of Bioscience Biotechnology	South Korea	25	0.91	2
20	Universidade Federal do Rio Grande	Brazil	25	0.91	1

P100 = number of papers with at least 100 citations.

concentration of research in the main institutions for public reasons. It was declared by 75.9% of all documents that there had funding for research, the support provided by associations to encourage development has been expressive for these institutions to develop research in this area (Konur, 2011).

2.3.5 Publishing year

Research in the area of wastewater treatment with consortium began to emerge in the early 1990s, with only 1 article being found in 1989. Fig. 2.1 shows the number of papers published over the years. The peak of publication in this area was reached in 2020 with 411 documents presented. However, only documents published until September 2021 (365 documents) were counted. This shows that research in this area is on the rise. The number of publications followed a sigmoid growth trend until 2020 ($R^2 = 0.98$). The accentuated

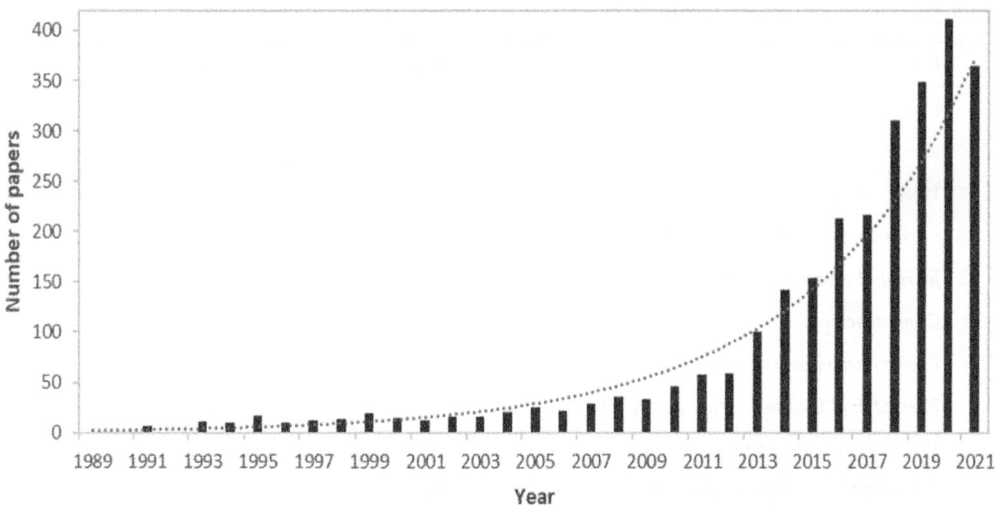

FIGURE 2.1   Publications distributed by years (1945–2021).

growth of documents since 2005 shows the importance of research in the area of wastewater treatment with microalgae consortia and is justified by the search for clean and sustainable technologies for the treatment of effluents (Khan et al., 2022).

2.3.6 Publishing journals

The analysis carried out on the WoS platform showed that 602 journals contributed with the documents found. Table 2.5 presents the 20 most influential journals. The 20 journals with the highest number of publications published 43.3% of the total papers. These 20 journals form the search pillar for documents in the wastewater treatment area with microalgae consortia. First in the rank is this *Bioresource Technology* which published 264 papers of which 22 had more than 100 citations. Followed by *Algal Research Biomass Biofuels and Bioproducts* and *Water Research* with 177 and 83 papers respectively. *Algal Research Biomass Biofuels and Bioproducts* published only 2 articles with more than 100 citations, however this is a newly created journal (2012) since *Water Research* was created in 1967, which has 11 articles with more than 100 citations. Featured journals have high impact factors 9642, 4401 and 11,236 for *Bioresource Technology*, *Algal Research Biomass Biofuels and Bioproducts*, and *Water Research*, respectively. Confirming the reliability of these journals.

2.3.7 Citation analysis

In this section, an analysis is made of the papers with the greatest impact on research on wastewater treatment with microalgae consortium. The authors of the 20 documents with the highest number of citations are shown in Table 2.6. The number of citations of a document reflects the importance of the topic, the quality and influence of the article. These documents were published between 1991 and 2018. The 2010 decade was the most

TABLE 2.5 The 20 journals with the highest number of publications.

	Journal	Paper number	Paper number %	P100
1	Bioresource Technology	264	9.60	22
2	Algal Research Biomass Biofuels and Bioproducts	177	6.43	2
3	Water Research	83	3.02	11
4	Science of the Total Environment	81	2.94	1
5	Journal of Applied Phycology	72	2.62	1
6	Chemosphere	61	2.22	3
7	Environmental Science and Pollution Research	48	1.74	0
8	International Biodeterioration Biodegradation	45	1.64	1
9	Water Science and Technology	44	1.60	0
10	Applied Microbiology and Biotechnology	39	1.42	2
11	Journal of Hazardous Materials	37	1.34	0
12	Journal of Environmental Management	36	1.31	0
13	Journal of Cleaner Production	35	1.27	2
14	Environmental Technology	31	1.13	1
15	Desalination and Water Treatment	25	0.91	0
16	Journal of Chemical Technology and Biotechnology	25	0.91	1
17	Renewable Sustainable Energy Reviews	24	0.87	4
18	Chemical Engineering Journal	22	0.80	1
19	Biochemical Engineering Journal	21	0.76	0
20	Ecological Engineering	21	0.76	0

P100 = number of papers with at least 100 citations.

proliferating with 15 documents. On the other hand, in the 2000s, only 4 documents were published.

The documents were represented by 7 research articles while the reviews were well represented with 13 documents. The review "Algal–bacterial processes for the treatment of hazardous contaminants: a review" had the highest number of citations (818).

There were 57 authors who contributed to the 20 most cited articles. Authors "Luz Bashan" and "Yoav Bashan" were the authors with the highest number of documents contributed with 3 articles. The most active countries were United States, India and Mexico with the highest number of articles (5, 4 and 4, respectively) among the 15 countries that contributed with the 20 most influential papers.

The institutions that contributed to this sample of 20 papers were 37. The institutions with the most influential documents are: The "Bashan Foundation," "Cranfield University," "Maharshi Dayanand Saraswati University," "University of Valladolid,"

TABLE 2.6 The 20 papers on wastewater treatment with the highest number of citations consortium.

	Paper	Journal	Year	Total number of citations
1	Muñoz and Guieysse (2006)	*Water Research*	2006	818
2	Chinnasamy et al. (2010)	*Bioresource Technology*	2010	479
3	Henderson et al. (2008)	*Water Research*	2008	437
4	de-Bashan and Bashan (2010)	*Bioresource Technology*	2010	432
5	Ramanan et al. (2016)	*Biotechnology Advances*	2016	428
6	Khan et al. (2018)	*Microbial Cell Fact*	2018	411
7	Bashan et al. (2013)	*Plant and Soil*	2013	402
8	Parmar et al. (2011)	*Bioresource Technology*	2011	336
9	Hart et al. (1991)	*Hydrobiologia*	1991	331
10	Wahlen et al. (2011)	*Bioresource Technology*	2011	309
11	Olguín (2012)	*Biotechnology Advances*	2012	264
12	Muñoz et al. (2015)	*Reviews in Environmental Science and Bio/Technology*	2015	246
13	Bhatnagar et al. (2011)	*Applied Energy*	2011	246
14	Oren (2010)	*Environmental Technology*	2010	246
15	Moreno-Garrido (2008)	*Bioresource Technology*	2008	245
16	Markou et al. (2014)	*Water Research*	2014	228
17	Subashchandrabose et al. (2011)	*Biotechnology Advances*	2011	228
18	De-Bashan et al. (2004)	*Water Research*	2004	225
19	Gonçalves et al. (2016)	*Algal Research*	2016	207
20	Henderson et al. (2010)	*Water Research*	2010	206

"University of Georgia" and "University System of Georgia." The 20 most influential articles were published in 10 journals. The most cited journals in the area were: *Water Research*, *Bioresource Technology*, and *Biotechnology Advances* which received 1916, 1802, and 922 citations in total, respectively.

2.4 Conclusion

The scientometric analysis carried out in this study, on research in the area of wastewater treatment with consortium of microalgae containing bacteria, fungi, and/or several species of microalgae, provided important data on the development of research in this area. Insights were provided on the quantity, language, and type of documents found, important authors, influential countries and research institutions, years of publication, and papers with the highest number of citations. The sample found revealed that 2743 documents were published from 1989 to 2021. Presenting a sigmoidal growth since the beginning of the research, it is expected to continue to follow the behavior and grow in the coming years due to the great importance of discoveries of clean and sustainable technologies for the treatment of wastewater. The results show that scientometric studies in this area have great potential to provide information on the progress of research on wastewater treatment with consortium, as well as solid technologies and the recognition of emerging methods. Especially for the provision of incentives, as there is interest especially on researchers, their institutions, and their most influential countries on research.

Appendix A

The keyword

"microalgae" or "microalga" or "micro-alga" or "micro-algae" or "chlorophycea" or "Chlorella" or "Dunaliella" or "Euglena" or "Haematococcus" or "Nannochloropsis" or "Ostreococcus" or "Scenedesmus" or "Tetraselmis"

"wastewater" or "effluent" or "waste water" or "waste-water" or "treatment" or "biotreatment" or "biodegradation" or "nutrient removal" or "nutrient recovery" or "phycoremediation" or "bioremediation"

"consortium" or "co-culture" or "consortia" or "co-cultivation" or "Fungi" or "fungus" or "bacterial" or "bacteria" or "Cyanobacteria"

References

Aketo, T., et al., 2020. Selection and characterization of microalgae with potential for nutrient removal from municipal wastewater and simultaneous lipid production. Journal of Bioscience and Bioengineering. Available from: https://doi.org/10.1016/J.JBIOSC.2019.12.004.

Bashan, Y., de-Bashan, L.E., Prabhu, S.R., Hernandez, J.-P., 2013. Advances in plant growth-promoting bacterial inoculant technology: formulations and practical perspectives (1998–2013). Plant and Soil 378, 1–33.

Bhatnagar, A., Chinnasamy, S., Singh, M., Das, K.C., 2011. Renewable biomass production by mixotrophic algae in the presence of various carbon sources and wastewaters. Applied Energy 88, 3425–3431.

Chia, S.R., et al., 2021. CO_2 mitigation and phycoremediation of industrial flue gas and wastewater via microalgae-bacteria consortium: Possibilities and challenges. Chemical Engineering Journal 425, 131436.

Chinnasamy, S., Bhatnagar, A., Hunt, R.W., Das, K.C., 2010. Microalgae cultivation in a wastewater dominated by carpet mill effluents for biofuel applications. Bioresource Technology 101, 3097–3105.

Dasan, Y.K., Lam, M.K., Yusup, S., Lim, J.W., Lee, K.T., 2019. Life cycle evaluation of microalgae biofuels production: effect of cultivation system on energy, carbon emission and cost balance analysis. The Science of the Total Environment 688, 112–128.

de Carvalho, J.C., et al., 2020. Microalgal biomass pretreatment for integrated processing into biofuels, food, and feed. Bioresource Technology 300, 122719.

de-Bashan, L.E., Bashan, Y., 2010. Immobilized microalgae for removing pollutants: Review of practical aspects. Bioresource Technology 101, 1611–1627.

De-Bashan, L.E., Hernandez, J.P., Morey, T., Bashan, Y., 2004. Microalgae growth-promoting bacteria as "helpers" for microalgae: a novel approach for removing ammonium and phosphorus from municipal wastewater. Water Research 38, 466–474.

Fontoura, J.T., Rolim, G.S., Farenzena, M., Gutterres, M., 2017. Influence of light intensity and tannery wastewater concentration on biomass production and nutrient removal by microalgae Scenedesmus sp. Process Safety and Environmental Protection 1, 355−362.

Gonçalves, A.L., Pires, J.C.M., Simões, M., 2016. A review on the use of microalgal consortia for wastewater treatment. Algal Research. Available from: https://doi.org/10.1016/j.algal.2016.11.008.

Hart, B.T., et al., 1991. A review of the salt sensitivity of the Australian freshwater biota. Hydrobiologia 210, 105−144.

Hena, S., Fatimah, S., Tabassum, S., 2015. Cultivation of algae consortium in a dairy farm wastewater for biodiesel production. Water Resources and Industry 10, 1−14.

Henderson, R.K., Baker, A., Parsons, S., et al., 2008. Characterisation of algogenic organic matter extracted from cyanobacteria, green algae and diatoms. Water Research 42. Available from: https://doi.org/10.1016/j.watres.2007.10.032.

Henderson, R.K., Parsons, S.A., Jefferson, B., 2010. The impact of differing cell and algogenic organic matter (AOM) characteristics on the coagulation and flotation of algae. Water Research 44, 3617−3624.

Kang, Y., Liu, R., 2021. Does the merger of universities promote their scientific research performance? Evidence from China. Research Policy 50, 104098.

Khan, M.I., Shin, J.H., Kim, J.D., et al., 2018. The promising future of microalgae: current status, challenges, and optimization of a sustainable and renewable industry for biofuels, feed, and other products. Microbial Cell Factories 17. Available from: https://doi.org/10.1186/s12934-018-0879-x.

Khan, S.A.R., Ponce, P., Yu, Z., Golpîra, H., Mathew, M., 2022. Environmental technology and wastewater treatment: strategies to achieve environmental sustainability. Chemosphere 286, 131532.

Konur, O., 2020b. The evaluation of the research on the biofuels: a scientometric approach | Publons. Handbook of Algal Science, Technology and Medicine. Elsevier.

Konur, O., 2020a. The scientometric analysis of the research on the algal foods. Handbook of Algal Science, Technology and Medicine. Elsevier, pp. 485−506. Available from: https://doi.org/10.1016/B978-0-12-818305-2.00031-0.

Konur, O., 2020c. The scientometric analysis of the research on the algal bioremediation. Handbook of Algal Science, Technology and Medicine. Elsevier, pp. 607−627. Available from: https://doi.org/10.1016/B978-0-12-818305-2.00038-3.

Konur, O., 2011. The scientometric evaluation of the research on the algae and bio-energy. Applied Energy 88, 3532−3540.

Markou, G., Vandamme, D., Muylaert, K., 2014. Microalgal and cyanobacterial cultivation: the supply of nutrients. Water Research 65, 186−202.

Mehariya, S., Goswami, R.K., Karthikeysan, O.P., Verma, P., 2021. Microalgae for high-value products: a way towards green nutraceutical and pharmaceutical compounds. Chemosphere 280, 130553.

Moreno-Garrido, I., 2008. Microalgae immobilization: current techniques and uses. Bioresource Technology 99, 3949−3964.

Muñoz, R., Guieysse, B., 2006. Algal−bacterial processes for the treatment of hazardous contaminants: a review. Water Research 40, 2799−2815.

Muñoz, R., Meier, L., Diaz, I., Jeison, D., 2015. A review on the state-of-the-art of physical/chemical and biological technologies for biogas upgrading. Reviews in Environmental Science and Bio/Technology 14, 727−759.

Murshid, S., Dhakshinamoorthy, G.P., 2021. Application of an immobilized microbial consortium for the treatment of pharmaceutical wastewater: Batch-wise and continuous studies. Chinese Journal of Chemical Engineering 29, 391−400.

Olguín, E.J., 2012. Dual purpose microalgae-bacteria-based systems that treat wastewater and produce biodiesel and chemical products within a Biorefinery. Biotechnology Advances 30, 1031−1046.

Oren, A., 2010. Industrial and environmental applications of halophilic microorganisms. Environmental Technology 31, 825−834.

Parmar, A., Singh, N.K., Pandey, A., Gnansounou, E., Madamwar, D., 2011. Cyanobacteria and microalgae: a positive prospect for biofuels. Bioresource Technology 102, 10163−10172.

Pena, A.C.C., Agustini, C.B., Trierweiler, L.F., Gutterres, M.V., 2020. Journal of Cleaner Production 263, 121618.

Pérez, L., Salgueiro, J.L., Maceiras, R., Cancela, Á., Sánchez, Á., 2016. Study of influence of pH and salinity on combined flocculation of Chaetoceros gracilis microalgae. Chemical Engineering Journal 286, 106−113.

Ramanan, R., Kim, B.H., Cho, D.H., Oh, H.M., Kim, H.S., 2016. Algae–bacteria interactions: Evolution, ecology and emerging applications. Biotechnology Advances 34, 14–29.

Shahid, A., et al., 2020. Cultivating microalgae in wastewater for biomass production, pollutant removal, and atmospheric carbon mitigation; a review. The Science of the Total Environment 704, 135303.

Solís-Salinas, C.E., et al., 2021. Long-term semi-continuous production of carbohydrate-enriched microalgae biomass cultivated in low-loaded domestic wastewater. The Science of the Total Environment 798, 149227.

Subashchandrabose, S.R., Ramakrishnan, B., Megharaj, M., Venkateswarlu, K., Naidu, R., 2011. Consortia of cyanobacteria/microalgae and bacteria: Biotechnological potential. Biotechnology Advances. Available from: https://doi.org/10.1016/j.biotechadv.2011.07.009.

Talapatra, N., Gautam, R., Mittal, V., Ghosh, U.K., 2021. A comparative study of the growth of microalgae-bacteria symbiotic consortium with the axenic culture of microalgae in dairy wastewater through extraction and quantification of chlorophyll. Materials Today: Proceedings. Available from: https://doi.org/10.1016/J.MATPR.2021.06.227.

Wahlen, B.D., Willis, R.M., Seefeldt, L.C., 2011. Biodiesel production by simultaneous extraction and conversion of total lipids from microalgae, cyanobacteria, and wild mixed-cultures. Bioresource Technology 102, 2724–2730.

Wang, J., et al., 2021. Construction of fungi-microalgae symbiotic system and adsorption study of heavy metal ions. Separation and Purification Technology 268, 118689.

Metabolic engineering of algal strains for enhancing wastewater treatment

Irteza Qayoom[1], Aashia Altaf[1], Suhail Bashir[2] and Zahid Bashir[1]

[1]Sri Pratap College Campus, Cluster University Srinagar, Srinagar, Jammu and Kashmir, India
[2]Al Noor Environment Consultants, Sharjah, United Arab Emirates

3.1 Introduction

Algae are nonvascular plants with a broad range of research purposes, including high species diversification, biofuel sources, and heavy metal adsorption, and are utilized as ingredients in health supplements after processing. Algal metabolism refers to the metabolic and transport processes through which algae absorb nutrients and convert them into the materials required for the organism's growth, reproduction, and protection of the organisms. Algal metabolism shares many similarities to that of other living species, but it also has certain differences. Environmental factors such as nutrition availability, temperature, and light extremes influence algal metabolism to a considerable extent. The macronutrients involved in metabolic processes in algae include nitrogen, carbon, and phosphorus. Among these, nitrogen is one of the main components of algae. In algae the regulation of transport and absorption varies greatly. The nitrogen resource utilized in the media drives microalgae culture, and their biomass is significant for third-generation biofuel production. For absorbing inorganic and organic nitrogen molecules, green microalgae have evolved a variety of regulatory mechanisms. The glutamine synthetase/glutamate synthase (GS/GOGAT) cycle is used to absorb ammonium, which is a primary nitrogen source under most conditions. However, under high ammonium conditions (e.g., *Chlorella* species) or when nitrogen is limited, some green microalgae use NADP-dependent glutamate dehydrogenase (GDH) to absorb ammonium (*Stichococcus bacillaris*, which possesses a high-affinity NADP-GDH). NAD-dependent GDH tends to function primarily in a heterotrophic environment and when amino acids are used as nitrogen sources, indicating that it has a catabolic role for this enzyme. At both high and low substrate concentrations, microalgae take up nitrogen faster

Valorisation of Microalgal Biomass and Wastewater Treatment
DOI: https://doi.org/10.1016/B978-0-323-91869-5.00003-X

33

per unit of biomass than macroalgae, while microalgae have a substantially higher affinity for nitrogen than macroalgae. Phosphorus (P) is a nonrenewable, vital nutrient that often restricts plant growth. *Chlorella* absorbs nitrogen and phosphorus reaches concentrations of 5.0%—10.1% nitrogen and 0.5%—1.3% phosphorus, respectively. Phosphorus could enter endogenous phosphorus as part of the microalga's photosynthetic metabolism or as an external flux. Phosphorus is an essential macronutrient for the physiological ecology of eukaryotic microalgae and cyanobacteria. Nutrient deprivation restricts cell division in some microalgae, resulting in the accumulation of organic carbon (e.g., in the form of triacylglycerols). Phosphorus is an essential component of the algal cell's structure and metabolism. It is needed in biomolecules, such as DNA, long-chain polyphosphates (polyPs), and phosphoproteins, all of which have a long life span. Phosphorus can be found in sugar molecules, RNAs, and cell membranes as phospholipids in biomolecule systems with shorter half-lives. The partitioning of photosynthetically assimilated carbon into pools of different macromolecules, such as proteins, lipids, carbohydrates, and nucleic acids, is known as carbon allocation in alga. The carbon flux for the synthesis of cell macromolecules is optimal under ideal conditions of light, nutritional availability, and temperature is in equilibrium with the energy flux provided by photosynthesis. Microalgae are unicellular photosynthetic microbes able to change atmospheric CO_2 into lipids and different high-valuable renewables. Pyrenoids are the carbon absorption sites of algae, and their structural properties differ depending on the species. Carbon concentrating mechanisms (CCMs) are the main processes in algal photosynthesis, metabolism, development, and biomass production, and uch of the research on CCMs has been done on model microalgal species. Algae can grow 1—2 cm per day, but if they absorb more carbon dioxide, they can grow considerably quicker.

3.2 Genetic modification of algae to enhance wastewater treatment

Potential uses of algae include bioremediation and wastewater treatment for agricultural runoff and other sewage wastes, as well as for removing toxic metals from industrial wastewater by uptake of heavy metals into their biomass, which stimulates the heavy metal stress response, resulting in heavy metal binding factors. It is also possible to increase the physiological features of algae by genetic and transgenesis alterations, which might boost the algae's utility in wastewater treatment and bioremediation even more. In the development of strong microalgal strains, genetic engineering is at the forefront. Microalgae may be developed as a renewable resource using techniques such as high-throughput screening approaches and genetic tools, as well as the use of the genome sequence. For wastewater treatment and bioremediation, this chapter explores the possibilities for genetic alterations and technologies that may be applied to algae.

3.2.1 Genetic engineering tools

Microalgae have been genetically engineered to increase their ability to utilize sunlight efficiently in laboratory trials. The decrease of the algal photosystem's chlorophyll antenna size is another topic that has attracted research attention. It has been estimated that algae with smaller chlorophyll antennas are 2–3 times more productive than those with larger antennas. To counteract the "packaging effect" and improve light dispersion inside the LHC (light-harvesting complex), a smaller antenna decreases the amount of pigment stored within the cell's LHC. Antenna proteins play an important function in dissipating surplus photon energy, therefore lowering their size and related proteins may not impair the photosystem's capacity to perform photosynthetically but might harm cell performance. As a result, species with a smaller antenna may be more susceptible to photodamage than others. Increasing the functioning of RuBisCO, a photosynthetic rate-limiting step is thought to improve light use efficiency. As a strategy to get around these restrictions, some have advocated genetic engineering of RuBisCO to boost enzymatic reactions and make them more selective to carbon dioxide rather than oxygen. To boost biomass production in high-rate algae pond systems (HRAPs), it has been proposed to introduce accessory pigments generated by other organisms, such as cyanobacteria's phycobilin pigment, into the desired species to enhance light absorption over the whole PAR spectrum.

3.2.2 Transformation technologies and selectable markers

Over the last decade, significant progress has been made in the field of microalgal genetics. Numerous microalgae genomes have been sequenced thus far, and many more are being sequenced, thanks to the establishment of Expressed Sequence Tag databases. Many genetic and molecular studies have focused on *Chlamydomonas reinhardtii*, which is a green alga. Most methods for transgenic expression and gene knockdown have been created for this algae species specifically. Diatoms and other algae, on the other hand, are becoming more and more popular in many industrial uses. C. *reinhardtii* nuclear transformation utilizing polyethylene glycol was first shown in the 1980s. Through the effective integration of the yeast arg4 locus, we were able to successfully complement an arginine-requiring, cell wall–deficient mutant. In the late 1980s, the biolistic transformation technique was used to successfully transport native genes to supplement auxotrophic growth in mutant C. *reinhardtii* and achieve sustained nuclear transformation. Transforming the nuclear genome of C. *reinhardtii* was most accomplished using the glass bead agitation and electroporation procedures first reported in the late 1990s. In other microalgal species, such as *Dunaliella*, the nuclear genome of C. *reinhardtii* has been effectively transformed by alternative means, such as the use of silicon carbide whiskers, Agrobacterium tumefaciens, and nanoparticles.

3.2.3 Genome editing

More and more progress has been achieved during recent years in the repertoire of microalgae genetic engineering tools to increase their usefulness as model organisms for scientific and/or industrial uses, notably in the remediation of wastewaters. Zinc-finger nucleases (ZFNs), meganucleases (MNs), transcription activator-like effector nucleases (TALEN), and CRISPR/Cas9 have emerged as green tools for genome editing in various microorganisms during the past decade. Cost, process complexity, and the capacity to do several adjustments at once all play a role in how widely these technologies may be used. The CRISPR/cas9-mediated genome-editing method has now become the ultramodern technique because of its simplicity and flexibility. Among others, ZFN-mediated genome editing was the first to have a significant impact on microalgae gene editing. Using engineered ZFNs, Sizova et al. and Greiner et al. targeted the *C. reinhardtii* COP$_3$ and COP$_4$ genes using designed ZFNs in their studies. Unique ZFNs with high specificity and affinity for the target sites are the most difficult to develop. Before carrying out the actual experiment, ZFNs must be verified using a gene-targeting selection technique. Targeting uridyl diphosphate (UDP)-glucose pyro phosphorylase in *Phaeodactylum tricornutum* with MNs and TALENs increased lipid accumulation, as well. Homologous recombination of the urease gene in *P. tricornutum* has been effectively disrupted using TALENs. A similar study by Serif et al. in P. *tricornutum* aimed to test if the uridine monophosphate synthase gene could be employed as an endogenous positive selectable marker for DNA-free genome editing. Jiang et al. used C. *reinhardtii* to show the CRISPR/cas9 system's efficacy in microalgal species for the first time. Using a codon optimized Cas9 gene and a single-guide RNA, this study effectively altered four genes (sgRNA). *P. tricornutum*, a model marine microalga, has also shown the capacity of the CRISPR system to be adaptable.

3.3 Genetic engineering in algae for enhanced nutrient removal

Rising global concern over increased pollution-related issues has resulted in the development of an approach toward advanced wastewater treatment and to achieve Sustainable Goal-6, it is also necessary to progress in wastewater management. Different sources of wastewater such as municipal, agricultural, and industrial wastewaters that contain a higher concentration of nutrients are particularly N and P as well as other trace metals and are present as secondary effluent gets discharged into the water bodies, thus negatively impacting the aquatic ecosystem (Preisner et al., 2021). Although there are conventional techniques available to deal with them, the ability to degrade these nutrients is limited. Furthermore, a long-term sustainable solution to wastewater management is required. Therefore it is necessary to cultivate algae strains that are adaptable and tolerant to remove nutrients from wastewater. This has been proven to be a promising method since algae have been recognized as an environmentally acceptable, economically sustainable, and inexhaustible source of carbohydrate, protein, and lipid, along with many therapeutically active enzymes, vitamins, sterols, and pigments (Kumar et al., 2021). Moreover, algae have shown efficacious removal of nutrients from wastewater as this environment is an ideal media for algae where algal cells can efficiently fix phosphorus and nitrogen content in wastewater and they also have a good nutrient absorption capacity (Choudhary et al., 2016; Jayaseelan, 2021) which is essential for its growth and metabolism. This in turn offers a great potential for sustainable by-products such as bioplastic, biofuels, and biofertilizers.

Chlorella genus (Wu et al., 2019), which includes *C. kessleri, C. pyrenidosa, C. sorokiniana, C. vulgaris*, and *C. emersonii*, has been studied extensively for its ability to remove nutrients from wastewater (Wu et al., 2019). Biotechnological interferences, transformations, and selection methods for a specific gene alteration can all be used to improve nutrient removal efficiency. Using genome editing and gene interference as methods of gene targeting is the most efficient method currently available. CRISPR-Cas9, ZFN, and TALEN are only a few of the new genome-editing techniques that have lately been used in gene modification. MicroRNA (miRNA) and silent RNA (siRNA), gene-interfering tools used in genetic modification, can assist achieve desirable microalgal phenotypes, either by stimulating or suppressing gene expression (Ng et al., 2017). A combination of molecular phylogeny and the use of omics technology to identify phenotypic and genotypic features in wastewater can help improve the removal of nutrients (Chamberlin et al., 2018). As omics approaches to algae are introduced, "Algomics" can provide insight into metabolic processes and dynamic changes in metabolic pathways as well as protein synthesis pathways and bioremediation pathways, and thus help unlock cell metabolism and physiology. A thorough understanding of these changes can ultimately lead to a better understanding of how cells function. (Patel et al., 2015, respectively) Since nitrogen metabolism is a major focus of microalgal metagenomics (Mishra et al., 2019), there has been relatively little research on individual microalgae utilized in wastewater treatment. Exogenous microalgae *Chlorella* were used in the work on Metagenomics investigation of the microbiota in swine lagoon effluent (Illumina Miseq platform). A CO_2 air bubbling phytoremediation mode with strong microalgae growth was shown to simultaneously remove N and P. This was the first research in a phytoremediation system that used metagenomics to examine the native microbial makeup of the soil (Ye et al., 2016). Proteomics-based research identified 2358 proteins with abundances of enzymes and proteins associated with nitrogen metabolism and assimilation, carbohydrates synthesis and utilization, amino acid recycling, evidence of oxidative stress, and little to the lipid biosynthesis of *C. reinhardtii* cultured in a wastewater containing nitrate and ammonia suitable for outdoor use (Patel et al., 2015). The findings of this study shed light on how microalgae respond to nitrogen inputs (nitrate/ammonia) and the enriched route, enzymes that may be genetically improved. Microalgae growth and nutrient removal may be better understood if more metabolic pathways were dissected.

3.4 Selection of marker genes

The area of genetic engineering holds immense promise for cultivating plants that have the traits we want them to have. Gene transfer can be used to introduce new features into plants that are hard to achieve through traditional breeding methods. The low frequency with which the injected DNA is taken up and integrated into the plant genome is the most significant technical barrier in existing gene transfer technologies. One in a thousand to one in a million times, this alien DNA is successfully incorporated. As a result, a marker gene is always cointroduced with the target gene. Only those plant cells that have the marker-encoded characteristic flourish under optimal growth conditions, thanks to selectable markers. Herbicide and antibiotic resistance have been the most often employed selectable marker genes. It has been argued that antibiotic-resistant genes, which can be easily produced and duplicated in bacteria, are the most contentious of these genes' many features. In the meantime, the biggest risk is that agricultural plants created with genes might end up contaminating harmful bacteria and causing disease. These situations and features create the specter of synthetic medications losing their therapeutic efficacy.

There has already been a rising medical worry about the growth of antibiotic-resistant bacteria and the limited supply of antibiotics.

Any new gene should be reviewed and dealt with in the same way as any other novel gene. Marker gene products need to be tested for allergenicity and potential toxicity, among other things. The impact of the marker gene's horizontal dissemination to unanticipated hosts should be evaluated regularly. It is necessary to analyze each of these marker genes on its own because of the differences in their features. Gene marker—based models for genetic evaluation may be developed through genomic selection, which uses molecular genetic markers to plan new breeding programs and build new breeding programs (Tuberosa, 2012). Complex features may be gained more quickly and cost-effectively through current plant breeding programs, thanks to advancements in genetics and breeding techniques. Genomic selection in agricultural plants relied heavily on the cost—benefit analysis. For its successful and effective implementation in crop species, the availability of genome-wide high-throughput, cost-effective, and flexible markers, having low ascertainment bias, suitable for large population size as well as for both model and nonmodel crop species with or without the reference genome sequence was the most important factor. Such issues were among the main drawbacks of prior marker systems. To achieve desired plant breeding results, the two primary methodologies of traditional and marker-assisted breeding (MAB) are employed and practiced. There are several drawbacks associated with conventional breeding programs, such as the lengthy period required to develop a new crop variety based on phenotypic selection and the high environmental noise that comes with it, which makes conventional breeding less effective for complex and low-heritable traits. The use of molecular markers for indirect selection of characteristics in crop species, little phenotypic information during training, and an effort to overcome the limits of conventional breeding are some of the advantages of advanced MAR.

3.5 Reporter and promoter genes

The characteristics of the reporter genes are extensive. Additionally, they aid in the study of gene expression and its regulation, and the movement of proteins. For gene expression analysis, reporter genes are utilized to encode enzymes that can be found in the genome. In gene expression investigations, the quality reporter genes have become a useful and improved instrument (Naylor, 1999). Researchers in domains, including molecular biology, biochemistry, pharmaceuticals, and biomedicine, rely on these reporter genes all the time. Scientists may use these high-quality reporter genes in cell cultures, animals, and plants to investigate and track different genes' activity. A limited percentage of people can benefit from gene therapy, but researchers can utilize reporter genes to determine whether cells have taken up and integrated the gene now under investigation into their genome. To ensure that just the reporter protein is produced, the reporter genes can be linked to other sequences.

The primary and some basic properties and events that the reporter genes can report:

1. the strength of promoters, whether native or modified for reverse genetic studies;
2. the efficiency of gene delivery systems;

3. the intracellular fate of a gene product, a result of protein traffic;
4. interaction of two proteins in the two-hybrid systems or of a protein and a nucleic acid in the one-hybrid system; and
5. the efficiency of translation initiation signals and the success of molecular cloning efforts.

When it comes to gene coding regions, cis-acting elements, which are specialized binding sites for proteins involved in the initiation and control of transcription, may be found in gene promoters. There are many types of promoters used in the field of biotechnology, etc. and include mainly constitutive, tissue-specific promoters, inducible promoters, and synthetic promoters. These quality and essential promoters dominate expressions largely in almost all the tissues and if not entirely, it is evident that these are independent of environmental and developmental factors.

3.6 Limitations in the field of genetically engineered algae

In genetic engineering, algae have become more important, although several obstacles and constraints need to be overcome. There are several issues associated with the creation of transgenic algae such as decreased gene expression and gene silencing, which stem from a cell's reaction to a virus or occasionally a foreign DNA and in response to the management of development challenges. When using DNA constructions from nonhomologous sources, problems occur. It is typically regarded as a disadvantage for translation if the DNA of the donor species is difficult to discover in the target organisms, and this harms expression rates because of the lack of DNA-related tRNA. It is crucial to know the codon use of most species in business. Examples of transformation of *Volvox carteri* and *C. reinhardtii* genes in *Streptoulloteichus hinduslanus* were identified as an important selectable marker. Genetic information for robust and economically acceptable strains of microalgae is not readily available, which limits the possibility for improving process economics through genetic engineering of microalgae. DNA synthesis and genetic modification tools and techniques have advanced rapidly in recent years, making it possible to build microalgae with more complicated functionalities than ever before. Even nevertheless, a lack of genetic strain design guidelines continues to impede advancement in this field. Once the genetically modified strains are produced, the safety of human health and the environment will be the determining factor in their economic success. CRISPR, a nontransgenic and marker-free genome editing technology, has the potential to transform microalgal bioengineering for the manufacture of non-GMO (genetically modified organisms) alga goods.

3.7 Future prospective

Genetically engineered algae, due to their complex and efficient complex metabolic capacity, are being continuously explored at a greater level for pharmaceuticals and other industrially important bioactivities. However, suboptimal yield and outcome of the bioactive of interest in robust wild-type strains are of perennial concern for their industrial applications. So, to deal with and address all such limitations, the improvement of strains through genetic engineering could play a great role. It is evident that the advanced tools for genetic engineering have emerged at a greater level, and they remain underused for microalgae as compared to other microorganisms. These genetically engineered algae have a lot of properties that need to be explored to deal with environmental constraints. Algae could be utilized in bioremediation and wastewater treatment to decrease nitrogen and phosphorus contents in agricultural wastes and to remove the toxic metals from industrial wastewater through the heavy metal uptake from the environment and stimulating a heavy metal stress response, thereby producing heavy metal binding factors and proteins.

There is still a lot that is needed to be explored. Future research will help the modern world to deal with and curb environmental pollution with the help of these efficient microorganisms. Genetically engineered algae are also utilized for concomitant CO_2 sequestration, wastewater treatment, and biomass production for high-volume low-value products.

3.8 Conclusions

Over the years, significant progress has been made to improve the catalog of available tools for genetic engineering in microalgae. CRISPR/cas9-mediated genome-editing system has emerged as the ultramodern tool due to its simplicity and versatility. Rising global concern over increased pollution-related issues has resulted in the development of an approach toward advanced wastewater treatment and to achieve Sustainable Goal-6, it is also necessary to progress in wastewater management. Algae have shown efficacious removal of nutrients from wastewater as this environment is an ideal media for algae. Gene-editing and gene-interfering can help in achieving desired microalgal phenotypes. The introduction of omics approaches in algae offers an insight into metabolic processes, dynamic changes in the metabolic pathways, protein synthesis pathways, bioremediation pathway, and the effect of different conditions on algal metabolism.

References

Chamberlin, J., Harrison, K., Zhang, W., 2018. Impact of nutrient availability on tertiary wastewater treatment by *Chlorella vulgaris*. Water Environment Research 90 (11), 2008–2016. Available from: https://doi.org/10.2175/106143017x15131012188114.

Choudhary, P., Prajapati, S.K., Malik, A., 2016. Screening native microalgal consortia for biomass production and nutrient removal from rural wastewaters for bioenergy applications. Ecological Engineering 91, 221–230. Available from: https://doi.org/10.1016/j.ecoleng.2015.11.056.

Jayaseelan, M., et al., 2021. Microalgal production of biofuels integrated with wastewater treatment. Sustainability. 13 (16), 8797. Available from: https://doi.org/10.3390/su13168797.

Mishra, A., et al., 2019. Omics approaches for microalgal applications: prospects and challenges. Bioresource Technology 291. Available from: https://doi.org/10.1016/j.biortech.2019.121890.

Ng, I.S., et al., 2017. Recent developments on genetic engineering of microalgae for biofuels and bio-based chemicals. Biotechnology Journal 12 (10). Available from: https://doi.org/10.1002/biot.201600644.

Patel, A.K., et al., 2015. Comparative shotgun proteomic analysis of wastewater-cultured microalgae: nitrogen sensing and carbon fixation for growth and nutrient removal in *Chlamydomonas reinhardtii*. Journal of Proteome Research 14 (8), 3051–3067. Available from: https://doi.org/10.1021/pr501316h.

Preisner, M., Neverova-Dziopak, E., Kowalewski, Z., 2021. Mitigation of eutrophication caused by wastewater discharge: a simulation-based approach. Ambio. Springer Science and Business Media LLC 50 (2), 413–424. Available from: https://doi.org/10.1007/s13280-020-01346-4.

Wu, T., et al., 2019. Sequencing and comparative analysis of three *Chlorella* genomes provide insights into strain-specific adaptation to wastewater. Scientific Reports 9 (1). Available from: https://doi.org/10.1038/s41598-019-45511-6.

Ye, J., et al., 2016. Metagenomic analysis of microbiota structure evolution in phytoremediation of a swine lagoon wastewater. Bioresource Technology 219, 439–444. Available from: https://doi.org/10.1016/j.biortech.2016.08.013.

Lab-scale to commercial-scale cultivation of microalgae

Ananya Roy Chowdhury[1] *and Achintya Das*[2]

[1]Department of Botany, Chakdaha College, Chakdaha, Nadia, India [2]Department of Physics, Mahadevananda Mahavidyalaya, Barrackpore, West Bengal, India

4.1 Introduction

The microalgae are a diverse group of cosmopolitan algae which may be of prokaryotic as well as dikaryotic type. They are predominantly aquatic; some can be edaphic, epilithic, lithophytic, cryophilic too. Their phylogenetic diversity is exhibited in their physiological characters as well as in the wide types of metabolites they produce. Microalgae, therefore, is a point of great commercial interest. The first commercial application of microalgae was started in the 1940s. Later in 1960, commercial production of *Chlorella* (heath food), a green unicellular freshwater alga, was started in Japan and Taiwan. After *Chlorella*, another filamentous blue-green alga, *Spirulina* was commercially produced in Mexico. After that, *Dunaliella salina* for beta carotene, *Haematococcus pluvialis* for astaxanthin, *Crypthecodinium cohnii* for long-chain PUFA, docosahexaenoic acid (DHA) was commercially cultivated in different countries of the world. Microalgae cultivation is aimed at the production of high-value products as its production cost is of high range. In addition to high-value secondary metabolites, microalgae are currently becoming a strong need because they can serve as the source of alternative fuel too. Microalgae-derived biodiesel will be a promising substitute for petroleum-derived fuels, therefore, not only for high nutritional value but also for strong fueling energy microalgae are of excellent commercial importance. So, to fulfill all these demands to mankind, the cultivation of microalgae has started from lab scale to commercial scale too. The present study aims to identify how lab-scale cultivation of microalgae can be transferred to large commercial scale cultivation, as well as the pros and cons of doing so.

Valorisation of Microalgal Biomass and Wastewater Treatment
DOI: https://doi.org/10.1016/B978-0-323-91869-5.00006-5

43

4.2 Cultivation of microalgae on a lab-scale

Several methods and use of basic culture were started long back in the 1800s and early 1900s. The algal culturing in laboratory scale is categorized into the following categories, such as 19th-century algal culturing, 20th-century algal culturing, conventional algal culturing, microalgal mass culturing, mariculture of seaweeds and cryopreservation, etc.

4.2.1 Nineteenth-century algal culturing

In 1850 Ferdinand Cohn first cultivated unicellular flagellate *Haematococcus* (Chlorophyceae) in his laboratory in Poland and the process he followed was termed as "cultivation." It was the first-ever published report of algal culture. He did not use any algae-specific culture medium and there was no algal culture isolation at all. Later on, in St. Petersburg first successful cultivation of *Chlorococcum infusionum* (Schrank), Meneghini, and *Protococcus viridis* C. Agardh were done using Knop's solution (Andersen and Culturing, 2005). Beijerinck first isolated free-living *Chlorella* and *Scenedesmus* and also isolated symbiotic green alga from Hydra (Andersen and Culturing, 2005). Later he also established axenic cultures of blue-green algae and diatoms. He also recognized that Anabaena can be grown in a completely nitrogen-free medium. At the end of the 19th century, it was established that agar and gelatin can be used as culture media for algal growth and sometimes the modifications can be done by adding other nutrients too.

4.2.2 Twentieth-century algal culturing

4.2.2.1 Conventional algal culturing

Zumstein, a student of Klebs and Benecke first isolated bacteria-free cultures of Euglena gracilis Klebs (Andersen and Culturing, 2005), Among different algae, Gymnodinium fucorum, Polytoma uvella Ehrenberg, *Carteria*, *Chlamydomonas*, *Chlorogonium*, and *Spondylomorum* (Chlorophyceae), *Chaetoceros*, *Skeletonema*, and *Thalassiosira*, Eudorina, Gonium, Acetabularia, *Scenedesmus*, *Navicula*, and Euglena, *Prymnesium parvum* Carter were isolated successfully and the cultures were established in the axenic condition in laboratory conditions by several phycologists. For obtaining bacteria-free algal growth, the use of antibiotics in the culture media was highlighted at this time. In 1948 first attempted the application of streptomycin to develop colorless mutant variants of Euglena (Andersen and Culturing, 2005). The addition of different metals in the culture media was also established. Among different metals, cobalt, copper, manganese, molybdenum, vanadium, and zinc were preferably added more in the culture media.

4.2.2.2 Microalgal mass culturing

The axenic culturing of *Chlorella* in dense laboratory culture was established (Andersen and Culturing, 2005). Later, in Germany and Japan, microalgal mass culturing was followed in the case of *Chlorella* cultivation. The main aim of microalgal mass culturing was to investigate the use of unicellular algae in space vehicles, particularly for air revitalization, waste treatment, food production etc.

4.2.2.3 *Mariculture of seaweeds*

Due to high nutritional value and fuel value, marine algae have been consumed by coastal people since ancient times. Among several seaweeds, *Porphyra* is one of the well-known macroalgae which is getting consumed by people in China, Japan, and the Western world too.

4.2.2.4 *Cryopreservation*

The storage of algae in liquid nitrogen is known as cryopreservation. There is no detailed report regarding cryopreservation of algae until the early 1960s, but further research is going on.

4.3 Commercial-scale production systems

Microalgae is a worldwide distributed, large and diverse group of chlorophyllous, photosynthetic organisms. They can produce diverse types of metabolites and hence, they can be a great source of bioproducts with high commercial value (Jerney and Spilling, 2020).

Commercial-scale cultivation of microalgae can be classified into two systems: open and closed types. The open systems can also be of four types: extensive unmixed shallow ponds, circular ponds with centrally rotating arms for mixing purposes, shallow sloping cascade system, and paddle wheel mixed raceway ponds. The closed photobioreactor is extensively used for large-scale commercial use, specifically tubular photobioreactors. In addition to these, standard fermenters are also used for the heterotrophic culture of microalgae. Pure culture of microalgae on a large scale is quite impossible except for those algal species that are grown heterotrophically.

4.3.1 Open cultivation system

Cultivation of microalgae in open ponds is the most ancient and simplest experimental way to culture algae. Nowadays, open systems are the most used technological approach for outdoor solar cultivation (Chaumont, 1993). Primarily raceway ponds or high-rate algal ponds were made in the 1950s for wastewater treatment purposes. Since the 1960s, the open raceway ponds were employed for the commercial production of microalgae and blue-green algae. The main merit of an open cultivation system is that it is cost-effective and easy to handle (Becker, 1994). Operation costs can be lowered by employing solar light as a primary energy source. Stagnant ponds are not well mixed; as a result, it facilitates a simple and cheap upscaling process. In the case of raceway ponds, it can be developed cost-effectively on a larger scale but turbulence is required in the form of a paddle wheel to keep the algal culture always moving (Pulz, 2001). This system of cultivation is mostly used for commercial purposes and there are lots of demerits too. Continuous regulation and control of pH, temperature, availability of light in the system are quite difficult to maintain properly. Besides this, there are lots of seasonal variations (Chisti, 2016; Brennan and Owende, 2010; Kumar et al., 2010; Wang et al., 2008; Ugwu et al., 2008) which may lead to variation in biomass production. During summer, water losses occur at a high rate because of evaporation and chances of contamination also increase by several algal growths which can lead to the destruction of desired algal culture (Christenson and Sims, 2011; Chen et al., 2011).

4.3.2 Closed cultivation system

Several photobioreactors are used in closed cultivation systems. It can be a laminar reactor, tubular reactor, hanging plastic sleeves or tank reactor (Kumar et al., 2010). As the construction and operating system of commercial-scale bioreactors demand a high cost, the large scale closed systems are getting employed for the production of algal biomass, particularly for the high-value products, for example, nutraceuticals and cosmetics.

Closed cultivation systems offer a better way of control for cultivation variables as compared to open systems, such as the lesser chance of microbial contamination, low involvement of grazers as compared to open ponds. Besides these, the productivity level of algal biomass is much higher in the case of closed systems (Posten, 2009).

Photobioreactors have some challenges, such as the exchange of gas and the need for cooling. Getting enough carbon dioxide for massive algal growth is a very important factor here (Carvalho et al., 2006).

4.4 Commercial microalgae and processes

Algae for health nutrients (Arthrospira and Chlorella): The chlorophyllous freshwater unicellular *Chlorella* and filamentous blue-green alga *Arthrospira (Spirulina)* (Tomaselli, 1997) were first cultivated algae on large scale. Both algae were primarily used as health food supplements and different types of bioproducts were also extracted from these algae. The *Chlorella* extract was used as a health tonic, *Arthospira* was used as a cosmetic colorant. The algal oil and protein—lipid rich extracts are also produced from *Chlorella protothecoides* and *Prototheca moriformis* (Borowitzka, 2018).

Production of Carotenoids (β-Carotene and Astaxanthin): The process for β-carotene production using *D. salina* and astaxanthin production using *H. pluvialis* is quite different because of the biological features of both these algae. These two algae accumulate high amounts of target carotenoids and other carotenoids in little quantity. *D.* salina can accumulate up to $\sim 14\%$ of dw β-carotene and H. pluvialis over 5% of dw astaxanthin (Ben-Amotz et al., 1988).

Production of long-chain polyunsaturated fatty acids: Microalgae is used as a potential source of lipid and long-chain poluunsaturated fatty acids (PUFAs). It contains >50% of dry weight as lipids and the lipids of different species contain higher amounts of PUFAs γ-linolenic acid (GLA, 18: 3n-6), arachidonic acid (AA, 20: 4n-6), DHA (22: 6n-3), and eicosapentaenoic acid (EPA) (20: 5n-3) (Stephens et al., 2010; Tababa et al., 2012). Due to the high lipid content, especially in the form of triglycerides, has drawn interest in the cultivation of microalgae as the chief source of renewable liquid fuels. The production of DHA by the heterotrophic culture of dinoflagellate *C. cohnii* is well established. Microalgal species under development include chlorophycean algae *Lobosphaera (Parietochloris) incisa* for amino acid and several diatoms for EPA (Cepák et al., 2014; Camacho-Rodríguez et al., 2014).

4.5 Challenges in scaling up

As the culture scale increases, the hydrodynamic conditions, light, and environment start to control algal growth. In the case of small-scale culture, it is comparatively easy for better mixing and supplying uniform light and proper chemical environment for algal cells. Nevertheless, in the case of large raceway ponds or photobioreactors, uniform mixing is quite difficult (Belay, 1997; Weissman et al., 1989; Olaizola, 2000; Jiménez et al., 2003).

Flow rate, turbulence, and mixing: Flow and proper mixing in large-scale culture systems differ from small laboratory-scale mini ponds and photobioreactors. For large raceway ponds, the flow rate is different along with the channels, and vertical mixing is generally observed in the vicinity of the paddlewheel where the flow is turbulent. The flow rate is generally higher at the ends than the inside of the pond in long channels, which can lead to settling and accumulation of algal cells at the outside of the bends (Dodd, 2017; Shimamatsu, 1987). The unequal mixing can easily be observed in the ponds by foam distribution on the water surface. To reduce this, flow controllers are installed to even out the flow surrounding the bends.

4.6 Commercial-scale outdoor culture

Algal cultures developed in raceway ponds are exposed under two different regimes (dark/light). In raceway ponds, the algal cells are transferred between full exposure to solar radiation (Grobbelaar, 1991, 1994; VONSHAK and GUY, 1992; Lu and Vonshak, 1999). The two light regimes create a unique physiological condition. Availability of proper wavelength of light controls the rate of photosynthesis. The temperature of the cultivation system is another parameter that controls commercial-scale outdoor cultivation to a major extent. Different algal genera prefer different temperature ranges for optimal growth (Vonshak et al., 2001, 2014). Genus-specific culture media as well as different nutrient components in the culture media are also one of the key factors controlling the mass cultivation of microalgae (Borowitzka and Borowitzka, 1988). After a certain period, the algal cells die, so, eventually, toxins are produced in the growth medium and recycling of culture medium is mandatory for continuing the algal growth efficiently (Borowitzka and Moheimani, 2013).

4.7 Strain selection

As algal culture is a new flourishing arena for nutrient production, the most promising strain selection is becoming an obvious part of cultivation. The outdoor method of cultivation is the most preferred way to select the strain. Different ecological conditions, weather, temperature range, pH changes control the growth and optimal mass production of algae; hence, the ideal strain selection for all environmental conditions is not so easy. Genetic modification or genetically engineered strain development aiming for optimal mass cultivation was introduced (Radakovits et al., 2010; Kilian et al., 2011; Guihéneuf et al., 2016; Nymark et al., 2016). These technologies incorporate genetic engineering, homologous recombination and CRISPR/Cas9 genome editing, etc. Little information is available regarding successful genetically engineered algal strains for commercial cultivation. The first test of phenotypic stability and risk associated with ecological conditions has been done in the United States using chlorophyte strain *Acutodesmus dimorphus*. This alga was genetically engineered by the addition of two genes, one for fatty acid biosynthesis enhancement and another gene for recombinant GFP expression (Szyjka et al., 2017). This genetically modified strain was field tested in almost 800 outdoor ponds for a period of 50 days, but it was not able to outcompete the native algal strains. So, identifying and selecting the ideal strain for optimum mass cultivation is under research and scientific investigation is going on.

4.8 Contaminants and diseases

All the large-scale cultures that may be raceway ponds or closed systems such as photobioreactors can easily be contaminated by other algal species, fungi, bacteria, and predatory protozoa. Microalgal species with major selective environmental requirements, such as *D. salina* (High salinity), *Chlorella* (high nutrients), *Arthospira* (high alkalinity), is suitable option to reduce the risk of contamination in culture but most of the algal species of interest are not extremophiles (Gutman et al., 2009; Carney and Lane, 2014). Besides these, bacterial contamination is also an emerging issue leading to the spoiling of the whole axenic culture of microalgae.

4.9 Online and daily monitoring

Visual checking daily is a mandatory way for continuous monitoring of algal culture. Changes in the color of the culture indicate contamination as well as deficiency. Simultaneously, if the algal cultures start to get adhered with ponds and photobioreactors or excessive foaming occurs, then it can be assumed that there must be some stress in that cultivation system. Irrespective of temperature and light control, monitoring of pH and oxygen concentration of the system is an immediate requirement for daily monitoring of algal culture. The variable fluorescence technique can also be an integral path for monitoring algal culture growth (Masojídek et al., 2010).

4.10 Regulations and standards

As most of the high-value algal products are used as nutraceuticals, human food supplements, pharmaceuticals, the maintenance of standards is a core part of the extensive cultivation of microalgae. The manufacture and standards regarding rules differ from country to country. For many markets and regulatory authorities, Good Manufacturing Practice certification and ISO 9001-200 and other ISO quality management system standards are essential. Besides these, some specific markets require certification by a recognized authority, such as Halal or Kosher or organic certification (Belay, 2008; Borowitzka, 2013; Gellenbeck, 2012; Grobbelaar, 2003; Ryan et al., 2010).

References

Andersen, R.A., Culturing, A., 2005. Algal culturing technique. Journal of Phycology 41, 906–908.DOI. Available from: https://doi.org/10.1111/j.1529-8817.2005.00114.x.

Becker, E., 1994. Microalgae: Biotechnology and Microbiology, first edit Cambridge University Press.

Belay, A., 1997. "Mass culture of Spirulina outdoors – The earthrise farms experience." In: Spirulina Platensis (Arthrospira): Physiology, Cell-biology and Biochemistry. Taylor and Francis, London, UK.

Belay, A., 2008. Spirulina (Arthrospira): production and quality assurance. In: Gershwin, M.E., Belay), A. (Eds.), Spirulina in Human Nutrition and Health. CRC Press, Boca Raton, FL, pp. 1–25.

Ben-Amotz, A., Lers, A., Avron, M., 1988. Stereoisomers of β-carotene and phytoene in the alga *Dunaliella bardawil*. Plant Physiology 86 (4), 1286–1291. Available from: https://doi.org/10.1104/pp.86.4.1286. Oxford University Press (OUP).

Borowitzka, M., 2018. Commercial-Scale Production of Microalgae for Bioproducts. Wiley, pp. 33–65. Available from: 10.1002/9783527801718.ch2.

Borowitzka, M.A., 2013. High-value products from microalgae-their development and commercialisation. Journal of Applied Phycology 25 (3), 743–756. Available from: https://doi.org/10.1007/s10811-013-9983-9. Australia.

Borowitzka, M.A., Borowitzka, L.J., 1988. Limits to growth and carotenogenesis in laboratory and largescale outdoor cultures of *Dunaliella salina*. Algal Biotechnology Elsevier Applied Science.

Borowitzka, M.A., Moheimani, N.R., 2013. Sustainable biofuels from algae. Mitigation and Adaptation Strategies for Global Change 18 (1), 13–25. Available from: https://doi.org/10.1007/s11027-010-9271-9. Australia: Kluwer Academic Publishers.

Brennan, L., Owende, P., 2010. Biofuels from microalgae—A review of technologies for production, processing, and extractions of biofuels and co-products. Renewable and Sustainable Energy Reviews. 14 (2), 557–577. Available from: https://doi.org/10.1016/j.rser.2009.10.009. Ireland.

Camacho-Rodríguez, J., et al., 2014. A quantitative study of eicosapentaenoic acid (EPA) production by Nannochloropsis gaditana for aquaculture as a function of dilution rate, temperature and average irradiance. Applied Microbiology and Biotechnology. Spain 98 (6), 2429–2440. Available from: https://doi.org/10.1007/s00253-013-5413-9.

Carney, L.T., Lane, T.W., 2014. Parasites in algae mass culture. Frontiers in Microbiology 5. Available from: https://doi.org/10.3389/fmicb.2014.00278. United States: Frontiers Research Foundation.

Carvalho, A.P., Meireles, L.A., Malcata, F.X., 2006. Microalgal reactors: A review of enclosed system designs and performances. Biotechnology Progress 22 (6), 1490–1506. Available from: https://doi.org/10.1021/bp060065r. Portugal.

Cepák, V., et al., 2014. Optimization of cultivation conditions for fatty acid composition and EPA production in the eustigmatophycean microalga Trachydiscus minutus. Journal of Applied Phycology 26 (1), 181–190. Available from: https://doi.org/10.1007/s10811-013-0119-z. Czech Republic: Kluwer Academic Publishers.

Chaumont, D., 1993. Biotechnology of algal biomass production: a review of systems for outdoor mass culture. Journal of Applied Phycology 5 (6), 593–604. Available from: https://doi.org/10.1007/BF02184638. France: Kluwer Academic Publishers.

Chen, C.Y., et al., 2011. Cultivation, photobioreactor design and harvesting of microalgae for biodiesel production: A critical review. Bioresource Technology 102 (1), 71–81. Available from: https://doi.org/10.1016/j.biortech.2010.06.159. Taiwan.

Chisti, Y., 2016. Large-Scale Production of Algal Biomass: Raceway Ponds. Springer Science and Business Media LLC, pp. 21–40. Available from: 10.1007/978-3-319-12334-9_2.

Christenson, L., Sims, R., 2011. Production and harvesting of microalgae for wastewater treatment, biofuels, and bioproducts. Biotechnology advances 29 (6), 686–702. Available from: https://doi.org/10.1016/j.biotechadv.2011.05.015. United States.

Dodd, J.C., 2017. Elements of pond design and construction. CRC Handbook of Microalgal Mass Culture 265–283. Available from: https://doi.org/10.1201/9780203712405. United States: CRC Press.

Gellenbeck, K.W., 2012. Utilization of algal materials for nutraceutical and cosmeceutical applications-what do manufacturers need to know? Journal of Applied Phycology 24 (3), 309–313. Available from: https://doi.org/10.1007/s10811-011-9722-z. United States.

Grobbelaar, J.U., 2003. Quality control and assurance: Crucial for the sustainability of the applied phycology industry. Journal of Applied Phycology. Available from: https://doi.org/10.1023/A:1023820711706. South Africa: Springer Netherlands.

Grobbelaar, J.U., 1991. The influence of light/dark cycles in mixed algal cultures on their productivity. Bioresource Technology 38 (2–3), 189–194. Available from: https://doi.org/10.1016/0960-8524(91)90153-B. South Africa.

Grobbelaar, J.U., 1994. Turbulence in mass algal cultures and the role of light/dark fluctuations. Journal of Applied Phycology 6 (3), 331–335. Available from: https://doi.org/10.1007/BF02181947. South Africa: Kluwer Academic Publishers.

Guihéneuf, F., Khan, A., Tran, L.S.P., 2016. Genetic engineering: A promising tool to engender physiological, biochemical, and molecular stress resilience in green microalgae. Frontiers in Plant Science 7 (2016). Available from: https://doi.org/10.3389/fpls.2016.00400. Ireland: Frontiers Media S.A.

Gutman, J., Zarka, A., Boussiba, S., 2009. The host-range of Paraphysoderma sedebokerensis, a chytrid that infects *Haematococcus pluvialis*. European Journal of Phycology. 44 (4), 509–514. Available from: https://doi.org/10.1080/09670260903161024. Israel.

Jerney, J., Spilling, K., 2020. Large scale cultivation of microalgae: open and closed systems,". Methods in Molecular Biology 1–8. Available from: https://doi.org/10.1007/7651_2018_130. Finland: Humana Press Inc.

Jiménez, C., et al., 2003. The feasibility of industrial production of Spirulina (Arthrospira) in Southern Spain. Aquaculture 217 (1–4), 179–190. Available from: https://doi.org/10.1016/S0044-8486(02)00118-7. Spain.

Kilian, O., et al., 2011. High-efficiency homologous recombination in the oil-producing alga *Nannochloropsis* sp. Proceedings of the National Academy of Sciences 108 (52), 21265–21269. Available from: https://doi.org/10.1073/pnas.1105861108. Proceedings of the National Academy of Sciences.

Kumar, A., et al., 2010. Enhanced CO_2 fixation and biofuel production via microalgae: Recent developments and future directions. Trends in Biotechnology 28 (7), 371–380. Available from: https://doi.org/10.1016/j.tibtech.2010.04.004. United States.

Lu, C., Vonshak, A., 1999. Photoinhibition in outdoor Spirulina platensis cultures assessed by polyphasic chlorophyll fluorescence transients. Journal of Applied Phycology 11 (4), 355–359. Available from: https://doi.org/10.1023/A:1008195927725. Israel: Springer Netherlands.

Masojídek, J., Vonshak, A., Torzillo, G., 2010. In: Suggett, D., Prášil, O., Borowitzka, M. (Eds.), Chlorophyll a Fluorescence in Aquatic Sciences: Methods and Applications. Developments in Applied Phycology, 4. Springer, Dordrecht, pp. 277–292. Available from: https://doi.org/10.1007/978-90-481-9268-7_13.

Nymark, M., et al., 2016. A CRISPR/Cas9 system adapted for gene editing in marine algae. Scientific Reports 6. Available from: https://doi.org/10.1038/srep24951. Norway: Nature Publishing Group.

Olaizola, M., 2000. Commercial production of astaxanthin from *Haematococcus pluvialis* using 25,000-liter outdoor photobioreactors. Journal of Applied Phycology. Available from: https://doi.org/10.1023/a:1008159127672. United States: Springer Netherlands.

Posten, C., 2009. Design principles of photo-bioreactors for cultivation of microalgae. Engineering in Life Sciences 9 (3), 165–177. Available from: https://doi.org/10.1002/elsc.200900003. Germany.

Pulz, O., 2001. Photobioreactors: Production systems for phototrophic microorganisms. Applied Microbiology and Biotechnology 57 (3), 287–293. Available from: https://doi.org/10.1007/s002530100702. Germany.

Radakovits, R., et al., 2010. Genetic engineering of algae for enhanced biofuel production. Eukaryotic Cell 9 (4), 486–501. Available from: https://doi.org/10.1128/EC.00364-09. United States.

Ryan, A.S., Zeller, S., Nelson, E.B., 2010. Safety evaluation of single cell oils and the regulatory requirements for use as food ingredients, Single Cell Oils: Microbial and Algal Oils, Second Edition Elsevier Inc., United States, pp. 317–350. Available from: 10.1016/B978-1-893997-73-8.50019-0.

Shimamatsu, H., 1987. A pond for edible Spirulina production and its hydraulic studies. Hydrobiologia 151–152 (1), 83–89. Available from: https://doi.org/10.1007/BF00046111. Japan: Kluwer Academic Publishers.

Stephens, E., et al., 2010. An economic and technical evaluation of microalgal biofuels. Nature Biotechnology 28 (2), 126–128. Available from: https://doi.org/10.1038/nbt0210-126. Australia.

Szyjka, S.J., et al., 2017. Evaluation of phenotype stability and ecological risk of a genetically engineered alga in open pond production. Algal Research 24, 378–386. Available from: https://doi.org/10.1016/j.algal.2017.04.006. United States: Elsevier B.V.

Tababa, H.G., Hirabayashi, S., Inubushi, K., 2012. Media optimization of *Parietochloris incisa* for arachidonic acid accumulation in an outdoor vertical tubular photobioreactor. Journal of Applied Phycology 24 (4), 887–895. Available from: https://doi.org/10.1007/s10811-011-9709-9. Japan: Kluwer Academic Publishers.

Tomaselli, L., 1997. "Morphology, ultrastructure and taxonomy of Arthrospira (Spirulina) maxima and Arthrospira (Spirulina) platensis." In: Spirulina Platensis (Arthrospira): Physiology, Cell-Biology and Biotechnology. Taylor and Francis, London, UK.

Ugwu, C.U., Aoyagi, H., Uchiyama, H., 2008. Photobioreactors for mass cultivation of algae. Bioresource Technology 99 (10), 4021–4028. Available from: https://doi.org/10.1016/j.biortech.2007.01.046. Japan.

Vonshak, A., et al., 2001. Sub-optimal morning temperature induces photoinhibition in dense outdoor cultures of the alga Monodus subterraneus (Eustigmatophyta). Plant, Cell and Environment 24 (10), 1113–1118. Available from: https://doi.org/10.1046/j.0016-8025.2001.00759.x. Israel.

Vonshak, A., et al., 2014. The effect of light availability on the photosynthetic activity and productivity of outdoor cultures of *Arthrospira platensis* (Spirulina). Journal of Applied Phycology 26 (3), 1309–1315. Available from: https://doi.org/10.1007/s10811-013-0133-1. Israel: Kluwer Academic Publishers.

VONSHAK, A., GUY, R., 1992. Photoadaptation, photoinhibition and productivity in the blue-green alga, Spirulina platensis grown outdoors. Plant, Cell and Environment 15 (5), 613–616. Available from: https://doi.org/10.1111/j.1365-3040.1992.tb01496.x. Wiley.

Wang, B., et al., 2008. CO2 bio-mitigation using microalgae. Applied Microbiology and Biotechnology 79 (5), 707–718. Available from: https://doi.org/10.1007/s00253-008-1518-y. Canada.

Weissman, J.C., Tillett, D.M., Goebel, R.P., 1989. Design and operation of an outdoor microalgae test facility. Final subcontractors report. SERI/STR 232-3569, 1–56.

Valorization of microalgal biomass for biofuels

Eleni Koutra[1], *Dimitris P. Zagklis*[1], *Konstantina Tsigkou*[1], *Sameh Samir Ali*[2,3] *and Michael Kornaros*[1]

[1]Laboratory of Biochemical Engineering and Environmental Technology (LBEET), Department of Chemical Engineering, University of Patras, Patras, Greece [2]Biofuels Institute, School of the Environment and Safety Engineering, Jiangsu University, Zhenjiang, P.R. China
[3]Botany Department, Faculty of Science, Tanta University, Tanta, Egypt

5.1 Introduction

Increasing population, industrialization, fossil fuel depletion, greenhouse gas emissions, climate change, and renewable energy demand represent major challenges of the modern world that call for immediate action and bio-based solutions toward sustainability despite the ongoing development. Currently, fossil-based fuels constitute the main source of energy, with CO_2 emissions from fossil combustion showing a 1.6% increase in 2017 (Khoo et al., 2020), while energy consumption from alternative sources, including biofuels and waste, represents only 9.5% (Aliyu et al., 2021). Between the available biomass sources, aquatic microorganisms, including macroalgae and microalgae, have been in the limelight of current scientific research to unravel their full potential. Microalgae represent a highly diverse group, mainly composed of photosynthetic microorganisms, that can thrive in almost every natural and engineered environment, under different modes of growth and a wide range of conditions, concurrently producing valuable biomass, rich in proteins, carbohydrates, lipids, and bioactive compounds, which can be further valorized toward the production of bioenergy and added-value compounds. Up to date, the commercially produced microalgae worldwide reach 10 million tons annually, mainly represented by *Arthrospira* (Spirulina), *Chlorella*, *Haematococcus*, and *Dunaliella*, and are intended for special applications or food and feed purposes (Muhammad et al., 2021; t' Lam et al., 2018). In terms of energy, microalgae-derived biofuels constitute the third generation of biofuels which offers significant advantages, including high biomass and lipid productivity, CO_2

sequestration, use of nonarable land, and bioremediation of wastewater (Sindhu et al., 2019), compared to conventional crops and nonedible substrates, such as lignocellulosic and waste biomass (Lee et al., 2021; Milano et al., 2016). However, up to date, microalgae-based biofuels have not been implemented at an industrial scale due to several technical and economic challenges. At the same time, a biorefinery approach with the concomitant valorization of all biomass fractions is largely considered a prerequisite for the sustainability of microalgal biofuels (t' Lam et al., 2018).

5.2 Microalgae-based biofuels

Microalgae biofuels such as biodiesel, biogas, biohydrogen, bioethanol, biobutanol, and biocrude oil are generated through biochemical and thermochemical processes (Fig. 5.1). In the case of biochemical applications, even though the advantages of lower energy demand and higher environmental friendliness are observed, the relatively low efficiencies usually discourage the scale-up of the processes. However, continuous research and process development lead to quite promising results. The main biofuels derived from the microalgae biomass valorization are presented next.

5.2.1 Biodiesel

Biodiesel is commonly known as the fatty acid methyl esters (FAMEs) mixture after transesterification, where fatty acids react with alcohols, such as methanol or ethanol. The typical biodiesel composition is usually characterized by 5−6 methyl ester types enabling

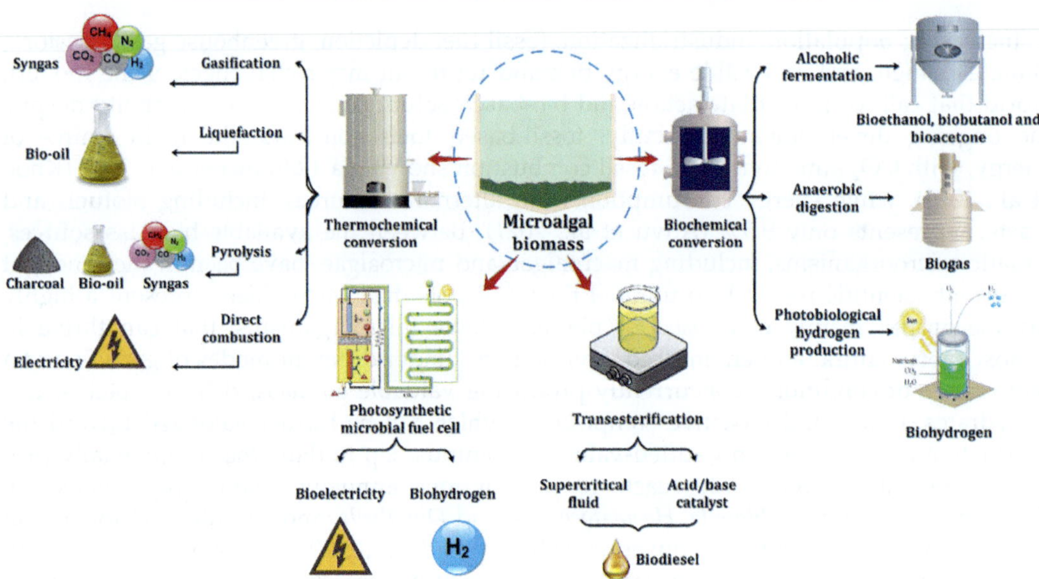

FIGURE 5.1   Microalgae-based biofuels.

conventional petroleum diesel replacement (Ananthi and Brindhadevi, 2021a). Specifically, in European Union, biodiesel and renewable diesel production accounted for almost 16 million liters for the year 2020 (Flach et al., 2020), making it 82% of the total produced biofuels (Yin et al., 2020), indicating not only the compatibility but also the value of such biofuel. The commonly determined methyl esters in microalgal biodiesel are saturated and unsaturated (C16:0, C18:0, and C18:1, C18:2, C18:3, respectively) (Ananthi and Brindhadevi, 2021a). Except for the lipids content and their composition, the biodiesel quality (viscosity, cetane number, flash point, oxidation stability, cold flow property, calorific value, density, etc.) is strongly affected by other components. The free fatty acid content should be less than 0.5% to reduce the catalyst used. In comparison, the water content should not exceed 0.05%, as water presence leads to soap and emulsion formation. Both phosphorus and sulfur content should be controlled as the catalytic converters for the emission control systems can be damaged (Chen et al., 2018a).

Concerning the fatty acid production and composition during microalgae cultivation, significant differences could be presented depending on physical (light, temperature, CO_2, pH, UV radiation, growth conditions) and chemical parameters (carbon, nitrogen, phosphorus, heavy metals, nanoparticles) (Ananthi and Raja, 2021b). During optimum cultivation conditions, the produced lipids are limited, and starvation cultivation strategies under nutrient stress conditions are proposed (Arguelles and Martinez-Goss, 2021). The microalgal strain selection for biodiesel production is also of paramount importance for the efficiency of the process due to the lipid productivity rate and the lipid content—for example, strains such as *Scenedesmus quadricauda*, *Chlorella vulgaris*, *Chlamydomonas reinhardtii*, *Nannochloropsis* sp. F&M-M49, *Nannochloropsis salina*, and *Parachlorella kessleri* are characterized by lipid productivity higher than $50 \, \text{mg} \, \text{L}^{-1} \, \text{d}^{-1}$ (Derakhshandeh et al., 2021; Yin et al., 2020), while *Chlorococcum pamirum*, *Chaetoceros muelleri*, *Chlorella emersonii*, *C. vulgaris*, *Desmodesmus* sp., *Nannochloropsis occulata*, etc. can accumulate more than 40% of lipids (dry weight biomass) (Chen et al., 2018a; Yin et al., 2020). Table 5.1 presents some examples of lipid production improvement during various stress conditions.

5.2.2 Biogas

Even though the anaerobic digestion (AD) process has been used since 1950 for wastewater treatment, its application for simultaneous energy production started only a few decades ago. AD is a biological process that takes place in four main stages, hydrolysis, acidogenesis, acetogenesis, and methanogenesis (Vargas-Estrada et al., 2021). Simple and more complex substrates can be decomposed during AD by an anaerobic mixed culture (bacteria and archaea). At the same time, a mixture of methane (60%−70%) and carbon dioxide (30%−40%), which is also known as biogas, is produced in the absence of oxygen. Biogas can be upgraded to biomethane through carbon dioxide removal or used in internal combustion engines (Kendir Çakmak and Ugurlu, 2020; Tsigkou et al., 2019). As microalgae biomass has been widely tested for various biofuels production, the case of AD exhibits the advantage of microalgal biomass conversion to biogas without dewatering, thus leading to significant energy savings (Xiao et al., 2019). However, several factors such as the rigid cell wall of microalgae, the C/N ratio, and the pH can strongly influence the efficiency of the specific

TABLE 5.1 Lipid production improvement by various stress conditions.

Factor	Strain	Stress conditions	Result	References
Light	*Phaeodactylum tricornutum*	Intensity incline from 50 to 300 μmol photons $m^{-2} s^{-1}$	39% and 101% increase in lipid and biomass content, respectively	Ananthi and Brindhadevi (2021a)
Temperature	*Coccomyxa subellipsoidea* C169	35°C	6% increase of dry weight triglyceride content	Allen et al. (2018)
CO_2	*Chlorella vulgaris* MSU AGM 14	8% CO_2	Biomass (23%) and lipid productivity (94%) promotion	Lakshmikandan et al. (2020)
pH	*Chlamydomonas reinhardtii*	pH 7.8 and 8.5	Lipid content increased 3 and 2.3 times respectively	Ochoa-Alfaro et al. (2019)
UV radiation	*Ettlia* sp.	UV-A and PAR	Lipid productivity and content were almost 43.7% and 33.7% higher	Seo et al. (2019)
Carbon	*Chlorella pyrenoidosa*	Acidified starch wastewater into anaerobic digestate of sludge (1:1, v/v)	Lipids production improvement by 3.2-fold (87.3 mg $L^{-1} d^{-1}$)	Tan et al. (2020)
Nitrogen	*Chaetoceros muelleri*	0.18 mM $NaNO_3$	1.6 times higher lipid content than the control	de Jesús-Campos et al. (2020)
Phosphorus	*C. reinhardtii*	0.4 mg L^{-1} P content in growing media	Maximum fatty acid content observation	Qari and Oves (2020)
Heavy metals	*Monoraphidium* sp. QLY-1	80 μM and 40 μM Cd	The highest lipid productivity (96.75 mg $\cdot L^{-1} \cdot d^{-1}$) and content (52.78%) were enhanced by 1.39- and 1.59-fold respectively	Zhao et al. (2019)
Nanoparticles	*C. pyrenoidosa*	20 mg IONPs per L and 30 mg IONPs per L IONPs: α-Fe_2O_3 nanoparticles	Highest biomass production of 2.16 g L^{-1} (lipid content of 13.77%) and lipid accumulation of 16.89%wt (biomass concentration of 2.01 g L^{-1}) respectively	Rana et al. (2020)

biological process. Yield limitations can be overcome due to several strategies, mainly concerning the biomass pretreatment, the nutrients balance, temperature, increase in hydraulic retention time (HRT), and decrease in organic loading rate (OLR) decrease.

5.2.2.1 Pretreatment

The intracellular compounds are liquified or released in the aqueous phase during various pretreatment methods targeting their concentration, accessibility, and biodegradability. However, pretreatment methods are usually characterized by higher energy demand

and thus increased cost and/or chemical addition (Kendir Çakmak and Ugurlu, 2020). Typical instances of microalgal biomass pretreatment refer to ultrasonic, microwave, bead milling, thermal, biological, and chemical methods. For example, after the thermal pretreatment of *Scenedesmus* sp. (120°C for 40 minutes) and *Chlorella* sp. (70°C–80°C for 30 minutes), the methane yield increased by 21%–27% and 37%–48%, respectively. In comparison, the enzymatic pretreatment of *Oocystis* sp. (laccase at 25°C for 20 minutes) and *Acutodesmus obliquus* (enzyme cocktail at 37°C for 24 hours) led to a 20% and 14% methane increase, respectively (Zabed et al., 2020).

5.2.2.2 Nutrients balance

C/N ratio is usually characterized as a factor of great importance for biogas production efficiency. The C/N variations could result in NH_3 increase and lead to AD performance limitation due to inhibition of methanogenic activity. According to the literature, the optimum C/N ratio is 25, and a common, cost-effective strategy for C/N improvement is AD codigestion with one or more substrates (Yukesh Kannah et al., 2021). For instance, the codigestion of raw sewage and harvested microalgal biomass in an up-flow anaerobic sludge blanket reactor led to 25% higher methane production (Vassalle et al., 2020).

5.2.2.3 Temperature

Temperature range is often reported as a factor that can significantly affect the microalgal biomass biodegradability and/or methane productivity, mainly due to the improved hydrolysis rate. However, the anaerobic inoculum should adapt before the thermophilic conditions as the lack of acclimated consortia could strongly affect and subsequently limit the reactor's performance and efficiency (Carrillo-Reyes et al., 2021).

5.2.2.4 Hydraulic retention time and organic loading rate

HRT is equal to the average time needed for a fixed biomass volume to pass through the reactor, while OLR is defined as the organic matter loaded daily into the reactor. Both low HRTs and high OLRs could cause process imbalances due to AD main-end products (volatile fatty acids and ethanol) accumulation, pH decrease and methanogens metabolism inhibition. Conventional AD occurs when applying an OLR between the range of $0.5–1.6 \, kg \, VS \, (m^3d)^{-1}$ and HRT of 20–30 days, while both parameters significantly depend on the biomass type, composition and reactor design (Zabed et al., 2020).

5.2.3 Biohydrogen

Biohydrogen is an ecological renewable fuel source with a high energy density compared to the other biofuels and a heating value of $141.65 \, MJ \, kg^{-1}$. Additionally, biohydrogen is recognized as the fuel of the future and as the only fuel that can be used in fuel cells without carbon dioxide (by-product) production. Many photobioreactors (PBRs) have been designed concerning biohydrogen production, such as membrane, fixed-bed, or continuous stirred-tank reactors. At the same time, the main requirement is energy supply in the form of carbohydrates, light or carbon dioxide, depending on the biological mechanisms. Various microbes such as photosynthetic and nonphotosynthetic algae, photosynthetic and dark fermentative bacteria, and cyanobacteria can produce hydrogen due to direct or indirect biophotolysis, dark fermentation and/or photofermentation (Anwar et al., 2019; Razu et al., 2019).

5.2.3.1 Biophotolysis

Photosynthetic microalgae such as *Scenedesmus*, *Chlorococcum*, *Chlorella*, *Botryococcus*, *Chlamydomonas*, and *Tetraspora* can produce hydrogen while the enzymes hydrogenase and nitrogenase are present. Microalgae photosynthesis system utilizes water electrons and sunlight while simple nutrients are provided, and atmospheric carbon dioxide is captured as a carbon source. Direct biophotolysis uses solar energy (sunlight) for splitting water into oxygen and hydrogen (Eq. 5.1), indicating a pretty attractive method; however, practical issues such as enzymatic activity or hydrogen yield inhibition could limit the mechanism. Concerning indirect biophotolysis (Eq. 5.2), hydrogen is produced from stored starch and glycogen following two steps: carbohydrates production, using light energy, and carbohydrates fermentation for hydrogen generation (Akhlaghi and Najafpour-Darzi, 2020; Anwar et al., 2019).

$$\text{Direct bio}-\text{photolysis}: 2H_2O + \text{solar energy} \rightarrow 2H_2 + O_2 \tag{5.1}$$

$$\text{Indirect bio}-\text{photolysis}: C_6H_{12}O_6 + 6H_2O \rightarrow 6CO_2 + 12H_2 \tag{5.2}$$

5.2.3.2 Dark fermentation

During dark fermentation, anaerobic fermentative consortia can decompose the organic matter of microalgal biomass into volatile fatty acids, ethanol, hydrogen, and carbon dioxide. Even though the scientific community has shown increased attention to hydrogen generation by dark fermentation of microalgal biomass, the yields are limited due to the prior need for pretreatment to decrease the carbohydrates' polymerization and increase microbial accessibility. As a feedstock for dark fermentation, various species such as *Saccharina*, *Chlorella*, and *Scenedesmus* have been tested without pretreatment, while physical, chemical, or biological pretreated biomass has exhibited higher yields. Equally to the case of biogas production through AD, during dark fermentation, several parameters like OLR, temperature, pH, and HRT are of great importance for the efficiency of the process (Nagarajan et al., 2020; Razu et al., 2019; Wang and Yin, 2018).

5.2.3.3 Photofermentation

Photofermentation (Eq. 5.3) is defined as the organic matter fermentative conversion into hydrogen and carbon dioxide, while solar energy is utilized as an energy source. Such a process is not inhibited by the partial pressure of hydrogen, eliminates environmental pollutants, and is capable of being implemented after dark fermentation to consume the produced volatile fatty acids or ethanol as feedstock. The limitation of photofermentation concerns mainly the nitrogen supply needs. According to the literature, *Rhodobacter*, *Rhodopseudomonas*, and *Rhodospirillum* have been widely employed in the photofermentation process (Razu et al., 2019; Salakkam et al., 2021).

$$\text{Photofermentation}: CH_3COOH + 2H_2O + \text{light} \rightarrow 2CO_2 + 4H_2 \tag{5.3}$$

5.2.4 Bioethanol

Among liquid biofuels, bioethanol, which is two-carbon alcohol, is characterized as the most widely used due to its ability to be deployed either blended with gasoline or as it is, as well as due to its biodegradability, sustainability, and low toxicity. According to the literature, microalgal biomass could be involved in bioethanol production via three possible routes, which are (1) hydrolysis combined with yeast fermentation, (2) dark fermentation, and (3) photofermentation (Bastos, 2018; Maia et al., 2020). During the hydrolysis–fermentation process, the first step is biomass pretreatment for cell structure breakdown and biomass hydrolysis, while fermentation of the treated biomass follows. Even though high efficiency could be achieved, the energy needs for the biomass pretreatment stage combined with the cost for the use of pure cultures may prohibit the process implementation. A typical example of hydrolysis–fermentation is the use of *Saccharomyces cerevisiae*, or *Pichia stipitis* strains for bioethanol production using carbohydrate-enriched microalgal biomass such as *Chlorella, Desmodesmus, Scenedesmus*, or mixed cultures after various pretreatment steps (acid, alkaline, or enzymatic hydrolysis) (Bastos, 2018; de Farias Silva et al., 2018). Concerning the process of dark fermentation, it is the same process that has already been extensively discussed in the case of biohydrogen production. Microalgae and cyanobacteria, including *C. vulgaris, Synechococcus* sp., *Oscillatoria limnetica, Chlamydomonas moewusii, C. reinhardtii, Cyanothece* sp., and *Chlorococcum littorale* have already been tested for bioethanol production. On the other hand, photofermentation requires genetic engineering of the microalgal biomass to redirect the current biochemical pathways, targeting higher efficiency and productivity (Bastos, 2018).

5.2.5 Biobutanol

Biobutanol, an alcohol with four atoms of carbon, is produced from biomass as feedstock and has many applications, including cosmetics, pharmaceuticals, and gasoline blending. Concerning its production, the bacterium *Clostridium acetobutylicum* is used for fermentation, while acetone, butanol, and ethanol (ABE) are generated. Such a process is also known as ABE fermentation because of its final products. According to the literature, the addition of *Clostridium beijerinckii* could help with higher sugar consumption and increased product yields (0.35 g of ABE per g of consumed sugar) (Timung et al., 2021). Microalgal biomass, such as *Chlorella zofingiensis* (Onay, 2020) and *C. vulgaris* (Arun et al., 2020), led to yields of 0.084 g biobutanol per g of microalgal biomass and 3.37 g biobutanol per L, respectively.

5.2.6 Biocrude oil

Biocrude oil is a synthetic fuel that could replace petroleum and can be produced during biomass valorization. Its composition significantly depends on the physicochemical characteristics of the feedstock. Two processes are widely proposed for biocrude oil production: pyrolysis and hydrothermal liquefaction (HTL). In the first case, not only biocrude oil but also biochar noncondensable pyrogas are produced. In the case of HTL, biocrude oil is the main product, while gases and residual solids and aqueous products are also generated. More specifically, during biomass pyrolysis, heat in the typical range of 300°C–700°C is applied at atmospheric pressure in the absence of oxygen and the presence or absence of a catalyst. Before pyrolysis implementation a drying stage of the microalgal biomass is required. On the other hand, the microalgal biomass can be valorized through HTL without drying. The process conditions refer to 200°C–380°C and 5–28 MPa pressure in the presence or absence of a catalyst. Disruption techniques for intracellular components release, such as milling, ultrasonication, or microwaves, favor both processes (Hu et al., 2019; Kazemi Shariat Panahi et al., 2019; Zhang et al., 2018).

Various microalgae species have been tested through HTL (*Pavlova, Nannochloropsis, Spirulina, Chlorella, Isochrysis, Botryococcus*, etc.) and pyrolysis (*Chlorella, Tetraselmis, Arthrospira, Dunaliella*, etc.) valorization (Gautam and Vinu, 2020). Some of the most recent published studies for HTL biocrude oil with promising results have mentioned mixed microbial biomass valorization with a yield of 21.7% w/w, at optimal conditions of 299.7°C, 16.1% w/v biomass loading, and duration of 65 minutes (Makut et al., 2020), while *Chlorella* sp. led to a yield of 31.8%–32.2% w/w after the addition of different catalysts. However, the presence of catalysts increased the low boiling fraction from 68% to 81% (Lu et al., 2020). Concerning the biocrude oil generation after wild-type, microalgal biomass pyrolysis, *Scenedesmus* and *Synechocystis* biomass resulted in yields of 35.3% and 21.1% w/w, respectively, at temperatures of 500°C and 600°C (Derakhshandeh et al., 2020).

5.3 Downstream processing for biofuels production

Downstream processing in microalgal technology includes several consecutive steps of biomass harvesting, cell disruption, product extraction, and occasional fractionation and purification, which mainly depend on the established process and the desired end-product (Fig. 5.2).

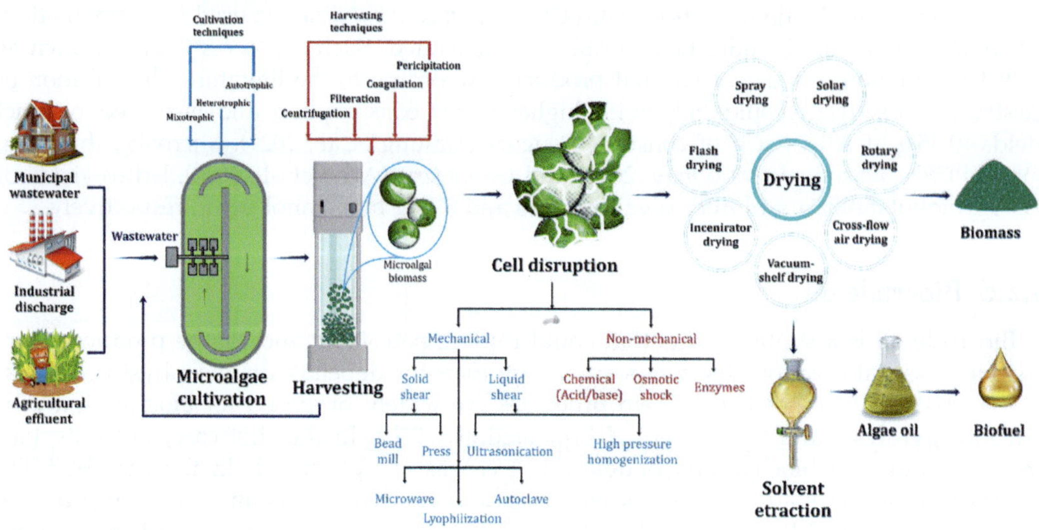

FIGURE 5.2   Downstream processing of microalgal biomass.

Among the most crucial factors, effectiveness, upscaling, and mildness must be carefully evaluated in downstream processing of microalgae, while in addition to low energy and cost requirements, they contribute to the sustainability of microalgal biorefinery (Vanthoor-Koopmans et al., 2013). Currently, multiple processing steps, along with the high cost of available technology, hinder large-scale applications. However, extensive research and several improvements in productivity of biomass and target compounds, technical advancements, application of recycling options, and integrated processes will contribute to the feasibility of microalgal biofuels in the future (Chen and Leng, 2018b).

5.3.1 Harvesting

Harvesting includes the recovery of microalgal cells, representing up to 30% of the total microalgae production cost. Therefore, in terms of sustainability, the high yield must be accompanied by low energy, operational, and maintenance requirements (Khoo et al., 2020). The effectiveness of harvesting mainly depends on cell characteristics, including size, shape, charge, and specific weight, while biomass recovery is also affected by cell concentration, as well as medium components and salinity (Aliyu et al., 2021). During biomass harvesting, microalgae are separated from the cultivation medium through mechanical, chemical, biological, or combined methods, each characterized by specific advantages and drawbacks. The most commonly used processes include centrifugation, flocculation, coagulation, filtration, sedimentation, and electrical methods, which can be applied either in a single step or after preconcentration of microalgal biomass in the case of dilute systems. Harvesting of *Scenedesmus obliquus* and *C. vulgaris* through flocculation has been tested under pH control, aluminum, $FeCl_3$, and chitosan addition, or air- and electro-flotation, showing a flocculation efficiency between 65% and 99% after 1 hour depending on the flocculant addition. However, NaOH-induced recovery proved to be the most environment-friendly and cost-effective harvesting method for biodiesel production on a large scale (Koley et al., 2017). Furthermore, similar lipid content values were observed in the case of *C. vulgaris* harvested by microfiltration and centrifugation, 26% and 28% respectively, though chitosan-based coagulation negatively affected biomass and lipid recovery (Ahmad et al., 2014). Also, *Penicillium* sp. in the pellet form served as an effective bioflocculant for *Chlorella* sp., reaching 98.2% harvesting efficiency in 2.5 hour, and under less carbon input in comparison with spores (Chen and Li, 2018c). In total, specific criteria determine the selection of the appropriate harvesting method, including biomass yield and quality, processing duration and cost, and applicability to different species (Singh and Patidar, 2018). Therefore, for biofuel production, flocculation is considered as the best option, owing to the easiness and transfer at industrial scale, low cost, time and space requirements. However, residual flocculants can contaminate the recovered biomass, making the method inappropriate for human or animal consumption in contrast to centrifugation or filtration which are preferable in the latter case (Singh and Patidar, 2018; t' Lam et al., 2018).

5.3.2 Drying

Upon biomass harvesting, moisture removal is necessary to retain microalgal quality. However, the applied method can either facilitate or damage the desired cellular components, while an additional cost is also imposed in the downstream processing (Aliyu et al., 2021; Show et al., 2017). Common drying methods include sun drying, spray drying, freeze-drying, and vacuum drying. However, the interaction between biomass harvesting and drying methods and the subsequent extraction step must be taken into careful consideration. *C. vulgaris* biomass drying at temperatures between 40°C and 140°C revealed a deterioration of cell morphological and biochemical characteristics compared to freeze-drying, while medium heat between 60°C and 80°C can best preserve biomass quality (Hosseinizand et al., 2018). Similarly, the application of drying temperatures between 50°C and 70°C retained the quality of *Chlorella pyrenoidosa* intended for bioethanol production (Megawati et al., 2020). In addition, the combined effect of two harvesting methods, centrifugation and flocculation, with *S. obliquus* biomass freezing, oven-, and freeze-drying has been studied, revealing the suitability of frozen biomass for biodiesel production, while FAME yield decreased upon the combination of flocculation and oven-drying (Oliveira et al., 2020). Lastly, freeze-drying proved to be more cost-effective than oven- and sun-drying. The combination of freeze-drying and cell disruption based on microwaves resulted in the highest lipid recovery of 25.4% by wastewater grown *S. obliquus* (Ansari et al., 2018).

5.3.3 Cell disruption

Applying disruptive techniques is a prerequisite for the effective recovery of microalgal compounds and biomass conversion to multiproducts. Cell disruption can be effectively performed through bead milling, homogenization, pulsed electric field, ultrasound, microwaves, enzymes, and chemical treatment (Günerken et al., 2015; Vanthoor-Koopmans et al., 2013), while bead milling and homogenization are between the most efficient methods for large-scale application (Khoo et al., 2020). However, the effectiveness of the applied methods is species-specific; therefore careful selection should be made based on the upstream process. In the case of lipid production from marine microalgae, the highest lipid recovery of *Nannochloropsis oceanica* was accomplished through microwave, while in the case of *Nannochloropsis gaditana* through the application of ultrasound, 49% and 21.7%, respectively (Quesada-Salas et al., 2021). In contrast, bead milling was the only effective disruptive method for the subsequent recovery of 12.6% lipid from *Tetraselmis suecica* biomass. Furthermore, ultrasonication, along with a novel disruptive method, hydrodynamic cavitation, was optimized for cell disruption of *C. pyrenoidosa*, revealing the high potential of the latter upon alkali pretreatment of microalgal biomass (Waghmare et al., 2019). Enzymatic pretreatment can also facilitate product recovery, as in the case of *C. reinhardtii* incubated with autolysin, resulting in effective degradation of cell walls and increased lipid and protein extraction, compared to ultrasonication or no disruption (Sierra et al., 2017).

5.3.4 Extraction

Target compounds from the disrupted biomass are subsequently extracted, commonly through the use of organic solvents. However, due to environmental and health concerns, along with the high cost of conventional solvents, more environment-friendly options have been investigated, including ionic liquids (ILs), supercritical CO_2 (SC-CO_2), deep eutectic solvents, and switchable solvents (Khoo et al., 2020; Lee et al., 2021). However, sequential processes significantly increase complexity, energy demand, and cost of downstream processing. Therefore process integration, such as cell disruption and concomitant extraction, can decrease processing steps, prevent product degradation, and recover the desired compound without impurities (Lee et al., 2021). Recently, lipid extraction was effectively performed from wet *C. vulgaris* biomass through a novel extraction process, including mild pressure accompanied by heat shock, increasing biodiesel yield by 26.7%, and energy output by 27.7%, compared to the conventional Bligh and Dyer method (Lakshmikandan et al., 2020). Furthermore, lipid extraction from wet *C. pyrenoidosa* biomass was performed by the green solvents 2-methyl tetrahydrofuran (2-MeTHF) and cyclopentyl methyl ether in addition to isoamyl alcohol, receiving 95.7 and 74.8 mg lipids g^{-1} biomass, respectively (de Jesus et al., 2019). However, up to date, high costs make their use uncompetitive to conventional solvents.

5.4 Life cycle analysis and techno-economic aspects of microalgal biofuels

The life cycle assessment (LCA) and techno-economic aspects of biofuel production from microalgae have been the focus of scientific research, alongside the technical aspects of these processes. After the exponential growth in microalgal biofuel publications from the mid-2000s, more and more researchers started examining these sides of the process, reaching around 100 publications per year in 2020, according to the Web of Science database (Web of Science, 2021) (Fig. 5.3). The economic potential of a new process or product

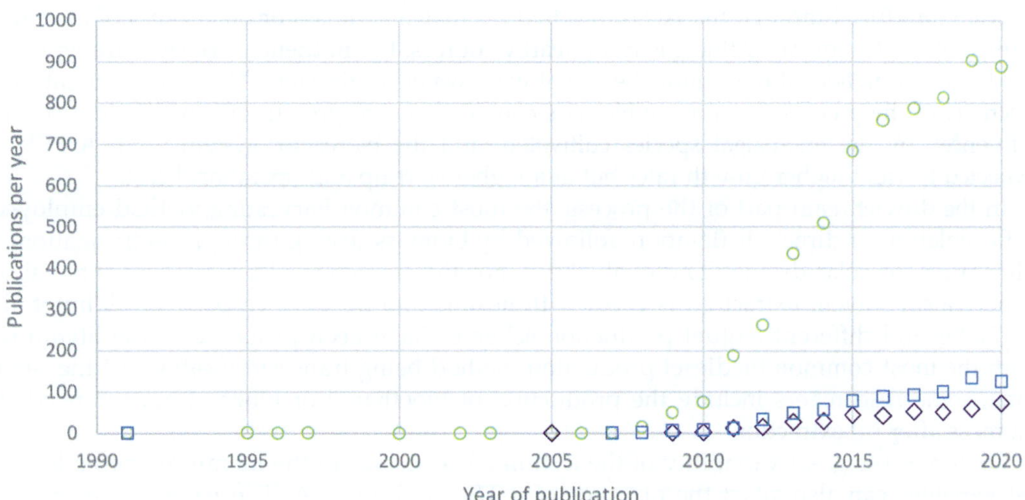

FIGURE 5.3 Number of publications per year in the field of biofuel production from microalgae, according to the Web of Science database.

is one of the most important factors that will determine its adoption, especially when it needs to displace a preexisting product in the market, like in the case of biofuels (Debnath et al., 2019). The increased environmental awareness and eco-friendly policies being adopted worldwide can give an extra edge to sustainable processes, even if they lag behind the conventional products, from an economic perspective. This is usually reinforced through some sort of penalty for using less environment-friendly products (like carbon tax on fossil fuels) (Ebadian et al., 2020). In short, the low environmental footprint of a product can alleviate, at least partially, the economic burden that a "greener" production process may incur.

Ideally, the design of a new process or product needs to be optimized, considering the technical, economic, and environmental aspects simultaneously, in conjunction with the region and period of implementation, and the weight of each of these aspects in the decision-making process. Because of the inherent difficulties of this kind of analysis, this is rarely the case (if ever). Typically, a researcher will carry out some technical optimization. The results will be used by the same or another team to calculate the economic potential of the process or its environmental impact through an LCA. Apart from the variables of the technical process examined in each case, the variables of the techno-economic analysis (TEA) and the LCA lead to notoriously high variability in the results reported in the literature, as is the case for biofuels production from microalgae. A brief overview of the variables mentioned above is presented in Fig. 5.4.

As it is to be expected, the technical characteristics of the biofuel production process strongly affect the results of any TEA or LCA. In the upstream part of microalgal cultivation, the most important variables are the sources of carbon and nutrients, the biomass growth rate and composition, and finally, the type of bioreactor used (Zhu et al., 2017). In most cases, atmospheric carbon dioxide is used as a carbon source for microalgal cultivation. Still, some studies cover these needs (at least partially) by using carbon dioxide from biogas, typically produced through the AD of residual microalgal biomass or other types of the agro-industrial waste stream, or flue gas from nearby sources. For nutrients, fertilizers are typically used, but researchers also examine the use of wastewater or digestate. The macronutrient productivity of the process depends on biomass composition and growth rate, which, in turn, are influenced by the microalgal species cultivated, and the bioreactor system used. A PBR is expected to yield higher growth rates but at a higher start-up and operational cost.

In the downstream part of the process, the most common harvesting method employed is flocculation or direct air flotation, followed by biomass drying through centrifugation, a filter press, or solar drying (Zhu et al., 2017). For the extraction of bio-oil, conventional or supercritical solvent extraction is used, with hexane and SC-CO$_2$, respectively (Zhu et al., 2017). Several different biofuel production schemes have been proposed in the literature, with the most common biodiesel production method being transesterification. At the same time, some researchers include the production of bioethanol or biogas from the residual biomass after oil extraction.

On top of the great variability of the examined processes in the literature, methodological variables can also affect the outcome of a TEA and an LCA. The parts of the process included in each analysis are one of them, for example, the inclusion of the bio-oil transformation to biofuels step in the TEA or the inclusion of carbon emissions of the infrastructure manufacturing in an LCA. The scale of a process affects the outcome of the TEA

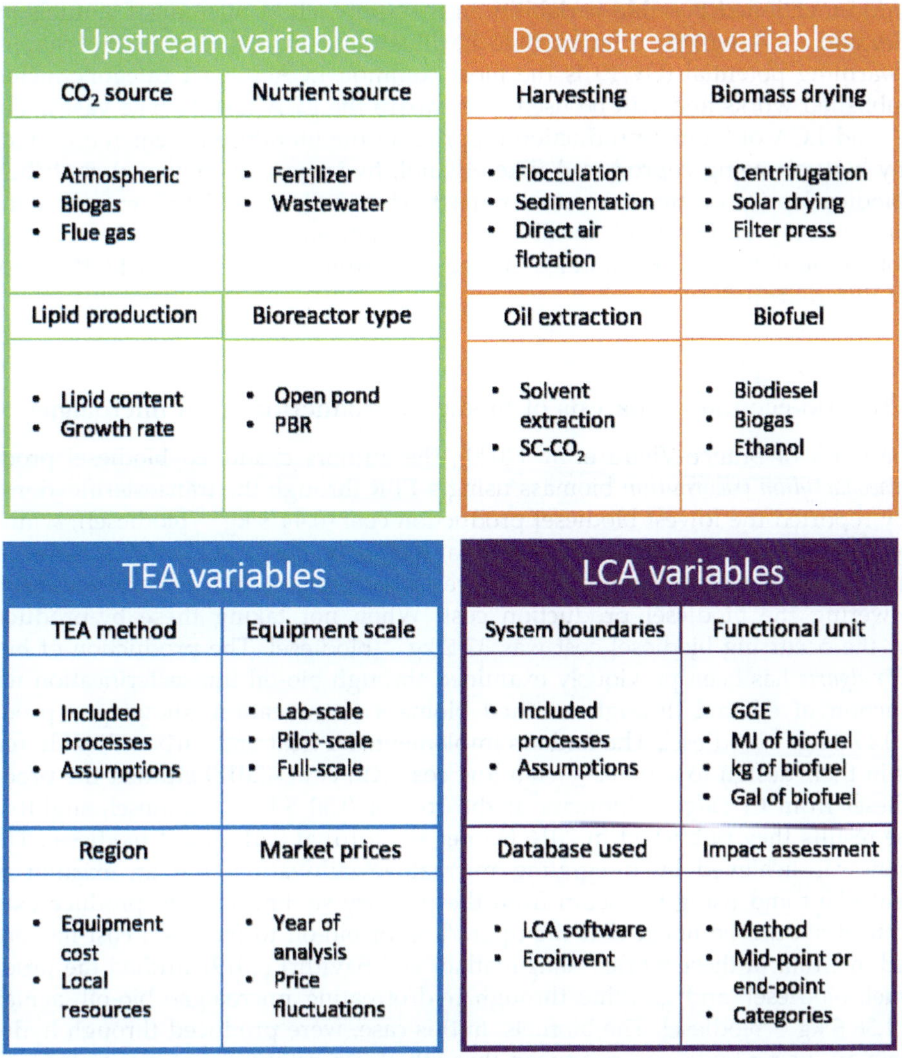

Upstream variables		Downstream variables	
CO$_2$ source	**Nutrient source**	**Harvesting**	**Biomass drying**
• Atmospheric • Biogas • Flue gas	• Fertilizer • Wastewater	• Flocculation • Sedimentation • Direct air flotation	• Centrifugation • Solar drying • Filter press
Lipid production	**Bioreactor type**	**Oil extraction**	**Biofuel**
• Lipid content • Growth rate	• Open pond • PBR	• Solvent extraction • SC-CO$_2$	• Biodiesel • Biogas • Ethanol

TEA variables		LCA variables	
TEA method	**Equipment scale**	**System boundaries**	**Functional unit**
• Included processes • Assumptions	• Lab-scale • Pilot-scale • Full-scale	• Included processes • Assumptions	• GGE • MJ of biofuel • kg of biofuel • Gal of biofuel
Region	**Market prices**	**Database used**	**Impact assessment**
• Equipment cost • Local resources	• Year of analysis • Price fluctuations	• LCA software • Ecoinvent	• Method • Mid-point or end-point • Categories

FIGURE 5.4   Variables affecting the results reported in literature for the TEA and LCA of biofuel production from microalgae. *LCA,* Life cycle analysis; *TEA,* techno-economic analysis.

through the economy of scale, with higher capacity equipment being cheaper per unit of capacity (to a certain point). The region and timeframe of the TEA analysis also affect the results through local market prices and inflation rates. In the case of LCA, a wide variety of functional units have been used in literature, like 1 MJ of biofuel energy or 1 kg of biofuel. Even if these units can be converted to a single unit for data comparison, this adds another layer of complexity to the process. The database used in LCA also affects the results, alongside the impact assessment method that may include different damage categories with abstract concepts.

The aforementioned possible variability in the approach of microalgal biofuels production TEA and LCA will be attempted to be illustrated in this part of the chapter. The global warming potential (GWP) is the most common LCA impact category included in most published works and will be used to compare the LCA results. The vast majority of the TEA and LCA of biofuel production reported in the literature is centered on biodiesel, with any biofuels being coproduced (like ethanol, hydrogen, or methane) contributing to lower biodiesel price or environmental impact. This analysis will follow a similar trend, primarily focusing on biodiesel. In all cases, the reported results have been converted to $ kg^{-1} of biodiesel for the TEA and kg CO_2 eq MJ^{-1} from biodiesel to facilitate comparison between different studies.

5.4.1 Techno-economic analysis of biodiesel production from microalgae

In the work of Branco-Vieira et al. (2020), the authors examined biodiesel production from *Phaeodactylum tricornutum* biomass using a PBR through the transesterification of bio-oil. They reported the lowest biodiesel production cost (0.44 $ kg^{-1} biodiesel), while interestingly, the analysis was carried out on a relatively small scale. This result can be partially attributed to the assumption that residual biomass and glycerol would also be sold, lowering the biodiesel production cost. When not taking these by-products into account, the occurring biodiesel cost was 37 $ kg^{-1} biodiesel. The production of biodiesel from *C. vulgaris* has been previously examined through bio-oil transesterification with the coproduction of ethanol through residual biomass fermentation, showing a production cost of 0.49 $ kg^{-1} biodiesel. The authors implemented a heat and carbon dioxide recovery scheme in their design to achieve such a low cost. Xin et al. (2018) studied the production of biodiesel from *C. vulgaris* biomass at the cost of 0.50 $ kg^{-1} biodiesel, slightly better than the results they published in 2016 (Xin et al., 2016) at 0.61 $ kg^{-1} biodiesel. This cost reduction was achieved by integrating microalgae cultivation into an existing sewage treatment plant and using the scum from the primary settling tank to produce extra biodiesel, but it should be noted that the upgrading of bio-oil to biodiesel cost has not been included in either of these works. Ranganathan and Savithri (2019) studied the production of jet fuel, biodiesel, and gasoline through hydrotreating microalgae bio-oil, achieving a cost of 1.36 $ kg^{-1} biodiesel. The biofuels, in this case, were produced through hydrotreating of bio-oil, using hydrogen produced through steam reforming of biogas from the AD of the residual biomass. In the work of Dutta et al. (2016), two cases were examined, with the base scenario describing the production of biodiesel, after hexane extraction of bio-oil, through transesterification, yielding a production cost of 3.33 $ kg^{-1} biodiesel. In the second scenario, hydrolysis fermentation and distillation of biomass were added before hexane extraction to produce bioethanol from the contained carbohydrates. The production of biodiesel was realized through hydrotreating, and finally, biogas was produced from the residual biomass. The second scenario exhibited a lower production cost of 1.37 $ kg^{-1} biodiesel. Hoffman et al. (2017) examined the cultivation of *N. salina* to produce biodiesel through hydroprocessing. The authors assessed two different schemes, with the first using a conventional raceway pond and biodiesel production through hydroproceeding yielding a production cost of 2.50 $ kg^{-1} biodiesel. In the second scenario, they examined the use of an algal turf scrubber system that led to lower production costs, at 1.88 \$ kg^{-1} biodiesel. Juneja and Murthy (2017) reported a production cost of 1.99 \$ kg^{-1} biodiesel for producing biodiesel from *C. vulgaris* biomass through hydrotreating and methane through the catalytic HTL of residual biomass. Finally, in the work of Sun et al. (2019), a production cost of 2.29 \$ kg^{-1} biodiesel was reported. The results mentioned above and the analysis variables reported by the authors are summarized in Table 5.2.

As biodiesel needs to displace an existing product (conventional diesel), it would be helpful to compare the results reported in the literature for biodiesel production cost from microalgae with conventional diesel prices. This can give an insight into how difficult it would be to adopt this product. For this purpose the results reported in Table 5.2 were combined with data from two recent literature reviews and are presented in Fig. 5.5. Duplicate results (datapoints included in more than one study) were removed, and all reported results were converted to \$ kg^{-1} of biodiesel. Because of the significant variation in the reported results, the *x*-axis illustrates the percentage of data with biodiesel production cost below the corresponding value on the *y*-axis.

Using the average price of diesel in the United States for the past 5 years (US Energy Information Administration, 2021), it is obvious that less than 20% of the reported data indicate a lower biodiesel production cost than the conventional diesel selling price. Biodiesel production cost varies between 0.44 and 11.24 \$ kg^{-1} biodiesel, with an average value of 2.56 \$ kg^{-1} biodiesel and a median value of 1.95 \$ kg^{-1} biodiesel. This great variation is to be expected when considering all the variables in the process and the TEA followed by each researcher. As a general remark, it seems that biodiesel would be difficult to compete with conventional diesel at the current technological state and market prices. This could change in the future, either through a technical breakthrough, increased diesel prices or eco-friendly policies that will cover part of the extra cost through some sort of subsidy.

5.4.2 Life cycle analysis of biodiesel production from microalgae

Juneja and Murthy (2017), in their work, also examined the environmental impact of the biofuel production process, calculating a GWP of -0.01 kg CO_2 eq MJ^{-1} of biodiesel, indicating that not only does the production of biodiesel through the proposed process have a low environmental footprint, but also led to the removal of carbon dioxide equivalents from the atmosphere through the displacement of more emission-intensive products. Wu et al. (2018) also carried an LCA to produce biodiesel and calculated a GWP of 0.04 kg CO_2 eq MJ^{-1}. Even if the net emissions of the process were found to be positive, their value is considered very low compared to fossil fuels, as will be later discussed. The LCA of the two processes examined by Dutta et al. (2016) resulted in lower emissions for the most cost-effective alternative that utilized residual biomass for the production of ethanol and methane alongside biodiesel, with 0.04 kg CO_2 eq MJ^{-1}, compared to the production of biodiesel through transesterification without residual biomass valorization, at 1.33 kg CO_2 eq MJ^{-1}. Mu et al. (2017) examined three alternative biofuel production scenarios, with the best LCA results occurring for the coproduction of methane and biodiesel through bio-oil transesterification and AD of residual biomass, at 0.28 kg CO_2 eq MJ^{-1}.

TABLE 5.2 Techno-economic analysis results of resent works examining the production of biofuels from microalgae.

		Upstream variables					Downstream variables				
CO_2	NUTR	Bioreactor	Harvesting	BM lipids (%)	BM yield ($g\ m^{-2}\ d^{-1}$)	BM drying	Oil extraction	Biofuel production	Scale (t BM y^{-1})	Cost ($\$ kg^{-1}$ BIOD)	References
A	F	PBR	N.S.	9	68	C	Milling, SC-CO2	BIOD: TE	$1.8\ 10^3$	0.44	Branco-Vieira et al. (2020)
A	F	OP	HYDRC	30	46	—	Cell disruption, SE	BIOD: TE, EtOH: FERM	$2.9\ 10^5$	0.49	Wu et al. (2018)
A	W	PBR	FLOC, SED	29	46	C	Hyd/mal dewat., pyrolysis	Not included	$3.5\ 10^3$	0.50	Xin et al. (2018)
A	W	PBR	FLOC, SED	29	46	C, SD	HDRL, SE	Not included	$3.3\ 10^6$	0.61	Xin et al. (2016)
A & B	W	OP	SED	54	30	—	HTL	Jet-fuel, BIOD, gasoline: HYDR	$2.1\ 10^5$	1.36	Ranganathan and Savithri (2019)
A	F	OP	SED, DAF	41	30	C	HDRL, SE	BIOD: HYDR, EtOH: FERM, CH_4: AD	$4.0\ 10^5$	1.37	Dutta et al. (2016)
FG	F	ATS	FLOC, DAF	10	20	C	HTL, SE	BIOD: HYDR	$1.5\ 10^5$	1.88	Hoffman et al. (2017)
A	W	OP	FLOC, SED	35	25	C	HTL	BIOD: HYDR, CH_4: HTL,	$2.8\ 10^4$	1.99	Juneja and Murthy (2017)
A	F	PBR	SED	38	23	FP	SE	BIOD	$8.4\ 10^3$	2.29	Sun et al. (2019)
FG	F	OP	SCR	10	20	—	HTL, SE	BIOD: HYDR	$4.4\ 10^5$	2.50	Hoffman et al. (2017)
A	F	OP	SED, DAF	41	30	C	SE	BIOD: TE	$4.0\ 10^5$	3.33	Dutta et al. (2016)

A, Atmospheric; *ATS*, algal turf scrubber; *B*, biogas; *BIOD*, biodiesel; *BM*, biomass; *C*, centrifugation; *DAF*, direct air flotation; *dewat.*, dewatering; *EtOH*, ethanol; *F*, fertilizers; *FERM*, fermentation; *FG*, flue gas; *FLOC*, flocculation; *FP*, filter press; *HDRL*, hydrolysis; *HTL*, hydrothermal liquefaction; *Hyd/mal*, hydrothermal; *HYDR*, hydrotreating; *N.S.*, not specified; *NUTR*, nutrient; *OP*, open pond; *PBR*, photobioreactor; *SC-CO₂*, supercritical carbon dioxide extraction; *SCR*, scraper; *SD*, solar drying; *SE*, solvent extraction; *SED*, sedimentation; *TE*, transesterification; *W*, wastewater.

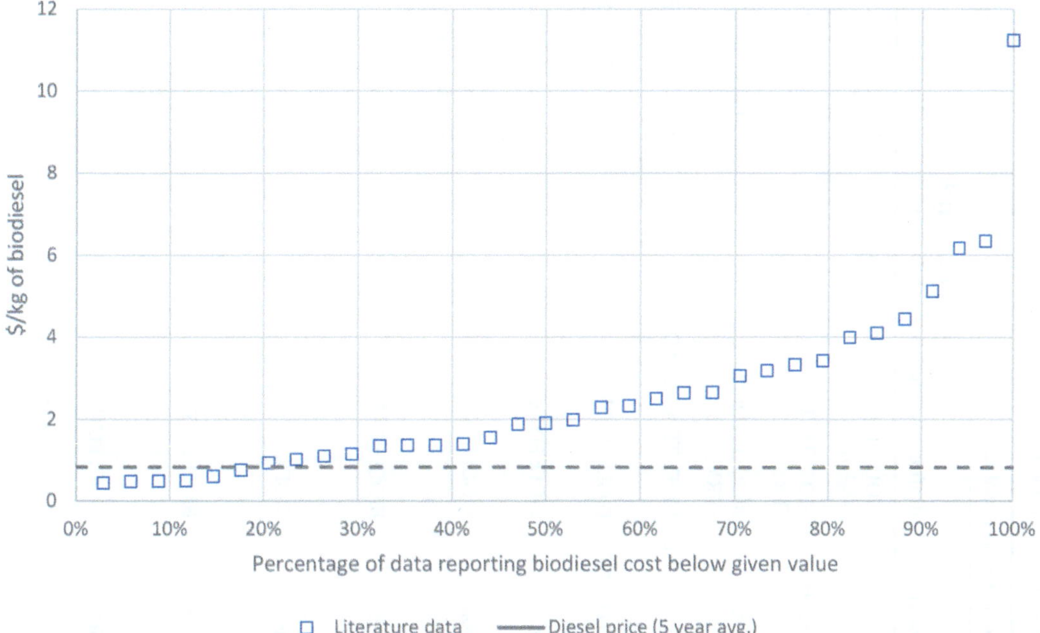

FIGURE 5.5 Comparison of literature data for the production cost of biodiesel from microalgae with diesel price. Source: *Data from Chen et al. (2018c); US Energy Information Administration (2021); Xin et al. (2016), Branco-Vieira et al. (2020), Wu et al. (2018), Xin et al. (2016, 2018), Ranganathan and Savithri (2019), Dutta et al. (2016), Hoffman et al. (2017), Juneja and Murthy (2017), Sun et al. (2019).*

The second scenario included the hydrolysis of biomass before bio-oil extraction and bio-diesel production through hydrothermal treatment with a GWP of 0.33 kg CO_2 eq MJ^{-1}. In comparison, the third scenario with no biomass hydrolysis had a GWP of 0.36 kg CO_2 eq MJ^{-1}. In the work of Saranya and Ramachandra (2020), the authors carried out an LCA for 12 different alternative scenarios for Isochrysis galbana cultivation and biodiesel production, with varying sources of nutrient for microalgae cultivation, different biomass harvesting and drying methods, and different bio-oil transformation techniques, with the GWP ranging from 0.55 to 1.55 kg CO_2 eq. MJ^{-1}. The most promising scenario was the use of solar drying and biocatalyzed transesterification. Finally, in the work of Branco-Vieira et al. (2020), mentioned for their TEA in the previous paragraph, they calculated a GWP of 5.74 kg CO_2 eq MJ^{-1}. The aforementioned results and the analysis variables reported by the authors are summarized in Table 5.3.

As in the case of TEA, it would be useful to compare the results of biodiesel LCA with other energy sources. To this end, data from two recent literature reviews were aggregated with the results presented in Table 5.3 and are summarized in Fig. 5.6. The GWP of coal, oil, natural gas and an average of most renewable energy sources were added for comparison (Edenhofer et al., 2011). All duplicates were removed, and results were converted to kg CO_2 eq MJ^{-1}.

TABLE 5.3 Life cycle analysis results of recent works examining the production of biofuels from microalgae.

		Upstream variables					Downstream variables			GWP (kg CO_2eq MJ^{-1} of BIOD)	References
CO_2	NUTR	Bioreactor	Harvesting	BM lipids (%)	BM yield ($g\,m^{-2}\,d^{-1}$)	BM drying	Oil extraction	Biofuel production			
A	W	OP	FLOC, SED	35	25	C	HTL	BIOD: HYDR, CH_4: catalytic HTL,		−0.01	Juneja and Murthy (2017)
A	F	OP	HYDRC	30	46	–	Cell disruption, SE	BIOD: TE, EtOH: FERM		0.04	Wu et al. (2018)
A	F	OP	SED, DAF	41	30	C	HDRL, SE	BIOD: HYDR, EtOH: FERM, CH_4: AD		0.04	Dutta et al. (2016)
FG & B	F	OP	FLOC, SED	17	7.2	C	Homogenization, SE	BIOD: TE, CH_4: AD		0.28	Mu et al. (2017)
FG	F	OP	FLOC, SED	17	7.2	C	HDRL, HTL, SE	BIOD: HYDR		0.33	Mu et al. (2017)
FG	F	OP	FLOC, SED	17	7.2	C	HTL, SE	BIOD: HYDR		0.36	Mu et al. (2017)
A	–	OP	MH	40	1.24	SD	Sonication	BIOD: BCTE		0.55	Saranya and Ramachandra (2020)
A	W	OP	MH	14	8	SD	Sonication	BIOD: BCTE		0.57	Saranya and Ramachandra (2020)
A	W	OP	MH	14	8	SD	Sonication	BIOD: ACTE		0.65	Saranya and Ramachandra (2020)
A	–	OP	MH	40	1.24	SD	Sonication	BIOD: ACTE		0.75	Saranya and Ramachandra (2020)
A	–	OP	MS	40	1.24	FP	Sonication	BIOD: BCTE		0.75	Saranya and Ramachandra (2020)

A	W	OP	MS	14	8	FP	Sonication	BIOD: BCTE	0.77	Saranya and Ramachandra (2020)
A	W	OP	MS	14	8	FP	Sonication	BIOD: ACTE	0.9	Saranya and Ramachandra (2020)
A	—	OP	MS	40	1.24	FP	Sonication	BIOD: ACTE	1	Saranya and Ramachandra (2020)
A	F	OP	MH	14	15	SD	Sonication	BIOD: BCTE	1.15	Saranya and Ramachandra (2020)
A	F	OP	MH	14	15	SD	Sonication	BIOD: ACTE	1.3	Saranya and Ramachandra (2020)
A	F	OP	SED, DAF	41	30	C	SE	BIOD: TE	1.33	Dutta et al. (2016)
A	F	OP	MS	14	15	FP	Sonication	BIOD: BCTE	1.4	Saranya and Ramachandra (2020)
A	F	OP	MS	14	15	FP	Sonication	BIOD: ACTE	1.55	Saranya and Ramachandra (2020)
A	F	PBR	N.S.	9	68	C	Milling, SC-CO$_2$	BIOD: TE	5.74	Branco-Vieira et al. (2020)

A, Atmospheric; *ACTE*, acid catalyzed transesterification; *AD*, anaerobic digestion; *B*, biogas; *BCTE*, biocatalyzed transesterification; *BIOD*, biodiesel; *BM*, biomass; *C*, centrifugation; *DAF*, direct air flotation; *EtOH*, ethanol; *F*, fertilizers; *FERM*, fermentation; *FG*, flue gas; *FLOC*, flocculation; *FP*, filter press; *HDRL*, hydrolysis; *HTL*, hydrothermal liquefaction; *HYDR*, hydrotreating; *MH*, manual harvesting; Ms, mechanical scrubbers; *N.S.*, not specified; *NUTR*, nutrient; *OP*, open pond; *PBR*, photobioreactor; *SD*, solar drying; *SE*, solvent extraction; *SED*, sedimentation; *TE*, transesterification; *W*, wastewater.

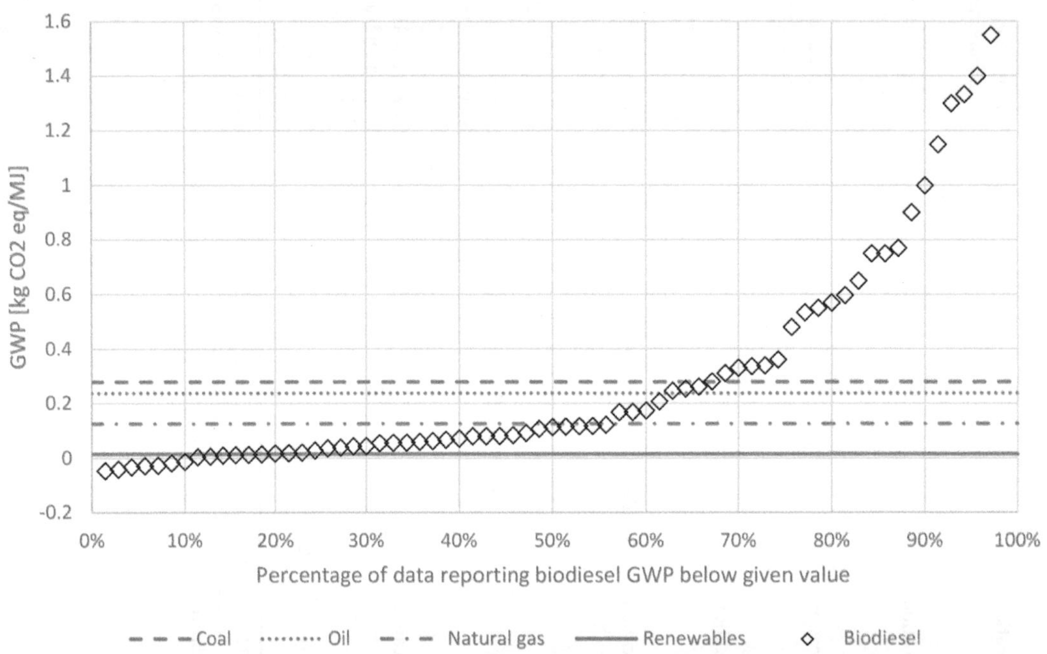

FIGURE 5.6 Comparison of literature data for the GWP of biodiesel production from microalgae with other energy sources' GWP. *GWP*, Global warming potential. Source: *Data from Chamkalani et al. (2020); Edenhofer et al. (2011); Morales et al. (2019); Valente et al. (2019), Juneja and Murthy (2017), Wu et al. (2018), Dutta et al. (2016), Mu et al. (2017), Saranya and Ramachandra (2020), Branco-Vieira et al. (2020).*

Once again, great variation in the reported results can be observed, but in the case of GWP, biodiesel seems to fare much better than fossil fuels, with 65% of the reported data having lower GWP than oil-based energy. It is also notable that other renewables have a lower environmental impact than 80% of biodiesel reported data. The GWP reported for biodiesel production from microalgae ranges from -0.05 to $5.74\,kg\ CO_2\ eq\ MJ^{-1}$ (two datapoints at 2.55 and $5.74\,kg\ CO_2\ eq\ MJ^{-1}$ are not shown because they would make the figure hard to read for the lower GWP values), with an average value of $0.39\,kg\ CO_2$ $eq\ MJ^{-1}$ and a median of $0.11\ kg\ CO_2\ eq\ MJ^{-1}$. Considering the median GWP value (a statistic value better describing nonnormal distributions), biodiesel production from microalgae fairs better than any fossil fuel included in Fig. 5.4 but is outperformed by other renewable energy sources.

As mentioned earlier, the adoption of biodiesel, according to the data presented herein, would be difficult from an economic point of view. Apart from a subsidy to financially back biodiesel production, a different approach that is implemented in many countries today is the fossil fuel carbon tax. Using the data from Figs. 5.5 and 5.6, a carbon tax for oil-based fossil fuels that would make biodiesel competitive can be calculated at 0.24 $\$\ kg^{-1}\ CO_2$. To put this number into context, the average carbon tax currently implemented in 18 European countries is $0.042\ \$\ kg^{-1}\ CO_2$, with the highest carbon tax being implemented in Sweden at $0.137\ \$\ kg^{-1}\ CO_2$ (Elke, 2021). These numbers indicate that there is still a long way to go until biodiesel can replace conventional diesel, but considering the depletion of fossil fuels and the ever-growing environmental awareness of stakeholders, this could change in the not-so-distant future.

5.5 Challenges and prospects for cost-effective biofuels

Biofuel production from microalgae has not been implemented at an industrial scale yet, since technological advancements and economic improvements need to be made. Under this scope, a multiproduct biorefinery approach can contribute to the sustainability of microalgal technology, while integrated processes and coproduction of biofuel and added-value products have currently been proposed (Sánchez-Bayo et al., 2020; Xu et al., 2019). Besides biofuels, microalgae represent an unlimited source of valuable compounds, including pigments (chlorophylls, carotenoids, and phycobilins), proteins, polyunsaturated fatty acids, vitamins and secondary metabolites, which can be further exploited for advanced applications (Koller et al., 2014). Furthermore, process economics and efficiency can be substantially improved through recycling options and the use of low-cost substrates, including CO_2 from flue gases, as well as various types of wastewater and effluents (Koutra et al., 2018). In this case, biomass processing for biofuels production is the best option since no specific requirements exist concerning contaminants (Delrue et al., 2016). In addition, mixotrophic cultivation offers significant advantages over autotrophic growth and can be based on low-cost organic substrates, while two-stage cultivation can also be performed, initially maximizing biomass growth and then followed by the desired product accumulation (Peng et al., 2020). Lastly, genetic and metabolic engineering tools can be applied to enhance species performance in terms of growth characteristics, lipid and other products accumulation, and specific adaptations to extreme and large-scale conditions (Peng et al., 2020).

5.6 Conclusions

Despite the dependence on fossil-based fuels, renewable energy sources have received increasing attention due to the numerous advantages directly affecting the natural environment and human prosperity. Under this scope, microalgae constitute the third-generation biomass that can be used for bioenergy purposes. Additionally, more valuable microalgal components can be further valorized toward various added-value products contributing to sustainability in the framework of a biorefinery approach. However, several challenges, including optimized large-scale cultivation, efficient downstream processing, and significant cost decrease, need to be met to make microalgal biofuels feasible soon.

References

Ahmad, A.L., et al., 2014. Comparison of harvesting methods for microalgae *Chlorella* sp. and its potential use as a biodiesel feedstock. Environmental Technology (United Kingdom) 35 (17), 2244–2253. Available from: https://doi.org/10.1080/09593330.2014.900117. Malaysia: Taylor and Francis Ltd.

Akhlaghi, N., Najafpour-Darzi, G., 2020. A comprehensive review on biological hydrogen production. International Journal of Hydrogen Energy 45 (43), 22492–22512. Available from: https://doi.org/10.1016/j.ijhydene.2020.06.182.

Aliyu, A., Lee, J.G.M., Harvey, A.P., 2021. Microalgae for biofuels via thermochemical conversion processes: a review of cultivation, harvesting and drying processes, and the associated opportunities for integrated production. Bioresource Technology Reports 14, 100676. Available from: https://doi.org/10.1016/j.biteb.2021.100676. Elsevier BV.

Allen, J.W., et al., 2018. Induction of oil accumulation by heat stress is metabolically distinct from N stress in the green microalgae Coccomyxa subellipsoidea C169. PLoS One 13 (9). Available from: https://doi.org/10.1371/journal.pone.0204505. United States: Public Library of Science.

Ananthi, V., Brindhadevi, K., et al., 2021a. Impact of abiotic factors on biodiesel production by microalgae. Fuel 284, 118962. Available from: https://doi.org/10.1016/j.fuel.2020.118962. Elsevier BV.

Ananthi, V., Raja, R., et al., 2021b. A realistic scenario on microalgae based biodiesel production: third generation biofuel. Fuel 284, 118965. Available from: https://doi.org/10.1016/j.fuel.2020.118965. Elsevier BV.

Ansari, F.A., et al., 2018. Evaluation of various cell drying and disruption techniques for sustainable metabolite extractions from microalgae grown in wastewater: a multivariate approach. Journal of Cleaner Production 182, 634–643. Available from: https://doi.org/10.1016/j.jclepro.2018.02.098. South Africa: Elsevier Ltd.

Anwar, M., et al., 2019. Recent advancement and strategy on bio-hydrogen production from photosynthetic microalgae. Bioresource Technology 292, 121972. Available from: https://doi.org/10.1016/j.biortech.2019.121972. Elsevier BV.

Arguelles, E.D., Martinez-Goss, M.R., 2021. Lipid accumulation and profiling in microalgae Chlorolobion sp. (BIOTECH 4031) and Chlorella sp. (BIOTECH 4026) during nitrogen starvation for biodiesel production. Journal of Applied Phycology 33 (1). Available from: https://doi.org/10.1007/s10811-020-02126-z. Philippines: Springer Science and Business Media B.V.

Arun, J., et al., 2020. A conceptual review on microalgae biorefinery through thermochemical and biological pathways: bio-circular approach on carbon capture and wastewater treatment. Bioresource Technology Reports 11. Available from: https://doi.org/10.1016/j.biteb.2020.100477. India: Elsevier Ltd.

Bastos, R.G., 2018. Biofuels from microalgae: bioethanol. Green Energy and Technology 229–246. Available from: https://doi.org/10.1007/978-3-319-69093-3_11. Brazil: Springer Verlag.

Branco-Vieira, M., et al., 2020. Economic analysis of microalgae biodiesel production in a small-scale facility. Energy Reports 6, 325–332. Available from: https://doi.org/10.1016/j.egyr.2020.11.156. Portugal: Elsevier Ltd.

Carrillo-Reyes, J., et al., 2021. Thermophilic biogas production from microalgae-bacteria aggregates: biogas yield, community variation and energy balance. Chemosphere 275. Available from: https://doi.org/10.1016/j.chemosphere.2021.129898. Mexico: Elsevier Ltd.

Chamkalani, A., et al., 2020. A critical review on life cycle analysis of algae biodiesel: current challenges and future prospects. Renewable and Sustainable Energy Reviews 134, 110143. Available from: https://doi.org/10.1016/j.rser.2020.110143. Elsevier BV.

Chen, Y.D., et al., 2018a. Waste biorefineries—integrating anaerobic digestion and microalgae cultivation for bioenergy production. Current Opinion in Biotechnology 50, 101–110. Available from: https://doi.org/10.1016/j.copbio.2017.11.017. China: Elsevier Ltd.

Chen, J., Leng, L., et al., 2018b. A comparative study between fungal pellet- and spore-assisted microalgae harvesting methods for algae bioflocculation. Bioresource Technology 259, 181–190. Available from: https://doi.org/10.1016/j.biortech.2018.03.040. China: Elsevier Ltd.

Chen, J., Li, J., et al., 2018c. The potential of microalgae in biodiesel production. Renewable and Sustainable Energy Reviews 90, 336–346. Available from: https://doi.org/10.1016/j.rser.2018.03.073. China: Elsevier Ltd.

de Farias Silva, C.E., Meneghello, D., Bertucco, A., 2018. A systematic study regarding hydrolysis and ethanol fermentation from microalgal biomass. Biocatalysis and Agricultural Biotechnology 14, 172–182. Available from: https://doi.org/10.1016/j.bcab.2018.02.016. Italy: Elsevier Ltd.

de Jesus, S.S., et al., 2019. Comparison of several methods for effective lipid extraction from wet microalgae using green solvents. Renewable Energy 143, 130–141. Available from: https://doi.org/10.1016/j.renene.2019.04.168. Brazil: Elsevier Ltd.

de Jesús-Campos, D., et al., 2020. Chemical composition, fatty acid profile and molecular changes derived from nitrogen stress in the diatom Chaetoceros muelleri. Aquaculture Reports 16. Available from: https://doi.org/10.1016/j.aqrep.2020.100281. Mexico: Elsevier B.V.

Debnath, D., et al., 2019. The future of biofuels in an electrifying global transportation sector: imperative, prospects and challenges. Applied Economic Perspectives and Policy 41 (4), 563–582. Available from: https://doi.org/10.1093/aepp/ppz023. United States: John Wiley and Sons Inc.

Delrue, F., et al., 2016. The environmental biorefinery: using microalgae to remediate wastewater, a win-win paradigm. Energies 9 (3). Available from: https://doi.org/10.3390/en9030132. France: MDPI AG.

Derakhshandeh, M., Ateş, F., Tezcan Un, U., 2020. Renewable bio-oil from pyrolysis of *Synechocystis* and *Scenedesmus* wild-type microalgae biomass. Bioenergy Research. Available from: https://doi.org/10.1007/s12155-020-10200-0. Turkey: Springer.

Derakhshandeh, M., Atici, T., Tezcan Un, U., 2021. Evaluation of wild-type *Microalgae* Species biomass as carbon dioxide sink and renewable energy resource. Waste and Biomass Valorization. Springer Science and Business Media LLC 12 (1), 105–121. Available from: https://doi.org/10.1007/s12649-020-00969-8.

Dutta, S., Neto, F., Coelho, M.C., 2016. Microalgae biofuels: a comparative study on techno-economic analysis & life-cycle assessment. Algal Research 20, 44–52. Available from: https://doi.org/10.1016/j.algal.2016.09.018. Portugal: Elsevier.

Ebadian, M., et al., 2020. Biofuels policies that have encouraged their production and use: an international perspective. Energy Policy 147, 111906. Available from: https://doi.org/10.1016/j.enpol.2020.111906. Elsevier BV.

Edenhofer, O., et al., 2011. Renewable energy sources and climate change mitigation: special report of the intergovernmental panel on climate change. Cambridge University Press, Cambridge, United Kingdom. Available at. Available from: https://www.ipcc.ch/report/renewable-energy-sources-and-climate-change-mitigation/.

Elke, A., 2021. Carbon taxes in Europe. <https://taxfoundation.org/carbon-taxes-in-europe-2021/> (accessed 14.08.21).

Flach, B., Lieberz, S., Bolla, S., 2020. Biofuels annual, USDA Foreign Agricultural Service. <https://apps.fas.usda.gov/newgainapi/api/Report/DownloadReportByFileName?fileName = Biofuels%20Annual_The%20Hague_European%20Union_06-29-2020>.

Gautam, R., Vinu, R., 2020. Reaction engineering and kinetics of algae conversion to biofuels and chemicals: via pyrolysis and hydrothermal liquefaction. Reaction Chemistry and Engineering. India: Royal Society of Chemistry 5 (8), 1320–1373. Available from: https://doi.org/10.1039/d0re00084a.

Günerken, E., et al., 2015. Cell disruption for microalgae biorefineries. Biotechnology Advances 33 (2), 243–260. Available from: https://doi.org/10.1016/j.biotechadv.2015.01.008. Belgium: Elsevier Inc.

Hoffman, J., et al., 2017. Techno-economic assessment of open microalgae production systems. Algal Research 23, 51–57. Available from: https://doi.org/10.1016/j.algal.2017.01.005. United States: Elsevier B.V.

Hosseinizand, H., Sokhansanj, S., Lim, C.J., 2018. Studying the drying mechanism of microalgae *Chlorella vulgaris* and the optimum drying temperature to preserve quality characteristics. Drying Technology 36 (9), 1049–1060. Available from: https://doi.org/10.1080/07373937.2017.1369986. Canada: Taylor and Francis Inc.

Hu, Y., et al., 2019. A review of recent developments of pre-treatment technologies and hydrothermal liquefaction of microalgae for bio-crude oil production. Renewable and Sustainable Energy Reviews 101, 476–492. Available from: https://doi.org/10.1016/j.rser.2018.11.037. Canada: Elsevier Ltd.

Juneja, A., Murthy, G.S., 2017. Evaluating the potential of renewable diesel production from algae cultured on wastewater: Techno-economic analysis and life cycle assessment. AIMS Energy 5 (2), 239–257. Available from: https://doi.org/10.3934/energy.2017.2.239. United States: AIMS Press.

Kazemi Shariat Panahi, H., et al., 2019. Recent updates on the production and upgrading of bio-crude oil from microalgae. Bioresource Technology Reports 7. Available from: https://doi.org/10.1016/j.biteb.2019.100216. Iran: Elsevier Ltd.

Kendir Çakmak, E., Ugurlu, A., 2020. Enhanced biogas production of red microalgae via enzymatic pretreatment and preliminary economic assessment. Algal Research 50, 101979. Available from: https://doi.org/10.1016/j.algal.2020.101979. Elsevier BV.

Khoo, K.S., et al., 2020. Recent advances in downstream processing of microalgae lipid recovery for biofuel production. Bioresource Technology 304. Available from: https://doi.org/10.1016/j.biortech.2020.122996. Malaysia: Elsevier Ltd.

Koley, S., et al., 2017. Development of a harvesting technique for large-scale microalgal harvesting for biodiesel production. RSC Advances 7 (12), 7227–7237. Available from: https://doi.org/10.1039/c6ra27286j. India: Royal Society of Chemistry.

Koller, M., Muhr, A., Braunegg, G., 2014. Microalgae as versatile cellular factories for valued products. Algal Research 6, 52–63. Available from: https://doi.org/10.1016/j.algal.2014.09.002. Austria: Elsevier.

Koutra, E., et al., 2018. Bio-based products from microalgae cultivated in digestates. Trends in Biotechnology 36 (8), 819–833. Available from: https://doi.org/10.1016/j.tibtech.2018.02.015. Greece: Elsevier Ltd.

Lakshmikandan, M., et al., 2020. Sustainable biomass production under CO_2 conditions and effective wet microalgae lipid extraction for biodiesel production. Journal of Cleaner Production 247. Available from: https://doi.org/10.1016/j.jclepro.2019.119398. China: Elsevier Ltd.

Lee, S.Y., et al., 2021. Techniques of lipid extraction from microalgae for biofuel production: a review. Environmental Chemistry Letters 19 (1), 231–251. Available from: https://doi.org/10.1007/s10311-020-01088-5. Malaysia: Springer Science and Business Media Deutschland GmbH.

Lu, J., et al., 2020. Catalytic hydrothermal liquefaction of microalgae over mesoporous silica-based materials with site-separated acids and bases. Fuel 279, 118529. Available from: https://doi.org/10.1016/j.fuel.2020.118529. Elsevier BV.

Maia, J.L.D., et al., 2020. Microalgae starch: a promising raw material for the bioethanol production. International Journal of Biological Macromolecules 165, 2739–2749. Available from: https://doi.org/10.1016/j.ijbiomac.2020.10.159. Brazil: Elsevier B.V.

Makut, B.B., Goswami, G., Das, D., 2020. Evaluation of bio-crude oil through hydrothermal liquefaction of microalgae-bacteria consortium grown in open pond using wastewater. Biomass Conversion and Biorefinery. Springer, India. Available from: http://doi.org/10.1007/s13399-020-00795-x.

Megawati, et al., 2020. Drying characteristics of *Chlorella pyrenoidosa* using oven and its evaluation for bio-ethanol production. Materials Science Forum. Trans Tech Publications Ltd, Indonesia. Available from: http://doi.org/10.4028/http://www.scientific.net/MSF.1007.1.

Milano, J., et al., 2016. Microalgae biofuels as an alternative to fossil fuel for power generation, Renewable and Sustainable Energy Reviews, 58. Elsevier Ltd, Malaysia, pp. 180–197. Available from: http://doi.org/10.1016/j.rser.2015.12.150.

Morales, M., et al., 2019. Life-Cycle Assessment of Microalgal-Based Biofuel. Elsevier BV, pp. 507–550. Available from: http://doi.org/10.1016/b978-0-444-64192-2.00020-2.

Mu, D., et al., 2017. Life cycle assessment and nutrient analysis of various processing pathways in algal biofuel production, Bioresource Technology, 230. Elsevier Ltd, United States, pp. 33–42. Available from: http://doi.org/10.1016/j.biortech.2016.12.108.

Muhammad, G., et al., 2021. Modern developmental aspects in the field of economical harvesting and biodiesel production from microalgae biomass. Renewable and Sustainable Energy Reviews. Elsevier Ltd, China, p. 135. Available from: http://doi.org/10.1016/j.rser.2020.110209.

Nagarajan, D., Chang, J.S., Lee, D.J., 2020. Pretreatment of microalgal biomass for efficient biohydrogen production – recent insights and future perspectives. Bioresource Technology. Elsevier Ltd, Taiwan, p. 302. Available from: http://doi.org/10.1016/j.biortech.2020.122871.

Ochoa-Alfaro, A.E., et al., 2019. pH effects on the lipid and fatty acids accumulation in *Chlamydomonas reinhardtii*. Biotechnology Progress 35.

Oliveira, C.Y.B.D., et al., 2020. A comparison of harvesting and drying methodologies on fatty acids composition of the green microalga *Scenedesmus obliquus*. Biomass and Bioenergy. Elsevier Ltd, Brazil, p. 132. Available from: http://doi.org/10.1016/j.biombioe.2019.105437.

Onay, M., 2020. The effects of indole-3-acetic acid and hydrogen peroxide on *Chlorella zofingiensis* CCALA 944 for bio-butanol production, Fuel, 273. Elsevier BV, p. 117795. Available from: http://doi.org/10.1016/j.fuel.2020.117795.

Peng, L., et al., 2020. Biofuel production from microalgae: a review. Environmental Chemistry Letters 18 (2), 285–297. Available from: https://doi.org/10.1007/s10311-019-00939-0. China: Springer.

Qari, H.A., Oves, M., 2020. Fatty acid synthesis by *Chlamydomonas reinhardtii* in phosphorus limitation. Journal of Bioenergetics and Biomembranes 52 (1), 27–38. Available from: https://doi.org/10.1007/s10863-019-09813-8. Saudi Arabia: Springer.

Quesada-Salas, M.C., et al., 2021. Article optimization and comparison of three cell disruption processes on lipid extraction from microalgae. Processes 9 (2), 1–20. Available from: https://doi.org/10.3390/pr9020369. France: MDPI AG.

Rana, M.S., et al., 2020. Effect of iron oxide nanoparticles on growth and biofuel potential of *Chlorella* spp. Algal Research. Elsevier B.V., India, p. 49. Available from: http://doi.org/10.1016/j.algal.2020.101942.

Ranganathan, P., Savithri, S., 2019. Techno-economic analysis of microalgae-based liquid fuels production from wastewater via hydrothermal liquefaction and hydroprocessing, Bioresource Technology, 284. Elsevier Ltd, India, pp. 256–265. Available from: http://doi.org/10.1016/j.biortech.2019.03.087.

Razu, M.H., Hossain, F., Khan, M., 2019. Advancement of bio-hydrogen production from microalgae. Microalgae Biotechnology for Development of Biofuel and Wastewater Treatment. Springer Singapore, Bangladesh, pp. 423–462. Available from: http://doi.org/10.1007/978-981-13-2264-8_17.

Salakkam, A., et al., 2021. Valorization of microalgal biomass for biohydrogen generation: a review, Bioresource Technology, 322. Elsevier BV, p. 124533. Available from: http://doi.org/10.1016/j.biortech.2020.124533.

Sánchez-Bayo, A., et al., 2020. Biodiesel and biogas production from *Isochrysis galbana* using dry and wet lipid extraction: a biorefinery approach, Renewable Energy, 146. Elsevier Ltd, Spain, pp. 188–195. Available from: http://doi.org/10.1016/j.renene.2019.06.148.

Saranya, G., Ramachandra, T.V., 2020. Life cycle assessment of biodiesel from estuarine microalgae. Energy Conversion and Management: X 8, 100065. Available from: https://doi.org/10.1016/j.ecmx.2020.100065.

Seo, S.H., et al., 2019. Maximizing biomass and lipid production in *Ettlia* sp. by ultraviolet stress in a continuous culture. Bioresource Technology. Elsevier Ltd, South Korea, p. 288. Available from: http://doi.org/10.1016/j.biortech.2019.121472.

Show, P.L., et al., 2017. A holistic approach to managing microalgae for biofuel applications. International Journal of Molecular Sciences 18 (1). Available from: https://doi.org/10.3390/ijms18010215. Malaysia: MDPI AG.

Sierra, L.S., Dixon, C.K., Wilken, L.R., 2017. Enzymatic cell disruption of the microalgae *Chlamydomonas reinhardtii* for lipid and protein extraction, Algal Research, 25. Elsevier B.V, United States, pp. 149–159. Available from: http://doi.org/10.1016/j.algal.2017.04.004.

Sindhu, R., et al., 2019. Biofuel production from biomass: toward sustainable development. Current Developments in Biotechnology and Bioengineering: Waste Treatment Processes for Energy Generation. Elsevier, India, pp. 79–92. Available from: http://doi.org/10.1016/B978-0-444-64083-3.00005-1.

Singh, G., Patidar, S.K., 2018. Microalgae harvesting techniques: a review, Journal of Environmental Management, 217. Academic Press, India, pp. 499–508. Available from: http://doi.org/10.1016/j.jenvman.2018.04.010.

Sun, J., et al., 2019. Microalgae biodiesel production in China: a preliminary economic analysis, Renewable and Sustainable Energy Reviews, 104. Elsevier Ltd, China, pp. 296–306. Available from: http://doi.org/10.1016/j.rser.2019.01.021.

t' Lam, G.P., et al., 2018. Multi-product microalgae biorefineries: from concept towards reality. Trends in Biotechnology 36 (?), 216–227. Available from: https://doi.org/10.1016/j.tibtech.2017.10.011. Netherlands: Elsevier Ltd.

Tan, X.B., et al., 2020. Lipids production and nutrients recycling by microalgae mixotrophic culture in anaerobic digestate of sludge using wasted organics as carbon source. Bioresource Technology. Elsevier Ltd, China, p. 297. Available from: http://doi.org/10.1016/j.biortech.2019.122379.

Timung, S., Singh, H., Annu, A., 2021. Bio-butanol as biofuels: the present and future scope. Wiley, pp. 467–485. Available from: http://doi.org/10.1002/9781119793038.ch13.

Tsigkou, K., Sakarika, M., Kornaros, M., 2019. Inoculum origin and waste solid content influence the biochemical methane potential of olive mill wastewater under mesophilic and thermophilic conditions, Biochemical Engineering Journal, 151. Elsevier BV, p. 107301. Available from: http://doi.org/10.1016/j.bej.2019.107301.

U.S. Energy Information Administration, (2021). <https://www.eia.gov/petroleum/gasdiesel/> (accessed 14.08.21).

Valente, A., Iribarren, D., Dufour, J., 2019. How do methodological choices affect the carbon footprint of microalgal biodiesel? A harmonised life cycle assessment, Journal of Cleaner Production, 207. Elsevier Ltd, Spain, pp. 560–568. Available from: http://doi.org/10.1016/j.jclepro.2018.10.020.

Vanthoor-Koopmans, M., et al., 2013. Biorefinery of microalgae for food and fuel, Bioresource Technology, 135. Elsevier Ltd, Netherlands, pp. 142–149. Available from: http://doi.org/10.1016/j.biortech.2012.10.135.

Vargas-Estrada, L., et al., 2021. A review on current trends in biogas production from miroalgae biomass and microalgae waste by anaerobic digestion and co-digestion. Bioenergy Research. Springer, Mexico. Available from: http://doi.org/10.1007/s12155-021-10276-2.

Vassalle, L., et al., 2020. Upflow anaerobic sludge blanket in microalgae-based sewage treatment: co-digestion for improving biogas production. Bioresource Technology. Elsevier Ltd, Brazil, p. 300. Available from: http://doi.org/10.1016/j.biortech.2019.122677.

Waghmare, A., et al., 2019. Hydrodynamic cavitation for energy efficient and scalable process of microalgae cell disruption, Algal Research, 40. Elsevier BV, p. 101496. Available from: http://doi.org/10.1016/j.algal.2019.101496.

Wang, J., Yin, Y., 2018. Fermentative hydrogen production using pretreated microalgal biomass as feedstock. Microbial Cell Factories 17 (1). Available from: https://doi.org/10.1186/s12934-018-0871-5. China: BioMed Central Ltd.

Web of Science, 2021. <https://www.webofscience.com/wos/woscc/basic-search> (accessed 14.08.21).

Wu, W., Lin, K.H., Chang, J.S., 2018. Economic and life-cycle greenhouse gas optimization of microalgae-to-biofuels chains, Bioresource Technology, 267. Elsevier Ltd, Taiwan, pp. 550–559. Available from: http://doi.org/10.1016/j.biortech.2018.07.083.

Xiao, C., et al., 2019. Exergy analyses of biogas production from microalgae biomass via anaerobic digestion, Bioresource Technology, 289. Elsevier BV, p. 121709. Available from: http://doi.org/10.1016/j.biortech.2019.121709.

Xin, C., et al., 2016. Comprehensive techno-economic analysis of wastewater-based algal biofuel production: a case study, Bioresource Technology, 211. Elsevier Ltd, China, pp. 584–593. Available from: http://doi.org/10.1016/j.biortech.2016.03.102.

Xin, C., et al., 2018. Waste-to-biofuel integrated system and its comprehensive techno-economic assessment in wastewater treatment plants, Bioresource Technology, 250. Elsevier Ltd, China, pp. 523–531. Available from: http://doi.org/10.1016/j.biortech.2017.11.040.

Xu, S., et al., 2019. Evaluation of bioethanol and biodiesel production from *Scenedesmus obliquus* grown in biodiesel waste glycerol: A sequential integrated route for enhanced energy recovery. Energy Conversion and Management. Elsevier Ltd, China, p. 197. Available from: http://doi.org/10.1016/j.enconman.2019.111907.

Yin, Z., et al., 2020. A comprehensive review on cultivation and harvesting of microalgae for biodiesel production: Environmental pollution control and future directions. Bioresource Technology. Elsevier Ltd, China, p. 301. Available from: http://doi.org/10.1016/j.biortech.2020.122804.

Yukesh Kannah, R., et al., 2021. A review on anaerobic digestion of energy and cost effective microalgae pretreatment for biogas production. Bioresource Technology. Elsevier Ltd, India, p. 332. Available from: http://doi.org/10.1016/j.biortech.2021.125055.

Zabed, H.M., et al., 2020. Biogas from microalgae: technologies, challenges and opportunities, Renewable and Sustainable Energy Reviews, 117. Elsevier BV, p. 109503. Available from: http://doi.org/10.1016/j.rser.2019.109503.

Zhang, B., et al., 2018. Effect of acidic, neutral and alkaline conditions on product distribution and biocrude oil chemistry from hydrothermal liquefaction of microalgae, Bioresource Technology, 270. Elsevier Ltd, China, pp. 129–137. Available from: http://doi.org/10.1016/j.biortech.2018.08.129.

Zhao, Y., et al., 2019. Influence of cadmium stress on the lipid production and cadmium bioresorption by *Monoraphidium* sp. QLY-1, Energy Conversion and Management, 188. Elsevier Ltd, China, pp. 76–85. Available from: http://doi.org/10.1016/j.enconman.2019.03.041.

Zhu, L., et al., 2017. Using microalgae to produce liquid transportation biodiesel: what is next? Renewable and Sustainable Energy Reviews, 78. Elsevier Ltd, China, pp. 391–400. Available from: http://doi.org/10.1016/j.rser.2017.04.089.

Further reading

Allen, J.W., et al., 2018. Induction of oil accumulation by heat stress is metabolically distinct from N stress in the green microalgae *Coccomyxa subellipsoidea* C169. PLoS One 13 (9). Available from: https://doi.org/10.1371/journal.pone.0204505. United States: Public Library of Science.

de Jesús-Campos, D., et al., 2020. Chemical composition, fatty acid profile and molecular changes derived from nitrogen stress in the diatom *Chaetoceros muelleri*. Aquaculture Reports. Elsevier B.V, Mexico, p. 16. Available from: http://doi.org/10.1016/j.aqrep.2020.100281.

Eppink, M.H.M., et al., 2019. From current algae products to future biorefinery practices: a review. Advances in Biochemical Engineering/Biotechnology. Springer Science and Business Media Deutschland GmbH, Netherlands, pp. 99–123. Available from: http://doi.org/10.1007/10_2016_64.

Jaiswal, K.K., et al., 2020. Ecological stress stimulus to improve microalgae biofuel generation: a review. Octa J. Biosci. 8 (1), 48–54.

Ochoa-Alfaro, A.E., et al., 2019. pH effects on the lipid and fatty acids accumulation in *Chlamydomonas reinhardtii*. Biotechnology Progress 35.

Qari, H.A., Oves, M., 2020. Fatty acid synthesis by *Chlamydomonas reinhardtii* in phosphorus limitation. Journal of Bioenergetics and Biomembranes 52 (1), 27–38. Available from: https://doi.org/10.1007/s10863-019-09813-8. Saudi Arabia: Springer.

Rana, M.S., et al., 2020. Effect of iron oxide nanoparticles on growth and biofuel potential of Chlorella spp. Algal Research. Elsevier B.V, India, p. 49. Available from: http://doi.org/10.1016/j.algal.2020.101942.

Tan, X.B., et al., 2020. Lipids production and nutrients recycling by microalgae mixotrophic culture in anaerobic digestate of sludge using wasted organics as carbon source. Bioresource Technology. Elsevier Ltd, China, p. 297. Available from: http://doi.org/10.1016/j.biortech.2019.122379.

Zhao, Y., et al., 2019. Influence of cadmium stress on the lipid production and cadmium bioresorption by *Monoraphidium* sp. QLY-1, Energy Conversion and Management, 188. Elsevier Ltd, China, pp. 76–85. Available from: http://doi.org/10.1016/j.enconman.2019.03.041.

Valorization of microalgal biomass for food

Ala'a H. Al-Muhtaseb[1], Farrukh Jamil[2], Asma Sarwer[2] and Suhaib Al-Maawali[1]

[1]Department of Petroleum and Chemical Engineering, College of Engineering, Sultan Qaboos University, Muscat, Oman [2]Department of Chemical Engineering, COMSATS University Islamabad (CUI), Lahore, Pakistan

6.1 Introduction

Depending on the cellular organization, algae are classified into macroalgae and microalgae. Microalgae can survive in a harsh environment. It can be grown in both water and land. It can grow with a minimal amount of nutrients and water when compared to the plant existing on the land. It can be grown in industrial wastewater using the nutrients already present in it, which makes the wastewater environmentally safe with the small quantity of water used (Bhalamurugan et al., 2018). Microalgae photosynthetic microorganisms have diverse species with a large group. Microalgae, in general, have some characteristics that are extremely useful for food/feed, biofuels, cosmetic, agricultural, and pharmaceutical industries. The high growth rate, ability to fix carbon dioxide, ability to be grown in soil or water, the ability to grow all over the year (according to climate), nonrequirement of pesticides/fertilizers, and biodegradability are some beneficial features, making it highly valuable for various industries (Blaga et al., 2018).

Microalgae are gaining attention due to their application in human and animal feed. It contains high-value nutrients, for example, proteins, pigments, and the presence of bioactive compounds, vitamins, sterols, minerals, and amino acids (Khanra et al., 2018). Because of its excellent nutritional profile, it has been used in areas of food and medicine, emerging as an alternative to use in industry-related aquaculture.

In addition, the microalgal biomass is a replacement for fish oil and fish meals. Currently, the fish industry relies on a fish meal containing high protein content and fish

Valorisation of Microalgal Biomass and Wastewater Treatment
DOI: https://doi.org/10.1016/B978-0-323-91869-5.00016-8

81

oil for feed. An increase in fish meal prices and excessive use of pelagic fishes for fish meals has initiated the trend to find a less expensive unconventional substitute.

Moreover, it is used to mitigate anthropogenic carbon dioxide and wastewater and produce many high valued low-volume products. According to research in this field, producing 1 kg of dry algal biomass consumes about 1.83 kg of carbon dioxide (Yadav et al., 2020). Its ability to capture carbon dioxide is ten times greater than its terrestrial counterparts (Kamyab et al., 2017).

The pigments produced by microalgae, aside from biomass, such as beta-carotene, chlorophyll, and carotenoids, are used in the food and cosmetic industries as colorants. Furthermore, algae strains like *Chlorella* sp., etc., are used in nutrient-rich supplements such as vitamin D2 (Blaga et al., 2018).

Microalgae are great sources of vegetable protein. Research related to nutritional elements on different microalgae describes that high quality and high amounts of proteins can be produced through microalgae, a source of essential amino acids. Recent studies elaborate that microalgae have the potential to be a substitute for the resource of recombinant protein (Khanra et al., 2018).

The chemical composition of microalgae varies depending on the species, environmental factors, and nutrient availability. These factors can affect the composition of main components, that is, proteins, lipids, and carbohydrates. The suitable strain selection of microalgae is according to the final objectives (production of food, feed, biofuel, etc.). Hence, the following characteristics must be evaluated:

1. Constant cultivation is possible either in a photobioreactor (PBR) (in a closed and controlled system) or in an open pond.
2. Constant and high quantities of valuable bioactive compounds are ensured.
3. Changes in climate and seasons do not affect the survival and growth of the microalgae.
4. Fouling inside the vessel is minimum.
5. Large-scale biomass production is possible.
6. Harvesting is easy and can be subjected to various methods of extraction.
7. Energy conversion rate and efficiency related to photosynthesis are high.

The nutritional factors that affect the growth are mainly C-source, nitrogen source, vitamins, minerals, and trace elements. Hence, manipulating any of these factors can enhance the productivity and selectivity of the desired product. In hydrocarbon and biomass production, the nitrogen and phosphorus ratio has a significant influence. Trace elements are necessary for the metabolic pathways, including the utilization of CO_2, phosphorus, nitrogen, and light. Thus nutrient concentration, high temperature, light intensity, high salinity, and substitutes of organic carbon induce stress in the cultivation process, which eventually increases the concentration of some compounds but affects the growth rate simultaneously (Blaga et al., 2018).

Nowadays, microalgae are an excellent unconventional resource for biomass production. But the energy requirements using the available technology are too high for biomass production for feed, food, fuel, and chemicals. The yield of microalgal biomass has to be increased for it to be a primary product, just like crops. Along with all these matters, the cost of production is also a barrier needed to economizing yet. To optimize the process, the yield of biomass is required to be increased. The use of PBRs, genetic engineering in metabolic pathways and selection of strains is a development in this field. But the advancement in downstream processing is not satisfactory, essential for reducing production costs (Vandamme et al., 2013).

Some industries produce microalgal biomass, such as Fuji Chemical, Cyanotic, Mera Pharmaceuticals, and Seambiotic, for value-added products in pharmaceuticals, feed, and cosmetics. Selection of harvesting and pigment extraction processes suitable for the type of microalgae used is also essential. The factors affecting are the type of solvent for extraction, features of the pigment itself, the time required for extraction, yield, cost, and ease of operation. The conventionally used processes for extraction in industries are percolation, soaking, pressurized liquid extraction, counter-current extraction, and soxhlet extraction. As described above, the downstream processes are not yet appropriately economized; these extraction processes also come with disadvantages such as the requirement of vast amounts of solvent, denaturation of molecules, environmental pollution, and health hazards. Thus more efficient and optimized extraction of biomolecules is needed, and advancement in technique and technology is the demand.

The extraction of protein from microalgae is carried out by alkaline, acidic, and aqua methods. Protein Recovery is made by ultrafiltration, precipitation, and centrifugation of chromatography techniques. The downstream process for proteins is also the same as pigments, as the extraction and purification are not researched widely. The major obstacles in upgrading the protein extraction process are diversity in the species of microalgae, cell structure variations, and the release of degrading enzymes. Some advanced techniques and technologies are being applied for extraction processes: ultrasound extraction, pulsed electric field, and microwave-assisted extraction (Khanra et al., 2018).

6.2 Products derived from microalgae

Refer Fig. 6.1.

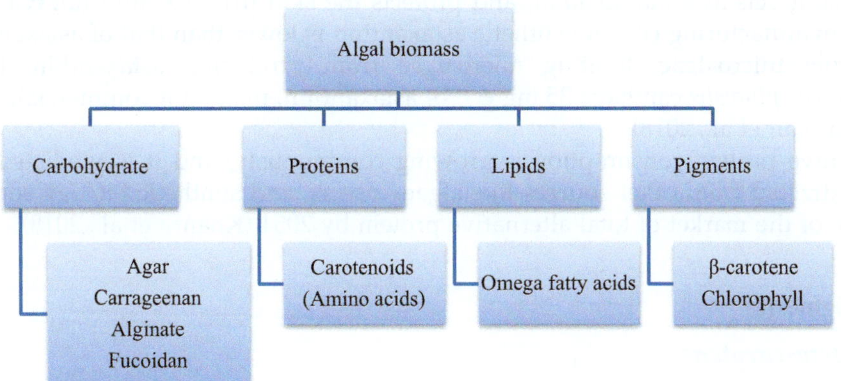

FIGURE 6.1   Classification of algal biomass based on the nature of molecules.

6.2.1 Proteins

Biomass derived from microalgae carries a good quantity of proteins and other components as mentioned in Fig. 6.1. All 20 essential amino acids can be synthesized through microalgae, making it an unconventional source for human nutrition, food ingredients, and medicine.

6.2.1.1 Carotenoids

The use of algae as a food source has been practiced for centuries due to its high protein value. Japan was the first country to manufacture *Chlorella* sp. commercially in 1960 for human consumption. Different microalgae species produce protein content ranging from 6% to 70% of their dry biomass.

The powerful carotenoids of human consumption are lutein, β-carotene, astaxanthin, lycopene, fucoxanthin, etc. When compared with terrestrial plants, microalgae contain higher carotenoid content. There are several advantages of obtaining carotenoid from microalgae due to additional nutritional value, antioxidant activity, and its use as a food colorant. Moreover, there are no requirements for storage of potential degradation of carotenoids in time, and there is a possibility to grow microalgae all year.

Carotenoids can be classified into primary and secondary carotenoids. This classification is based on the difference in the cultivation process. The primary carotenoids are subjected to a one-step cultivation process because of the degradation of carotenoids under stress. At the same time, the secondary carotenoids are produced through a two-step cultivation process. In the first step, growth and multiplication occur in the optimized conditions, while in the second step, extreme stress conditions are applied, that is, reduced nitrogen and intense light.

Lutein carotenoids are found in fruits and vegetables, but the most important sources are egg yolk and maize. It is obtained from Marigold flower petals. At the same time, microalgae can produce higher amounts of lutein carotenoid. The labor and the area required for the cultivation of marigold flowers are more elevated than cultivating microalgae. *Muriellopsis* sp. and *Scenedesmus almeriensis* are common sources of lutein carotenoid.

Astaxanthin is a carotenoid that is considered to be a high-value product available in the market. It acts as an antioxidant and protects the skin from harmful ultraviolet radiation. The manufacturing cost of synthetic astaxanthin is lower than that of astaxanthin produced from microalgae, limiting microalgae from producing astaxanthin. However, *Haematococcus pluvialis* can have 35 mg g^{-1} of astaxanthin, making it commercially possible (Bhalamurugan et al., 2018).

Alternative protein consumption is growing continuously, and it is predicted that the protein extracted from other sources like algae, insects, and synthetic biology sources will cover 50% of the market of total alternative protein by 2054 (Khanra et al., 2018).

6.2.2 Pigments

6.2.2.1 Beta-carotene

The pigment like beta-carotene is primarily a carotenoid, lipid-soluble natural pigment—the resources of beta-carotene algae, plants, and photosynthetic bacteria. Pigments play a vital role in the process of photosynthesis. It is used as a coloring agent, antioxidant, and vitamin A supplement. Beta-carotene is related to a group of carotenoid pigments. It was the first molecule of carotenoid which was commercialized. The primary source of natural beta-carotene is carrot. Recently, microalgae have been used as a substitute natural source for commercially producing beta-carotene. The preferred source to produce beta-carotene is *Dunaliella salina* because of its highest carotenoid content, approximately 10% of dry biomass. Other microalgae species commonly used in manufacturing beta-carotene are *Dunaliella bradawl* and *S. almeriensis* (Guedes et al., 2011).

6.2.2.2 Chlorophyll

Chlorophyll is a green-colored pigment naturally present in plants, cyanobacteria, and algae. It is an essential pigment for photosynthesis. The most popular and primary source of chlorophyll manufacturing is *Chlorella* sp. The *Chlorella* sp. contains approximately 7% of the biomass, several times more than the chlorophyll in spirulina.

6.2.3 Carbohydrates

Several products are used for food purposes derived from different microalgae species composed of polysaccharides. The activities of polysaccharides are strongly influenced by the glycosidic linkages and the position and content of sulfate groups depending on the growing conditions and extraction processes used without degrading the procedure for deriving a polysaccharide.

6.2.3.1 Agar

Polysaccharides derived from microalgae are nontoxic, renewable materials, and biodegradable. Agar is a complex mixture of polysaccharides. Other molecules make the cell wall of marine red algae. It is composed of two sugar molecules: agarose and agaropectin. Agarose is the gelling part of the agar molecule. Agaropectin is composed of disaccharide molecules. Sulfate content in agarose, which is 0.15%, is shallow compared to agaropectin (5%–8%). Therefore, it is a gelatinous material and found its application in the food industry and mostly in the bakery because of its water-retaining capacity and thickening agent. On an industrial scale, two species of algae, delirium and gracilaria, are known resources of agar at an industrial scale.

6.2.3.2 Carrageenan

Carrageenan is a group of hydrophilic polysaccharides naturally occurring in Rhodophyta, such as *Eucheuma* sp. These molecules can increase viscosity and gel formation; they are composed of sulfated galactans. Carrageenan is the polymer of carbohydrates and protein molecules. Because of its water gel applications are used in food and pharmaceutical industries for fruit juices, fruit gels, marmalades, and gummy candy manufacturing. Several red strains are used to produce carrageenan, such as *Gigartina* sp., *Hypnea* sp., *Chondrus* sp., and *Eucheuma* sp. The United States is on the list of most prominent carrageenan manufacturing individuals, producing over 30 carrageenan products for food applications such as desserts, meat, beverages, and dairy.

6.2.3.3 *Alginate*

The structure of alginates is psycho-colloid. Alginates have applications in the food, cosmetics, and pharmaceutical industries. In food industries, they are used as thickening agents. Alginate gives proper appearance and uniformity to dairy products because they possess moisture retention properties (Khanra et al., 2018). They have also been used in preprepared food, for instance, pie fillings, ready-mix cakes, and salad dressing.

6.2.3.4 *Fucoidan*

Fucoidan is a polysaccharide macromolecule made up of fructose sugar. Its yield varies with seasonal variations, the sea's depth and algae species. It is a useful bioactive material for medicine due to its antivirus, antioxidant, antitumor, and anticoagulant properties. It is not a part of human food or animal feed.

6.2.4 Lipids

Microalgae contain 50%–75% of dry biomass and therefore are excellent sources of lipids. Which makes them valuable sources of long-chain fatty acids that can be used in the medical field or as a feedstock for biodiesel production (Khanra et al., 2018) (Table 6.1).

6.3 Microalgae as human food

Humans need nutrients that are provided through food intended to respond to hunger. The world's population is overgrowing and is expected to grow by 9.5 billion in 2050. Therefore, the required amount of food will increase as well as the type of food and the

TABLE 6.1   Microalgal strains and obtained products.

Microalgae	Product
Dunaliella salina (Rabbani et al., 1998) *Scenedesmus almeriensis* (Macías-Sánchez et al., 2010)	Lutein β-carotene
Tetraselmis suecica (Garcia et al., 2018)	Protein
Scenedesmus obliquus (Koley et al., 2018)	Protein and carbohydrates
Nannochloropsis sp. (Gerde et al., 2013)	Proteins
Isochrysis sp. (Lacour et al., 2012)	Carotenoids
S. obliquus (Ansari et al., 2015)	β-CaroteneChlorophyllCarbohydrate
Nannochloropsis gaditana (Sánchez-Camargo et al., 2018)	Chlorophyll
Haematococcus pluvialis (Huang et al., 2006)	β-Carotene
Chlorella sp. (Lee et al., 2015; Laurens et al., 2015)	Carbohydrate
Scenedesmus sp. (Yang et al., 2010)	Carbohydrates and lipids

corresponding benefits to diet will also increase. Microalga biomass is an excellent source of proteins, carbohydrates, and lipids. Humans have used microalgae as a food source in China, Africa, Japan, and Mexico because of their protein content, essential fatty acids, and vitamins. For food purposes, the most commonly available microalgae are *Chlorella* and Spirulina; their properties like faster growth rates. Both the microalgae are being used as the coloring agent, beverages manufacturing like microalgal sour milk, health drinks, and microalgal green tea.

The use of microalgae as a protein source in the diet is very poorly developed in the world except in some countries because it requires the development of new values among people, and much attention is needed to optimize the production cost and process stability, food safety, and customer acceptance. Other technical issues like extraction, refining, and palatability are a significant concern for their relative composition in food products (Henchion et al., 2017).

In recent years, the awareness of food and its impact on health has increased, making the demand for functional food higher. Functional foods improve wellbeing, physiological activities, and consumer health by reducing the risk of diseases. So the regular intake of available food through diet can enhance the quality of life, reducing the budget for healthcare. Higher protein content, polysaccharides, fatty acids, pigment, vitamins, volatile compounds, minerals and sterol provide the functional value for microalgae to be utilized as food.

Microalgal polyunsaturated fatty acids have been used as supplements in the food industry. Eicosapentaenoic acid (EPA) and docosahexaenoic acid (DHA) are related to reducing arthritis, cardiovascular effusions, hypertension, etc. DHA is also responsible for developing and functioning of the nervous system. The pigments present in the microalgae Primary application in food as colorants; however, they have several health advantages, for example, anticancer, antioxidant, anti-inflammatory, anti-aging, and antidepressant features. The human body cannot synthesize the pigments, so the only source of these pigments is dietary intake. Astaxanthin and beta-carotene can be responsible for more than 80% of the total carotenoids in the cell. These two pigments have a strong market demand. For Industrial bioprocessing, *H. pluvialis* and *D. salina* are considered good sources of microalgal biomass due to the high-level *endocellular* accumulation of carotenoids and high market demand.

Seafoods like crabs, shrimp, lobsters, and trout are a great source of astaxanthin in the diet, but the quantity is deficient, which is insufficient for human beings. Astaxanthin from *H. pluvialis* has an amount 90-fold greater than astaxanthin produced synthetically. The astaxanthin derived from *H. pluvialis* is conjugated to proteins and is acidified with fatty acids, making them able to provide stability to the molecule against oxidation. At the same time, the synthetically produced astaxanthin is produced in the free form. Strict regulations and increased awareness among people about synthetic derivatives greatly enhance the market demand for natural astaxanthin. Only *H. pluvialis*, among other sources of algal astaxanthin, had been approved for the human diet as a supplement. *Chlorella zofingiensis* and *Chlorococcum* sp. are used to produce feed for aquatic animals.

Deep research and evaluation are needed to study the interaction of microalgae with food like protein, water content, fat, pH, oxygen concentration, and preservation. Some technical issues on the extraction of biomass and adapting it to the market's specifications as final food have been very much solved. These additives can be produced in alternative forms such as powder bread, biscuits and pasta, liquid for drinks and beverages, and capsules and tablets for supplements.

Sensory aspects can determine the acceptance of these foods by consumers. For example, powders obtained from microalgae cultured either from traditional open ponds or indoor PBRs show a dark green color due to the presence of chlorophyll, and the taste is also very unpleasant, which makes the consumer compromise their use in formulation for food. Studies show that these unappealing features can be reduced by bleaching the microalgal biomass. The strategy is that these functional foods can be used as an additive but not as basic food and can find a way around the difficulties in changing food and eating habits (Camacho et al., 2019).

This approach uses microalgal biomass in different compositions with cassava flour and wheat flour, enhancing the nutritional quality related to fibers, proteins, lipids, and minerals. Andrade et al. (2018) describe that dried spirulina was added to 22 Indian food recipes and was declared satisfactory for sensory aspects. Standard semolina Spaghetti was compared with the Spaghetti having additive biomass of *Chlorella vulgaris* and *Spirulina maxima*. The comparison that the nutritional value and quality related to sensory aspects were improved when adding microalgal-derived additives. Moreover, changes regarding cooking and properties related to texture were also not observed (Andrade et al., 2018).

Bhalamurugan et al. (2018) present the risk factors of using microalgal biomass as food for human beings. A study was conducted on rats for four generations who were raised consuming 10% of *Dunaliella*. The results depicted no significant change in gross Pathology. However, chronic inflammation was decreased, and no adverse effects were noticed, indicating that it is safe for conception for human beings. Studies related to nutrition present there consuming algae to a certain amount had no adverse effects even when the intake was prolonged. An increase in body weight was seen when feeding *Scenedesmus obliquus* to children and adults with no other adverse effects. The study was then carried out for seriously or slightly malnourished infants feeding them *S. obliquus*. The results showed increased body weight due to the addition of algal biomass compared to the infants fed a regular diet. The conclusion was that an algal diet could enhance health without adverse effects.

Research by Kagan and Matulka (2015) also represents that using *Nannochloropsis oculata* freeze-dried biomass in pasta and cookies can enrich them with Omega 3 fatty acids, which are DHA and EPA and is very promising for food-related industries.

Hence, the increasing manufacturing of microalgae species and biomass presents a possibility of using microalgae for food purposes and can be a solution for increasing the world's population. A significant concern for incorporating microalgal biomass in human food is that it carries large amounts of nucleic acid. Nucleic acid, upon metabolic degradation, might cause kidney or gout stones (García et al., 2017).

6.4 Parameters influencing the production of biomass

The accumulation of specific types of biomass and the productivity of certain chemical substances are detrimental factors to the efficiency and effectiveness of the strain of microalgae chosen. These parameters can be affected by the following conditions: pH, Temperature,

Culture medium used, growth phase, the intensity of light, available nitrogen concentration, and the harvesting method.

The most commonly used microalgae, for example, *Chlorella, Neochloris, Haematococcus, Nannochloropsis,* and *Botryococcus,* have an optimum temperature ranging from 15°C to 35°C. The strains of *Chlorella sorokiniana* can grow at high temperatures, that is, 43°C and are also resistant to the high concentration of carbon dioxide and NO. Temperature for cultivation for various microalgae biomass yields is recommended between 27°C and 30°C. If the cultivation temperature is increased to 35°C, the yield decreases.

Change in hydrogen potential (pH) determines the solubility of carbon dioxide and minerals in the medium and directly affects the microalgae; therefore, it is important in the cultivation process. The alteration in pH can be seen due to changes in temperature, the metabolic activity of cells, quantity of carbon dioxide dissolved, composition, etc. The growth rate is affected by pH, and different microalgae strains have different tolerance levels for pH. Using buffer solution can reduce pH fluctuations, but this will add up to the overall production cost, making its use infeasible (Dolganyuk et al., 2020).

Cavonius et al. (2015) studied the pH shift process on *N. oculata* to extract protein-rich food ingredients. Two versions of the pH shift process were carried out at pH 7 with the first separation step. The second was carried out after alkalization to pH 10. The main constituent of the Elgin growth media was seawater. It was used to reduce the consumption of resources. The process was repeated at different pH levels to obtain protein and fatty acids. For the pH7 process, the protein was about to 86%, while for the pH 10 process, it was 72%. The total fatty acid recovered at pH 7 was 85%, while at pH 10, the recovery was 73%. They also studied the profile for amino acids and polypeptide content. Polypeptide content was slightly affected by pH shift, whether the solubilization was at pH 7 or 10.

At high pH values, the effect was positive due to auto flocculation, which allows microalgae to settle, eliminating the use of chemical flocculants. But, a disadvantage of a high pH value is that it hinders the process of photosynthesis due to the nitrate accumulation and eventually the death of microalgae (Iasimone et al., 2018).

Guedes et al. (2011) presented the effects of temperature and pH on the growth and antioxidant content of specific microalgae, *S. obliquus.* Results depicted that the lowest impact was seen at pH 6 and 25°C, while the highest was at pH 8. The results show that pH's effect on biomass depends on temperature. The probable reason for this interaction might be the environmental temperature that limits several metabolic reactions; for example, with the change in temperature balance of reaction rate is changed; therefore, The reactions occurring with the help of enzymes become rate-limiting, which has an impact due to enzymes with clearly distinct pH optimization, arising from a different amino acid composition of the active site. It can be seen that the time needed to reach the required status is a function of both temperature and pH. Growth acceleration can be stressful compared to stress due to nutrient change usually applied to activate secondary metabolism. With the increase in pH, the antioxidant content production also increased at all temperatures. The significantly actual results were obtained at 30°C at pH 7 and 8. The effect is significant and positive as at the highest temperature, the amount of antioxidant content was highest without considering pH, showing that temperature had a much more substantial impact than pH between pH 7 and 8, around 30°C. The significant pH impact noticed was that at any temperature with pH 6, the production was the lowest.

Another essential factor in biomass is nitrogen concentration. Menegol et al. (2017) showed that a critical factor affecting microalgae growth, particularly chlorophyll and protein synthesis, is nitrogen concentration. The concentration of nitrogen can highly influence biomass concentration. They exerted different nitrogen concentrations into the cultivation medium to grow *Chlorella luteoviridis*. The culture for microalgae growth with the highest concentration of nitrogen gave the highest amount of biomass which was 2.3 times higher than the lowest concentration of nitrogen. Nutritional changes are linked with stress conditions, and nitrogen deficiency can cause stress conditions to decrease biomass formation. The highest productivity can be due to the characteristics of *Heterochlorella luteoviridis*, but it can also be due to the performance of the PBR, allowing the highest specific growth rates.

They also presented a study of different temperature conditions along with various nitrogen concentrations to check the production of carbohydrates, lipids, and proteins. Results depicted that there was no change in carbohydrate content at low temperatures despite any concentration of nitrogen. On the other hand, with higher temperature and nitrogen concentration, the carbohydrate content produced was increased. A higher nitrogen concentration was suitable for creating higher protein content. In contrast to the carbohydrate conditions, the yield of protein content was higher at low temperatures and increased nitrogen concentration. This shows that yield of protein content is more dependent on nitrogen concentration than temperature. This behavior can be illustrated through the following example: When *Isochrysis* sp. and prymnesiophyte biomass was exerted to high temperatures, the protein structure was broken down, and interference with enzyme regulators was seen. Therefore in the conditions like nitrogen starvation, the protein content will be decreased, showing that nitrogen is essential for protein synthesis. This is due to the reduced metabolism of microalgae under nitrogen starvation conditions, which leads to the biosynthesis of carbohydrates and lipids.

The synthesis of carbohydrates requires less energy; therefore, they are synthesized before lipids respond to environmental stress. Under nitrogen deficiency for an extended period, the photosynthesis efficiency decreases. Then cells convert carbohydrates into energy. This is the reason for the low carbohydrate content under nitrogen starvation. Therefore continuous nitrogen deficiency during microalgae cultivation leads to higher lipid content. The study concluded that lipid content decreased with higher nitrogen concentration. At the same time, the temperature effect on *H. luteoviridis* lipids synthesis was not detected.

Gonçalves et al. (2016) studied the effects of temperature and average irradiance of daylight on the growth of microalgae and uptake rates of nutrients. Increased biomass productivity increased with the increased supply of light and removal of nutrients. The most suitable temperature was 25°C. At this temperature, an increase in photosynthetic activity was observed. *Synechocystis Salina C. vulgaris*, and *Microcystis aeruginosa* were most effective for biomass production. Corresponding to the irradiance effect, the yield of beta-carotene increased at low irradiance along with less nitrate supply (Sánchez-Camargo et al., 2018). However, intense light exposure can lead to photodamaged microalgae developing a pale green color with a size reduction and indicating the decrease in chlorophyll and impacting cell development (Nurachman et al., 2015).

Another study shows that the lowest conservation of nutrients and highest light intensity positively affect microalgae cultivation (Iasimone et al., 2018).

With nongrowth-limiting environmental conditions and enough nutritional supplies, Mixing to make the flow turbulent can be an essential factor in obtaining a high biomass yield. Mixing has three significant effects, including

1. Prevention of the formation of gaseous and nutritional gradient: stirring increases the Likelihood of carbon dioxide capture from surroundings, facilitating the transfer of oxygen (biosynthesized) from liquid to the gaseous phase, stimulating the process of photosynthesis in the culture.
2. Prevention of algal sedimentation: Algal biomass sedimentation or accumulation in the dead zone can significantly negatively affect productivity due to developing anoxic biomass decay products. These zones might have protozoa or invertebrates proliferate. These things can cause severe damage leading to the failure of ponds very fast.
3. Moving cells in variation with quality and quantity of light lower the chances of photoinhibition or self-darkening (nonuniform distribution of light can lead to self-darkening). This can happen due to high cell concentrations because the cells at the surface block the light at the surface, reducing the intensity with depth.

Effective stirring is of great significance for obtaining high cell concentrations. Another advantage of induced turbulence is the elimination of thermal separation. Mixing is done by moving the liquid in a container, channel or tube. The turbulence resulting from the blending can be directly determined by the degree of mixing in the reactor. Turbulence caused by studying increases the growth rate of photosynthetic microorganisms. Turbulence levels higher than optimum can cause sharp decreasing growth due to cell damage. Thus the stirring should be carried out while keeping into account the hydrodynamic stress (Grobbelaar, 2010; Dolganyuk et al., 2020).

The thermal resistance of microalgae upon heating can be assessed through thermogravimetric analysis, which can be helpful in food processing. It is observed that microalgae undergo a progression of changes upon heating, depending on the group they belong to. It can be described in different stages and is related to the biomolecules in algal cells. In every step, weight loss occurs; in the first stage, weight loss occurs between 40°C and 200°C−250°C. In this range, free water and water loosely bound the molecules to happen. This process progressively destroys the cell structure and phenomenon as changes in lipids structure and thermal unfolding of protein occur. The second stage's most significant weight loss occurs at the onset of temperatures above 221°C. For example, Carotenogenic algae have thermal stability. After those two stages, the residue is slowly decomposed, producing a nondegradable material. This stage occurs above 900°C (Batista et al., 2013).

Carbon dioxide is required for the growth of microalgae as a carbon source. A limiting factor in productivity can be the insufficient carbon dioxide supply. Daylight carbon dioxide must be continuously supplied, and it can be monitored through pH measurement. Carbon dioxide utilization for carbon sources for microalgae is safe. Moreover, it provides benefits to the ecosystem. Carbon dioxide emitted from power plants can be utilized for this purpose. The optimal concentration for carbon dioxide ranges between 0.038% and 10% for most species; higher concentrations of carbon dioxide can cause harmful effects on the growth of *Chlorella* sp. as it grows at a concentration below 5%. Some species of microalgae can survive under high concentrations. But, this inhibits carbon fixation and the rate of biomass production (Eloka-Eboka and Inambao, 2017). Ambient air enriched with carbon dioxide is favorable for producing oily content. Only the carbon source alone is accountable for 60% of the nutrients/minerals required (Singh and Patidar, 2020).

Another critical factor affecting the growth of microalgae is salinity. For biosynthesis of lipids in microalgal cells, energy storage can be stimulated by the accumulation of molecules for Osmo protecting and scavenging reactive oxygen species. Stress due to salinity is also dependent on the species and strain. In some species, excess salt concentrations can lower the efficiency of the photosynthesis process, reducing the amount of biomass production. The types of microalgae found in freshwater have very little tolerance for salinity. Microalgae can also be classified according to their resistance to salinity. The types of species which can survive only in shallow salinity water are called oligohaline. The species which can be cultivated in moderate levels of salinity are called mesogaline. The species that can develop with high salinity levels are called polyhaline (Cavonius et al., 2015; Chen et al., 2017).

One of the crucial parameters for specified microalgal biomass characteristics is biomass composition harvesting time. Conducting an economic assessment is necessary, which includes the practical variations in the biomass composition (Sui et al., 2020).

6.5 Processing

Microalgae is a large group containing autotrophic organisms, growing through the process of photosynthesis same as the plants growing on the land. They are unicellular in structure (not always), which allows them the conversion of energy from the sun into chemical energy (Harun et al., 2010). Growing microalgae is a straightforward and an easy process, particularly when phytoplankton is observed naturally blooming. The energy source is required light energy for the growth of autotrophic microalgae or organic compounds for heterotrophic microalgal growth (Grobbelaar, 2010). Microalgal biomass processing to convert it into valuable products consists of several stages, that is, unit operations. Cultivating microalgae is a relatively new process, and a massive set of data utilizing a wide range of species is required.

Moreover, developing a new process also requires analytical and integrative approaches. Analytical approaches are needed to get information and optimize the processing steps. At the same time, integration occurs due to the flux of the products in each step as the requirement and outputs of each stage affect the other processes.

After the cultivation process, the microalgal biomass obtained is very dilute, containing low concentrations, that is, 0.5–5 g L^{-1} (Lafarga, 2020). Suspension concentration is significant in lessening the processing volumes and equipment sizes downstream. Different techniques and technologies can be used to concentrate the culture, for example, sedimentation, filtration, centrifuge and flocculating agents.

The resulting slurry or paste has a concentration of about 2%–3% solid with simple sedimentation. The numbers turn to 10% reliable when centrifuges are used. Intracellular water is still present in biomass even after vacuum or pressure filtration. Generally, the highest efficiency is 30% solid content. In the next step, cell concentrate, if required, can be dried. The reason is that complications arise with water as it makes extraction of carotenoids and lipids difficult with nonpolar solvents; for instance, the direct transesterification process is sensitive to moisture.

On the other hand, drying requires energy for face change, and increased temperature can also alter the composition of biomass and structure. Still, eventually, drying helps to stabilize biomass in subsequent processes. The concentrate obtained can be stored either frozen or cold for wet handling of biomass; some pretreatment techniques such as wet extraction and hydrothermal liquefaction. Drying is not required to extract specific components for food or feed purposes, but conditioning or modification can be done. This process can help increase the properties of biomolecules, such as accessibility or digestibility. The conditioned biomass is then fractionated. The first step is the solubilization of lipid, followed by the next steps, protein, and carbohydrate solubilization (de Carvalho et al., 2020).

Sometimes, further purification is required because the extraction of the product obtained is a mixture of different compounds. For instance, during the extraction of lipids, when a low polarity solvent mixture is used like methanol and chloroform, it gives lipid fractions as polar and nonpolar lipids, which can be further segregated in the proteins and triglycerides. Hot ethanol, a more polar solvent, can also be used to extract proteins and carbohydrates and disrupt biomass processing. After this process, separating solid with water gives minerals and protein extract. According to requirements, extract in raw form (without purification) can be dried as a final product example of this process is the extraction of protein from *Arthrospira* sp. (Bhalamurugan et al., 2018). Sometimes, purification is also required for speciality products such as peptides (de Carvalho et al., 2020).

6.6 Cultivation

Depending on the mode of cultivation, the process of cultivation is divided into three types.

6.6.1 Batch

In this mode of cultivation, all the nutrients required are added to the cultivation medium before starting the cultivation. During the batch cultivation process, the growth of microbiology stops because of limited depletion of substrate or due to the accumulation of growth-inhibiting products. The environment changes continuously, resulting in the alteration in the composition of cells and the existence of growth phases (lag, exponential, stationary, and death phase), consequently causing fluctuations in productivity. Currently, microalga cultivation is primarily based on batch systems. This is due to the simple operation, less risk of contamination, and reduced maintenance cost.

6.6.2 Fed-batch

In this cultivation system, nutrients in the cultivation medium are added continuously or periodically. Still, products and cells are removed at the end of each cycle. This type of culture results in variable volume and dilution rates. As fed-batch is operated between batch and continuous mode, it is suitable for metabolite production that is not linked with substrate inhibition.

6.6.3 Continuous

Unlike batch cultivation systems, continuous systems have controlled cultivation environments and conditions, enabling biomass composition at a fixed rate. A constant cultivation system requires less space, high volumetric productivity, and fewer unprofitable periods. These systems are open systems; the fresh medium is fat continuously along with the continuous removal of product. The volume of the medium inside the bioreactor remains, more or less, unchanged. Theoretically, it is possible to keep the culture in an exponential phase by the relationship of limiting substrate availability and microbial growth. Still, problems related to cell aging stop the growth eventually.

6.7 Cultivation systems

The most commonly used technique for microalgae cultivation is an open system and a closed system.

6.7.1 Open system

The most commonly used system for the cultivation of microalgae is open systems. Open systems are used in many forms and shapes with particular pros and cons. The most widely used systems for microalgal cultivation are raceway ponds, circular ponds, big shallow ponds and lagoons, and inclined systems. Open systems are most commonly used for large-scale cultivation because they are easy to build, extend their lifetime, have low cost, and have a larger production capacity. In open pond systems, factors affecting the growth cannot be completely controlled, that is, light for photosynthesis and climate. Therefore pond location plays a critical role in successful cultivation because the system becomes a function of the environment. There are several technical challenges to properly controlling the parameters affecting the growth, for example, evaporation. Another issue is the growth of predators and other heterotrophs, causing contamination in the pond. This issue arises due to an uncontrolled environment and lack of sterile conditions. Thus artificial or natural ponds have limitations due to naturally growing organisms.

Using open pond systems due to the severe environments is a challenge. It is limited to a range of fast-growing microalgae species and is naturally resistant to environmental conditions, like *Dunaliella*, which can grow in high salinity levels (Borovkov et al., 2020), while *Chlorella* can be cultivated in high nutrient concentrations. For the production of carotenoids, *D. salina* can be cultivated in open ponds. Moreover, *Muriellopsis* sp. is cultivated to produce biomass rich in lutein, which can be further used as pigments in food, and as additives in feed for aquaculture and poultry farming. The yield obtained was comparable to the closed systems (Blanco et al., 2007).

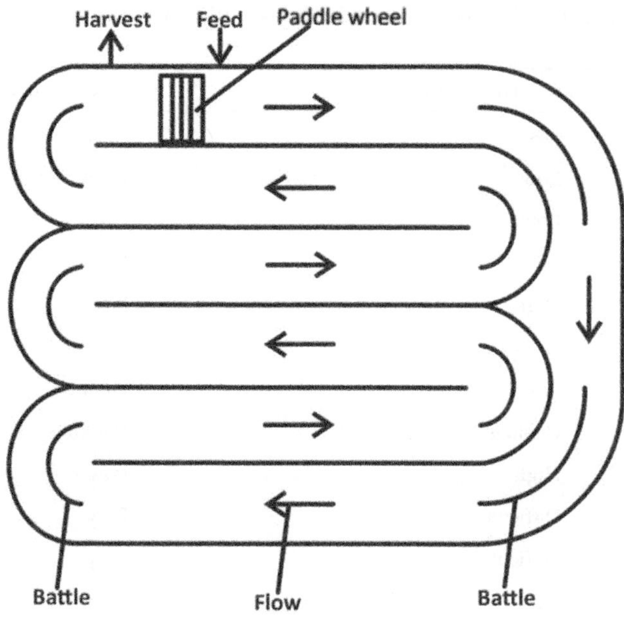

FIGURE 6.2   Raceway pond.

Circular ponds and raceway ponds are also widely used for commercially manufacturing *Chlorella* (Guccione et al., 2014). They are composed of a central agitator, which causes mechanical restrictions. It also requires high energy for agitation, which makes it economically unfeasible to use on a large scale.

The most commonly used open ponds are oval-shaped raceway ponds with a closed loop-like formation. The depth of the pond is kept between 0.2 and 0.5 m. A continuously moving paddle wheel induces the mixing effect while avoiding sedimentation (Fig. 6.2).

C cascade systems reach high cell density (Masojídek et al., 2015). Culture suspension flows through the surface, and gravitational force induces turbulence. Due to the high turbulence, thin culture layers can be cultivated, providing high cell concentrations compared to raceway ponds.

6.7.2 Closed systems

In closed systems, the direct interaction between atmosphere and culture is prohibited. Closed systems for microalgal cultivation mainly comprise closed PBRs. Close PBRs are more costly than open ponds, but the yield is more desirable (Harun et al., 2010). Owing to a contamination-free environment, essential parameters affecting the cultivation process can be controlled to allow the growth of monoseptic cultures (Geada et al., 2017). The variables, which can be maintained while cultivating microalgae, are as follows (Table 6.2):

1. The depth of the culture can be controlled. Control over light intensity is a complicated process, but it gives better control of distribution and attenuation when compared to open ponds.

TABLE 6.2　Advantages and disadvantages of open and closed systems (Grobbelaar, 2010; Harun et al., 2010; Geada et al., 2017; Xu et al., 2009).

Parameters	Closed system	Open system
Process control	Easy	Difficult
Required area	Low	High
Water loss	Low	High
Risk of contamination	Low	High
Capital cost	High	Low
Operational cost	High	Low
Harvesting cost	Low	High
Temperature	Cooling is required	Variable
Biomass yield	High	Low
Commercialization	Difficult	Easy
Cleaning and maintenance	Difficult	Easy
Weather dependence	None	High
Light utilization	Efficient	Poor
CO_2 losses	Low	High
Oxygen concentration	High	Low

2. Through mixing, the consequent turbulence can be controlled. A higher turbulence rate can be obtained in PBRs. On the other hand, in open raceway ponds, the flow is majorly laminar.
3. In PBRs, deficit zones can be avoided and better control carbon dioxide and nutrient concentration control.
4. The biomass concentration can be a determinant factor of light intensity inside the culture.
5. There is a variety of modes of operation like batch, fed-batch a, continuous systems.
6. Temperature and pH are controlled within the limits.
7. Evaporation of growth medium is reduced when compared to open ponds.
8. The risk of contamination is less; thus the pharmaceutical compounds can also be obtained from biomass cultivated from closed PBRs (Grobbelaar, 2010; Wang et al., 2012).

Different designs of the bioreactor are used, such as those (Blaga et al., 2018; García et al., 2017; Harun et al., 2010) (Fig. 6.3; Tables 6.3 and 6.4).

1. Tubular PBRs (Wang et al., 2012): They are manufactured with different structures and shapes such as manifold, spiral, serpentine, and conical-shaped.
2. Flat PBR: They are composed of glass plates for alveolar panels.
3. Stirred tank.
4. Airlift.

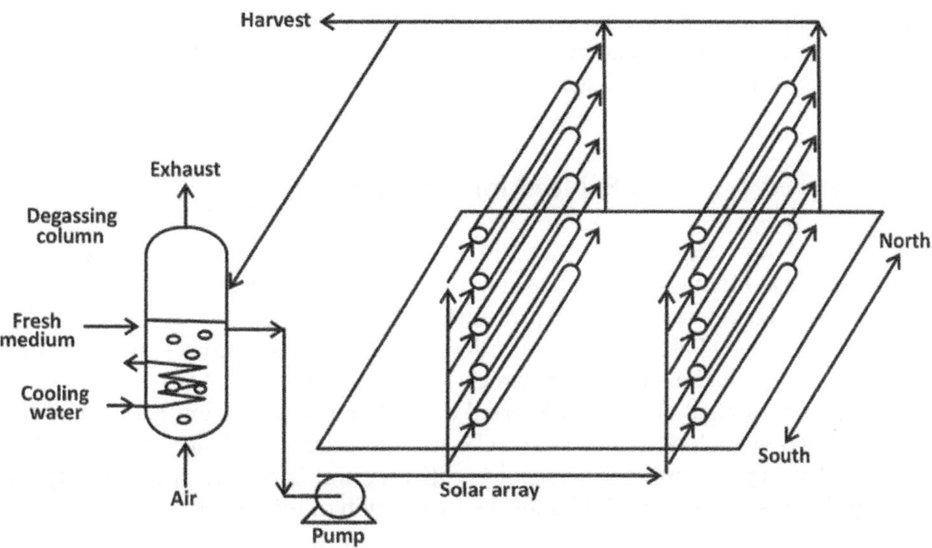

FIGURE 6.3 Tubular reactor diagram.

TABLE 6.3 Tubular photobioreactor advantages and limitations (Geada et al., 2017).

Advantages	Disadvantages
Good biomass productivity	Fouling
Illumination surface area	Significant wall growth
Economical	Large space is required
Suitable outdoor cultures	Risk of photoinhibition effect
	CO_2, O_2, and pH gradient in tubes

TABLE 6.4 Flat photobioreactor advantages and limitations (Geada et al., 2017).

Advantages	Disadvantages
Cost-effective	Significant wall growth
Maintenance and cleaning are easy	Photosynthetic efficiency is low
Good biomass productivities	Temperature control is difficult
Illumination surface area is large	Upgrading to a bigger scale is difficult
Oxygen buildup is low	

The development in the field of microalgal cultivation is related to the commercial availability of PBRs designed to operate under defined conditions with shallow contamination risk (García et al., 2017). Manirafasha et al. (2016) state that the design of the bioreactor will be effective when the energy consumption is minimized while increasing productivity and eventually reducing the overall cost of manufacturing. Other challenges reported are growth kinetics, photosynthetic efficiency, and the mass transfer rate (Blaga et al., 2018; Hallenbeck et al., 2016).

6.8 Need for pretreatment

Pretreatment of microalgae is required just like any other biomass. But size reduction, like in lignocellulose biomass, is unnecessary in this case. Because it is already microscopic and the size range is between 5 and 50 μm. Inside the cells, molecules are protected by cell walls and membranes. Pretreatment of microalgal biomass requires uncovering these molecules of interest for the subsequent processes, either digestion or extraction.

The pretreatment of algal biomass is interlinked with fractionation and harvesting. It can be served as food feed or fuel, or its fraction can be used to produce valuable chemicals based on different uses. Therefore biomass processing is different and based upon its final application. But three significant steps are required after the production of microbiology.

1. Biomass recovery: biometric ovary is usually done through flocculation followed by sedimentation, filtration or centrifugation.
2. Biomass pretreatment: treatment is necessary to increase digestibility or expose the required molecules.
3. Isolation and purification of product: this process is precise depending on the fraction or biomolecule of interest, and conventional processing strategies are followed for this step.

Steps 1 and 3 are pretty similar, but in step 2, biomass pretreatment is different as this process involves structuring the Polymers inside the cell envelope. This step requires high energy, making it expensive. This step consists of a structure and composition alteration transformation, eventually affecting the product. The studies in this area mainly concern the digestion of biomass and biorefineries. Therefore the required products are mostly lipids but to increase the cost-effectiveness of the process, products such as bio pigments, vitamins, minerals and peptides are also obtained.

6.9 Pretreatment

There is a vast diversity of microalgal species. The drive product leads to various pretreatments: the pretreatment required for microalgal-derived biomass for food applications drying, cell disruption, and size reduction. The pretreatment is required after cultivation and before the extraction of microalgae.

6.9.1 Drying

Drinking free treatment stabilizes biomass because lower water content reduces chemical and microbial decomposition. Water removal depends on the heat transfer rate and diffusion of water molecules. For the removal of moisture content generally, two methods are used:

TABLE 6.5 Advantages and disadvantages of methods of drying (de Carvalho et al., 2020).

Methods of drying	Advantages	Disadvantages
Solar	Low cost	Degradation of biomass due to slow processing It depends on weather conditions The required area is large
Spray	The process is fast Biomass is obtained in powdered form	Highly energy-intensive Large air volume is required Degradation of biomass can occur
Microwave	The process is fast Control is easier	Safety concerns regarding biomass composition
Conduction (tray, drum)	The process is fast	Degradation of biomass occurs
Convective (tunnel, tray oven)	commercially available technology	At higher temperatures, degradation can occur
Fluidized bed	The process is fast Biomass is obtained in powdered form	Highly energy-intensive Large air volume is required Degradation of biomass can occur
Freeze-drying	The biomass obtained is of high quality	The process is slow

thermal drying and freeze-drying. A wide range of drying temperatures can be used for thermal drying, that is, from 25°C to 60°C, while the reported time is about 18 to 48 hours. But most commonly, overnight duration along with 60°C temperature is used. Freeze-drying temperature ranges between -50°C and -80°C; usually, 24–48 hour time duration is used. To prevent the product from spoiling, the freezing phase is needed to be maintained. Freeze-drying is commonly used when the end product is thermos liable or a living system. This way, biomolecules have their structure intact. The freeze-drying temperature is energy-consuming and takes more time, increasing the cost compared to thermal drying.

Usually, 1%–10% water content remains after drying. Required concentrations of residual water are needed to be evaluated in dry biomass, taking the following steps downstream into account. For example, the case of protein-bound water can increase stability. Moreover, when the end product is enzymes or vitamins, slow drying gently is needed to produce porous and viable biomass without much alteration in the molecules. Therefore it is not a problem for nutritional biomass (Table 6.5).

6.9.2 Mechanical cell disruption

6.9.2.1 High-pressure homogenizer

At high velocity, the suspension is passed through a homogenizer valve in high-pressure homogenizers. They are due to the high shear deceleration of liquid and the velocity gradient that occurs in the system, deforming and disrupting the cells. To generate high velocities, high pressure is required in narrow passages. Conventionally the forces used to range between 500 and 1500 bar.

6.9.2.2 Sonication

Sonication is generally used in labs to extract nucleic acid, lipids, and protein characterization. Energy requirements are very high. Currently, large-scale industrial devices are available, allowing continuous mode operation. The disadvantage of this technique is that it can pose destructive effects on microalgal cells due to acoustic cavitation.

6.9.2.3 Bead milling

Bead milling is the simplest, reliable, and fast method, but the energy requirement is very high and depends on the microalgal species used. It can be used in large-scale applications because of its high efficiency in single-pass operation, and a large workforce is not required. The major disadvantage is that it produces heat, and cell disruption is very high, which complicates the purification steps downstream.

6.9.2.4 Steam explosion

Through steam explosion, cell disruption occurs so that the intracellular components are removed easily. Biomass is exposed at a high temperature ranging from 160°C to 260°C and high vapor pressure between 1.03 and 3.45 MP, which is then depressurized to bring the temperature to ambient temperature leading to the disruption of cells. This process is relatively new and is being researched because of its potential for commercial applications.

6.9.3 Nonmechanical/chemical cell disruption

6.9.3.1 Acid/alkali cell disruption

Chemical cell disruption methods are scalable, for example, alkali or acid treatment, but an economic evaluation is necessary according to the process requirement. These processes may overlap when targeted fractionation is intended. For example, hydrolysis of cell walls using alkali or acid produces fractions that are rich in carbohydrate content.

6.9.3.2 Enzyme

Permeation of cell walls of microbiology can be done through enzymes. Different types of enzymes respond differently to various species of microalgae. These processes are expensive and cannot be used for low-cost products. Because enzymes can target specific cell wall components, they can be used in lower concentrations. They can be integrated into other processes to facilitate cell disruption and decrease the energy requirements, making the process economically feasible. The time required is about 24–72 hours. However, it is integrated with other preprocessing methods like autoclaving, milling, etc.

6.9.3.3 Size reduction

Before the extraction process, size reduction can help increase efficiency by increasing the surface area for biomass to solve interaction. However, tiny particle size can affect the higher absorption capability of lipids (Khanra et al., 2018; de Carvalho et al., 2020) adversely (Table 6.6).

Some other methods for biomass cell disruption are (Table 6.7):

TABLE 6.6 Summary of cell disruption techniques and their principles (de Carvalho et al., 2020).

Methods	Principle
High-pressure homogenization	Suspension is passed through the homogenizer valve at high velocity due to High shear velocity gradient occurs, disrupting the cell wall
Sonication	The cavitation of microbubbles causes high shear in the cell wall
Grinding	Collisions and attrition
Milling	By fluid agitation along with hard solids, high shear is generated
Alkaline treatment	Hydrolysis of cell wall occurs due to very high pH
Acid treatment	Hydrolysis of cell wall occurs due to low pH
Enzymatic	Enzymes target specific cell wall components and then weaken or dissolve them.
Thermal treatment (steam explosion)	Heating at a high temperature disrupts the cell wall

TABLE 6.7 Methods for cell disruption (de Carvalho et al., 2020).

Methods	Principle
Freeze thawing	Ice crystals damage cell walls and cell membranes
Osmotic shock	Biomass is immersed in a hypotonic solution, and osmosis occurs
Pulsed electric field	Pores are formed due to the rearrangement of membranes by the application of a high-intensity electric field
Solvent solubilization	Solvents destabilize cell membranes
Detergent solubilization	Cell membranes are dissolved in detergents

6.10 Harvesting/dewatering techniques

Different microalgae dewatering technologies and techniques are as follows:

6.10.1 Flocculation

Flocculation has been chosen as an optimized method for harvesting microalgae at a low cost, and different flocculation technologies are also developed. Flocculation is a process in which agglomeration of suspended solids/cells occurs. Flocculation is either induced (with the addition of flocculating agents) or naturally happens (due to the production of the flocculating chemical by microalgal cells). Auto flocculation occurs due to the increase in pH. Sometimes, it can be a consequence of the interaction of contaminating species with microalgae. Microalgal cells form a relatively stable suspension. Every cell in the suspension carries a negative charge (due to functional group ionization) (Zeng, Guo et al., 2016). The presence of negative charge repulsion prevents agglomeration and sedimentation. Flocculation is carried out using physical, chemical, and bio-based methods.

1. **Physical flocculation**: Physical population is carried out to prevent contamination, which can be induced when chemical flocculants are added. Technologies used for flocculation are electrolytic coagulation, ultrasound electro-flocculation, and magnetic separation.
2. The addition of flocculating agents carries chemical flocculation. There are three flocculating agents: inorganic polymers, inorganic flocculants, and organic polymers. Using chemical flocculation is feasible for bulk biomass production. Using inorganic flocculants can alter the composition of cells, which can affect the final application of the biomass, such as in food or feed. Organic flocculants are biodegradable and safe, but their application is restricted due to pH dependence. Organic Polymers are expensive. A significant disadvantage of proper flocculation chemicals is their contamination and secondary pollution, which makes them applicable to small microalgae species.
3. **Bio-flocculation**: This flocculation method uses extracellular polymeric substances or microorganisms. It is environmentally friendly and is being utilized in wastewater treatment already. Three types of bio-based flocculation are microbial bio-flocculation, microorganisms associated with bio-flocculation and self-flocculation (Wan et al., 2015; Vandamme et al., 2013).

6.10.2 Gravity sedimentation

It is one of the cheapest flocculation techniques for separating solid and liquid and has found its application in water treatment processes; however, as far as microalgae are concerned, until the particle size of microalgae is increased, this process cannot be applicable. Moreover, dewatering is required separately through filtration, centrifugation or evaporation. It is a slow process, but it can be integrated with flocculation.

6.10.3 Filtration

Fundamentally the process of filtration uses semipermeable material, retaining the solid content on its surface while the fluid passes through it. The fluid is driven by pressure difference. But the use of filtration to dewater microalgae is not visible as the microalgal cells accumulate on the surface and act as a resistance for further fluid flow, thus reducing the speed of the process. When more enormous algal strains are used, filter press should be used under pressure or vacuum, for example, *Spirulina platensis*. Pressure filtration is considered an energy-efficient process for dewatering (Harun et al., 2010). Fouling of the filtration medium also occurs due to the presence of extracellular polymers secreted by cells. This phenomenon also affects the filtration rate with time (Zeng, Guo et al., 2016).

6.10.4 Evaporation/drying

The drying process requires heat, which can damage the cells or products from microalgae, restricting it to heat-resistant materials. This process is not promising for live-cell harvesting. Sensitive materials can be subjected to evaporation. But both of the processes need a more significant amount of energy than other dewatering processes. Freeze-drying is preferred because it does not denature the microalgal cells or products (Zeng, Guo et al., 2016).

6.11 Extraction

6.11.1 Protein extraction

The protein content in biomass depends on the microalgae species used, Painting from 30% to 70% dry biomass. Removing other molecules like nucleic acid, lipid, pigment, and carbohydrate is necessary to extract protein. There are various methods to extract protein, including organic solvent extraction and sonication.

6.11.1.1 Organic solvent extraction

Solvent extraction is a conventional process, and solvents used in the process are hexane, ethanol, and methanol. For extraction of UTM and reaction theme, hexane is famous for the treatment because it converts both products into their free forms and chlorophyll and carboxylic acids are left in the aqueous phase. A precipitate of trichloroacetic acid/acetone is commonly used for small amounts of protein. In this process, denaturation and precipitation of the cell's protein depend upon the low pH level and net negative charge on trichloroacetic acid and organic acetone. Protein denaturation stops the proteolytic and other activities of enzymes that degrade protein.

Phong et al. (2016) choose *Chlamydomonas* sp. for four protein content extractions using different solvents to evaluate and compare them. The solvent in the study was water, methanol, ethanol and n-propanol. 0.1 g of biomass was added to the 10 mL of each solvent. The protein concentration per dry weight obtained was 3% for water, 2.7% for propanol, 1.9% for ethanol, and 0.7% for ethanol. The results show that water can be an excellent solvent for downstream processing of microalgal biomass because of the higher yield of protein than any other solvent.

The solvent extraction method is quite a work intensive process. Using organic solvents can be potentially toxic and induce alteration in the spatial arrangement and structure. Due to Extreme conditions like oxygen and light intensity, important biotic components can be decomposed. Moreover, using methanol and chloroform commercially for extraction have a severe disadvantage due to potential hazard to human health and the environment (Alam et al., 2020).

Miazek et al. (2017) describe in their study that the impact of using organic solvent on microalgae depends on the strain of microalgae used, the concentration of solvent and the type of solvent used. *Synechocystis* sp. and *Synechococcus elongatus* were used to test hexane, ethanol and butanol. The result depicted that hexane was the highly toxic of the three solvents. Butanol comes second, while ethanol was the least toxic solvent among the three.

6.11.1.2 Sonication

In ultrasound sonication for extraction, 20–100 MHz ultrasound waves are used. Waves are not what causes the degradation process improvement of availability of proteins present in microalgal biomass, but due to the development, growth, and collapse of bubbles, which is called acoustic cavitation. At extreme temperature and pressure, microscopic regions are formed due to the imploding bubbles, which result in the chemical excitation of the liquid and its contents, reading the breakdown and degradation of the intended compound. The advantages of ultrasound-assisted extraction are nonthermal properties, less solvent required, and less time needed for the processing while producing high purity products, which eventually reduces the process required downstream (Bleakley and Hayes, 2017; Alam et al., 2020).

Free treatment through ultrasound resulted in increased protein content in the extraction process of *Ascophyllum nodosum* with alkaline and acidic treatment. The protein content extracted was 27% higher in alkaline treatment while 540% with acid treatment (Kadam et al., 2017).

Other technologies are pulsed electric field lysis, microwave-assisted extraction, Super and subcritical extraction (CO_2) etc. (Fig. 6.4).

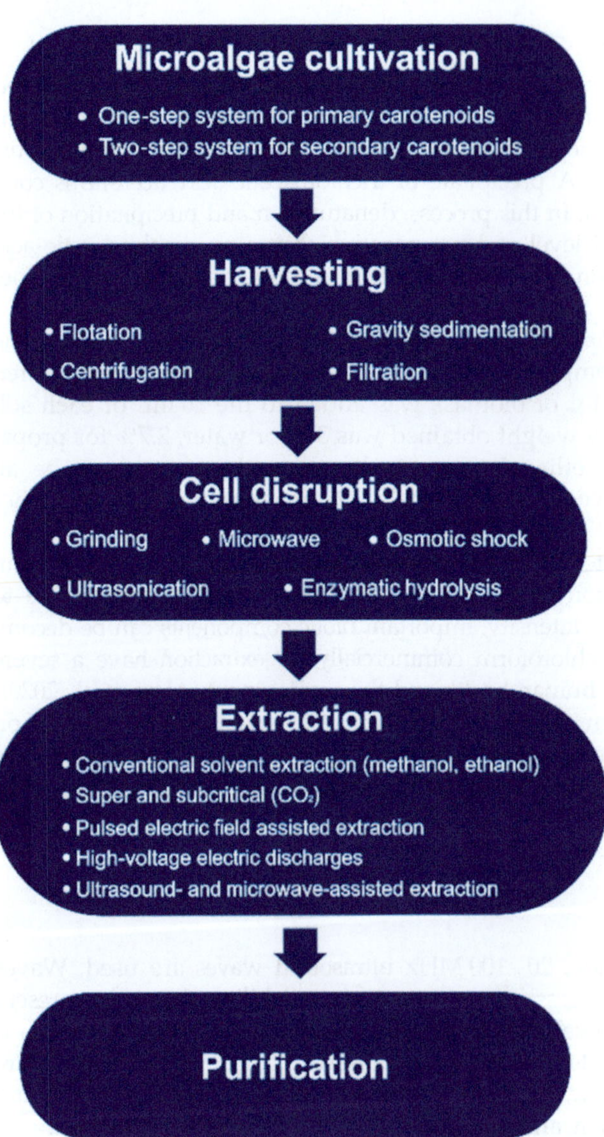

FIGURE 6.4   Process diagram of carotenoids production.

6.11.2 Protein purification

For the purification of protein chromatography techniques, for example, Ion exchange, affinity size exclusion, hydrophobic interaction etc., are used. Chromatography is used in labs (Alam et al., 2020; Khanra et al., 2018). Purification membrane filtration techniques, such as ultrafiltration and cross-flow filtration, are also used. The requirements of the filtration process are minor, provide high yield, and the overall process is cost-effective. One thing that needs to be taken into account is the required size of the membrane, which is a crucial factor of a specific protein (Buyel et al., 2015) (Fig. 6.5).

FIGURE 6.5 Process diagram of carotenoid purification.

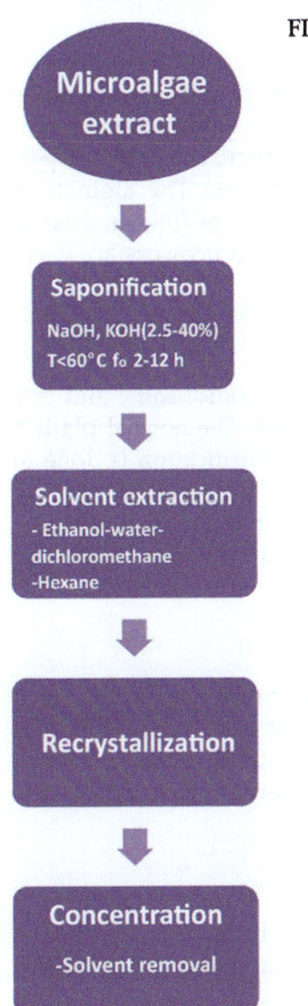

6.11.3 Carbohydrate extraction

Available polysaccharide extraction processes are alkali extraction, alcohol, potassium chloride, and drum drying.

To extract agar, biomass obtained from microalgae is boiled in water. Acid is added to control the pH between 6.3 and 6.5 and to enhance the efficiency of the extraction process. Pressure also plays a critical role in agar extraction by increasing yield and reducing the time required. The optimum conditions can vary with the strain of microalgae used as well. For purification, filtration is needed. The residue biomass is removed, and the hot filtrate is cooled to form the gel. To remove the color of the gel and enhance the quality, the gel is sometimes bleached as well. The remaining water is extracted using pressure to squeeze or a freeze-thaw process.

For alginate extraction, Particle size reduction of biomass is required before extraction to get better efficiency in the extraction process. After the decrease in particle size, biomass is hydrated with formaldehyde solution into insoluble alginic acid. To remove external salt acid, treatment is also required. Water is added to the biomass, increasing the temperature to 80°C. The pH 10 is maintained by adding sodium carbonate. The alginate is extracted from biomass into the liquid phase, and a rotary vacuum drum filter is used to filter the alginate solution. For the extraction of carrageenan, types of processors are used, which are:

6.11.3.1 Alcoholic process

The biomass in dry form is washed with water to remove shells, sand, salts, and any articles. In alkaline conditions, the process of extraction is carried out. The normal pH is 8 or 9. Then the desired product is transferred to the solution phase. Purification is done to remove other insoluble materials through a pressure filter. To obtain the product from the solution, vacuum evaporation is done. After which organic solvent, that is, Isopropyl alcohol, is used to precipitate. This process is majorly done for *Chondrus crispus* and *Gigartina* sp. (Hernandez-Carmona et al., 2013).

6.11.3.2 Potassium chloride process

The extraction process is similar to the alcohol process. The extraction is done in a solution phase. But instead of alcohol 1% of weight/volume, KCl is used to precipitate it at low temperature and is carried out by the freeze-thaw process similar to the Agar process. This process is majorly used for Kappa carrageenan (Hernandez-Carmona et al., 2013).

6.11.3.3 Drum drying process

Instead of precipitation, recovery of carrageenan is carried out by a drum dryer. To obtain better efficiency, vacuum conditions can be applied during the evaporation, which breaks down carrageenan due to the hot environment (Hernandez-Carmona et al., 2013) (Fig. 6.6).

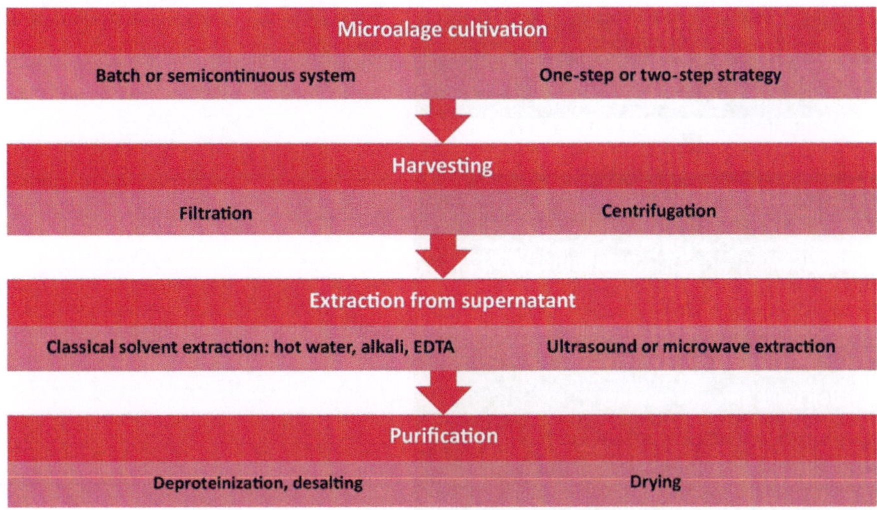

FIGURE 6.6   Carbohydrate process flow diagram.

6.11.4 Lipid extraction

For the extraction of lipids, wet and dry biomass can be used—the preferred method in recovering lipid content in the dry biomass extraction method. Owing to the requirement of high energy for drying, and overall extraction of lipids, therefore, at a large scale, this method is not suitable. Consequently, algal biomass in the wet form is used to extract lipids. But the high water content prevents the recovery of liquid by solvent, but the pretreatment of wet algal biomass can overcome this problem. The pretreatment techniques are described in a previous section. For the extraction of lipids, seven techniques were used, mainly organic solvent extraction/ionic liquid-assisted method, supercritical carbon dioxide, and transesterification method.

6.11.4.1 Solvent extraction

Hexane, chloroform, and methane are used for organic solvent extraction because this process has low vapor pressure, is nonflammable, and has high chemical and thermal stabilities. Other than organic solvents, inorganic solvents can also be used. Both nonpolar and polar solvents can extract polar and neutral lipids. Using only nonpolar solvents can only remove neutral lipids. Extraction is performed in two steps to obtain both nonpolar and polar lipids. After extraction, it can be connected into other categories based on its polarity, length of the carbon chain, and saturation level of the corresponding fatty acid (Sarkar et al., 2020).

Other methods like pyrolysis and transesterification are mainly used for lipids' bio-oil production (Fig. 6.7).

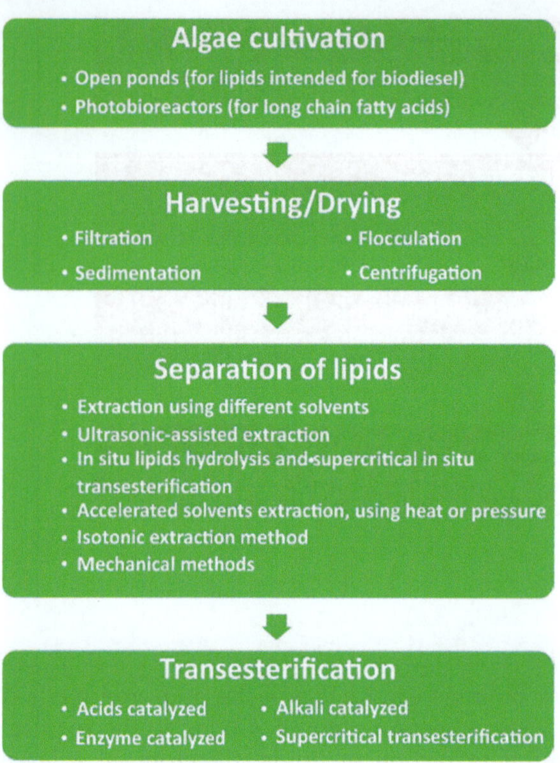

FIGURE 6.7 Process diagram of fatty acids and lipids.

6.12 Conclusions

To safely use microalgae in the food sector, the significant biodiversity of microalgae offered by nature is needed to be explored and evaluated in terms of their properties and sustainability. Microalgae have significant technological importance due to high-value chemicals' availability and wide applications in different areas. They can produce proteins, unsaturated fatty acids, lipids, polysaccharides, pigments, etc. Biologically active compounds extracted from microalgae have antiviral, antioxidant, antibacterial, antihypertensive, and immune system stimulation. Thus, it is proven as a source of both functional and nutritional additives. Only a few species of microalgae are available for human consumption. The most important compound that can be obtained from microalgae is carotenoids. Microalgae strains should be selected very carefully, following the selection criteria. If the cultivation and growth conditions are optimized, research in downstream processing is still needed. Pretreatment processes must be thoroughly evaluated and optimized to make downstream processing more efficient and sustainable in yield and selectivity. Cell dispersion must be studied as a pretreatment technique because it makes the downstream process more energy efficient while increasing the yield. In downstream processing, a significant problem is the use of organic toxic solvent extraction processes. Water extraction should be studied more than organic solvents, which can pose serious hazards.

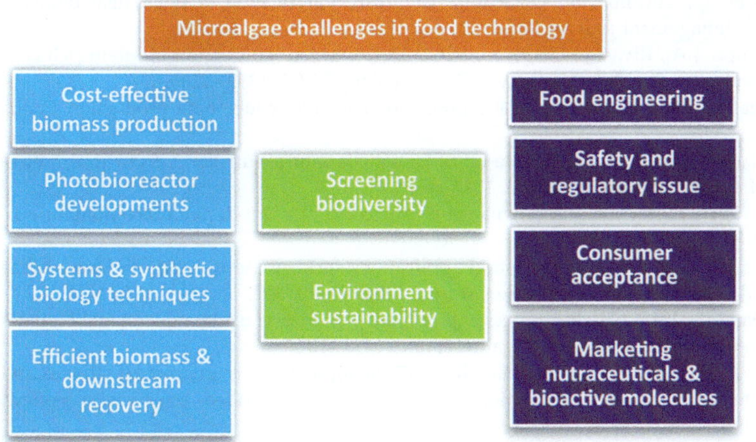

FIGURE 6.8 Overall challenges in utilization of microalgae.

6.13 Future perspective

The following factors should be considered to make microalgal biomass-derived products suitable for large-scale commercialization and, eventually, human life (Fig. 6.8):

1. The influence of cultivation conditions on the bioactive compounds in the cell should be assessed.
2. The research should be directed toward the fractionation process to obtain a wide range of products of different nature from microalgae along with the minimum number of steps, process integration, and reduced chemical requirement.
3. Genetic mutation may be required to increase a specific compound.
4. Complete understanding of process design and intracellular activities to make them commercially available, along with using each by-product for economic feasibility.
5. Most of the energy is required for downstream processing; therefore, technological development is necessary to reduce energy consumption.
6. Improvement of PBRs (closed) for their commercial availability.
7. Environmental, medical, and economic risks should be assessed.

References

Alam, M.A., Xu, J.-L., Wang, Z., 2020. Microalgae biotechnology for food, health and high value products. Springer.

Andrade, L., Andrade, C., Dias, M., Nascimento, C., Mendes, M., 2018. Chlorella and spirulina microalgae as sources of functional foods. Nutraceuticals, and Food Supplements 6 (1), 45—58.

Ansari, F.A., Shriwastav, A., Gupta, S.K., Rawat, I., Guldhe, A., Bux, F., 2015. Lipid extracted algae as a source for protein and reduced sugar: a step closer to the biorefinery. Bioresource Technology 179, 559—564.

Batista, A.P., Gouveia, L., Bandarra, N.M., Franco, J.M., Raymundo, A., 2013. Comparison of microalgal biomass profiles as novel functional ingredient for food products. Algal Research 2 (2), 164—173.

Bhalamurugan, G.L., Valerie, O., Mark, L., 2018. Valuable bioproducts obtained from microalgal biomass and their commercial applications: A review. Environmental Engineering Research 23 (3), 229—241.

Blaga, A.C., Cascaval, D., Kloetzer, L., Tucaliuc, A., Galaction, A.I., 2018. Valorization of microalgal biomass. Environmental Engineering & Management Journal 17 (4).

Blanco, A.M., Moreno, J., Del Campo, J.A., Rivas, J., Guerrero, M.G., 2007. Outdoor cultivation of lutein-rich cells of Muriellopsis sp. in open ponds. Applied Microbiology & Biotechnology 73 (6), 1259–1266.

Bleakley, S., Hayes, M., 2017. Algal proteins: extraction, application, and challenges concerning production. Foods 6 (5), 33.

Borovkov, A.B., Gudvilovich, I.N., Avsiyan, A.L., 2020. Scale-up of Dunaliella salina cultivation: from strain selection to open ponds. Journal of Applied Phycology 32 (3), 1545–1558.

Buyel, J., Twyman, R., Fischer, R., 2015. Extraction and downstream processing of plant-derived recombinant proteins. Biotechnology advances 33 (6), 902–913.

Camacho, F., Macedo, A., Malcata, F., 2019. Potential industrial applications and commercialization of microalgae in the functional food and feed industries: A short review. Marine drugs 17 (6), 312.

Cavonius, L.R., Albers, E., Undeland, I., 2015. pH-shift processing of Nannochloropsis oculata microalgal biomass to obtain a protein-enriched food or feed ingredient. Algal research 11, 95–102.

Chen, B., Wan, C., Mehmood, M.A., Chang, J.-S., Bai, F., Zhao, X., 2017. Manipulating environmental stresses and stress tolerance of microalgae for enhanced production of lipids and value-added products—a review. Bioresource technology 244, 1198–1206.

de Carvalho, J.C., Magalhães Jr, A.I., de Melo Pereira, G.V., Medeiros, A.B.P., Sydney, E.B., Rodrigues, C., Aulestia, D.T.M., de Souza Vandenberghe, L.P., Soccol, V.T., Soccol, C.R., 2020. Microalgal biomass pretreatment for integrated processing into biofuels, food, and feed. Bioresource technology 300, 122719.

Dolganyuk, V., Belova, D., Babich, O., Prosekov, A., Ivanova, S., Katserov, D., Patyukov, N., Sukhikh, S., 2020. Microalgae: a promising source of valuable bioproducts. Biomolecules 10 (8), 1153.

Eloka-Eboka, A.C., Inambao, F.L., 2017. Effects of CO2 sequestration on lipid and biomass productivity in microalgal biomass production. Applied Energy 195, 1100–1111.

Garcia, E.S., Van Leeuwen, J., Safi, C., Sijtsma, L., Eppink, M.H., Wijffels, R.H., van den Berg, C., 2018. Selective and energy efficient extraction of functional proteins from microalgae for food applications. Bioresource technology 268, 197–203.

García, J.L., de Vicente, M., Galán, B., 2017. Microalgae, old sustainable food and fashion nutraceuticals. Microbial Biotechnology 10 (5), 1017–1024.

Geada, P., Vasconcelos, V., Vicente, A., Fernandes, B., 2017. Microalgal biomass cultivation. Algal Green Chemistry. , pp. 257–284. Elsevier.

Gerde, J.A., Wang, T., Yao, L., Jung, S., Johnson, L.A., Lamsal, B., 2013. Optimizing protein isolation from defatted and non-defatted Nannochloropsis microalgae biomass. Algal. Research 2 (2), 145–153.

Gonçalves, A.L., Pires, J.C., Simoes, M., 2016. The effects of light and temperature on microalgal growth and nutrient removal: an experimental and mathematical approach. RSC advances 6 (27), 22896–22907.

Grobbelaar, J.U., 2010. Microalgal biomass production: challenges and realities. Photosynthesis research 106 (1), 135–144.

Guccione, A., Biondi, N., Sampietro, G., Rodolfi, L., Bassi, N., Tredici, M.R., 2014. Chlorella for protein and biofuels: from strain selection to outdoor cultivation in a Green Wall Panel photobioreactor. Biotechnology for Biofuels 7 (1), 1–12.

Guedes, A.C., Amaro, H.M., Pereira, R.D. and Malcata, F.X. (2011). Effects of temperature and pH on growth and antioxidant content of the microalga Scenedesmus obliquus. 27(5): 1218–1224.

Hallenbeck, P., Grogger, M., Mraz, M., Veverka, D., 2016. Solar biofuels production with microalgae. Applied Energy 179, 136–145.

Harun, R., Singh, M., Forde, G.M., Danquah, M.K. J.R. and Reviews, S.E. (2010). Bioprocess engineering of microalgae to produce a variety of consumer products. 14(3): 1037–1047.

Henchion, M., Hayes, M., Mullen, A.M., Fenelon, M., Tiwari, B., 2017. Future protein supply and demand: strategies and factors influencing a sustainable equilibrium. Foods 6 (7), 53.

Hernandez-Carmona, G., Freile-Pelegrín, Y., Hernández-Garibay, E., 2013. Conventional and alternative technologies for the extraction of algal polysaccharides. Functional ingredients from algae for foods and nutraceuticals. Elsevier, pp. 475–516.

Huang, J.-C., Chen, F., Sandmann, G., 2006. Stress-related differential expression of multiple β-carotene ketolase genes in the unicellular green alga Haematococcus pluvialis. Journal of biotechnology 122 (2), 176–185.

Iasimone, F., Panico, A., De Felice, V., Fantasma, F., Iorizzi, M., Pirozzi, F., 2018. Effect of light intensity and nutrients supply on microalgae cultivated in urban wastewater: Biomass production, lipids accumulation and settleability characteristics. Journal of environmental management 223, 1078–1085.

Kadam, S.U., Álvarez, C., Tiwari, B.K., O'Donnell, C.P., 2017. Extraction and characterization of protein from Irish brown seaweed Ascophyllum nodosum. Food Research International 99, 1021–1027.

Kagan, M.L., Matulka, R.A., 2015. Safety assessment of the microalgae Nannochloropsis oculata. Toxicology Reports 2, 617–623.

Kamyab, H., Chelliapan, S., Shahbazian-Yassar, R., Din, M.F.M., Khademi, T., Kumar, A., Rezania, S., 2017. Evaluation of lipid content in microalgae biomass using palm oil mill effluent (Pome). Jom 69 (8), 1361–1367.

Khanra, S., Mondal, M., Halder, G., Tiwari, O., Gayen, K., Bhowmick, T.K., 2018. Downstream processing of microalgae for pigments, protein and carbohydrate in industrial application: A review. Food & bioproducts processing 110, 60–84.

Koley, S., Khadase, M.S., Mathimani, T., Raheman, H., Mallick, N., 2018. Catalytic and non-catalytic hydrothermal processing of Scenedesmus obliquus biomass for bio-crude production–A sustainable energy perspective. Energy Conversion & Management 163, 111–121.

Lacour, T., Sciandra, A., Talec, A., Mayzaud, P., Bernard, O., 2012. Diel variations of carbohydrates and neutral lipids in nitrogen-sufficient and nitrogen-starved cyclostat cultures of isochrysis sp. 1. Journal of Phycology 48 (4), 966–975.

Lafarga, T., 2020. Cultured microalgae and compounds derived thereof for food applications: Strain selection and cultivation, drying, and processing strategies. Food Reviews International 36 (6), 559–583.

Laurens, L., Nagle, N., Davis, R., Sweeney, N., Van Wychen, S., Lowell, A., Pienkos, P., 2015. Acid-catalyzed algal biomass pretreatment for integrated lipid and carbohydrate-based biofuels production. Green Chemistry 17 (2), 1145–1158.

Lee, O.K., Oh, Y.-K., Lee, E.Y., 2015. Bioethanol production from carbohydrate-enriched residual biomass obtained after lipid extraction of Chlorella sp. KR-1. Bioresource technology 196, 22–27.

Macías-Sánchez, M., Fernandez-Sevilla, J., Fernández, F.A., García, M.C., Grima, E.M., 2010. Supercritical fluid extraction of carotenoids from Scenedesmus almeriensis. Food Chemistry 123 (3), 928–935.

Manirafasha, E., Ndikubwimana, T., Zeng, X., Lu, Y., Jing, K., 2016. Phycobiliprotein: Potential microalgae derived pharmaceutical and biological reagent. Biochemical Engineering Journal 109, 282–296.

Masojídek, J., Sergejevová, M., Malapascua, J.R., Kopecký, J., 2015. Thin-layer systems for mass cultivation of microalgae: flat panels and sloping cascades. Algal biorefineries. Springer, pp. 237–261.

Menegol, T., Diprat, A.B., Rodrigues, E., Rech, R., 2017. Effect of temperature and nitrogen concentration on biomass composition of Heterochlorella luteoviridis. Food Science & Technology 37 (SPE), 28–37

Miazek, K., Kratky, L., Sulc, R., Jirout, T., Aguedo, M., Richel, A., Goffin, D., 2017. Effect of organic solvents on microalgae growth, metabolism and industrial bioproduct extraction: a review. International Journal of Molecular Sciences 18 (7), 1429.

Nurachman, Z., Hartini, H., Rahmaniyah, W.R., Kurnia, D., Hidayat, R., Prijamboedi, B., Suendo, V., Ratnaningsih, E., Panggabean, L.M.G., Nurbaiti, S., 2015. Tropical marine Chlorella sp. PP1 as a source of photosynthetic pigments for dye-sensitized solar cells. Algal research 10, 25–32.

Phong, W.N., Le, C.F., Show, P.L., Lam, H.L., Ling, T.C., 2016.). Evaluation of different solvent types on the extraction of proteins from microalgae. Chemical Engineering Transactions 52, 1063–1068.

Rabbani, S., Beyer, P., Lintig, J.V., Hugueney, P., Kleinig, H., 1998. Induced β-carotene synthesis driven by triacylglycerol deposition in the unicellular alga Dunaliella bardawil. Plant Physiology 116 (4), 1239–1248.

Sánchez-Camargo, Ad.P., Pleite, N., Mendiola, J.A., Cifuentes, A., Herrero, M., Gilbert-López, B., Ibáñez, E., 2018. Development of green extraction processes for Nannochloropsis gaditana biomass valorization. Electrophoresis 39 (15), 1875–1883.

Sarkar, S., Manna, M.S., Bhowmick, T.K., Gayen, K., 2020. Priority-based multiple products from microalgae: Review on techniques and strategies. Critical reviews in biotechnology 40 (5), 590–607.

Singh, G., Patidar, S., 2020. Development and Applications of Attached Growth System for Microalgae Biomass Production. BioEnergy Research 1–14.

Sui, Y., Jiang, Y., Moretti, M., Vlaeminck, S.E., 2020. Harvesting time and biomass composition affect the economics of microalgae production. Journal of Cleaner Production 259, 120782.

Vandamme, D., Foubert, I., Muylaert, K., 2013. Flocculation as a low-cost method for harvesting microalgae for bulk biomass production. Trends in biotechnology 31 (4), 233–239.

Wan, C., Alam, M.A., Zhao, X.-Q., Zhang, X.-Y., Guo, S.-L., Ho, S.-H., Chang, J.-S., Bai, F.-W., 2015. Current progress and future prospect of microalgal biomass harvest using various flocculation technologies. Bioresource technology 184, 251–257.

Wang, B., Lan, C.Q., Horsman, M., 2012. Closed photobioreactors for production of microalgal biomasses. Biotechnology advances 30 (4), 904–912.

Xu, L., Weathers, P.J., Xiong, X.R., Liu, C.Z., 2009. Microalgal bioreactors: challenges and opportunities. Engineering in Life Sciences 9 (3), 178–189.

Yadav, G., Meena, D.K., Sahoo, A.K., Das, B.K., Sen, R., 2020. Effective valorization of microalgal biomass for the production of nutritional fish-feed supplements. Journal of Cleaner Production 243, 118697.

Yang, Z., Guo, R., Xu, X., Fan, X., Li, X., 2010. Enhanced hydrogen production from lipid-extracted microalgal biomass residues through pretreatment. International journal of hydrogen energy 35 (18), 9618–9623.

Zeng, X., Guo, X., Su, G., Danquah, M.K., Chen, X.D., Lin, L., Lu, Y., 2016. Harvesting of microalgal biomass. Algae Biotechnology. Springer, pp. 77–89.

Valorization of microalgal biomass for fertilizers and nanoparticles

Umarin Jomnonkhaow[1,2], *Sureewan Sittijunda*[3] *and Alissara Reungsang*[1,2,4]

[1]Department of Biotechnology, Faculty of Technology, Khon Kaen University, Khon Kaen, Thailand [2]Research Group for Development of Microbial Hydrogen Production Process from Biomass, Khon Kaen University, Khon Kaen, Thailand [3]Faculty of Environment and Resource Studies, Mahidol University, Nakhon Pathom, Thailand [4]Academy of Science, Royal Society of Thailand, Bangkok, Thailand

7.1 Introduction

Algal organisms are photosynthetic organisms that live in a wide range of environments. These microorganisms have a short generation time and multiply exponentially under favorable environmental conditions (Parsaeimehr & Lutzu, 2016). They are categorized into macro and microalgae-based on their size and morphology. Macroalgae are macroscopic and are classified into three major groups: green algae (Chlorophyta), red algae (Rhodophyta), and brown algae (Phaeophyceae) (Pereira, 2021), for example, seaweed is a macroalga with numerous cells that organize into structures that resemble higher plant roots, stems, and leaves (Salehi et al., 2019). In contrast, microalgae are microscopic photosynthetic organisms that are primarily unicellular (Roy & Mohanty, 2021; Salakkam et al., 2021). They can be grown under autotrophic, heterotrophic, and mixotrophic conditions. Autotrophic algae use photosynthesis to capture sunlight and fix the inorganic carbon from atmospheric CO_2, which is then assimilated in the form of reserve food materials in their bodies as carbohydrates (Pandey et al., 2020; Parsaeimehr & Lutzu, 2016). Heterotrophic algae can use organic molecules such as sugars, organic acids, and wastewater as energy and carbon sources and convert them into their building blocks, such as proteins, fats, and oils (Duran et al., 2021; Salehi et al., 2019). Finally, some algal species can use both atmospheric CO_2 and organic molecules from the environment, a process known as mixotrophy.

113

Fritsch and Oliver (1907) were the first to report the use of algal biofertilizers when they used blue-green algae in paddy fields for improving soil fertility through atmospheric nitrogen fixation. Furthermore, several studies have found that biofertilizers can improve carbon fixation along with nitrogen (N), phosphorus (P), potassium (K), and other soil elements (Chatterjee et al., 2017; Guo et al., 2020). In addition, the secretion of exopolysaccharide (EPS) and phytohormones from algae increased crop production and yield (Fig. 7.1). Therefore using algae as biofertilizers is a potential approach for sustainable management in modern agriculture. On the other hand, excessive use of chemical fertilizers has resulted in soil hardening, salinization, nutrient reduction, macroelement imbalance, N, P, K ratio, groundwater pollution, and plant health problems (Chatterjee et al., 2017; Guo et al., 2020; Pandey et al., 2020; Parsaeimehr & Lutzu, 2016). In terms of human health, chemical fertilizers have direct as well as indirect toxic effects. High levels of nitrate and nitrites in chemical fertilizer cause diseases, such as, hemoglobin disorder, Alzheimer's disease, diabetes, and nonalcoholic steatohepatitis (De La Monte et al., 2009). Therefore using algal biofertilizers can help us achieve sustainable consumption and production patterns (the 12th Sustainable Development Goal).

Nanotechnology is one of the most active research fields that has become an emerging multidisciplinary field, including agriculture, medicine, and the environment. Compared

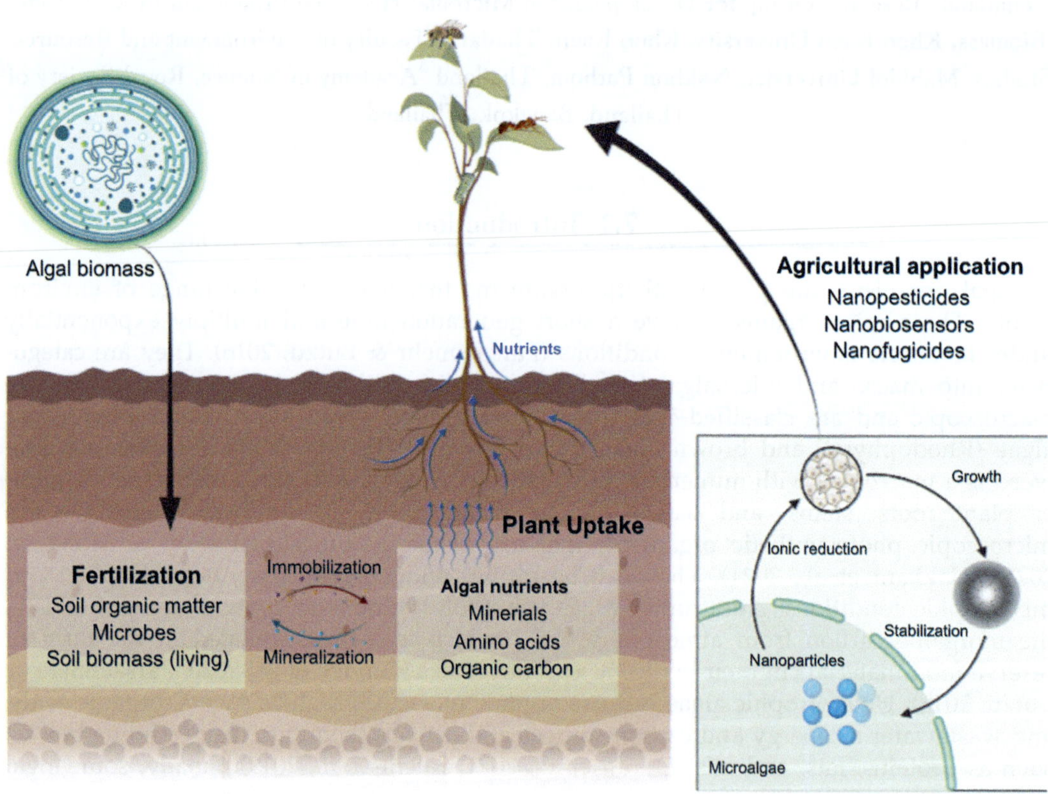

FIGURE 7.1   Algal utilization as biofertilizer and biological synthesis of nanoparticle.

to physical and chemical methods, nanoparticle (NP) synthesis via biosynthetic pathways is simpler, cost-effective, and environment-friendly (Pandey et al., 2020). NP synthesis from biological resources is a green technology for sustainable production. A wide range of NPs, such as silver (Ag), gold (Au), palladium (Pd), iron (Fe), and zinc oxide (ZnO), have been synthesized using biological resources (Khalafi et al., 2019; Khanna et al., 2019; Shankar et al., 2016) such as bacteria, fungi, plant extracts, cyanobacteria, and algae (Fawcett et al., 2017; Saravanan et al., 2021).

The biosynthesis of NPs using microalgae is a developing and challenging field. Numerous methodologies have been developed to synthesize metallic (gold, silver, and copper oxide) NPs from aqueous salt solutions using microalgae (Pandey et al., 2020; Sun et al., 2018). Synthesizing NPs using microalgae is simple because the process can be performed at ambient temperature, and the process requires only the metal ion solution to act as the precursor (Jacob et al., 2021; Moraes et al., 2021; Zayadi & Abu Bakar, 2020). For instance, silver nitrate was used as the metal ion precursor for the biosynthesis of silver nanoparticles (AgNPs) from microalgae. At the same time, bioactive compounds in algal cells act as the reducing and capping agents (Chokshi et al., 2016). In the chemical synthesis of AgNPs, silver nitrate, ethylene glycol, and poly(vinyl pyrrolidone) were used as the metal ion precursor, reducing agent, and capping agent, respectively (Sun & Xia, 2002). In addition, microalgae have various characteristics to be outstanding biological resources such as high photosynthetic efficiency, growth under extreme environments such as high salinity, extreme temperature, and heavy metal stress (Leong & Chang, 2020). It has been reported that microalgae have multiple mechanisms for self-protection from high heavy metal stress, such as heavy metal immobilization, gene regulation, chelation, and production of antioxidants and reducing enzymes to reduce heavy metals via redox reactions (Leong & Chang, 2020). These abilities of heavy metal tolerance and changing the metal ions via redox reactions can be used in NP synthesis. Algal NPs can be applied in biomedical applications such as anticancer, antibacterial, antifungal, and biosensing. Moreover, it can be used as an antibiofilm agent against multidrug-resistant bacteria.

This chapter discusses the impact of algal biofertilizers on soil properties, enzyme activity, nutrient availability, and the mechanisms involved in plant growth enhancement. Furthermore, an overview of the use of microalgae to synthesize NPs is provided, including the fundamental approaches to NP synthesis, the biosynthesis of NPs via intracellular and extracellular synthesis routes, and synthesis of NPs (Ag and Au) using various microalgal species. In addition, the application of NPs synthesized from microalgae has been discussed. Furthermore, research prospects for algal biofertilizers and NP synthesis using microalgae are also discussed.

7.2 Biofertilizer from microalgal biomass

Biofertilizers contain living organisms that promote plant growth and increase nutrient availability when applied to the plant surface or soil (Ginni et al., 2020). In modern agricultural systems, algae can be used as biofertilizers to replace the excessive use of chemical fertilizers (Guo et al., 2020). Several algal types, such as cyanobacteria, green algae, red algae, and brown algae, are preferred as biofertilizers (Table 7.1).

TABLE 7.1 Types of algal biofertilizers.

Algal biofertilizer	Application	Contributions	Crop plants	References
Cyanobacteria/blue-green algae	– Suspension		*Matricaria chamomilla* L.,	Zarezadeh et al. (2020)
Nostoc carneum		1. Nitrogen fixation	*Pisum sativum,*	Osman et al. (2010)
Wollea vaginicola		2. Supply of nutrients 3. Carbon fixation	*Oryza sp.,*	Bocchi and Malgioglio (2010)
Nostoc punctiforme		4. Produce plant growth-promoting hormone (auxin and cytokinin)	*Solanum lycopersicum,*	Vaishampayan et al. (2001)
Nostoc entophytum *Oscillatoria angustissima*		5. Produce exopolysaccharide	*Zea mays,*	Singh (1961)
Tolypothrix Westiella,			*Triticum*	
Nostoc muscorum, Nostoc sp.,			*Gossypium*	Watanabe et al. (1978)
Anabaena,				
Camptylonema,				
Oscillatoria				
Cylindrospermum,				
Arthrospira				
Anabaenopsis,				
Chlorogloea,				
Aphanothece, Aulosira,				
Calothrix,				
Chlorogloeopsis,				
Green algae			*Eruca vesicaria* (L.),	Hassan et al. (2021)
Ulva lactuca	– Liquid extracted biofertilizer	1. Produce phytohormones	*Medicago truncatula,*	
Chlorella sp.	– Live algal cells – Liquid biofertilizer	2. Produce algal exopolysaccharide 3. Nitrogen fixation	*Cyamopsis tetragonoloba*	Gitau et al. (2021)
Chlamydomonas reinhardtii		4. Increase phosphorus availability	*Cucumis sativus,*	
Chlorococcum sp.		5. Supply nutrient 6. Increase photosynthetic pigments,	*Vigna radiata,*	Deepika and MubarakAli (2020)
Scendesmus spp.		7. Increase protein content, amino acids, reducing sugar, ascorbic acid, and nitrate reductase activity.	*Solanum lycopersicum,*	Hamouda (2013)
		8. Source for macro and micro elements (significantly high N.P.K. content)	*Capsicum annuum,*	McHugh (2003) Khan et al. (2021b)

Red algae			*Z. mays* L.,	McHugh (2003)
Laurencia obtusa	– Dry powder	1. Source for macro and micro elements such as potassium, nitrogen, and phosphorus	*Eruca vesicaria* (L.)	Hamouda (2013)
Corallina elongata	– Liquid extracted biofertilizer			Hassan et al. (2021)
Jania rubens		2. Neutralize acid soil		
		3. Source for micro elements		
Phymatolithon calcareum				
Lithothamnion corallioides				
Pterocladiella capillacea (S.G. Gmelin)				
Brown algae	– Liquid extracted biofertilizer		*Solanum melongena,*	Rahayu et al. (2021)
Stoechospermum marginatum		1. Supply nutrient source and trace element	*Cyamopsis tetragonoloba,*	Ramya et al. (2010)
Durvillea potatorum,		2. Produce auxin, gibberellin, and cytokinin	*Phaseolus vulgaris,*	Ramya et al. (2015)
Sargassum wightii		3. Increase vitamin C and fiber content	*Picea, Triticum*	Sharma et al. (2014)
Macrocystis pyrifera		4. Produce algal exopolysaccharide as the active plant defend		
Ascophyllum sp.				
Ecklonia maxima,				
Laminaria digitata, Ascophylum nodosum, Fucus serratus,				
Himanthalia elongate				

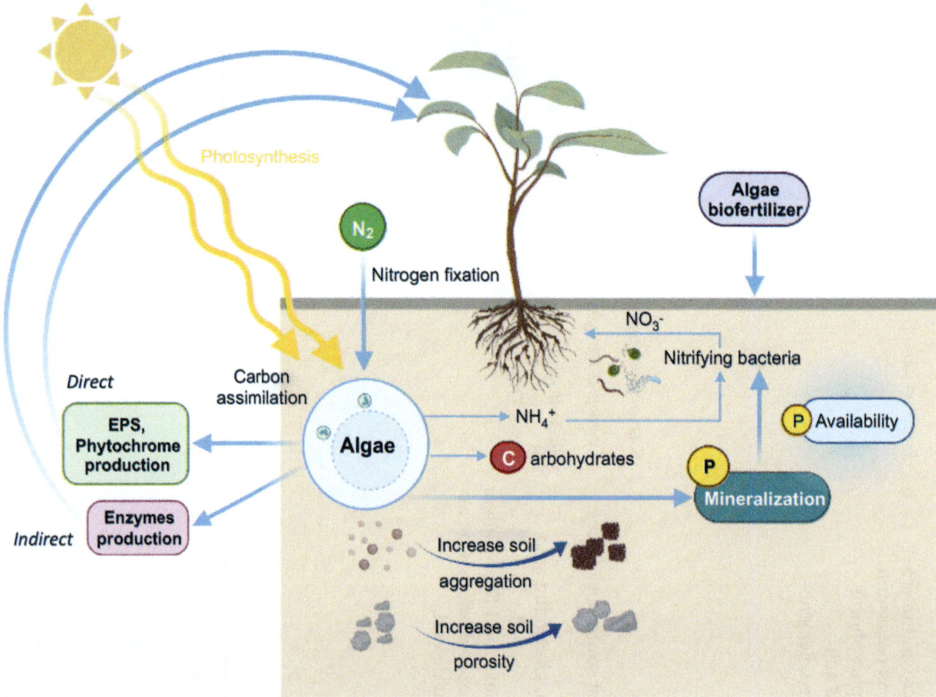

FIGURE 7.2 Algal biofertilizers enhance nitrogen fixation, carbon assimilation, phosphorus solubilization, enzyme activity, EPS, and plant growth-promoting hormone production.

Living or extracted algal biomass, either in liquid or solid forms, can be used as biofertilizers. However, as a part of the zero-waste concept, using extracted algal biomass as a biofertilizer is an interesting approach for a long-term solution. Algal biofertilizers can improve soil properties and crop yield through various mechanisms, including nitrogen fixation, carbon assimilation, and phosphorus solubilization. Moreover, it enhances crop yield due to metabolites, such as EPS and plant growth-promoting hormone (PGPH) production (Fig. 7.2). The following sections review the specifics of each contribution.

7.2.1 Algal biofertilizer influence soil properties

Soil properties, including soil aggregation, porosity, pH, and fertility, are critical factors affecting crop growth. The use of chemical fertilizers is a practical way to improve these properties. However, over the long term, excessive use of chemical fertilizers harms these soil properties. As a result, algal biofertilizers are an appealing approach for mitigating these negative impacts. Several studies have found that green algae and cyanobacteria can secrete EPS into their environment (Babiak & Krzemińska, 2021; Liu et al., 2016), improving organic matter (OM) content in the soil and preventing soil erosion. OM content is significant to soil's physicochemical and biochemical properties, such as its buffering ability, absorptive capacity, biodiversity, and biological activity (Guo et al., 2020; Nayak et al.,

2019; Pandey et al., 2020). A significant increase in OM was observed when *Chlorella vulgaris* and *Spirulina platensis* were used as liquid biofertilizers. As a result, the levels of carbohydrates, indoles, and phenols in the leaves and roots of maize plants increased significantly (Dineshkumar et al., 2019). Using algal biomass as a biofertilizer in liquid or dry form can improve soil fertility by increasing OM, potassium, and nitrogen levels in the soil (Table 7.1). Monosaccharides such as fucose, galactose, glucose, xylose, rhamnose, arabinose, and ribose are the primary EPS produced by green algae, red algae, and cyanobacteria (Hu et al., 2003; Liu et al., 2016; Maksimova et al., 2004; Mishra & Jha, 2009), while uronic acid, glucuronic acid, galacturonic acid, amino acids, and sulfate are the minor components (Hu et al., 2003; Liu et al., 2016; Maksimova et al., 2004). The presence of a negative charge causes the EPS to become sticky. The polysaccharide-based EPS is so sticky that it acts as a soil-particle-binding agent, preventing soil erosion (Babiak & Krzemińska, 2021; Dineshkumar et al., 2019). Due to its anionic nature, EPS has an affinity toward positively-charged ions, most notably metal ions. This strategy benefits the bioremediation of heavy metal contaminated soil or water because it is cost-effective and environmentally safe (Igiri et al., 2018).

Furthermore, EPS can also aid in the aggregation of soil particles, resulting in the minimum pore size required for root penetration and nutrient adsorption. For example, the application of dry *C. vulgaris* and *S. platensis* as biofertilizers improves soil aggregation and porosity due to the adhesive properties of proteoglycans in microalgae (Dineshkumar et al., 2019). *Chlorella* sp. combined with vermiculite as a biofertilizer significantly affected the soil aggregate, stability, and organic carbon (OC) content. An increase in soil stability and OC content has been linked to the photosynthetic activity of algal species (Yilmaz & Sönmez, 2017). The application of cyanobacteria to alkali soils resulted in significant improvements in soil aggregation, OM, and total N and P availability (Aziz and Hashem, 2003). Moreover, EPS also plays a significant role in water storage because of the hygroscopic properties of polysaccharides. Thus EPS can boost the water retention capacity of the soil.

7.2.2 Impact on enzyme activity and nutrient availability

Algal biofertilizer application can increase plant nutrient content through the life cycle of algae, that is, nutrients in the microalgal cells are secreted from them after their death. In general, algal biomass contains different contents of the main organic groups (carbohydrates, proteins, and lipids). Algal biomass is mainly made of protein, followed by lipids and carbohydrates (Salakkam et al., 2021). Phosphorus ranging from 1% to 2% (Yaakob et al., 2021) is another micronutrient in algal cells. These macro- and micronutrients in algal biomass act as nutrient supplements in soil. As a result, soil fertility increases remarkably (Guo et al., 2020; Sharma et al., 2021). Macro- and micronutrients can be released into the ground through soil microbial degradation or secretion after cell death (Chatterjee et al., 2017; Guo et al., 2020). For example, the dry biomass of *Chlorella* sp. had a 3.30%−6.40% (w/w) of nitrogen content that can be converted into NO_3^- by nitrifying soil bacteria around the plant roots leading to NO_3^- assimilation in the plants.

Similarly, the use of *Chlorella minutissima* as a biofertilizer in the cultivation of baby corn improves dehydrogenase activity (Sharma et al., 2021), resulting in significantly increased soil OM and nitrogen availability, with 5% and 1% probability levels, respectively. Consequently, the length of baby corn was 1.24 times higher than control (without algal biomass) (Sharma et al., 2021). Dehydrogenase is an important enzyme involved in the biological oxidation of soil OM by transferring hydrogen from organic substrates to inorganic acceptors (Zhang et al., 2010). It is also used as an indicator of overall soil microbial activity, soil health, and fertility because of its impact on the physicochemical and biochemical properties of the soil. However, under flooded or anaerobic conditions, there can be very high dehydrogenase activity, which affects the transport and availability of Fe to plants due to its utilization as a terminal electron acceptor (Das & Varma, 2010). Furthermore, seasonal changes in soil moisture content, soil aeration, and soil pH affect soil dehydrogenase activity, making accurate soil condition assessment difficult.

Urease is another essential soil enzyme that participates in the hydrolysis of urea to ammonia (Eq. 7.1) (Mobley & Hausinger, 1989).

$$CO(NH_2)_2 + H_2O + 2H^+ \rightarrow \rightarrow \rightarrow CO_2 + NH_4^+ \tag{7.1}$$

As ammonia gets further oxidized to nitrate by the soil bacteria, this enzyme indicates nitrogen availability. In general, green algae and cyanobacteria (*Nostoc muscorum* and *Tolypothrix tenuis*) produce urease (Singh et al., 2016). For example, urease activity increased when green algae (*C. minutissima*) were used as biofertilizers in the soil where spinach was grown. This resulted in a significant (p < 0.01) increase in nitrogen availability in the ground. As a result, the spinach's fresh biomass, leaf, and root lengths increased (Sharma et al., 2021).

Nitrate reductase is an enzyme that reduces NO^-_3 to the organic form (NO^-_2) within the plant. This enzyme activity reflects the level of nitrogen activity in leaves (Beevers & Hageman, 1969; Lane et al., 1975). Oosterhuis and Bate (1983) found that the changes in nitrate reductase activity in leaves correlated with changes in petiole nitrate concentration. Thus nitrate reductase assay is a reliable indicator of the nitrogen status in plant leaves. Furthermore, green algae as a biofertilizer improved nitrate reductase by 83% compared to the control experiment (without green algae) in the spinach pots. Hence, a significant increase in nitrogen availability in the soil after spinach harvesting was also observed (Sharma et al., 2021). It is worth noting that the nitrate reductase assay is too expensive and time-consuming to assess nitrogen levels in plants on a commercial scale. The other methods for determining nitrogen levels in the plant, such as the Kjeldahl method, an automated carbon hydrogen nitrogen analyzer (McGeehan & Naylor, 1988), and a soil-plant analysis development meter, are more suitable for commercial use due to their shorter time requirements and lower cost.

7.2.3 Nutrient improvement through nitrogen fixation, carbon assimilation, phosphorus solubilization, and plant growth-promoting hormone production

Additional approaches for improving plant nutritional content by algal biofertilizers include nitrogen and carbon fixation and phosphorus solubilization. Nitrogen fixation, carbon assimilation, and phosphorus solubilization are shown in Fig. 7.2. Nitrogen fixation is the process by which nitrogen-fixing microorganisms convert atmospheric nitrogen into ammonia. Nitrogen fixation occurs in heterocysts, which are special cells containing the nitrogenase enzyme, the most critical enzyme in the conversion of atmospheric nitrogen into ammonia (Hanrahan and Chan, 2005). This process requires ferredoxin as an electron donor and ATP as an energy source (Poudel et al., 2018). Among algal species, cyanobacteria have a specialized unit called heterocyst that absorbs atmospheric nitrogen and converts it to ammonium (NH_3) by nitrogenase enzyme. Nitrogenase activity in heterocystous cyanobacteria is light-dependent (Guo et al., 2020; Aziz and Hashem, 2003; Pandey et al., 2020; Singh et al., 2016). Nonheterocystous cyanobacteria such as *Plectonema*, *Oscillatoria*, *Aphanothece*, and *Gloeocapsa* also fix atmospheric nitrogen (Issa et al., 2014). Since the nitrogenase complex is very sensitive to oxygen, anaerobic conditions are required for biological N_2 fixation. As a result, inadequate oxygen and light levels are unfavorable for optimizing the nitrogen fixation process. Nevertheless, several studies have shown that inoculation with a cyanobacterial consortium enhances the soil nitrogen content of crops (Chatterjee et al., 2017; Guo et al., 2020; Pandey et al., 2020; Sharma et al., 2021). For example, Grzesik et al. (2017) proved that inoculating the foliar spray of *Anabaena* PCC7120 increases the nitrogen content in the soil. In addition, the inoculation of green algae (*Chlorella* sp.) in willow plants increased the amount of nitrogen. The plant takes up fixed nitrogen forms in soil using nitrifying bacteria located in the plant root.

In addition to nitrogen fixation, "carbon assimilation" occurs when microalgal biofertilizers are used. Carbon assimilation is the process by which living organisms transform inorganic carbon (CO_2) into organic molecules (Chatterjee et al., 2017; Guo et al., 2020). Microalgae and cyanobacteria are essential OM sources because of their ability to photosynthesize to convert atmospheric carbon dioxide into organic algal biomass (Vale et al., 2020). Thus the assimilation of carbon dioxide by algae can significantly increase the carbohydrate content of algal biomass (Chatterjee et al., 2017; Guo et al., 2020). As a result, the algal biomass remaining in the soil can decompose and improve the organic content. Moreover, EPS secreted by algae can boost soil carbon levels and encourage other microflora and fauna (Costa et al., 2018).

Another essential role of microalgal biofertilizers is the solubilization of phosphorus. Phosphorus is necessary for plant growth, reproduction, and metabolic activity (Solovchenko et al., 2021). As a result, supplementing crop cultivation with chemical P fertilizer is a quick and straightforward method. However, chemical P fertilizer cannot be retained for long periods because metal-cation complexes precipitate it and rapidly become fixed in soils. This causes long-term environmental effects, such as eutrophication, soil fertility depletion, and carbon footprint (Kalayu, 2019). In general, almost 80% of organic and inorganic phosphorus in soil is in an insoluble form because of its binding to the soil constituents that plants cannot absorb (Pérez Corona et al., 1996). Insoluble phosphorus is converted into soluble phosphorus by a few efficient soil bacteria (Kalayu, 2019). This process may release phosphorus too fast for the plants to absorb (Solovchenko et al., 2021). Research reported that green algae and cyanobacteria could convert insoluble phosphate complexes into soluble forms (Kalayu, 2019). For example, cyanobacteria such as *Anabaena* and *Nostoc* can convert complex phosphates such as aluminum phosphate

(AlPO$_4$) and ferric orthophosphate (FePO$_4$) into a soluble form (Yibeltie et al., 2018). Therefore microalgae play a significant role in the availability of phosphorus to plant roots. Thus, in addition to increasing phosphorus availability, the use of phosphorus-rich algal biofertilizers is an attractive approach.

In general, algae can remove various types of wastewater pollution, particularly nitrogen and phosphorus, through a process known as eutrophication (Wurtsbaugh et al., 2019). These characteristic make it possible to produce the algae for the purpose of removing phosphorus and then utilize algae that contain phosphorus as P-rich fertilizer. According to Powell et al. (2011), *Scenedesmus* sp. can remove phosphorus from waste stabilization ponds and accumulate it as polyphosphate, after which it is further slowly decomposed by soil microbial phosphatase. Thus, based on their findings, the alga acted as a P-rich slow-release fertilizer due to the slow decomposition of polyphosphate by soil microbial phosphatases. Furthermore, the bioavailability of phosphorus released from algal cells to the soil is proportional to the plant uptake rate (Coppens et al., 2016). Simultaneously, the resumption of cell division caused the polyphosphate to decrease. As a result, harvesting polyphosphate-rich algal biomass over time is a critical step for biofertilizer application (Solovchenko et al., 2019).

Adding a natural or synthetic plant growth hormone is an attractive approach to increase crop yield and productivity while also controlling weeds in agricultural cultivation (Gonçalves, 2021; Guo et al., 2020; Ronga et al., 2019; Singh et al., 2016). However, the potential risk of leakage into the surrounding area and water bodies is a primary environmental concern. Green algae and cyanobacteria can produce PGPH (auxin, cytokinins, and gibberellins) and secrete them into the environment (Gonçalves, 2021; Guo et al., 2020; Ronga et al., 2019; Singh et al., 2016). Many cyanobacterial strains, *Phormidium* SM-14, *Phormidium* SM-15 *Nostoc calcicola*, and *Anabaena vaginicola* have been reported to produce indole-3-acetic acid (I.A.A.) (Mazhar and Hasnain, 2000; Hashtroudi et al., 2012). I.A.A. is an auxin class hormone that induces cell elongation and cell division. Moreover, I.A.A. serves as a signaling molecule essential for developing plant organs and coordination of growth in large amounts. Wheat growth was increased when using *Nostoc calcicola* and *Anabaena vaginicola* as biofertilizers because of indole-3-propionic acid (I.P.A), I.A.A., and indole-3-butyric acid (I.B.A.) hormone secretion (Mazhar and Hasnain, 2000). In vitro experiments revealed that cyanobacterial growth-promoting substances of *Anabaena variabilis* caused 100% seed germination in *Zea mays* and *Sorghum bicolor* grains. In comparison, *N. calcicola* caused 90% germination (Suresh et al., 2019). Cytokinin is a hormone that is related to the growth of auxiliary buds and enhances seed yield. Furthermore, it induces the mobilization of nutrients and acts as a plant defense by inducing resistance against phytopathogenic agents. For example, the cytokinins produced from *Dunaliella. salina*, *Acutodesmus obliquus*, *C. vulgaris*, and *Gracilaria caudata* stimulated the accumulation of photoprotective pigments in these organisms, resulting in higher photosynthetic rates and, consequently, higher biomass productivity (Mousavi et al., 2016; Piotrowska & Czerpak, 2009; Renuka et al., 2017; Souza & Yokoya, 2016). Gibberellins are involved in cell elongation and division, similar to auxin, and induce pollen germination. For example, gibberellins derived from *Chlorella pyrenoidosa*, *C. vulgaris*, and *Microcystis aeruginosa* have been linked to increased cell numbers, sizes, and faster growth rates in microalgae and cyanobacteria (Du et al., 2015; González-Garcinuño et al., 2016; Falkowska et al., 2011; Pan et al.,

2008). Hence, the gibberellin produced by microalgae and cyanobacteria regulate seed germination, leaf expansion, stem elongation, and flower development in the same way that plant gibberellins do.

7.3 Nanoparticles from microalgal biomass

The synthesis of NPs is fundamentally classified into two approaches: top-down and bottom-up (Khanna et al., 2019; Saravanan et al., 2021). In the top-down approach, the process involves breaking bulk components into reduced-size particles. The top-down approach usually requires microfabrication techniques where external energy is used to cut, grind, and shape particles into desired size and shape (Khanna et al., 2019). Top-down approaches include mechanical ball milling (Zhang & Wen, 2020), etching (Nikravan et al., 2018), laser ablation (Rashid et al., 2021), electrical explosion (Tanaka et al., 2018), and sputtering (Verma et al., 2018). The bottom-up approach refers to molecular nanotechnology involving the aggregation of atoms, molecules, or small particles to form NPs. In this method, single self-assembled molecules are created to produce complex conformations as NPs (Khanna et al., 2019; Saravanan et al., 2021). Thus the NPs formed are clustered from smaller units. The advantage of the bottom-up approach is that the synthesized NPs are stable and can be formed in a crystalline structure (Kakakhel et al., 2021). Bottom-up approaches include supercritical fluid synthesis (Jiang et al., 2020), spinning (Mohammadi et al., 2014), sol–gel processes (Khan et al., 2021a), laser pyrolysis (D'Amato et al., 2013), chemical vapor deposition (Hwang et al., 2020), chemical reduction (Lokman et al., 2021), and biosynthesis.

The use of chemicals for NP synthesis is hazardous and causes several environmental issues, while physical methods are energy-intensive and require high energy and input costs (Asmathunisha & Kathiresan, 2013; Patil & Kim, 2018; Saravanan et al., 2021). Biological methods, referred to as biosynthesis, are more attractive than chemical and physical methods because they are cost-effective, have low energy input, are easily scaled up for large-scale synthesis, and are environment-friendly (Shankar et al., 2016). Moreover, the NPs achieved from biosynthesis can be applied in biomedical applications and clinical fields because they are free of contamination as the biosynthesis process does not utilize any toxic chemicals (Saravanan et al., 2021). Biological methods of NP synthesis include the application of various plants (Rodriguez-Felix et al., 2021) and microbes such as bacteria (Anthony et al., 2014), algae (Abdul Razack et al., 2016), fungi, and yeast (Ananthi et al., 2018).

7.3.1 Biosynthesis of nanoparticles using microalgae

Microalgae are well known to serve as integrated biofactories that can simultaneously grow in heavy metal-polluted environments. Thus microalgae can be potentially used for the biosynthesis of metal NPs. Microalgal cells use metal ions at low concentrations to function as cofactors in enzymatic reactions. However, high concentrations of metal ions have been reported to have adverse effects on their growth and metabolism (Jacob et al.,

2021). Under metal stress conditions, microalgae have multiple strategies of unique self-protection mechanisms against toxicity of metals, such as metal immobilization, exclusion, and chelation of antioxidants, and reducing enzymes that reduce metals by redox reactions (Gómez-Jacinto et al., 2015; Leong & Chang, 2020). The interaction of metal ions and microalgae under high metal concentrations has been beneficially used for the biosynthesis of NPs (Vargas-Estrada et al., 2020). Moreover, microalgae are a potential source of biomolecules consisting of carbohydrates; proteins; fats; mineral oil; aromatic compounds; bioactive compounds such as tocopherols and polyphenols; and pigments such as chlorophyll, phycobilins, and carotenoids, which are capable of acting as reducing agents in algal cells for the synthesis of NPs without producing toxic by-products (Khanna et al., 2019; Michalak & Chojnacka, 2015; Saravanan et al., 2021). In addition, using algal cells for NP synthesis is possible for economic scaling and easy disruption processes. For these reasons, microalgae have received much attention as a potential source of NP synthesis (Saravanan et al., 2021).

Previous studies have proposed three preparation processes for NP synthesis from microalgae: (1) algae extraction, (2) metal precursor solution preparation, and (3) incubation of a mixture of algae extract and metal precursor solution (Saravanan et al., 2021; Uzair et al., 2020). Algae extracts composed of active compounds have been identified as reducing and capping agents (Khanna et al., 2019). During the incubation, the reactions between the algae extract and the metal precursor solution can be initially detected by visual observation as a function of the color change of the reaction mixture. The three following stages could describe the mechanism of NP biosynthesis: (1) nucleation in which the negatively charged metal ions reduce metallic cations (M^+) at the initiation reaction; (2) seed growth stage where more NPs (M^0) is produced and metal ions (M^+) get reduced on the surface of existing nuclei, causing the reaction to accelerate as the diameter of particles increases with time; and (3) the stabilization phase, whereas the reaction proceeds the reaction rate decreases and becomes slow due to the limiting concentrations of the reactants or the reducing agents (Bao & Lan, 2018). During the stabilization phase, stable shapes and sizes of the NPs are formed. The general mechanism of NP synthesis is shown in Fig. 7.3. The ability of microalgae to synthesize NPs occurs through two routes of synthesis: intracellular and extracellular synthesis (Khanna et al., 2019; Patil & Kim, 2018; Sharma et al., 2016). The general methods for NP synthesis through intracellular and extracellular synthesis are shown in Fig. 7.4.

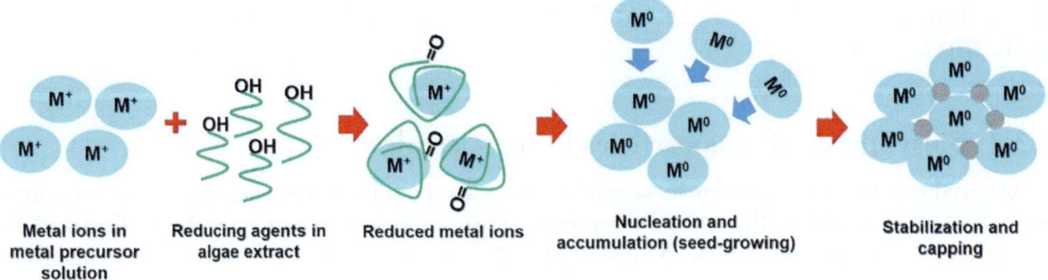

FIGURE 7.3   Schematic depicting the general mechanism for nanoparticles synthesis.

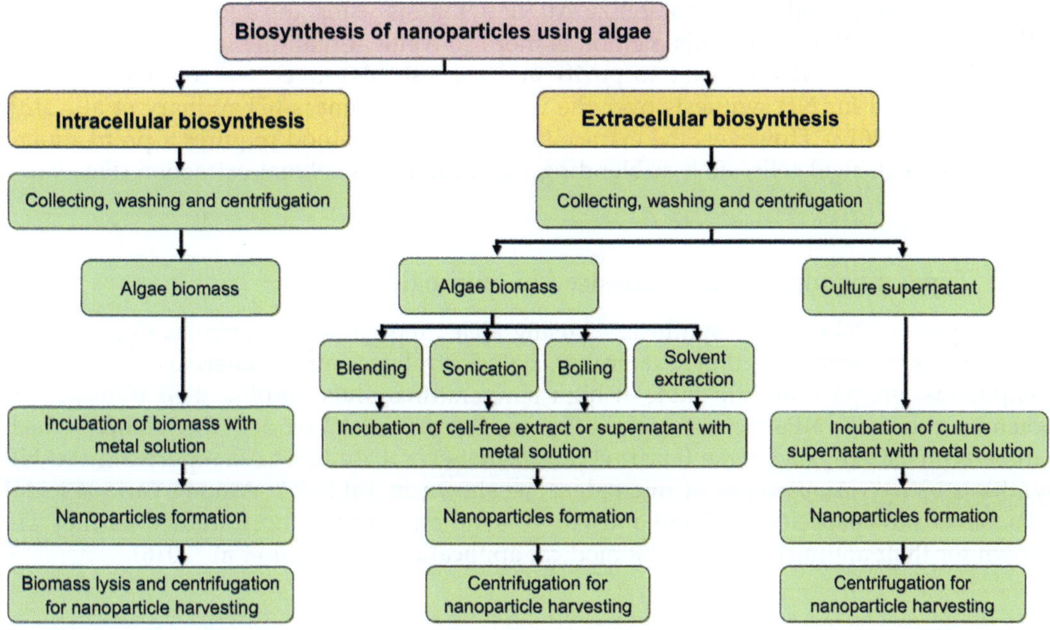

FIGURE 7.4 General methods for nanoparticles synthesis through intracellular and extracellular synthesis.

7.3.1.1 Intracellular synthesis

This process takes place inside the algal cell. It does not require the pretreatment for breaking or disrupting of the microalgal cell, such as blending, sonication, and boiling because processes related to metabolic pathways are responsible for syntheses, such as photosynthesis, respiration, and nitrogen fixation (Khanna et al., 2019; Sharma et al., 2016). The positively-charged metal ions that carry out the intracellular synthesis of NPs trapped on the surface of the cell wall and subsequently transported into the microbial cell containing negatively charged enzymes or proteins. The metal ions are reduced inside the cell to form small nuclei and later form different morphological NPs, then release the stabilized metal NPs into the culture medium (Patil & Kim, 2018; Shankar et al., 2016; Sharma et al., 2016). The study by Senapati et al. (2012) reported that the intracellular synthesis of gold nanoparticles (AuNPs) in the algae *Tetraselmis kochinensis* was formed by reducing metal ions at the surface of the algae cell and cytoplasmic membrane.

7.3.1.2 Extracellular synthesis

Extracellular synthesis refers to processes that occur outside the algal cell. Here, metal ions attached to the surface of algal cells and their metabolites are reduced by reducing agents such as polysaccharides, peptides, various proteins, enzymes, reducing sugars, pigments, and other reducing factors. Subsequently, reduced metal ions are stabilized and capped in NPs (Chaudhary et al., 2020; Khanna et al., 2019; Sharma et al., 2016). Singh et al. (2019) demonstrated that the functional groups in flavonoids, polyphenols, tannins, and proteins might reduce and stabilize AuNP synthesis using the aqueous extract of *D. salina*. The extracellular synthesis method is more convenient than the intracellular synthesis method because NPs are easy to purify and harvest. Moreover, the cell-free extract is highly exploited for NP synthesis over the use of whole biomass (Chaudhary et al., 2020; Shankar et al., 2016). However, the extracellular synthesis method requires a pretreatment process to break algal cells, such as blending, sonication, and solvent extraction (Fig. 7.4).

7.3.2 Type of nanoparticle synthesized by microalgae

Two types of NPs can be synthesized using algae: inorganic and organic. Organic NPs (chitosan, poly-ε-lysine, cation quaternary polyelectrolytes, and quaternary ammonium compounds) are generally unstable at high temperatures, thus limiting their translational potential. Inorganic NPs include metal (Ag, Au, Pd) and metal oxides (CuO, ZnO), which can be synthesized using algae (Uzair et al., 2020). Several studies have reported metal NP synthesis using various types of microalgae, as shown in Table 7.2. Among various metal NPs, silver nanoparticles (AgNPs) and AuNP synthesis are the most studied as they are known for their antimicrobial and biomedical applications (Shankar et al., 2016).

7.3.2.1 Silver (Ag) nanoparticles

It has been reported that more than 20 species of green microalgae are efficient in AgNP synthesis. Almost all green microalgae synthesize AgNPs using the extracellular mode (Chaudhary et al., 2020). Chokshi et al. (2016) studied the synthesis of AgNPs using the extract of dried de-oil microalga *Acutodesmus dimorphus*. The algae extract was incubated with 1 mM $AgNO_3$ solution for 24 hours at room temperature. The formation of AgNPs was detected by visual observation, where the colorless reaction solution changed to brownish. The amide linkage, functional groups, and proteins may be related to reducing silver ions to synthesize AgNPs. The analysis showed that the AgNPs produced are spherical in the range of 2–20 nm. Bao and Lan (2018) reported that that chlorophylls were the essential compounds mediating the reduction of Ag^+ to Ag^0 for AgNP synthesis using the sonicated extract of *Neochloris oleoabundans*. In addition, the results suggested that proteins may serve as stabilizers in the synthesis of AgNPs. In another study, ethanol extracts of *Chlorella* sp., *Lyngbya putealis*, and *Scenedesmus vacuolatus* were used to reduce silver metal ions into an Ag/AgCl nanohybrid material. These results revealed an average Ag/AgCl particle size of 90.6, 136.2, and 241.8 nm synthesized by *Chlorella* sp., *S. vacuolatus*, and *L. putealis*, respectively. *Chlorella* sp. was selected to test the effect on the concentration of $AgNO_3$ solution (0.1–1 mM). The results concluded that 0.5 mM of $AgNO_3$ is the optimal concentration for Ag/AgCl synthesis. The formation of Ag/AgCl at a low concentration of 0.1 mM $AgNO_3$ was found to be unstable, while at a high concentration of 1 mM $AgNO_3$, Ag/AgCl impurity was present (Kashyap et al., 2019). From these studies, it can be concluded that proteins and chlorophylls could act as reducing agents. The species of algae affect the size of AgNPs, and the concentration of metal ions affects the formation and purity of AgNPs.

TABLE 7.2 Synthesis of nanoparicles using microalgae.

Nanoparticles							
Type	Shape	Size (nm)	Algae species	Reactive algal-materials	Metal ions solution	Applications	References
Ag	Spherical	2–20	*Acutodesmus dimorphus*	De-oiled algae extract	1 mM AgNO$_3$	Antioxidant	Chokshi et al. (2016)
Ag	n.d.	n.d.	*Neochloris oleoabundans*	Algae extract	1 mM AgNO$_3$	n.d.	Bao and Lan (2018)
Ag	Spherical	15.67 ± 1	*Botryococcus braunii*	Cell-free culture liquid	1 mM AgNO$_3$	Antibacterial	Patel et al. (2015)
Ag	Spherical	19.28 ± 1	*Coelastrum* sp. 143–1	Cell-free culture liquid	1 mM AgNO$_3$	Antibacterial	Patel et al. (2015)
Ag/AgCl	Spherical	10–20	*Chlorella* sp.	Algae extract	0.5 mM AgNO$_3$	Antibacterial	Kashyap et al. (2019)
Au	Spherical	22.4	*Dunaliella salina*	Algae extract	1 mM HAuCl$_4$ · xH$_2$O	Anticancer	Singh et al. (2019)
Au	Spherical	14.65 ± 3.95	*C. vulgaris*	Powdered algal solution	0.2 g HAuCl$_4$/g-algae	Antioxidant and catalytic	Zayadi and Abu Bakar (2020)
Au	Spherical	50–60	*D. salina*	Algal suspension	1.5 mM HAuCl$_4$ · 3H$_2$O	Antibacterial	Basiratnia et al. (2021)
Fe	Spherical	20–50	*Chlorococcum* sp. MM11	Biomass	0.1 M FeCl$_3$	Bioremediation	Subramaniyam et al. (2015)
Pd	Spherical	May-20	*C. vulgaris*	Algae extract	1 mM PdCl$_2$	n.d.	Arsiya et al. (2017)
ZnO	Hexagonal	20	*Chlorella* sp.	Algae extract	80 mL Zn(CH$_3$COO)$_2$ · 2H$_2$O/ 20 mL algal extract	Bioremediation	Khalafi et al. (2019)

n.d.: no data.

7.3.2.2 Gold (Au) nanoparticles

Several microalgae have been reported to be efficient in the synthesis of AuNPs. Zayadi and Abu Bakar (2020) used a simple method for AuNP synthesis using *C. vulgaris*. The powdered algal solution was incubated with $HAuCl_4$ (0.2 g $HAuCl_4$/g-algae) at room temperature for 3 hours, and the successful synthesis of AuNPs was visually confirmed by a change of the mixing solution from light yellow to wine red. The spherical shape with the particle size of 14.65 ± 3.95 nm of AuNPs was detected. The researchers concluded that only using gold salt ($HAuCl_4$), distilled water, and natural resources (algae) was enough to produce biostabilized colloidal AuNPs. Singh et al. (2019) investigated the synthesis of AuNPs using an aqueous extract of *D. salina*. The effect of sunlight exposure (0–90 min), aqueous extract of algae (5%–40%), and $HAuCl_4 \cdot xH_2O$ concentration (0.5–3 mM) were investigated. The results showed that the formation of AuNPs was observed by the change of reaction mixture from yellow to dark purple. The optimal conditions for sunlight exposure, aqueous extract of algae, and $HAuCl_4 \cdot xH_2O$ concentrations were 75 minutes, 30%, and 1 mM, respectively. Exposure to sunlight for longer than 75 minutes did not yield any corresponding increase in the synthesis of AuNPs. The size of AuNPs depends on the amount of bioreductant. An increase in the amount of aqueous extract of algal, a bioreluctant, results in an enhancement in AuNP synthesis and lead to the synthesis of the small size of AuNPs. However, an increase in inoculum beyond 30% resulted in the overgrowth of AuNPs that settled in the bottle. The results on the varying $HAuCl_4 \cdot xH_2O$ concentration revealed that increasing the concentration from 0.5 to 1 mM enhanced the synthesis of AuNPs. In comparison, a continuous increase in the concentration to 3 mM led to the overgrowth and aggregation of the AuNPs. The concentration of gold metal ion solutions for AgNP synthesis was also studied by Basiratnia et al. (2021). During biosynthesis, the concentration of $HAuCl_4 \cdot 3H_2O$ was varied between 0.5 and 2 mM. The results showed that a concentration of 1.5 mM $HAuCl_4 \cdot 3H_2O$ is sufficient to synthesize AuNPs. Moreover, these results also indicate that pH affects the formation of these NPs. At alkaline pH, gold ions are hydrolyzed to form stable species of hydroxides, preventing the entry of other ions. Furthermore, AuNP formation did not occur at a low pH (2–4). Therefore AuNP synthesis was optimal at pH 6. From these studies, it can be summarized that the formation of AuNPs depends on the amount of bioreductant, pH, and gold ion concentration.

7.3.3 Applications of nanoparticles from microalgal biomass

Recently, algae-mediated NP synthesis has attracted attention owing to its broad application in antimicrobial, anticancer, and bioremediation functions. AgNPs have been proven to be effective antimicrobial agents even at low concentrations. They can also inhibit the growth of antibiotic-resistant bacteria (Jena et al., 2014). The characteristics of AgNPs, such as size and shape, and electrical and magnetic properties, influence their functions, such as sanitizing properties, insecticides, detergents, disinfectants, and pesticides. The antimicrobial activity involves the AgNPs adhering to the surface of the bacterial cell membrane, affecting its permeability and respiratory functions. The interaction of AgNPs with water or tissue fluid led to the release of silver ions which penetrate the cell membrane and interact with electron-donor functional groups (phosphates, thiols, and indoles). These electron-donor functional groups contain sulfur and phosphorus, which have a high affinity for silver ions. This interaction causes the inactivation of proteins and loss of membrane permeability, resulting in cell death (Terra et al., 2019). Kashyap et al. (2019) studied the antimicrobial activity of Ag/AgCl synthesized by *Chlorella* sp. using different concentrations of $AgNO_3$ (0.1−1 mM). The synthesized Ag/AgCl was tested on *Bacillus subtilis*, *Bacillus sphaericus*, *Bacillus* pasteurii, and *Escherichia coli* (DH5-α). The results indicated that a zone of microbial inhibition was found with different bacterial strains, proving that Ag/AgCl has an antimicrobial function. In addition, the concentration of 0.5 mM $AgNO_3$ was sufficient for antibacterial effect.

AuNPs have been used in various biomedical applications, such as biosensors, targeted drug delivery, and anticancer activity. AuNPs synthesized from halotolerant microalgae, *D. salina*, were tested for their anticancer activity in a cancer cell line and kill cancer cells. Therefore AuNPs synthesized by microalgae may be potential candidates for anticancer drugs (Singh et al., 2019). In addition, AuNPs synthesized from *C. vulgaris* were tested for their antioxidant activity using 2,2-diphenyl-1-picrylhydrazyl (DPPH) and hydroxyl radical assays. The AuNPs possessed DPPH and hydroxyl scavenging activities of 42.06% and 35.39%, respectively. The antioxidants might be adsorbed onto the surface of AuNPs during the biosynthesis and scavenging of hydroxyl radicals. Bioactive compounds, including carotenoids, proteins, and phenolic compounds, may serve as microalgae antioxidants (Zayadi & Abu Bakar, 2020).

Iron NPs are the most studied and used nanomaterials for remediation of chromium and other contaminants because of their high reducing properties, high surface area, and small size. Iron NPs synthesized from the soil microalga *Chlorococcum* sp. MM11 were tested for their efficiency in reducing Cr(VI) in wastewater containing Cr(VI) of 4 mg L^{-1}. The results revealed that 92% of Cr(VI) was reduced. In addition, the study concluded that the application of 2 μg L^{-1} of iron NPs is sufficient to treat 4 mg L^{-1} of Cr(VI) (Subramaniyam et al., 2015).

7.4 Challenges and perspective

Algae are fast-growing microorganisms found in freshwater, saltwater, and wastewater and play an essential role in creating a sustainable environment for living beings. Algal functions, such as photosynthesis, nitrogen fixation, phosphorus solubilization, and the production of biochemical substances (EPS and phytohormones), are essential for agricultural sustainability. Using wastewater-enriched nutrients as a source for crop cultivation via algae is an appealing approach to reducing wastewater, minimizing the use of chemical fertilizers, and protecting against the adverse effects of chemical fertilizer runoff. One of the most critical tasks is improving nutrient assimilation and controlling the potential risk of algal biofertilizer leakage to the surrounding area and water bodies to minimize environmental problems. In terms of sustainability, combining algal biomass for biofuels and agricultural biofertilizers could be a promising way to overcome commercialization barriers. In addition, the ability of algae to synthesize metal NPs suggests that it will be helpful in the future in medical applications, phytoremediation of toxic or metal substances, and as an antibiofilm agent against multidrug-resistant bacteria. However, different algal species have different effects on the formation and size of NPs, which may be due to differences in biomaterials in their algal cells, which act as reducing agents and stabilizing agents during the biosynthesis step. As a result, future research should concentrate on other microalgae species because selecting suitable and efficient algal species has a significant impact on the formation of NPs.

Furthermore, research on scaling up to an industrial scale is critical. Expanding studies on these applications are also very important, especially for biomedical applications. The investigation of safety and its possible application in humans still needs to be explored, especially when AuNPs are used in medical applications such as anticancer drugs, drug delivery, and antioxidant function. Since NPs have a minimal size, it is tough to recover the NPs from products and the environment. Therefore the behavior of NPs in the environment must be investigated to prevent the buildup of the NPs waste in the ecosystem.

Acknowledgments

The authors would like to acknowledge financial support from the Research and Graduate Studies Division, Khon Kaen University, and a TRF Senior Research Scholar (Grant No. RTA6280001). We are very grateful to Dr. Chonticha Mamimin for her graphical design.

References

Aziz, M.A., Hashem, M.A., 2003. Role of cyanobacteria in improving fertility of saline soil. Pakistan Journal of Biological Sciences 6.

Abdul Razack, S., Duraiarasan, S., Mani, V., 2016. Biosynthesis of silver nanoparticle and its application in cell wall disruption to release carbohydrate and lipid from *C. vulgaris* for biofuel production, Biotechnology Reports, 11. Elsevier B.V, India, pp. 70–76. Available from: http://doi.org/10.1016/j.btre.2016.07.001.

Ananthi, V., et al., 2018. Comparison of integrated sustainable biodiesel and antibacterial nano silver production by microalgal and yeast isolates. Journal of Photochemistry and Photobiology B: Biology 186, 232–242. Available from: https://doi.org/10.1016/j.jphotobiol.2018.07.021. India: Elsevier B.V.

Anthony, K.J.P., Murugan, M., Gurunathan, S., 2014. Biosynthesis of silver nanoparticles from the culture supernatant of Bacillus marisflavi and their potential antibacterial activity. Journal of Industrial and Engineering Chemistry 20 (4), 1505–1510. Available from: https://doi.org/10.1016/j.jiec.2013.07.039. India: Korean Society of Industrial Engineering Chemistry.

Arsiya, F., Sayadi, M.H., Sobhani, S., 2017. Green synthesis of palladium nanoparticles using *Chlorella vulgaris*. Materials Letters 186, 113–115. Available from: https://doi.org/10.1016/j.matlet.2016.09.101. Iran: Elsevier B.V.

Asmathunisha, N., Kathiresan, K., 2013. A review on biosynthesis of nanoparticles by marine organisms. Colloids and Surfaces B: Biointerfaces 103, 283–287. Available from: https://doi.org/10.1016/j.colsurfb.2012.10.030. India.

Babiak, W., Krzemińska, I., 2021. Extracellular polymeric substances (EPS) as microalgal bioproducts: a review of factors affecting EPS synthesis and application in flocculation processes. Energies 14 (13), 4007. Available from: https://doi.org/10.3390/en14134007. MDPI AG.

Bao, Z., Lan, C.Q., 2018. Mechanism of light-dependent biosynthesis of silver nanoparticles mediated by cell extract of *Neochloris oleoabundans*. Colloids and Surfaces B: Biointerfaces 170, 251–257. Available from: https://doi.org/10.1016/j.colsurfb.2018.06.001. Canada: Elsevier B.V.

Basiratnia, E., et al., 2021. Biological synthesis of gold nanoparticles from suspensions of green microalga *Dunaliella salina* and their antibacterial potential. BioNanoScience . Available from: https://doi.org/10.1007/s12668-021-00897-4Iran: Springer.

Beevers, L., Hageman, R.H., 1969. Nitrate reduction in higher plants. Annual Review of Plant Physiology 495–522. Available from: https://doi.org/10.1146/annurev.pp.20.060169.002431. Annual Reviews.

Bocchi, S., Malgioglio, A., 2010. Azolla-Anabaena as a biofertilizer for rice paddy fields in the Po Valley, a temperate rice area in Northern Italy. International Journal of Agronomy 1–5. Available from: https://doi.org/10.1155/2010/152158. Hindawi Limited.

Chatterjee, A., et al., 2017. Role of algae as a biofertilizer. In Algal Green Chemistry: Recent Progress in Biotechnology. Elsevier, India, pp. 189–200. Available from: http://doi.org/10.1016/B978-0-444-63784-0.00010-2.

Chaudhary, R., et al., 2020. An overview of the algae-mediated biosynthesis of nanoparticles and their biomedical applications. Biomolecules 1498. Available from: https://doi.org/10.3390/biom10111498. MDPI AG.

Chokshi, K., et al., 2016. Green synthesis, characterization and antioxidant potential of silver nanoparticles biosynthesized from de-oiled biomass of thermotolerant oleaginous microalgae: *Acutodesmus dimorphus*. RSC Advances 6 (76), 72269–72274. Available from: https://doi.org/10.1039/c6ra15322d. India: Royal Society of Chemistry.

Coppens, J., et al., 2016. The use of microalgae as a high-value organic slow-release fertilizer results in tomatoes with increased carotenoid and sugar levels. Journal of Applied Phycology 28 (4), 2367–2377. Available from: https://doi.org/10.1007/s10811-015-0775-2. Belgium: Springer Netherlands.

Costa, O.Y.A., Raaijmakers, J.M., Kuramae, E.E., 2018. Microbial extracellular polymeric substances: ecological function and impact on soil aggregation. Frontiers in Microbiology . Available from: https://doi.org/10.3389/fmicb.2018.01636Frontiers Media SA.

D'Amato, R., et al., 2013. Synthesis of ceramic nanoparticles by laser pyrolysis: from research to applications. Journal of Analytical and Applied Pyrolysis 104, 461–469. Available from: https://doi.org/10.1016/j.jaap.2013.05.026. Italy: Elsevier B.V.

Das, S.K., Varma, A., 2010. Role of enzymes in maintaining soil health. Springer Science and Business Media LLC, pp. 25–42. Available from: http://doi.org/10.1007/978-3-642-14225-3_2.

De La Monte, S.M., et al., 2009. Epidemilogical trends strongly suggest exposures as etiologic agents in the pathogenesis of sporadic alzheimer's disease, diabetes mellitus, and non-alcoholic steatohepatitis. Journal of Alzheimer's Disease 17 (3), 519–529. Available from: https://doi.org/10.3233/JAD-2009-1070. United States: IOS Press.

Deepika, P., MubarakAli, D., 2020. Production and assessment of microalgal liquid fertilizer for the enhanced growth of four crop plants. Biocatalysis and Agricultural Biotechnology 28, 101701. Available from: https://doi.org/10.1016/j.bcab.2020.101701. Elsevier BV.

Dineshkumar, R., et al., 2019. The impact of using microalgae as biofertilizer in maize (*Zea mays* L.). Waste and Biomass Valorization 10 (5), 1101–1110. Available from: https://doi.org/10.1007/s12649-017-0123-7. India: Springer Netherlands.

Du, K., et al., 2015. Enhanced growth and lipid production of *Chlorella pyrenoldosa* by plant growth regulator GA3. Fresenius Environmental Bulletin 24 (10B), 3414–3419 China: Parlar Scientific Publications. Available from: http://www.psp-parlar.de/feb_auswahl.asp.

Duran, S.K., Kumar, P., Sandhu, S.S., 2021. A review on microalgae strains, cultivation, harvesting, biodiesel conversion and engine implementation. Biofuels 12 (1), 91–102. Available from: https://doi.org/10.1080/17597269.2018.1457314. India: Taylor and Francis Ltd.

Falkowska, M., et al., 2011. The effect of gibberellic acid (GA3) on growth, metal biosorption and metabolism of the green algae *Chlorella vulgaris* (chlorophyceae) beijerinck exposed to cadmium and lead stress. Polish Journal of Environmental Studies 20 (1), 53–59. Poland.

Fritsch, F.E., Oliver, F.W., 1907. A general consideration of the subaerial and fresh water algal flora of ceylon. a contribution to the study of tropical algal ecology. Part I. subaerial algoe and algoe of the Inland fresh waters, Proceedings of the Royal Society of London.

Ginni, G., Adish, K.S., Mohamed, U.T.M., Pakonyi, P., Banu, J.R., 2020. Integrated biorefineries of food waste. Food waste to valuable resources. In: Banu, J.R., Kumar, G., Gunasekaran, M., Kavitha, S. (Eds.). Academic Press, pp. 275–298.

Gitau, M.M., et al., 2021. Strain-specific biostimulant effects of *Chlorella* and *Chlamydomonas* green microalgae on medicago truncatula. Plants 10 (6). Available from: https://doi.org/10.3390/plants10061060. Hungary: MDPI AG.

Gómez-Jacinto, V., et al., 2015. Elucidation of the defence mechanism in microalgae *Chlorella sorokiniana* under mercury exposure. Identification of Hg-phytochelatins. Chemico-Biological Interactions 238, 82–90. Available from: https://doi.org/10.1016/j.cbi.2015.06.013. Spain: Elsevier Ireland Ltd.

Gonçalves, A.L., 2021. The use of microalgae and cyanobacteria in the improvement of agricultural practices: a review on their biofertilising, biostimulating and biopesticide roles. Applied Sciences 11 (2), 871. Available from: https://doi.org/10.3390/app11020871. MDPI AG.

González-Garcinuño, Á., et al., 2016. Understanding and optimizing the addition of phytohormones in the culture of microalgae for lipid production. Biotechnology Progress 32 (5), 1203–1211. Available from: https://doi.org/10.1002/btpr.2312. Spain: John Wiley and Sons Inc.

Grzesik, M., Romanowska-Duda, Z., Kalaji, H.M., 2017. Effectiveness of cyanobacteria and green algae in enhancing the photosynthetic performance and growth of willow (*Salix viminalis* L.) plants under limited synthetic fertilizers application. Photosynthetica 55 (3), 510–521. Available from: https://doi.org/10.1007/s11099-017-0716-1. Poland: Kluwer Academic Publishers.

Guo, S., et al., 2020. Microalgae as biofertilizer in modern agriculture. Microalgae biotechnology for food, health and high value products. Springer Singapore, China, pp. 397–411. Available from: http://doi.org/10.1007/978-981-15-0169-2_12.

Hamouda, R., 2013. Effect of some red marine algae as biofertilizers on growth of maize (Zea mayz L.) plants. International Food Research Journal 20.

Hanrahan, G., Chan, G., 2005. Nitrogen. Encyclopedia of analytical science. In: Worsfold, P., Townshend, A., Poole, C. (Eds.). Elsevier, Oxford, pp. 191–196.

Hashtroudi, et al., 2012. Endogenous auxin in plant growth-promoting cyanobacteria-*Anabaena vaginicola* and *Nostoc calcicola*. Journal of Applied Phycology 25.

Hassan, S.M., et al., 2021. The potential of a new commercial seaweed extract in stimulating morpho-agronomic and bioactive properties of *Eruca vesicaria* (L.) cav. Sustainability (Switzerland) 13 (8). Available from: https://doi.org/10.3390/su13084485. Egypt: MDPI AG.

Hu, C., et al., 2003. Extracellular carbohydrate polymers from five desert soil algae with different cohesion in the stabilization of fine sand grain. Carbohydrate Polymers. China 54 (1), 33–42. Available from: https://doi.org/10.1016/S0144-8617(03)00135-8.

Hwang, S.H., et al., 2020. Size and flux of carbon nanoparticles synthesized by Ar + CH4 multi-hollow plasma chemical vapor deposition. Diamond and Related Materials 109. Available from: https://doi.org/10.1016/j.diamond.2020.108050. Japan: Elsevier Ltd.

Igiri, B.E., et al., 2018. Toxicity and bioremediation of heavy metals contaminated ecosystem from tannery wastewater: a review. Journal of Toxicology 2018. Available from: https://doi.org/10.1155/2018/2568038. Nigeria: Hindawi Limited.

Issa, A., Abd-Alla, M., Ohyama, T., 2014. Nitrogen fixing cyanobacteria: Future prospect. IntechOpen, pp. 23–48.

Jacob, J.M., et al., 2021. Microalgae: A prospective low cost green alternative for nanoparticle synthesis. Current Opinion in Environmental Science and Health. Elsevier B.V, India, p. 20. Available from: http://doi.org/10.1016/j.coesh.2019.12.005.

Jena, J., et al., 2014. Microalga *Scenedesmus* sp.: a potential low-cost green machine for silver nanoparticle synthesis. Journal of Microbiology and Biotechnology 24 (4), 522–533. Available from: https://doi.org/10.4014/jmb.1306.06014. India: Korean Society for Microbiolog and Biotechnology.

Jiang, H., et al., 2020. Synthesis and catalytic performance of highly dispersed platinum nanoparticles supported on alumina via supercritical fluid deposition. The Journal of Supercritical Fluids 166, 105014. Available from: https://doi.org/10.1016/j.supflu.2020.105014. Elsevier BV.

Kakakhel, M.A., et al., 2021. Green synthesis of silver nanoparticles and their shortcomings, animal blood a potential source for silver nanoparticles: A review. Journal of Hazardous Materials Advances 1, 100005. Available from: https://doi.org/10.1016/j.hazadv.2021.100005. Elsevier BV.

Kalayu, G., 2019. Phosphate solubilizing microorganisms: promising approach as biofertilizers. International Journal of Agronomy 2019, 1–7. Available from: https://doi.org/10.1155/2019/4917256. Hindawi Limited.

Kashyap, M., et al., 2019. Screening of microalgae for biosynthesis and optimization of Ag/AgCl nano hybrids having antibacterial effect. RSC Advances 9 (44), 25583–25591. Available from: https://doi.org/10.1039/c9ra04451e. India: Royal Society of Chemistry.

Khalafi, T., Buazar, F., Ghanemi, K., 2019. Phycosynthesis and enhanced photocatalytic activity of zinc oxide nanoparticles toward organosulfur pollutants. Scientific Reports 9 (1). Available from: https://doi.org/10.1038/s41598-019-43368-3. Iran: Nature Publishing Group.

Khan, I., et al., 2021a. Synthesis, characterization and magnetic properties of ε-Fe2O3 nanoparticles prepared by sol-gel method. Journal of Magnetism and Magnetic Materials 538, 168264. Available from: https://doi.org/10.1016/j.jmmm.2021.168264. Elsevier BV.

Khan, M.J., et al., 2021b. Insights into diatom microalgal farming for treatment of wastewater and pretreatment of algal cells by ultrasonication for value creation. Environmental Research 201. Available from: https://doi.org/10.1016/j.envres.2021.111550. India: Academic Press Inc.

Khanna, P., Kaur, A., Goyal, D., 2019. Algae-based metallic nanoparticles: Synthesis, characterization and applications. Journal of Microbiological Methods 163, 105656. Available from: https://doi.org/10.1016/j.mimet.2019.105656. Elsevier BV.

Lane, H. et al., 1975. Some observations on nitrate reduction in cotton, Proc Beltwide Cotton Prod Res Conf.

Leong, Y.K., Chang, J.-S., 2020. Bioremediation of heavy metals using microalgae: Recent advances and mechanisms. Bioresource Technology 303, 122886. Available from: https://doi.org/10.1016/j.biortech.2020.122886. Elsevier BV.

Liu, L., Pohnert, G., Wei, D., 2016. Extracellular metabolites from industrial microalgae and their biotechnological potential. Marine Drugs 14 (10). Available from: https://doi.org/10.3390/md14100191. China: MDPI AG.

Lokman, M.Q., et al., 2021. Synthesis of silver nanoparticles using chemical reduction techniques for Q-switcher at 1.5 μm region. Optik 244. Available from: https://doi.org/10.1016/j.ijleo.2021.167621. Malaysia: Elsevier GmbH.

Maksimova, I.V., Bratkovskaya, L.B., Plekhanov, S.E., 2004. Extracellular carbohydrates and polysaccharides of the alga *Chlorella pyrenoidosa* chick S-39. Biology Bulletin (2), 175–181. Available from: https://doi.org/10.1023/B:BIBU.0000022474.43555.ec. Springer Nature.

Mazhar, S., Hasnain, S., 2000. Screening of native plant growth promoting cyanobacteria and their impact on *Triticum aestivum* var, growth. African Journal of Agricultural Research 6.

McGeehan, S.L., Naylor, D.V., 1988. Automated instrumental analysis of carbon and nitrogen in plant and soil samples. Communications in Soil Science and Plant Analysis 19 (4), 493–505. Available from: https://doi.org/10.1080/00103628809367953. United States.

McHugh, D.J., 2003. A guide to the seaweed industry. Food and Agriculture Organization of the United Nations, Rome, Italy. Available from: https://www.fao.org/3/y4765e/y4765e00.htm.

Michalak, I., Chojnacka, K., 2015. Algae as production systems of bioactive compounds. Engineering in Life Sciences 15 (2), 160–176. Available from: https://doi.org/10.1002/elsc.201400191. Poland: Wiley-VCH Verlag.

Mishra, A., Jha, B., 2009. Isolation and characterization of extracellular polymeric substances from micro-algae *Dunaliella salina* under salt stress. Bioresource Technology. India 100 (13), 3382–3386. Available from: https://doi.org/10.1016/j.biortech.2009.02.006.

Mobley, H.L.T., Hausinger, R.P., 1989. Microbial ureases: significance, regulation, and molecular characterization. Microbiological Reviews. 53 (1), 85–108. Available from: https://doi.org/10.1128/mmbr.53.1.85-108.1989. United States.

Mohammadi, S., Harvey, A., Boodhoo, K.V.K., 2014. Synthesis of TiO$_2$ nanoparticles in a spinning disc reactor. Chemical Engineering Journal 258, 171–184. Available from: https://doi.org/10.1016/j.cej.2014.07.042. United Kingdom: Elsevier.

Moraes, J.G., Pereira, L., dos, S., Sampaio, V.L.F., 2021. Além do corpo escalpelado: o compromisso da psicologia diante da região amazônica. Atena Editora 177–188. Available from: https://doi.org/10.22533/at.ed.39421300317.

Mousavi, P., et al., 2016. Investigating the effects of phytohormones on growth and β-carotene production in a naturally isolates stain of *Dunaliella salina*. Journal of Applied Pharmaceutical Science 6 (8), 164–171. Available from: https://doi.org/10.7324/JAPS.2016.60826. Iran: Open Science Publishers LLP Inc.

Nayak, M., Swain, D.K., Sen, R., 2019. Strategic valorization of de-oiled microalgal biomass waste as biofertilizer for sustainable and improved agriculture of rice (*Oryza sativa* L.) crop. Science of the Total Environment 682, 475–484. Available from: https://doi.org/10.1016/j.scitotenv.2019.05.123. India: Elsevier B.V.

Nikravan, G., Haddadi-Asl, V., Salami-Kalajahi, M., 2018. Synthesis of dual temperature – and pH-responsive yolk-shell nanoparticles by conventional etching and new deswelling approaches: DOX release behavior. Colloids and Surfaces B: Biointerfaces 165, 1–8. Available from: https://doi.org/10.1016/j.colsurfb.2018.02.010. Iran: Elsevier B.V.

Oosterhuis, D.M., Bate, G.C., 1983. Nitrogen uptake of field-grown cotton. II. Nitrate reductase activity and petiole nitrate concentration as indicators of plant nitrogen status. Experimental Agriculture. Zimbabwe 19 (1), 103–109. Available from: https://doi.org/10.1017/S0014479700010565.

Osman, M.E.H., et al., 2010. Effect of two species of cyanobacteria as biofertilizers on some metabolic activities, growth, and yield of pea plant. Biology and Fertility of Soils. Egypt 46 (8), 861–875. Available from: https://doi.org/10.1007/s00374-010-0491-7.

Pan, X., et al., 2008. Effects of gibberellin A3 on growth and microcystin production in *Microcystis aeruginosa* (cyanophyta). Journal of Plant Physiology 165 (16), 1691–1697. Available from: https://doi.org/10.1016/j.jplph.2007.08.012.

Pandey, S.N., Verma, I., Kumar, M., 2020. Cyanobacteria: potential source of biofertilizer and synthesizer of metallic nanoparticles. Elsevier BV, pp. 351–367. Available from: http://doi.org/10.1016/b978-0-12-819311-2.00023-1.

Parsaeimehr, A., Lutzu, G.A., 2016. Algae as a novel source of antimicrobial compounds. Elsevier BV, pp. 377–396. Available from: http://doi.org/10.1016/b978-0-12-803642-6.00018-6.

Patel, V., et al., 2015. Screening of cyanobacteria and microalgae for their ability to synthesize silver nanoparticles with antibacterial activity. Biotechnology Reports 5 (1), 112–119. Available from: https://doi.org/10.1016/j.btre.2014.12.001. India: Elsevier B.V.

Patil, M.P., Kim, G.D., 2018. Marine microorganisms for synthesis of metallic nanoparticles and their biomedical applications. Colloids and Surfaces B: Biointerfaces 172, 487–495. Available from: https://doi.org/10.1016/j.colsurfb.2018.09.007. South Korea: Elsevier B.V.

Pereira, L., 2021. Macroalgae. Encyclopedia 1 (1), 177–188. Available from: https://doi.org/10.3390/encyclopedia1010017.

Pérez Corona, M.E., Van Der Klundert, I., Verhoeven, J.T.A., 1996. Availability of organic and inorganic phosphorus compounds as phosphorus sources for Carex species. New Phytologist 133 (2), 225–231. Available from: https://doi.org/10.1111/j.1469-8137.1996.tb01889.x. Wiley.

Piotrowska, A., Czerpak, R., 2009. Cellular response of light/dark-grown green alga *Chlorella vulgaris* Beijerinck (Chlorophyceae) to exogenous adenine- and phenylurea-type cytokinins. Acta Physiologiae Plantarum. Poland 31 (3), 573–585. Available from: https://doi.org/10.1007/s11738-008-0267-y.

Poudel, S., et al., 2018. Electron transfer to nitrogenase in different genomic and metabolic backgrounds. Journal of Bacteriology 200 (10). Available from: https://doi.org/10.1128/JB.00757-17. United States: American Society for Microbiology.

Powell, N., et al., 2011. Phosphate release from waste stabilisation pond sludge: Significance and fate of polyphosphate. Water Science and Technology 63 (8), 1689–1694. Available from: https://doi.org/10.2166/wst.2011.336. New Zealand.

Rahayu, S.T., Rosliani, R., and Prathama, M., 2021. Enhancing bush beans quality by applying brown algae (*Ascophyllum* sp) organic fertilizer, IOP Conference Series: Earth and Environmental Science. IOP Publishing, 724(1), p. 012001. doi: 10.1088/1755-1315/724/1/012001.

Ramya, S., Nagaraj, S., Narayanan, V., 2010. Biofertilizing efficiency of brown and green algae on growth, biochemical and yield parameters of *Cyamopsis tetragonolaba* (L.) Taub. Rec Res Sci Tech 2, 45–52.

Ramya, S.S., Vijayanand, N., Rathinavel, S., 2015. Foliar application of liquid biofertilizer of brown alga *Stoechospermum marginatum* on growth, biochemical and yield of *Solanum melongena*. International Journal of Recycling of Organic Waste in Agriculture 4 (3), 167–173. Available from: https://doi.org/10.1007/s40093-015-0096-0. India: Springer Berlin Heidelberg.

Rashid, T.M., et al., 2021. Synthesis and characterization of Au:ZnO (core:shell) nanoparticles via laser ablation. Optik 244. Available from: https://doi.org/10.1016/j.ijleo.2021.167569. Iraq: Elsevier GmbH.

Renuka, N., et al., 2017. Evaluating the potential of cytokinins for biomass and lipid enhancement in microalga *Acutodesmus obliquus* under nitrogen stress. Energy Conversion and Management 140, 14–23. Available from: https://doi.org/10.1016/j.enconman.2017.02.065. South Africa: Elsevier Ltd.

Rodriguez-Felix, F., et al., 2021. Sustainable-green synthesis of silver nanoparticles using safflower (*Carthamus tinctorius* L.) waste extract and its antibacterial activity. Heliyon 7 (4).

Ronga, D., et al., 2019. Microalgal biostimulants and biofertilisers in crop productions. Agronomy 9 (4),. Available from: https://doi.org/10.3390/agronomy9040192. Italy: MDPI AG.

Roy, M., Mohanty, K., 2021. Valorization of de-oiled microalgal biomass as a carbon-based heterogeneous catalyst for a sustainable biodiesel production. Bioresource Technology 337, 125424. Available from: https://doi.org/10.1016/j.biortech.2021.125424. Elsevier BV.

Salakkam, A., et al., 2021. Valorization of microalgal biomass for biohydrogen generation: a review. Bioresource Technology 322, 124533. Available from: https://doi.org/10.1016/j.biortech.2020.124533. Elsevier BV.

Salehi, B., et al., 2019. Current trends on seaweeds: looking at chemical composition, phytopharmacology, and cosmetic applications. Molecules (Basel, Switzerland) 24 (22). Available from: https://doi.org/10.3390/molecules24224182. Iran: MDPI AG.

Saravanan, A., et al., 2021. A review on biosynthesis of metal nanoparticles and its environmental applications. Chemosphere 264, 128580. Available from: https://doi.org/10.1016/j.chemosphere.2020.128580. Elsevier BV.

Senapati, S., et al., 2012. Intracellular synthesis of gold nanoparticles using alga *Tetraselmis kochinensis*. Materials Letters 79, 116–118. Available from: https://doi.org/10.1016/j.matlet.2012.04.009. India.

Shankar, P.D., et al., 2016. A review on the biosynthesis of metallic nanoparticles (gold and silver) using bio-components of microalgae: formation mechanism and applications. Enzyme and Microbial Technology 95, 28–44. Available from: https://doi.org/10.1016/j.enzmictec.2016.10.015. India: Elsevier Inc.

Sharma, H.S.S., et al., 2014. Plant biostimulants: a review on the processing of macroalgae and use of extracts for crop management to reduce abiotic and biotic stresses. Journal of Applied Phycology 26 (1), 465–490. Available from: https://doi.org/10.1007/s10811-013-0101-9. United Kingdom: Kluwer Academic Publishers.

Sharma, A., et al., 2016. Algae as crucial organisms in advancing nanotechnology: A systematic review. Journal of Applied Phycology 28 (3), 1759–1774. Available from: https://doi.org/10.1007/s10811-015-0715-1. India: Springer Netherlands.

Sharma, G.K., et al., 2021. Circular economy fertilization: Phycoremediated algal biomass as biofertilizers for sustainable crop production. Journal of Environmental Management 287. Available from: https://doi.org/10.1016/j.jenvman.2021.112295. India: Academic Press.

Singh, R.N., 1961. Role of blue-green algae in nitrogen economy of Indian agriculture. Indian Council of Agricultural Research, New Delhi.

Singh, N.K., Dhar, D.W., Tabassum, R., 2016. Role of cyanobacteria in crop protection, Proceedings of the National Academy of Sciences India Section B - Biological Sciences. India: Springer India, 86(1), pp. 1–8. doi: 10.1007/s40011-014-0445-1.

Singh, A.K., et al., 2019. Green synthesis of gold nanoparticles from *Dunaliella salina*, its characterization and in vitro anticancer activity on breast cancer cell line. Journal of Drug Delivery Science and Technology 51, 164–176. Available from: https://doi.org/10.1016/j.jddst.2019.02.023. India: Editions de Sante.

Solovchenko, A., et al., 2019. Phosphorus starvation and luxury uptake in green microalgae revisited. Algal Research 43, 101651. Available from: https://doi.org/10.1016/j.algal.2019.101651. Elsevier BV.

Solovchenko, A., Zaitsev, P., Zotov, V., 2021. Phosphorus biofertilizer from microalgae. Elsevier BV, pp. 57–68. Available from: http://doi.org/10.1016/b978-0-12-821667-5.00022-1.

Souza, J.M.C., Yokoya, N.S., 2016. Effects of cytokinins on physiological and biochemical responses of the agar-producing red alga *Gracilaria caudata* (*Gracilariales*, Rhodophyta). Journal of Applied Phycology 28 (6), 3491–3499. Available from: https://doi.org/10.1007/s10811-016-0885-5. Brazil: Springer Netherlands.

Subramaniyam, V., et al., 2015. *Chlorococcum* sp. MM11—a novel phyco-nanofactory for the synthesis of iron nanoparticles. Journal of Applied Phycology 27 (5), 1861–1869. Available from: https://doi.org/10.1007/s10811-014-0492-2. Australia: Kluwer Academic Publishers.

Sun, Y., Xia, Y., 2002. Shape-controlled synthesis of gold and silver nanoparticles. Science (New York, N.Y.) 298 (5601), 2176–2179. Available from: https://doi.org/10.1126/science.1077229. United States.

Suresh, A., et al., 2019. Evaluation and characterization of the plant growth promoting potentials of two heterocystous cyanobacteria for improving food grains growth, Biocatalysis and Agricultural Biotechnology, 17. Elsevier Ltd, India, pp. 647–652. Available from: http://doi.org/10.1016/j.bcab.2019.01.002.

Tanaka, S., et al., 2018. Synthesis of metastable cubic tungsten carbides by electrical explosion of tungsten wire in liquid paraffin. Advanced Powder Technology 29 (10), 2447–2455. Available from: https://doi.org/10.1016/j.apt.2018.06.025. Japan: Elsevier B.V.

Terra, A.L.M., et al., 2019. Microalgae biosynthesis of silver nanoparticles for application in the control of agricultural pathogens. Journal of Environmental Science and Health - Part B Pesticides, Food Contaminants, and Agricultural Wastes 54 (8), 709–716. Available from: https://doi.org/10.1080/03601234.2019.1631098. Brazil: Taylor and Francis Inc.

Uzair, B., et al., 2020. Green and cost-effective synthesis of metallic nanoparticles by algae: safe methods for translational medicine. Bioengineering 129. Available from: https://doi.org/10.3390/bioengineering7040129. MDPI AG.

Vaishampayan, A., et al., 2001. Cyanobacterial biofertilizers in rice agriculture. Botanical Review 67 (4), 453–516. Available from: https://doi.org/10.1007/BF02857893. India.

Vale, M.A., et al., 2020. CO$_2$ capture using microalgae. Elsevier BV, pp. 381–405. Available from: 10.1016/b978-0-12-819657-1.00017-7.

Vargas-Estrada, L., et al., 2020. Role of nanoparticles on microalgal cultivation: a review. Fuel 280. Available from: https://doi.org/10.1016/j.fuel.2020.118598. Mexico: Elsevier Ltd.

Verma, M., Kumar, V., Katoch, A., 2018. Sputtering based synthesis of CuO nanoparticles and their structural, thermal and optical studies. Materials Science in Semiconductor Processing 76, 55–60. Available from: https://doi.org/10.1016/j.mssp.2017.12.018. India: Elsevier Ltd.

Watanabe, I., Lee, K.-K., de Guzman, M., 1978. Seasonal change of N$_2$ fixing rate in rice field assayed by in situ acetylene reduction technique. Soil Science and Plant Nutrition 24 (4), 465–471. Available from: https://doi.org/10.1080/00380768.1978.10433126. Informa UK Limited.

Wurtsbaugh, W.A., Paerl, H.W., Dodds, W.K., 2019. Nutrients, eutrophication and harmful algal blooms along the freshwater to marine continuum. WIREs Water 6 (5). Available from: https://doi.org/10.1002/wat2.1373. Wiley.

Yaakob, M.A., et al., 2021. Influence of nitrogen and phosphorus on microalgal growth, biomass, lipid, and fatty acid production: an overview. Cells. India: NLM (Medline) 10 (2). Available from: https://doi.org/10.3390/cells10020393.

Yibeltie, G., Sahile, S., Box, P., 2018. The role of cyanobacteria on agriculture. Journal of Natural Sciences Research 8 (23).

Yilmaz, E., Sönmez, M., 2017. The role of organic/bio–fertilizer amendment on aggregate stability and organic carbon content in different aggregate scales. Soil and Tillage Research 168, 118–124. Available from: https://doi.org/10.1016/j.still.2017.01.003. Turkey: Elsevier B.V.

Zarezadeh, S., et al., 2020. Effects of cyanobacterial suspensions as bio-fertilizers on growth factors and the essential oil composition of chamomile, *Matricaria chamomilla* L. Journal of Applied Phycology 32 (2), 1231–1241. Available from: https://doi.org/10.1007/s10811-019-02028-9. Iran: Springer.

Zayadi, R.A., Abu Bakar, F., 2020. Comparative study on stability, antioxidant and catalytic activities of bio-stabilized colloidal gold nanoparticles using microalgae and cyanobacteria. Journal of Environmental Chemical Engineering 8 (4), 103843. Available from: https://doi.org/10.1016/j.jece.2020.103843. Elsevier BV.

Zhang, N., et al., 2010. Pedogenic carbonate and soil dehydrogenase activity in response to soil organic matter in *Artemisia ordosica* community. Pedosphere 20 (2), 229–235. Available from: https://doi.org/10.1016/S1002-0160(10)60010-0. China: Soil Science Society of China.

Zhang, Z., Wen, G., 2020. Synthesis and characterization of carbon-encapsulated magnetite, martensite and iron nanoparticles by high-energy ball milling method. Materials Characterization 167, 110502. Available from: https://doi.org/10.1016/j.matchar.2020.110502. Elsevier BV.

Life cycle assessment of wastewater treatment by microalgae

Christy B.K. Sangma and Rokozeno Chalie-u
ICAR Research Complex for NEH Region, Nagaland Centre, Dimapur, Nagaland, India

8.1 Background

Water is one of the important sources of livelihood and survival for every living being on earth. Water has never been scarce on the earth as 75% of the earth's surface is covered with water, consisting of 97.5% salt water and 2.5% fresh water. The fast-growing economy and the population explosion in recent years with the changing climatic conditions have limited the supply of fresh water with exhaustive depletion of water resources, reduced precipitation, and increased wastewater generation. The global wastewater production as per the estimates ranges from 358.0×10^9 to 450×10^9 m^3 year^{-1} (units is million-m^3 per year) which was quantified based on the reported data and modeling of return flow (in WaterGAP3—Water Global Analysis and Prognosis) and the number of urban population (Flörke et al., 2013; Qadir et al., 2020). Of the total wastewater collected the wastewater treatment ranges from 186.6×10^9 to 189.3×10^9 m^3 year^{-1} of the produced wastewater, and the rest goes back to the environment untreated (Jones et al., 2021). The wastewater treatment estimates are based on the standardized national level recording of merely 42 countries of point source and not considering the nonpoint source (point source: industries, households, services, and agriculture, and nonpoint source: unmonitored run-off from urban and agricultural land) and representing only 18% of the global population (70—80 L *approx.* of wastewater per capita and 29—64 L treated water per capita), as the standardized records are not available for all the countries (Habitat and WHO, 2021).

Every wastewater treatment generally follows two basic steps: (1) primary and (2) secondary processes, with several subunits having specific purposes. The primary treatment consists of physical techniques, namely, screening, sedimentation, filtration, flow equalization, and clarification. After the primary treatment, water enters the aeration tank in which nutrients and minerals are added and microbial action takes place in the presence of air (aerobic oxidation), causing degradation of organic matter. In secondary treatment the sludge (biosolids) and the water (liquid) are separated then move to the chlorination tank where disinfection, filtration and final disposal to the environment takes place. There are several types of standard wastewater treatment plants (WWTPs), and these are conventional activated sludge (CAS) system, CAS systems with filtration (CAS-F), immersed membrane bioreactor (MBR), including aerobic and anaerobic types, external MBR, upflow anaerobic sludge blanket react (UASB), sequencing batch reactor, and other land-based treatment technologies (Ortiz et al., 2007).

Besides the conventional wastewater treatment, algal ponds with microalgae biomass are also used to treat urban wastewater. These algal ponds are of two types, open and closed types. The open ponds are shallow, and paddlewheel and microalgae are used to absorb nutrients and produce oxygen, which is then utilized by heterotrophic bacteria to oxidize organic matter improving the water quality (Olguín et al., 2003; Cragg et al., 2014). Open pond includes stirred reactor (continuous stirred tank reactors, CSTR), raceway pond (mostly used for photoautotrophic microalgae cultivation), and inclined surface and the closed photobioreactor (PBR) consisting of tubular PBR, flat plate reactors, bag systems bioreactors, and annular reactor (Pulz, 2001; Posten, 2009). In closed-type PBR the environment and processes can be controlled based on the requirement of the type of microalgae adding up the capital investment, but there are several operational issues like overheating, fouling, and limitation on gas exchange, whereas the open-type ponds the operational cost is higher and are extremely affected by environmental factors like the temperature, humidity, invasion of bacteria, other algae species, etc. (Benemann, 2008). These methods of wastewater treatments compared to the conventional types are low cost, require less maintenance, are ecologically safer, and consume less energy but need high solar radiation for the growth of microalgae and larger area for the treatment plant (Molinos-Senante et al., 2014; Garfí et al., 2017). Another disadvantage of conventional wastewater treatment over microalgal treatment is that the incomplete utilization of natural resources leads to more secondary pollutions. The predominant microalgae cultures used for wastewater treatment and removal of nutrients are *Chlorella* sp. (Lee and Lee, 2001), *Scenedesmus* (Martínez et al., 2000), *Spirulina* (Olguín et al., 2003), *Nannochloris* (Jiménez-Pérez et al., 2004), *Botryococcus braunii* (An et al., 2003), and *Cyanobacteria phormidium* (Dumas et al., 1998). The microalgal biomass grown in these ponds can be harvested and reused for biofuel since it contains high energy value, also as a biofertilizer that is known to enhance plant growth and other nonfood bioproducts (Deviller et al., 2004; Coppens et al., 2016; Chew et al., 2017).

In recent times with global concern for environmental health and sustainability, wastewater treatment has become a mandatory act by government regulations. Wastewater treatments and the valorizations of waste streams have a considerable positive impact on the environment by reducing the sewage nutrient load in natural waters (Garcia et al., 2000; Rothermel et al., 2013). These environmental impacts are studied by employing the life cycle assessment (LCA) tools that are the standardized decision support system, for quantifying the different impact categories for providing a product or a service, which might have negative and positive impacts. LCA is also increasingly used as a support for manufacturing commercial products and their environmental performance. The use of LCA tools for wastewater treatment and their impact assessment is started very recently (Corominas et al., 2013, Zang et al., 2015). These LCA tools consider the impacts associated with the operation of the system for all types of WWTPs (both conventional and algal ponds) throughout its life cycle from cradle to grave and facilitate in choosing the best method of wastewater treatment out of many. The LCA studies follow ISO standards (International Organization for Standardization) with a baseline framework consisting of four phases, goal and scope, life cycle inventory analysis (LCI), life cycle impact assessment (LCIA), and interpretation of results (ISO 14040 International Standard, 2006; Ferreira et al., 2017).

8.2 Wastewater: composition, environmental impacts, and phases of wastewater treatment

8.2.1 Composition

The wastewater is previously utilized water from residents (domestic), farming exercises (agriculture), industries, surface overflow or storm water, any sewer inflow or sewer penetration, etc. The composition of wastewater may differ from place to place and from community to community. However, all kinds of wastewater typically consist of the following groups (Table 8.1) which made up their compositions (Hussain et al., 2002) and they are effluent organic matter; nutrients such nitrogen (N), phosphorus (P), and potassium (K); inorganic matter or dissolved minerals; toxic compounds [arsenic (Ar), cadmium (Cd), chromium (Cr), copper (Cu), lead (Pb), mercury (Hg), zinc (Zn), etc.]; and pathogens. The common pathogenic microorganisms in wastewater are viruses, bacteria, protozoa, etc. (Table 8.1).

8.2.2 Environmental impacts

The process of urbanization in the recent decades depicts an increased level of effluents discharge. The release of such raw and improperly treated wastewater into different water bodies or surroundings poses both short- and long-term threats to the environment, which is a global concern. The impact on the environment depends on the concentration and composition of wastewater, the frequency of wastewater released entering surface water sources (Akpor and Muchie, 2011).

The bacterial breakdown of organic solids present in wastewater and the oxidation of chemicals in it can consume much of the dissolved oxygen in the receiving water bodies (Borchardt and Statzner, 1990). A low level of dissolved oxygen affects the living organisms, especially fish by making them more susceptible to diseases, growth retardation, inability to swim, migration, and in extreme cases leads to death. Long-term effect includes changes in species composition (Welch, 1992; Chambers and Mills, 1996). As with any other organisms, aquatic animals also have a distinct ecological preference, tolerance to a certain temperature and pressure level. Ill-treated or raw wastewater released into the water bodies leads to physical changes, including temperature, because wastewater effluents are warmer (Welch, 1992). Another environmental impact of untreated wastewater is the phenomenon of bioaccumulation and biomagnification of contaminants. Owing to the phenomenon of bioaccumulation, certain substances that are barely measurable or are found in low concentrations in water can be found in high concentrations in the tissues of plants and animals. In other cases, through the process of biomagnification, the concentrations of some of the

TABLE 8.1   General composition and pathogens of wastewater and their concentration level.

Constituents	Concentrations (mg L^{-1})			Constituents	Concentrations (mg L^{-1})		
	Weak	Medium	Strong		Weak	Medium	Strong
Nitrogen	20	40	85	Fixed	20	55	75
Phosphorus	6	9	28	Volatile	80	165	275
Chloride	30	50	100	TDS	250	500	850
Total Organic Carbon	80	160	290	Fixed	145	300	525
Alkalinity (as CaCO$_3$)	50	100	200	Volatile	105	200	325
Sulfate	20	30	50	BOD5 at 20°C	205	200	250
TS	300	500	1100	COD	250	500	1000
SS	100	200	350	Grease	50	100	150

Pathogens present in wastewater

Pathogen	Species	Concentration per liter
Viruses	Enteroviruses	5000
	Adenoviruses	31.6
	Noroviruses	31.6
	Rotavirus	316.2
Bacteria	*Salmonella* sp.	1000–7000
	Shigella sp.	7000
	Vibrio cholerae	1000
	Escherichia coli	3,162,277
Protozoa	*Entamoeba histolytica*	4500
Helminths	*Ascaris lumbricoides*	600
	Hookworms	32
	Schistosoma mansoni	1
	Taenia saginata	10
	Trichuris trichiura	120

BOD, Biological oxygen demand; *COD*, chemical oxygen demand; *SS*, suspended solids; *TDS*, total dissolved; *TS*, total solids.

contaminants may be increased dramatically through a passage in the food chain that is prey to predators (Chambers and Mills, 1996). Because of the processes of bioaccumulation and biomagnification, very low concentrations of certain substances in wastewater are of concern. For instance, substances such as organochlorine pesticides and heavy metals can be detrimental even though they are present in low concentrations. In addition, eutrophication of water sources can lead to nutrient enrichment effects. Nutrient-induced production of aquatic plants in receiving water bodies has detrimental consequences as follows—(1) they interfere with recreational and esthetic water use through algal clumps, odors and decoloration of the water; (2) extensive growth of rooted aquatic life interferes with navigation, aeration, and channel capacity; (3) dead macrophytes and phytoplankton settled at the bottom of water body stimulate microbial breakdown, a process which requires oxygen, thus causing oxygen depletion; (4) extreme oxygen depletion can lead to the death of desirable aquatic life; (5) siliceous diatoms and filamentous algae may clog water treatment plant filters and result in reduced backwashing; and (6) algal blooms may shade and submerge aquatic vegetation, thus reducing or eliminating photosynthesis and productivity (Alm, 2003; Mbewele, 2006; McCasland et al., 2008). Though nutrients like nitrogen and phosphorus are beneficial to aquatic life in small amounts, they contribute to eutrophication when their content increases. Generally, the net effect of eutrophication on an ecosystem is usually an increase of a few plant types and a decline in the number and variety of other plant and animal species in the system (Environmental Canada, 1999).

Among all constituents nitrate leached from the wastewater poses one of the greatest threats to groundwater health (Bond, 1998). Groundwater that gets contaminated with nitrate can be treated or reduced by matching plant production systems to effluent characteristics. Besides the leaching of nitrate, groundwater contaminated with pharmaceutically active compounds and endocrine-disrupting chemicals, pathogens, salts, and nutrients also deteriorates the quality of groundwater and surface water (Stagnitti, 1999; Fent et al., 2006). Surface water contaminations by wastewater have led to extensive ecological degradation such as a decline in water quality and availability, intense flooding, loss of species, and changes in the distribution and structure of the aquatic biota (Oberdorff et al., 2002).

Wastewater contains many toxic compounds, which goes undetected due to the lack of analytical techniques and the increasing number of compounds that are being produced and discharged to sewers. Domestic wastewater normally has low contents of toxic organic compounds, but concentrations can increase if it receives industrial discharges, agricultural runoff (containing pesticides and their residues), leaks from storage tanks or pipes (that contain products such as fuels), leachates from polluted soils, confinement sites, and landfills and air pollutants deposited in rain. Among those compounds are industrial compounds (phthalates, biphenyl, p-nonylphenol, polychlorinated biphenyls (PCBs) and tributyltin), pesticides (atrazine, simazine, methoxychlor, 2, 4-D, DDT, dieldrin, endosulfan, and lindane), petroleum components, disinfection by-products or their precursor's hormones (from humans, such as 7-ethinylestradiol, or plants, such as 17-α-estradiol, estriol), and pharmaceuticals. These pollutants may have carcinogenic, teratogenic, and or mutagenic effects.

The impact of wastewater on the production of greenhouses can be subtle yet crucial enough not to be overlooked. It is reported that huge emissions of greenhouse gas (GHG) are found to be associated with the reuse of wastewater. This issue is not just exclusive to wastewater; some other promulgated solutions to water shortages, such as desalination, are considerably more energy-hungry than wastewater reuse (Cohen, 2006). It does nonetheless highlight that in the haste to reap environmental benefits through reusing wastewater for irrigation, detrimental impacts could easily be overlooked.

Wastewater irrigation to agricultural crops poses several risks to human health, due to the presence of pathogenic microorganisms (Fegen et al., 1998; Toze, 2006), organic chemicals such as endocrine-disrupting compounds, pharmaceutically active compounds (Moore and Chapman, 2003) and heavy metals (Chang et al., 1996; Bahri, 1998). Of all these, diseases caused by pathogenic microorganisms, bacteria, viruses, and protozoa, are the most common health hazards associated with wastewater (Environmental Canada, 1999; Chigor et al., 2013). The factors that contribute the most to causing many waterborne diseases are the pathogenic microbes. The presence of such in wastewater can cause chronic diseases which can have long-term effects, including degenerative heart disease and stomach ulcers. The numerous pathogenic microorganisms found in wastewater are bacteria species *Salmonella, Shigella*; and enteropathogenic *Escherichia coli*, protozoans—*Cryptosporidium parvum, Giardia intestinalis*, and *Entamoeba histolytica*; viruses—adenovirus, poliovirus, hepatitis-A virus, and rotavirus and parasitic worms—*Ascaris lumbricoides, Necator americanus*, and *Trichuris trichiura* (Toze, 2006; Yates and Gerba, 1997). Sieving through, viruses are among the most potentially hazardous pollutants in wastewater. They are generally more resistant to treatment, more infectious, more difficult to detect and require smaller doses to cause infections (Toze, 1997; Okoh, et al., 2007). Wastewater consists of vast quantities of bacteria, most of which are harmless to man. However, pathogenic forms that cause diseases, such as typhoid, dysentery, and other intestinal disorders may be present in wastewater (Table 8.2). The tests for total coliform and fecal coli from nonpathogenic bacteria are used to indicate the presence of pathogenic bacteria.

Humans excrete are said to contain more than 100 different types of enteric viruses capable of producing infection or disease to humans (Rose and Gerba, 1991; Absar, 2005). The presence and subsequent consumption of toxic algae or organisms that feed on them also lead to serious harm to humans and other terrestrial animals. The resulting toxins can

TABLE 8.2 Common diseases associated with pathogenic microbes in wastewater.

Microbes	Diseases	Causal organisms	References
Bacteria	Typhoid	*Salmonella* sp.	Yates and Gerba (1997); Metcalf and Eddy (2003)
	Dysentery	*Shigella* sp. and *Entamoeba histolytica*	
	Gastroenteritis	Enteropathogenic *Escherichia coli, Pseudomonas* strains	
	Skin and tissue infection	*Leptospirosis, Vibrio*	Absar (2005)
	Cholera	*Vibrio cholerae*	
Viruses	Diarrhea, vomiting	Adenovirus, rotavirus, *Cryptosporidium parvum, Giardia duodenalis*, and *Trichuris trichiura*	Yates and Gerba (1997)
	Anemia	*Necator americanus*	
Protozoa	Gastrointestinal, immune systems disorder	*C. parvum, Cyclospora*, and *Giardia lamblia*	Ingraham and Ingraham (1995)

cause gastroenteritis, liver damage, nervous system impairment, and skin irritation. In other cases, liver cancer in humans is thought to be associated with exposure to cyanobacterial toxins through the drinking water line and exposure to these toxins has usually been through contaminated drinking water or recreational water contact (WHO, 2006).

8.2.3 Phases of wastewater treatment

Considering the need for water for agriculture, industries, and domestic use as well, there has been a rising issue of water shortage throughout the world. In this regard, wastewater is considered a suitable alternative or substitute available to counteract water scarcity. However, it is a prerequisite step to know the composition of wastewater before reuse to avoid any kind of ill effects from its usage. Satisfactory recycling of wastewater is essential to ensure enough care for the public health, its environment, and water resources (Fahad et al., 2019). Wastewater treatment is done in two simple types—first, toxic or unwanted substances are removed from the water by cleaning, iron and manganese removal, sterilization, desalination, or softening; secondly, substances are specifically supplemented to improve the quality and influence parameters. These steps can be broken down into primary or mechanical phase, secondary or biological stage, tertiary, or chemical phase and nanofiltration (NF) (Mostafa, 2015; Samer, 2015). Before the commencement of these four stages, it is imperative to conduct a pretreatment phase; this stage consists of materials removed from the wastewater which would have otherwise damaged or clogged the pipes. The treatment processes in WWTPs are divided into the following stages.

8.2.3.1 Primary/mechanical wastewater treatment phase

The objective here is to separate the organic matter and sludge from the rest of the water and remove 20%–30% of the contained solids. This is carried out by guiding the wastewater into a screening plant, where sieve drums filter out coarse impurities step by step, after which the screened debris is dewatered and disposed of in an incineration plant. A sedimentation tank is used to remove the coarse particles (stones, sands, or glass, etc.). In this process a distinction is made between the aerated and nonaerated long sand collector and the round sand collector. The aerated sand collector removes fats and oils from the wastewater; the round sand collector separates substances from the wastewater with centrifugal force and sucks them away. The next stage is the primary wastewater treatment tank, where a low flow velocity is required such that finer particles get settled at the bottom of the water or on the surface. The sludge produced by sedimentation is called primary sludge, consisting of organic material. The primary sludge is pushed from the bottom into a fresh sludge hopper by a scraper. Then a pump transports the fresh sludge to what is known as a digestion tower, where the process of digestion takes about 4 weeks. In the digestion tower, methane gas is produced and converted into electricity in a block heating plant and can be used to supply the plant with energy; the remaining product is an odorless sludge. During the physical or mechanical treatment, biological oxygen demand (BOD) is reduced by about 5%–50%, 50%–70% of the total suspended solids (SS) and 65% of the oil and grease and facilitates subsequent treatment processes (Kesari et al., 2011; Jayalekshmi et al., 2021). On average, 30%–40% of the pollutants are removed by the first stage (Rivas et al., 2010).

8.2.3.2 *Secondary/biological wastewater treatment phase*

The mechanically cleaned wastewater proceeds to second-phase treatment in a circulation tank. In-tank the conditions are created for microorganisms to break down the organic substances present in the water. The bacteria from activated sludge flocks float freely in the water and therefore this process is also referred to as the activated sludge process. The activated sludge settles at the bottom of the purified water, where it can be separated from the clear water by mechanical clearing devices at the bottom and part of it is transferred to the digestion tower as additional biomass. The other part of the sludge, also known as "return sludge," is returned to the aeration tank. This process accounts for approximately 90% of cleaned wastewater. In a study by Abou-Elela et al. (2010), an activated sludge reactor with single *Staphylococcus xylosus* inoculum was used to treat pickled-vegetable plant sewage which concluded with nearly 90% chemical oxygen demand (COD) removal efficiency. Once the water has reached the legally prescribed quality, it can be returned to the water cycle.

The use of suspended biological methods for bioremediation of wastewater and resources recovery is an eco-friendly and cost-efficient way. Different types of contaminants in sewage sludge would be removed simultaneously by mixed microbial consortia. In the past few years, biofilm technology has gained much attention and has been widely applied to the breakdown of pollutants from waste sludge (Miranda et al., 2017). Wastewater treatment technology based on microbial biofilms is also considered an efficient and inexpensive alternative. Two different aerobic biofilm reactors, moving bed biofilm reactors and aerobic submerged fixed bed reactors, have found a niche in the treatment of municipal and agricultural wastewaters due to their promising advantages such as stability, reusability, low maintenance, and increased reaction rates (Ghimire and Wang, 2019). The varying species in biofilms, such as bacteria, fungi, algae, and yeast, form micro colony clusters enclosed within a self-produced glue-like extracellular polymeric substances (EPS) matrix (Dang and Lovell, 2016; Machineni et al., 2017). The biofilm can remove pollutant metals such as copper, lead, and zinc when metal ions are bound to the EPS matrix (Teitzel and Parsek, 2003).

8.2.3.3 *Tertiary/chemical wastewater phase*

The purpose of this stage is to further improve the quality of the water before it is safely released and reused. This treatment involves the use of chemicals to filter water and removes any inorganic substances (namely, N, P, toxic or heavy metals), viruses, bacteria, etc. The process includes neutralization, disinfection, flocculation, and precipitation. Neutralization is used to produce the desired pH value by adding an acid, for example, hydrochloric acid (HCl), or a base (e.g., milk of lime). In the process of disinfection, pathogens are killed by adding either chlorine or chlorine dioxide. The chemical precipitation or flocculation process is associated with the removal of phosphates. The phosphate precipitation is partly triggered by the addition of aluminum or iron salts in the sand collector or the secondary wastewater treatment tank. The metal-phosphate flocs that are formed during this secondary clarification are then taken out of the wastewater together with the activated sludge.

The chemical wastewater treatment includes processes such as chemical precipitation, ozonation, adsorption, photocatalytic oxidation, ion exchange, hydrodynamic cavitation, and ozone oxidation (Mahvi et al., 2008; Rafati et al., 2010; Chou et al., 2011; Musmarra et al., 2016;

Oppong et al., 2019; Yin et al., 2019). The chemical precipitation of heavy metals takes place by reaction with certain chemicals to form insoluble precipitates and the resultant precipitates are removed by sedimentation or filtration. The most commonly used chemical coagulants in wastewater treatment are alum [$Al_2(SO_4)_3H_2O$], ferric chloride ($FeCl_3 \cdot 6H_2O$), ferric sulfate [$Fe_2(SO_4)_3$], ferrous sulfate ($FeSO_4 \cdot 7H_2O$), and lime [$Ca(OH)_2$] (Sincero and Sincero, 2002; Chakinala et al., 2009). In recent years, advanced oxidation processes have emerged as a promising wastewater treatment technology for oxidation of various contaminants, including recalcitrant organic dyes, toxic chemicals derived from fixing agents, detergents, and salts. An ultrasound-enhanced electrochemical oxidation process was reported to be efficient for the thorough removal of dyes and demineralization of textile wastewater (Zhu et al., 2018).

8.2.3.4 Nanofiltration/membrane process wastewater phase

In the final wastewater treatment stage, membrane and filter processes are used. In NF, water is passed under pressure through a membrane that retains even the smallest dissolved particles. The pollutants retained during filtration, and NF is filtered into the sludge treatment in the form of filter sludge via the primary wastewater treatment tank. The water now reaches the last area of the WWTP, the treated water storage tank. Water samples are taken again here, and the water quality is checked. The purified water is only returned to the water cycle when the legally prescribed parameters have been met.

Membrane filtration is one of the most adopted modern wastewater treatments. It includes NF, ultrafiltration, reverse osmosis (RO), and electrodialysis.

8.2.4 Nanofiltration

NF is one of the widely used modern techniques for wastewater treatment and has successfully replaced RO membranes in many applications owing to its lower energy consumption and higher flux rates (Bowen and Mukhtar, 1996; Bowen and Welfoot, 2002). NF is a potential technique to eliminate heavy metals such as nickel, chromium, copper, and arsenic from wastewater (Semerjian and Ayoub, 2003; Maximous et al., 2010; Abhang et al., 2013). NF also does not require any heating or cooling of feed, for instance, distillation, or any mechanical stirring which ultimately reduces the cost of separation effectively.

8.2.5 Ultrafiltration

The use of *ultrafiltration* is a relatively recent technique that works at low trans membrane pressures for the removal of dissolved and colloidal material commonly used in many industrial applications such as food or pharmaceutical industries (Best et al., 2001). The membrane pore sizes are larger than dissolved metal ions which add to their advantage in easily passing through the membrane.

8.2.6 Reverse osmosis

The *RO* process uses a semipermeable membrane, allowing the liquid that is being purified to pass through it while rejecting the contaminants. RO is one of the techniques able to remove a wide range of dissolved species from water and accounts for more than 20% of the world's desalination capacity (Ben et al., 1993).

8.2.7 Biosorption

Biosorption is a physicochemical process defined as the ability of biological materials to accumulate heavy metals from wastewater through the metabolically mediated pathway of uptake. It is widely used in activated carbon filter, removes effluents containing toxic metals, removes contaminants even in dilute solution, can filter air and water by letting pollutants bind to their porous and high surface area structure. Conventional techniques like coagulation, electrocoagulation, electro-floatation, and electrodeposition have been used for the removal of heavy metals from the wastewater (Jayalekshmi et al., 2021). The process offers many advantages such as simple operation, no additional nutrient, low quantity of sludge, high efficiency, regeneration of biosorbent, and no increase in COD of water. The challenges faced are like those faced by membrane filtration technology, including the cost and stability of the biosorbent (membrane), the decrease in binding sites (fouling), and poor understanding and general reluctance to adopt new technologies (Kanamarlapudi et al., 2018).

8.2.8 Hydrogels

Hydrogels are the most recent tactic in handling wastewater. They have a three-dimensional network structure and the ability to incorporate different functional groups into polymeric networks, thereby upscaling their advantage as selected adsorbents to remove heavy metal ions from wastewater (Erdem et al., 2004; Eloussaief et al., 2009, 2011; Peng et al., 2012).

Of late *microalgae* are becoming a potential alternative in wastewater treatment (Fig. 8.1). A total decrease in the levels of COD (86%), total nitrogen (93%), and total phosphorus (83%) was observed after using algae in the municipal wastewater consortium. It is also capable of eliminating toxic substances such as selenium, zinc, and arsenic from the aquatic environment by accumulating those substances.

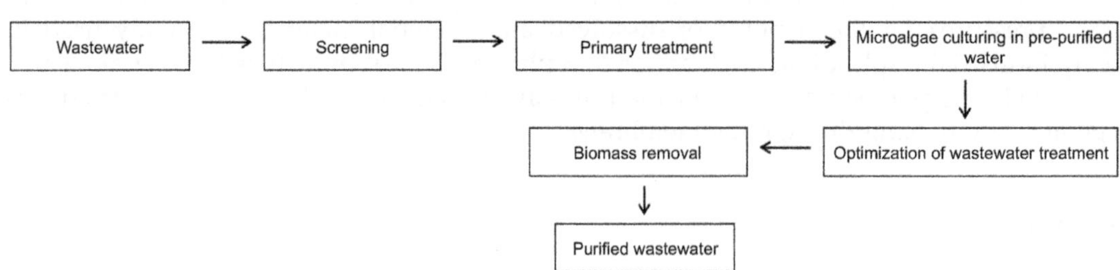

FIGURE 8.1   Microalgae wastewater treatment flowchart.

8.3 Algal features and their potentiality in wastewater treatment

8.3.1 Macro versus microalgae

Algae can be classified into two categories: macroalgae or seaweeds (multicellular) and microalgae. Macroalgae can be classified into three broad groups, based on their pigmentation: brown seaweed (Phaeophyceae), red seaweed (Rhodophyceae), and green seaweed (Chlorophyceae) (Chen et al., 2015). Microalgae are generally unicellular including both eukaryotic microalgae and prokaryotic cyanobacteria, also called phytoplankton, microscopic beings with a size smaller than 0.4 mm in diameter. The important classes of microalgae in terms of abundance are the diatoms (Bacillariophyceae), the green algae (Chlorophyceae), and the golden algae (Chrysophyceae). Their mechanism of survival in both types of algae is photosynthetic like terrestrial plants and aquatic in nature. They require CO_2, nutrients, and sunlight to convert solar energy to biomass. They have very high growth and reproduction rate (with microalgae doubling their biomass in a period of 3.5 hour during the exponential growth phase), high biomass production rate biomass production ($2\,g\,L^{-1}\,day^{-1}$ microalgae dry biomass), biodegradable and can grow in wastewaters (Chisti, 2007; Spolaore et al., 2006; Yuan et al., 2012) and productivity per unit of land is several thousand times more than the terrestrial crops. Their efficiency of photosynthesis is higher than that of terrestrial plants and ranges from 3% to 8% photosynthetic efficiencies, compared with 0.5% for many terrestrial plants. Macroalgae in comparison with microalgae have higher biomass densities, higher amount of carbohydrates and protein but do not produce significant quantities of lipid (<5% of dry weight) as compared to microalgae, where the lipid concentration (50%−70%) can be increased by optimizing the growth conditions. Microalgae are low in lignin content, and high number of proteins and oil and other compounds, which make them suitable for the synthesis of high-value products.

8.3.2 Potential of microalgae in wastewater treatment and production of high-value products

For the last 10 years, microalgae have been gaining importance in terms of wastewater treatment (otherwise called "phycoremediation") and the production of high-value products (Table 8.3). Microalgae especially *Chlorella* sp., *Scenedesmus* sp., *Cosmarium* sp. etc. have the important feature which is their ability to survive in brackish, saline, or wastewater loaded with high nutrient content and scavenge them as the source of nutrients and carbon for their growth, and fix CO_2 and evolve oxygen through photosynthesis (Min et al., 2011; Yuan et al., 2012). For profuse growth, microalgae require all kinds of nutrients, namely, carbon (C), nitrogen (N), phosphorous (P) which are macronutrients, and a variety of minerals essential micro-nutrients such as potassium (K), calcium (Ca), magnesium (Mg), iron (Fe), copper (Cu), and manganese (Mn), and other metals for their growth and metabolism. Many of these nutrients are present in the form of pollutants in wastewaters. So, microalgae are one of the best options for sustainable WWTPs with twofold or threefold benefit, for example, wastewater treatment + biofuel (biodiesel, bioethanol, biooil, biogas, etc.), production unit + production of biofertilizers from the sludge obtained

TABLE 8.3 Microalgae species, type of wastewater treatment, oil content in microalgae, and type of product generated from microalgae.

Microalgae spp.	Type of wastewater	Treatment	Oil content (% of dry wt)	Types of product
Chlorella sp.	Hog waste slurry	Removal rate: BOD, COD, SS, TN and TP: 91–95 + ; 60%–100% NH4; 78%–96% PO4	5–63	Biomethane
Combination of *Chlorella protothecoides, Scenedesmus obliquus,* and *Chlorella vulgaris*	Beef packaging wastewater treatment	91% COD, 67% TN, and 69% TP- PO_4^{3-}	5–58	–
Combination of *Chlorella* and *Stigeoclonium* sp.	Urban wastewater	Removal rate: NH4-N + : 93%–91%. COD: 62%–65%	*Chlorella* sp. 5–63	Biomethane
A mixture of *Chlamydomonas subcaudata* (92%), *Anabaena* sp. (7.5%), and *Nitzschia* sp. (0.5%)	Slaughterhouse wastewater	COD: 86%–92%; NH_4 -N^+: 79%–80%; soluble P: 71%–91%		Biomethane
Nitzschia sp.	Secondary effluent	2.2 mg L^{-1} N and 0.15 mg L^{-1} P	28–50	–
Botryococcus braunii	–	–	25–75	–
Chaetoceros calcitrans	–	–	15–40	–
S. obliquus (SO) and Chlorella zofingiensis (CZ)	Dairy treated wastewater	58,49% N and 23,94% P for SO and 97.5% N and 51.7% P for CZ	–	–
Dunaliella spp.	–	–	6–71	–
Spirulina with bacteria	Municipal wastewater-filtered	Total soluble N: 86%. NH4 + -N: 100%	4–17	Biocrude oil
Benthic pennate diatoms	Urban wastewater	N and P in biomass: 5.79% and 3.02%, respectively	–	Lipid
Haematococcus pluvialis	–	–	25	–
Chlorella pyrenoidosa with a bacteria strain belonged to *Klebsiella sp*	Municipal wastewater	The bacteria strain showed P accumulation at 9.48 L^{-1} mg^{-1} in 48 h	–	Lipid
Tetraselmis suecica	–	–	15–32	–
Thalassiosira pseudonana	–	–	21–31	–
Scenedesmus sp. ASK22	Urban wastewater and diary effluent	COD 90.50%, removal rate 292.85 mg d^{-1}; NO_3 100.00%, removal rate 13.56 mg d^{-1}; P 91.24%, removal rate 6.30 mg d^{-1}	2–55	Lipid

Isochrysis spp.	—	—	7–40	—
Desmodesmus sp., *Oscillatoria*, and *Arthrospira*	Facultative lagoon plant	83% N and 60% P	—	—
Monallanthus salina	—	—	20	—
Chlorococcum humicola, Selenastrum sp., *C. vulgaris*	Domestic wastewater	—	—	—
Nannochloris spp.	—	—	6–63	—
Parachlorella kessleri-1	Lake sewage	TN: 81%; TP: 98%; Mg: 84%; COD: 69%; BOD: 68%; TOC: 48%.	—	Lipid
Schizochytrium spp	—	—	50–77	—
Chlorella minutissima	Urban wastewater	Removal rates: N: 100%; P: 91%; C: 85%	—	Lipid
Nannochloropsis spp.	—	—	12–68	—
Oscillatoria sp., *C. vulgaris, Chlamydomonas* sp., *Scenedesmus chlorelloides*	—	Heavy metal removal	24–94	—
Neochloris oleoabundans	—	—	29–65	—
C. vulgaris, C. pyrenoidosa, Diplosphaera sp. MM1, *Scenedesmus* sp.	Industrial wastewater	—	—	—
Pavlova spp.	—	—	31–36	—
Scenedesmus obliquus	Piggery wastewater	—	—	—
Phaeodactylum spp.	—	—	20–57	—
C. vulgaris (UTEX 265), *Euglena gracilis* (SAG 1224), *Oocystis* sp., *Chlorella minutissima, Chlorella sorokiniana, Scenedesmus bijuga*	Agricultural wastewater	—	—	—

BOD, Biochemical Oxygen Demand; *COD*, Chemical Oxygen Demand; *SS*, Suspended Solids; *TP*, Total Phosphorus; *TN*, Total Nitrogen; *NH4*, Ammonium; *PO43+*, Phosphate; *N*, Nitrogen; *TOC*, Total organic carbon; *P*, Phosphorus; *M*, Magnesium. (Garrett and Allen, 1976; Arashiro et al., 2019; Hernandez, 2016; Zhou et al., 2013; Marella et al., 2019; Wang et al., 2019; Pandey et al., 2019; Singh et al., 2017; De Bhowmick et al., 2019; Boelee et al., 2011; Aslan and Kapdan, 2006; Zhang et al., 2014; Abou-Shanab et al., 2013; Huo et al., 2012; Prandini et al., 2016; Komolafe et al. 2014; (Abou-Shanab, 2013; Arashiro, 2019; Aslan and Kapdan, 2006; Boelee, 2011; De Bhowmick et al., 2019; Garrett and Allen, 1976; Hernández, 2016; Huo, 2012; Komolafe, 2014; Marella, 2019; Pandey et al., 2019; Prandini, 2016; Singh, 2017; Wang, 2019; Zhang, 2014; Zhou, 2013; Sakaguchi et al., 1981; Das et al., 2017; Riaño et al., 2011, Singh et al., 2011).

from the anaerobic digester which is a rich source of nutrients for agricultural reuse. Bioplastics from microalgae are also gaining popularity in the present time. The other advantage of microalgae is their ability to absorb the CO_2 gas for building their biomass and reducing the CO_2 gas emissions to the environment.

Microalgae are of different types, namely, photoautotrophic, heterotrophic, and mixotrophic in terms of carbon utilization. So, according to the complexity of the wastewater, the treatment requires the selected species of microalgae. The microalgae selected for the targeted treatment should have the criteria like fast growth, absorb many nutrients, utilize the low concentration of nutrients, easy to harvest, tolerant to the changing environment and resistant to other microorganisms' contamination, etc. (Wang et al., 2017). The different classified microalgae are as follows.

8.3.2.1 Photoautotrophic microalgae

Examples of phototrophic microalgae are *Phormidium* sp., *Oscillatoria* sp., *Stigeoclonium* sp. with biomass productivity of 122 ± 140 mg L^{-1} day^{-1}, and *Arthrospira* (Spirulina) with biomass productivity of 9.0 g m^{-2} day^{-1} (Dunn et al., 2013; Van Den Hende et al., 2012). In WWTPs for photoautotrophic microalgae, light is utilized as an energy source and CO_2 (inorganic carbon) is the main carbon supply (Salama et al., 2017). In this case, the bioreactors are placed vertically to provide and enhance the strength of light to microalgal anabolic reactions. The microalgal biomass productivity rate is more in vertical type than nonvertical bioreactor (Hu et al., 1996). Here the waste flue gas like CO_2 is inserted in the wastewater treatment process to improve the growth of microalgae as it acts as the source of inorganic carbon to them.

8.3.2.2 Heterotrophic microalgae

Organic carbon is the main source of carbon for heterotrophic microalgae and light is not required (Salama et al., 2017), so the wastewater treatments with this type of microalgae should be designed in such a way that the density of cells can be increase and more number can be accommodated for better results. The lipid production and recovery are reported to be higher in heterotrophic microalgae (39.5%) with glycerol than the autotrophic and mixotrophic types (10.5%) (Nzayisenga et al., 2018). Examples for heterotrophic microalgae are *Auxenochlorella protothecoides* UMN280 (biomass productivity of 1.12 g L^{-1}), *Phormidium autumnale* (630 g m^{-3} day^{-1}), *Gonium* sp. (0.53 g L^{-1}), etc. (Zhou et al., 2012; Rodrigues et al., 2014; Boduroğlu et al., 2014).

8.3.2.3 Mixotrophic microalgae

Mixotrophic types in wastewater treatment require a reduction in photoinhibition which can improve their growth rate (Salama et al., 2017). This type of microalgae requires both organic and inorganic carbon sources, as some grow autotrophically during daytime and heterotrophically at night or in insufficient sunlight due to high cell density. Glycerol is the better source of organic carbon and was found to enhance the growth rate of mixotrophs, compared to ethanol and acetate in the reactors (Sforza et al., 2010). Examples of mixotrophic microalgae are *Chlorella pyrenoidosa* (0.3 g L^{-1}), *Chlorella vulgaris* (0.251 g L^{-1}), etc. (Wang et al., 2012; Gao et al., 2014) (Table 8.3).

8.3.3 Factors affecting the microalgae growth

Various physical, chemical, and biological factors affect the cell growth of microalgae.

8.3.3.1 Physical factors

Factors like light and temperature are the abiotic or physical factors that are very critical for the growth of microalgae. As microalgae are mostly phototrophs, light conditions, both duration and intensity of light directly, affect the growth and photosynthesis. So, it is important to reach light for algae in mass culture production like raceway ponds and algal ponds. So, to remove the light limitations in algal cultures is to reduce the depth of the culture medium and keep the depth to between 15 and 50 cm, and during winter due to the lower light condition, keep the depth shallower, that is, lower than 20-cm depth (Fontes et al., 1987). Even though the microalgae require light, there are reports of photoinhibition due to the intense light. Temperature is also one of the crucial parameters for the growth of microalgae. Most microalgae species can tolerate temperatures between 16°C and 27°C. The high temperature increases the growth of algae up to a certain range, but after the critical temperature growth ceases.

8.3.3.2 Chemical factors

pH of the growth medium is also an important factor that affects the growth of microalgae. The highest growth rate of microalgae was observed when the medium is at a constant pH value of 7.0. The pH range for most cultured algal species is between 7 and 9, with the optimum range being 8.2−8.7. Salinity is another factor that affects the growth of microalgae. Marine microalgae are tolerant to the saline culture medium, but the best algae-growing condition for most species is at a salinity level that is slightly lower than that of their native habitat, which is obtained by diluting seawater with tap water. Salinities of $20-24$ g L^{-1} are optimum for the growth condition of most of the microalgae.

Carbon, nitrogen, and phosphorus are crucial components of microalgae nutrients. The growth of microalgae is more likely to be affected by carbon content rather than the other nutrients. In wastewaters, these nutrients occur in the form of CO_2 (carbon dioxide), NH_4 (ammonia), NO_2 (nitrite), NO_3 (nitrate), and PO_4 (orthophosphate). Microalgae utilized CO_2 as the inorganic carbon source during the fixation. Some species can utilize carbonates like Na_2CO_3 and $NaHCO_3$ for their cell growth (Wang et al., 2010), whereas the species like green algae *Chlorella* and *Scenedesmus* can shift their mode of carbon nutrition from autotrophy to heterotrophy depending on the carbon source (Becker, 1994). The element nitrogen mainly comes from sewage and 50% phosphorous comes from the detergents. The widely used microalgae cultures for nutrient removal are *Chlorella* (Lee and Lee, 2001), *Scenedesmus* (Martinez et al., 2000), and *Spirulina* (Olgun et al., 2003). *Scenedesmus* sp. is very common in all kinds of freshwater bodies, which play an important role in the purification of eutrophic waters (Mohamed, 1994).

The wastewaters suitable for algal growth are, namely, municipal wastewater, agricultural wastewater, eutrophic water bodies, streams, and effluents from the food processing industry, like breweries or dairies, etc. Industrial wastewater may be harmful which requires pretreatment to remove hazardous materials like wastewater from the medical sector, textile industry (chemicals in the effluent, low nutrient content), chemical, metallurgical, and mineral processing industry, etc. Some substances present in these wastewaters like heavy metals [cadmium (Cd), lead (Pb), mercury (Hg), zinc (Zn), copper (Cu) etc.] herbicides, pesticides, and chemicals in detergents can inhibit cell growth. A high concentration of ammonia that exceeds 20-mg NH_4-N at high pH and temperature is toxic for algal growth (Quiroz Arita et al., 2015).

8.3.3.3 Biological factors

The biological factors include the competition among the different species of microalgae, and the infections from viruses, protozoa, rotifers, and other pathogens like parasites, predators, etc. Some microalgae spp. also produce substances toxic to themselves during their metabolism, which eventually accumulates to high concentrations of toxic levels and inhibits their growth (autoinhibition).

8.3.3.4 Operational factors

Operational factors consisting of mixing of cells, dilution factor, depth of culture medium, the addition of bicarbonate, the distance between the algae and the light source, optimum pH adjustments and harvesting frequency, etc. affect the cell growth of microalgae.

8.4 Life cycle assessment

As such, wastewater treatment is not environment-friendly and economical process since a lot of energy and chemicals are involved, which require proper analysis on the benefits and damage it can cause to the environment. One such tool for carrying out such analysis is LCA, first applied in the field of wastewater treatment and reuse in the 1990s (Corominas et al., 2013), which is the compilation and evaluation of the outputs and their potential impacts on the environment throughout its life cycle based on the data obtained. The method can be widely used for environmental footprint studies for different products, technologies, and services. The LCA helps in validating the best method based on sustainability or its impacts on the environment. It works as a decision-making tool in environment management (Martínez-Blanco et al., 2016; Garfí et al., 2017). The LCA studies follow ISO 14040:2006 (International Organization for Standardization) baseline framework consisting of four phases, determining the goal and scope, LCI, LCIA, and interpretation (Fig. 8.2). The LCA considers the consumption of resources and energy and the generation of emissions and waste during the operation of treatment plants (Ferreira et al., 2017; ISO, 2006). Considerable studies have been carried out for LCA worldwide, covering municipal and industrial WWTPs (Lundin et al., 2000; Machado et al., 2007).

8.4.1 Goal and scope

The goal and scope of LCA must be predetermined based on the data quality and availability. The purpose of the goal and scope is to assess the environmental effects for the

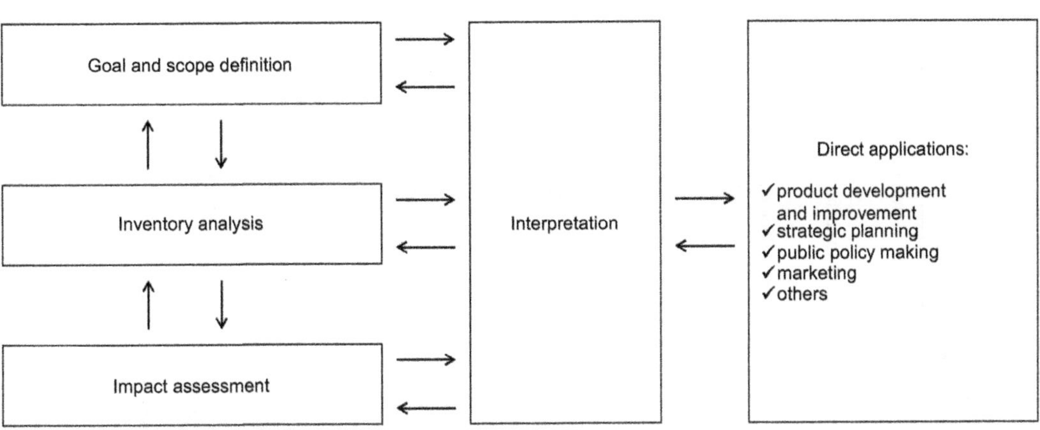

FIGURE 8.2   Four-phase framework of LCA (ISO 14040:2006). *LCA*, Life cycle assessment.

reuse of the treated water, for example, whether the water is to be used for irrigation purposes in agriculture or other purposes. The goal will include the *functional units* and the *system boundaries*. The functional unit is the water quality basic information, namely, COD, BOD, pH, and turbidity or the functional unit may adopt the use of people equivalent during analysis, that is, $1 \, m^3$ of potable water at the point of distribution or number of the population served. The system boundaries are time and spatial boundaries. Time boundary studies should define the input and output to be considered and should be as broad as possible and separate phases are considered the construction, operation (effluent quality assessment, energy requirements, etc.), maintenance, and demolition phases (end of life). In natural systems other than man-made structures, like wetlands, ponds, UASB systems, etc. are considered under the construction phase in the LCA analysis. The spatial boundaries include the operational treatment processes like pretreatment, primary treatment, secondary treatment, sludge collection, anaerobic digestion, thermal drying, etc.

8.4.1.1 Life cycle inventory analysis

The life cycle inventory analysis accumulates the data (background or foreground) or the database for analysis, using specific criteria or data quality matrices, in this phase (Corominas et al., 2013). This phase is the critical point in the LCA studies. In most of the LCI studies, Ecoinvent or Gabi is used as the database; Korean Ministry of Environment database (Piao et al., 2016), European life cycle reference database (Niero et al., 2014), LIPASTO database (Lehtoranta et al., 2014), etc. are the examples of commonly used databases (Table 8.4). As far as possible, the primary dataset should be considered during the LCI analysis since the wastewater quality may differ from time to time. As a general the following points should be considered in LCI studies: (1) database with version; (2) modifications made to the database used to fit the goals; (3) data from primary sources or estimated; (4) methodology for data estimations; (5) data quality; and (6) process identification in the database chosen. The LCI stage involves a four-step methodology, data collection, normalization, allocation, and data evaluation. But ISO 14040 standard

TABLE 8.4 Some databases supporting the life cycle assessment (LCA) and life cycle impact assessment (LCIA) tools and their uses (From: Lehtinen et al., 2011; Maekelae and Auvinen, 2009).

Tool name	Supplier	Supports LCI and LCIA	Supports full LCA	Main database	Special area
Ecoinvent waste disposal inventory tools v1.0	Doka Life Cycle Assessments (Doka Okobilanzen)	Yes	—	Ecoinvent database	Waste management
LIPASTO	VTT Technical Research Centre of Finland Ltd.				Unit emission database
The European reference Life Cycle Database (ECLD)	Institute for Environment and Sustainability (IES)	Yes			Energy carriers, transport, and waste management
BEDEC *banc* database	Construction Technology Institute of Catalonia				The constructive budgets to obtain the inventories for each WWTP construction and sewer system construction
GEMIS version 4.4	Oeko-Institut (Institute for Applied Ecology), Darmstadt office	Yes			Energy transport, recycling, and waste treatment
WRATE	UK Environment Agency		Yes		Municipal waste management systems
USES-LCA	Radboud University Nijmegen	Yes			Toxic impacts between substances
SALCA-tools	Agroscope Reckenholz-Tänikon Research Station ART	Yes			Agriculture
BEES 4.0	National Institute of Standards and Technology (NIST)		Yes	Bees database	Construction industry
LEGEP 1.2	LEGEP Software GmbH		Yes	LEGEP database	Construction industry
Eco Bat2.1	Haute Ecole d'Ingénierie et de Gestion du Canton de Vaud	Yes		Eco-Bat database	Construction industry

LCI, Life cycle inventory analysis; *WWTP*, wastewater treatment plant.
Lehtinen, et al., 2011. "A review of LCA methods and tools and their suitability for," SMEs. Biochemistry, 1; Maekelae, K., Auvinen, H., 2009. LIPASTO-Transport Emission Database: Report No. VTT-SYMP-262 VTT. Technical Research Centre of Finland, Espoo (Finland).

includes normalization in the LCIA phase, so the methodology depends on the standards used for the analysis.

8.4.1.2 Life cycle impact assessment

The main purpose of the impact assessment is to interpret the impact categories of inventory analysis. It requires *selection and definition* (goal) of impact categories, *classification* (grouping of impact categories based on their relevance) and *characterization* (category indicator results). In LCIA methodology studies, two approaches were evaluated: midpoint (gives the damage categories through modeling) and end-point analyses (the top-down approach where environmental burdens are measured—e.g., ecosystem quality, human health, and resources). These approaches have indicators or parameters and subcategories depending on the software used. Most of the studies used eutrophication as the midpoint category. Other categories (Table 8.5) include climate change or global warming, ozone layer depletion potential, acidification, and toxicity, odor, abiotic depletion elements and bacterial oxygen depletion, freshwater aquatic ecotoxicity potential, human toxicity potential, terrestrial ecotoxicity potential, and photochemical ozone creation potential, etc. (Bai et al., 2018; Cadena et al., 2018). In some studies, sensitivity (energy, transportation, materials use, resource recovery, main emissions, dosages used, or chemicals used in the process, e.g., sludge quantity, wastewater quality, treatment efficiency, and transport type) and uncertainty analysis can be incorporated at this stage. At the end of the analysis, a guideline is proposed, based on the problems identified.

LCIA is usually carried out with different types of software (Table 8.6), namely, SimaPro software, Gabi, OpenLCA, and Umberto. Other software are JEMAI-Pro 2.1.2 (Limphitakphong et al., 2016), Daycent (Miller-Robbie et al., 2017), Easetech (Fang et al., 2016), and Wastewater-Energy Sustainability Tool (Holloway et al., 2016), etc. The impact assessment transforms the mathematical data to environmental effect equivalent (significant pollution type or unit) via the factor multiplication.

8.4.1.3 Interpretation

Interpretation is the final assessment and the last phase of LCA. This phase consists of (1) identification and summarization of the significant issues based on the results of LCI and LCIA phases; (2) evaluation of sensitivity and consistency; and (3) conclusions, limitations, recommendations, and decision-making.

8.5 Life cycle assessment of wastewater treatment by microalgae

Sustainability of wastewater treatment is at the forefront of discussion, due to the population explosion, increasing pollution, resource efficiency, and conservation of nonrenewable resources, safety, and discharge regulations, climate change, etc. (Guest et al., 2009). The sustainability of WWTPs is measured based on the impact whether positive or negative, they generated on the environment. The wastewater treatment as such is not environment-friendly. They consume a lot of chemicals during the process of purification and produce secondary pollution due to the production of reagents and consequent energy consumption. The wastewater treatment with microalgae can significantly reduce

TABLE 8.5 Indicators for life cycle impact assessment and global normalization factors for emissions and resource extraction according to European Commission.

Indicator	Impact category	Subcategories	Unit	Reference value
Global warming potential	Global warming	Carbon dioxide (CO_2), nitrous oxides (N2O), methane (CH4), chlorofluorocarbons (CFCs)	kg CO_2-Eq (equivalent)	$5.79 \times 10^{+13}$
Ozone depleting potential	Ozone depletion	Chlorofluorocarbons (CFCs), hydrochlorofluorocarbons (HCFCs), halons, methyl bromide (CH3Br), or bromofluorocarbons	kg CFC-11 Eq	$1.61 \times 10^{+08}$
Acidification potential	Acidification general (terrestrial acidification included)	Sulfur oxides (SOx), nitrogen oxides (NOx), ammonia (NH4), hydrochloric acid (HCl), hydrogen (H +) ion	kg SO_2-Eq, Mole of H + Eq.	$3.83 \times 10^{+11}$
Eutrophication potential	Eutrophication (terrestrial, marine and freshwater)	Phosphate (PO_4), nitrogen dioxide (NO_2), nitrates, ammonia (NH4)	kg PO_4-Eq, kg N- Eq	$1.95 \times 10^{+11}$
Photochemical ozone creation potential	Photochemical smog	Nonmethane hydrocarbon (NMHC)	kg ethane (C_2H_6)-Eq, kg O3 (ozone) Eq	$2.80 \times 10^{+11}$
Human health	Human toxicity	Carcinogens, noncarcinogens, respiratory inorganics, ionizing radiation, ozone layer depletion, respiratory organics, heavy metals, particulate matter formation	kg 1,4-DCB (dichlorobenzene)-Eq, CTU (comparative toxic unit), PM10-Eq	—
Ecosystem quality	Freshwater aquatic ecotoxicity Marine aquatic ecotoxicity, terrestrial ecotoxicity	Heavy metals (lead (Pb), cadmium (Cd), chromium (Cr), zinc (Zn), copper (Cu), mercury (Hg) and nickel (Ni) LC50 (lethal concentration)	kg 1,4-DCB-Eq, CTU	$8.15 \times 10^{+13}$
Abiotic resource depletion potential	(mineral + fossil + renewable) Fossil depletion, metal depletion, water depletion	—	kg antimony (Sb)-Eq,kg Oil Eq, kg Fe (iron) Eq, m3	-
Energy resource	—	—	MJ	$4.50 \times 10^{+14}$
[a]Water footprint	Blue water, green water	—	m^3	$7.91 \times 10^{+13}$
Land use	—	—	ha	$9.64 \times 10^{+15}$

[a]Water footprint—the sum of water used during all stages of microalgal culturing and production; Blue water footprint—the amount of water incorporated and determined by evaporation rate; Green water footprint—the volume of water consumed during the process.
Jolliet et al. (2003); Itsubo and Inaba (2012); Goedkoop et al., 2013; Zang et al. (2015).

TABLE 8.6 Some of life cycle impact assessment (LCIA) models or tools.

S. no	LCA models	Impact category	References
1	SimaPro software (by the PréConsultants Company, a Dutch company) Versions: SimaPro 7.3.3 and SimaPro 8.0.3	Can calculate the environmental impacts; detect hot spots systematically and transparently. SimaPro give results in midpoint or end-point impact categories. Also, analyze only single impacts as carbon footprint or water use. SimaPro is used in industry, consultancies, universities and many research centers in different countries	PRé Consultants (2010)
2	CML (by the Institute of Environmental Science of the University of Leiden) Baseline versions: CML 2000 baseline and CML-IA baseline	CML 2000, Ecological scarcity (end-point): Environmental damages, CML-IA baseline 4.1: Eutrophication, human toxicity, freshwater aquatic ecotoxicity. GHG emissions	Guinée et al. (2001a,b), Barjoveanu (2013), O'Connor et al. (2014)
	Nonbaseline version: CML nonbaseline methods (CML 2001-midpoint and CML-IA nonbaseline)	Extended versions of the baseline methods contain the baseline categories plus alternative impact categories recommended for extended LCA studies. Impact categories: abiotic depletion, acidification, eutrophication, global warming, ozone layer depletion, human toxicity, freshwater aquatic ecotoxicity, marine aquatic ecotoxicity, terrestrial ecotoxicity, and photochemical oxidation	Guinée et al. (2001a,b), Guinee (2002); Acero et al. (2016)
3	ReCiPe (by RIVM, CML, PRé Consultants and Radboud Universiteit Nijmegen. (Most recommended method at present time). This method was created with the combination of two older methods CML and Eco-indicator 99	This method differentiates between 2 levels of indicators, the 18 midpoint categories, and the 3 end-point categories. It contains characterization factors for different substances, as well as factors for normalization from Europe and from all over the world. Human health, ecosystems, climate change, fossil depletion, freshwater eutrophication, human toxicity, particulate matter, terrestrial acidification, and resources (impact categories)	Goedkoop et al. (2009), Risch et al. (2014); Buonocore et al. (2018)
4	TRACI method [Tool for the reduction and assessment of chemical and other environmental impacts (TRACI) is a midpoint method]. Version: TRACI 2.1	Characterization of environmental stressors. Impacts such as global warming, acidification, carcinogens, noncarcinogens, respiratory effects, eutrophication, ozone depletion, ecotoxicity, and smog	Bare (2002); Amini et al. (2015)
5	GaBi is LCA modeling software by the German company Thinkstep. It includes its database and processes if primary data is not available	Ecoinvent and US LCI databases can be used with the software. In GaBi, processes inside the chosen system boundaries are modeled into a plan. Processes are connected with flows of energy and materials. After modeling the entire system with LCI into GaBi, the software calculates results for LCIA	Thinkstep (2019)

(Continued)

TABLE 8.6   (Continued)

S. no	LCA models	Impact category	References
6	OpenLCA (openLCA is open-source and free software for Sustainability and Life Cycle Assessment) Version: v2.1.2	Different impact categories include 43 different LCIA methods	https://www.openlca.org/
7	Excel-based Carbon Footprint Calculation Tool by Swedish Water Development (used in comparison with modeling results from GaBi)	Calculates carbon footprints for all functions contributing to emissions	Gustavsson and Tumlin (2013)
8	USES-LCA	Global warming, acidification, human toxicity, land use, cost	Hong et al. (2009)
9	Eco-indicator99	Environmental damages	Fuchs et al. (2011)
10	Impact 2002 +	Human health, ecosystem, climate change, resources	Thibodeau et al. (2014)
11	LIME	Freshwater use, global warming, acidification, eutrophication	Lam et al. (2015)
12	Umberto (developed by the IFU, Hamburg) The Umberto NXT family consists of four software solutions: Versions: Umberto NXT CO2 for calculating carbon footprints, Umberto NXT Efficiency for optimizing production processes from a cost perspective, Umberto NXT Universal for eco-efficiency analyses (consists of ecoinvent dataset for conducting the LCA study), and Umberto NXT LCA	The LCA modeling of this study was carried out in 5 phases: water collection, sludge activation, treatment, purification, and re-distribution. Indian electricity mix and diesel generators are utilized to incorporate the energy inputs of the study. The well-known ReCiPe method for both midpoint and end-point assessment was used. The midpoint assessment provides the result for various damage categories. The end-point category, e.g., human health is affected by ozone depletion, human toxicity, ionizing radiation, smog, particulate matter, and climate change which are measured in the midpoint categories. Similarly, the ecosystem at the end-point is affected by terrestrial ecotoxicity, terrestrial acidification, land occupation, marine and freshwater ecotoxicity and eutrophication, and climate change at the midpoint	Swiss Centre for Life Cycle Inventories, Eco-invent database v3.0, databases, 2015
13	JEMAI-LCA Pro is developed by JEMAI. Version: 2.1	Environmental impacts	Limphitakphong et al. (2016)

the pollution by using the nutrients in wastewater (through uptake by cells) as well as absorption of CO_2 present in the flue gases produced during the combustion processes (Casazza et al., 2016). The main advantage of using algae for wastewater treatment is the low cost of operation and the possibility of recycling assimilated nutrients into high-value products and avoiding a sludge handling problem, and the discharge of oxygenated effluent into the water body.

The wastewater treatment in algal ponds also employs LCA methodology to study the environmental impacts of the systems throughout the whole process chain, from raw material extraction, microalgal cultivation, waste disposal, and up to the discharge of purified water to the water bodies. The LCA methodology for wastewater treatments by microalgae also follows ISO standards (ISO, 2006) with four main stages: (1) goal and scope definition, (2) inventory analysis, (3) impacts assessment, (4) interpretation of the results.

8.5.1 Goal and scope

The goal is to evaluate the environmental impact of microalgal ponds for wastewater treatment and the recovery of resources. The goal can be determined based on different resource recovery scenarios, namely, (1) microalgal wastewater treatment with simultaneous energy recovery through biofuels production, (2) microalgal wastewater treatment with nutrient recovery through biofertilizer production. The goal can also be determined based on the type of cultivation systems used for microalgae, namely, (1) raceway ponds, (2) tubular PBR, (3) flat plate PBR, (4) fermenter, etc.

The functional unit for this study depends on the type of goal set for LCA. When the different microalgae cultivation systems are compared, the functional unit can be set at 1 m^3 of treated water [e.g., in high-rate algal pond (HRAP) system] or the functional unit of 1 kg of produced biomass dry weight in 22% DW slurry in the case when different species are used for the wastewater treatment (e.g., for *Nannochloropsis* sp.). The system boundaries or cradle-to-grave boundaries include construction, operation, and maintenance (Deprá et al., 2020) of the WWTP. The system boundary is usually considered for a certain period, for example, 20-year period. In some LCA studies the system boundaries were divided into the foreground and background processes (Garfí et al., 2017; Pérez-López et al., 2017; Rahman et al., 2016). The foreground systems were classified into four subsystems, (1) cleaning of the reactor, (2) preparation of the culture medium, (3) cultivation, (4) flocculation, (5) biomass concentration, and (6) drying. Input and output flow of materials (namely, construction materials and chemicals) and energy resources (heat and electricity), direct GHG emissions and NH_4^+ volatilization associated with wastewater treatment were also included in the boundaries. Treated water discharged, sludge disposal (i.e., incineration) into the environment, direct emissions to soil (heavy metals) in terms of biofertilizers and water used were also considered (Collet et al., 2011; ISO, 2006; Sfez et al., 2015).

8.5.1.1 Cleaning of reactor

This is carried out to ensure the complete removal of competing algae and protozoa or any pathogen. It also ensures the cleaning of previously cultured and encrusted microalgae biomass. For this, tap water (6 m^3) is pumped or air gun at a pressure of 4−6 bar is shot to a silo through the tubing and the biomass is collected at the end, without the chemical products addition. Then it is sterilized with sodium hypochlorite (NaClO) (2 mg L^{-1}) and passed through activated carbon filters to remove the hypochlorite. In the second rinse, 3% of a disinfection agent (containing hydrogen peroxide) is added, whereas in the third one, 0.5 g L^{-1} plastic beads for biofilm removal is used and a vacuum cleaning system for 1−2 hour after the last washing step to remove all water.

8.5.1.2 *Preparation of culture medium*

Based on the type of strain selected for the type of wastewater, the culture medium is prepared. Previously cultivated microalgae are taken as the resource used and the reactants are the composition of the culture medium and the dissolved organic salts (chemicals). Taking the example of *Nannochloropsis* sp., which requires natural seawater, is the main nutrient source in this case. To avoid contamination, the culture medium here is sterilized by adding hypochlorite (5 mg L^{-1}) and removing the chlorine with active carbon. Then, it was passed through a cascade filter (10 μm, 5 μm, and 1 μm) and supplied to the systems. Additionally, a culture medium with $NaNO_3$ as the main nitrogen source was supplemented to the reactor. The nitrate solution consisted of 212 g L^{-1} $NaNO_3$, 11.5 g L^{-1} KH_2PO_4, 3 g mL^{-1} Na_2 ethylenediaminetetraacetic acid (EDTA), 50 mL of trace mineral stock solution and 17.5 mL NaOH 4 M to adjust pH is used for *Nannochloropsis* sp., likewise, the culture medium changes based on the species of microalgae's requirement. For the tubular PBRs a dosage of 10 mL of nitrate solution per L of seawater was added in the final culture medium, whereas 2 mL L^{-1} were used in the medium of the open raceway pond (ORP). In some other studies, Chu liquid medium (Chu, 1974) was also used by modifying the composition as (g L^{-1}): $NaNO_3$ (0.25), $CaCl_2 \cdot 2H_2O$ (0.025), $MgSO_4 \cdot 7H_2O$ (0.075), K_2HPO_4 (0.075), KH_2PO_4 (0.175), NaCl (0.025), EDTA (0.05), KOH (0.031), $FeSO_4 \cdot 7H_2O$ (4.98 10^{-3}), H_3BO_3(11.42 10^{-3}), $ZnSO_4 \cdot 7H_2O$ (8.82 10^{-6}), $MnCl_2 \cdot 4H2O$ (1.44 10^{-6}), $NaMoO_4 \cdot 2H_2O$ (1.19 10^{-6}), $CuSO_4 \cdot 5H_2O$ (1.57 10^{-6}), $Co(NO_3)_2 \cdot 6H_2O$ (0.49 10^{-6}).

8.5.1.3 *Cultivation*

This stage consisted of a semicontinuous process in which the microalgae biomass was operated with a fixed daily dilution rate by adding the wastewater and harvesting (de Vree et al., 2015). The light requirement is provided based on the strains used in the culture medium. Some strains do not require any light source. The culture medium is maintained at the temperature of 20°C and 30°C with the use of heaters or chillers based on the requirement. The energy resource utilized in this stage is the use of electrical energy for circulation of the biomass to maintain temperature and the operation of the compressor for aeration. The possible emission in this stage is oxygen (O_2) produced during photosynthesis and other gases which are evolved during the biological activities of microalgae. To maintain purity and controlled contamination, the cultures were checked under a microscope several times per week (Bosma et al., 2014; Sandnes et al., 2005). The predominant microalgal species usually used are Chlorophyceae mainly *Chlorella* sp., *Scenedesmus* sp., and Cyanophyceae. In some PBRs, heterogenous biomasses are also maintained consisting of different species of the algal cells, protozoa, and small metazoan.

The following are the different types of cultivation systems of microalgae:

8.5.2 Open pond system

The open pond system, namely, natural lakes, lagoons, or ponds are the oldest and most common method of culturing microalgae since the year 1950. In an open pond system, raceway pond (resembling racetrack or lane format) and turf scrubbers are the most commonly used. (1) ORP also referred to as HRAP or aerobic ponds can be employed both as open (aerobic) and closed type (anaerobic) (Hase et al., 2000). In this system a pond with typically 1-ft depth (30–100 cm) and an area ranging from one to several acres are adopted, wherein the algae are exposed to the natural solar radiation and in the process, they convert this energy into biomass. The raceway pond uses paddle wheels to keep the algae circulating and to provide even distribution of nutrients as well as sunlight. The system offers advantages such as low cost, simple design, and lower operating and maintenance costs. The system, however, results in a risk of contamination by other microbes such as bacteria and phycophages; parasites, zooplanktons, and unwanted algal species; weather dependence; low reproducibility and biomass concentration; evaporative losses and requirement of large areas for operating (Ugwu et al., 2008). (2) *Turf Scrubber*, on the other hand, uses clusters of several filamentous algae species effective in wastewater treatment. Here the community of microalgae grows attached to a screen through which the wastewaters flow and absorb the compounds (Ray et al., 2015; Mulbry et al., 2010).

8.5.2.1 Closed system or photobioreactors

PBRs are the closed systems of algal culturing with a controlled environment and sterile conditions. This system provides adequate light exposure using natural or artificial ways to the algae for the process of photosynthesis. Because the system is closed and is under a controlled environment, the growth of algae can be regulated based on different growth parameters like the pH, temperature, light or quality, and quantity of nutrients. Closed systems have been developed to overcome the shortcomings of the open pond system. The PBRs are made up of glass/plexiglas/transparent polyvinyl chloride (PVC) material with internal or external illumination and provided with controlled gas exchange and circulation of media, the size and type of which vary with strain type and end-use of biomass and scalability. Closed bioreactors can be categorized into various types—*flat plate, vertical or inclined tubular, horizontal or serpentine tubular, airlift, helical, bubble column, membrane or hybrid* (Qiang and Richmond, 1996; Rubio et al., 1999; Ugwu et al., 2008). The advantages of the PBRs system are equipped with a light for photosynthetic activity, low contamination risk, and higher microalgae biomass yield. The major drawback is the high operating cost as well as the cost of initial investment; the system is also high maintenance.

8.5.2.2 Flocculation

Here sodium hydroxide (NaOH) or ferric chloride (FeCl$_3$) is used for thickening the microalgae biomass suspended in the culture medium. The chemicals used plus the electrical energy consumed for the homogenization in the flocculation tank come under the resource utilized in LCA analysis. The residual liquids generated can be reused in the other stages if they are treated properly.

8.5.2.3 Biomass concentration

The final biomass concentration varied for a given period depending on the cultivation system and seasonally due to the different weather conditions. With due course of time, the biomass in the PBR changes in terms of concentration and composition. So, to obtain a defined biomass concentration (22% DW), regardless of the reactor system and season, the microfiltration and centrifugation are used (1 kg of produced biomass, and 4.55 kg of slurry i.e. 22% DW is obtained.

Biomass concentration or harvesting refers to the process of separating microalgae from water for subsequent usage or production. The process of harvesting microalgae is, however, a challenging one because of the small algal cell size, which generally ranges from 1 to 20 μm and their nature of existence in nature in suspended form (Lam and Lee, 2012; Suali and Sarbatly, 2012). The method followed is carried out in two steps—(1) *bulk harvesting* to separate microalgae from bulk suspension through gravity sedimentation, flocculation, and flotation (2) *thickening* to concentrate the microalgae slurry after bulk harvesting using techniques such as centrifugation and filtration (Brennan and Owende, 2010; Chen et al., 2011). However, to obtain a higher rate of microalgal separation with low cost, two or more harvesting techniques combined could be employed. For instance, the significant downstream processes carrying the combined effects of flocculation, followed by sedimentation with centrifugation could be employed for cost-efficient (Schlesinger et al., 2012).

The types of harvesting systems are described in the following sections.

8.5.3 Gravity sedimentation

This technique of algal harvesting follows Stokes' law of sedimentation, where the algae are then allowed to settle naturally using gravity and density after the water is agitated. The sedimentation of SS is determined by the density and radius of microalgae. This method is low cost with no additional requirement for chemical or physical treatment; the drawback however is the long time it takes to settle. The best outcome of microalgal harvesting through gravity sedimentation was achieved through lamella-type separators with the recovery of 1.6% TSS (total SS) and sedimentation tanks recovery of 3% TSS (Show and Lee, 2013).

8.5.4 Flocculation

Flocculation is an algal harvesting process, in which scattered units are collected to form huge units and thus settle down by different types of flocculants or coagulants (Uduman et al., 2010). Microalgae cells found in the dispersed state are negatively charged and have a density near the growth medium. Thus ions in the chemical flocculants can interact with microalgal cells, resulting in the efficient harvest of microalgae (Singh and Patidar, 2018; Yin et al., 2020). Flocculation method of algal harvesting occurs through four mechanisms in combination or singly—(1) charge neutralization process where the charged microalgal cell surface is adsorbed by the oppositely charged ions, polymers or colloids; (2) electrostatic patch mechanism in which charge of the microalgal cell surface is bounded by the oppositely charged polymers thus resulting in patches to connect opposite charged microalgal cells and causes flocculation; (3) bridging process in which charged colloids or polymers binds together the surface of the two different microalgal cells to form the bridge, and lastly, (4) sweeping flocculation process in which microalgal cells are entrapped by the precipitation minerals present in the cultivation media. The main inorganic flocculants are $Fe_2(SO_4)_3$, $FeCl_3$, $Al_2(SO_4)_3$, $AlCl_3$, $ZnSO_4$, $ZnCl_2$, $CaSO_4$, $CaCl_2$, $MgSO_4$, $MgCl_2$, $(NH_4)_2SO_4$, and NH_4Cl (Papazi et al., 2010; Mathimani and Mallick, 2018).

8.5.5 Flotation

Flotation is the process whereby bubbles are ejected into the suspension to adhere to solid or suspended particles and thereafter lift to the surface of the liquid, achieving the process of skimming off or separating (Barros et al., 2015; Laamanen et al., 2016). Flotation is also called inverted sedimentation in which particles are separated with air bubbles using lifting force provided by air (Christenson and Sims, 2011). It is considered faster and more efficient than the sedimentation process because of the self-floating characteristics of microalgal cells with low density (Singh et al., 2011; Hanotu et al., 2012).

8.5.6 Centrifugation

Centrifugation is an advanced form of gravity sedimentation where centrifugal force replaces gravity in separating microalgae from their growth medium. The process of centrifugal separation depends on factors such as the cell settling characteristics, that is, cell size and negligible density difference of microalgal cells to their culture medium, cell slurry retention time in the centrifuge (Singh and Patidar, 2018). It is considered the fastest algal harvesting method, but also the most expensive owing to its high energy consumption, a high capital investment which limits its application to high-value products, such as highly unsaturated fatty acids, pharmaceuticals, and other commodities (Molina et al., 2003; Christenson and Sims, 2011; Rawat et al., 2011).

8.5.7 Filtration

Filtration is usually a dewatering process used to separate solids from liquids or gasses by interposing a medium where only fluid can pass and it is normally applied following flocculation to improve harvesting efficiency (Enamala et al., 2018). In this process, algae suspensions are run through a filter or a porous membrane, which retains algae slurry, while water is passed through a filter. The microalgal deposits on the filtration membrane tend to grow thicker throughout the process, increasing resistance and decreasing filtration flux upon a constant pressure drop (Show and Lee, 2013). The different types of filtrations used include microfiltration, macro-filtration, ultrafiltration, vacuum filtration, dead-end filtration, pressure filtration, and tangential flow filtration (Enamala et al., 2018). As the name suggests, microfiltration is used for harvesting smaller algal cells ranging from 0.1 to 10 mm (Li et al., 2011), whereas macro-filtration is used for larger cell size, particularly to filter flocs (Mathimani and Mallick, 2018). A recovery of about 70%−89% of freshwater algae is obtained through tangential flow filtration (Pragya et al., 2013). Despite microalgal cells of very low densities can be harvested by this method, membrane filtration is not commonly applied in large-scale processes (Molina et al., 2003).

8.5.8 Drying

It is the last stage in microalgae biomass production. The resource consume here is the use of electrical energy for the operation of the pressing filter and spray drier.

8.5.8.1 Inventory analysis

The inventory data collected are the significant inputs and outputs of the system. In this stage the primary data are collected for the foreground system. For the recording of the primary data, multiparametric probes are used for the determination of pH, oxygen diffusion (OD), NO_3, NH_4^+, chlorophyll-a, temperature, turbidity, etc. COD, total nitrogen (N), and phosphorus (P) are measured traditionally. The inputs and the outputs are quantified based on the volume of water at the cleaning stage. During the preparation of culture medium, the average dilution rate with water for preparation of chemicals, the energy requirements for monitoring system, mixing, aeration and temperature control, the building materials for the reactor, etc. (Fig. 8.3) are considered. A life span of 10 or 20 years was considered for the building materials depending on their properties and function. The input for the biomass concentration is the energy consumption for microfiltration and centrifuge, which is calculated based on the total volume of medium to separate from biomass to achieve the final 22% DW concentration. In all subsystems the solid wastes were assumed to be disposed of in either sanitary or inert landfills, whereas the resulting wastewater was collected in the general sewage system and treated in a conventional WWTP. An average transport distance of 200 km was considered for chemicals and building materials and 50 km was estimated for wastes. The inventory data related to the background system were obtained from the Ecoinvent database (Frischknecht et al., 2007). These inputs include the production of the chemicals required for the cleaning and the nutrients for the culture medium, as well as the production of electricity used throughout the stages of the processes, the manufacture of the building materials for each reactor, and the waste disposal (Pérez-López et al., 2014).

In the case of HRAP systems coupled with biogas and biofertilizer production, inventory analysis for foreground data includes construction materials and operations based on the detailed engineering designs performed. Treated wastewater characteristics were estimated considering the removal efficiencies and experimental results, NH_4^+ volatilization was estimated through nitrogen mass balance, NH_3 and N_2O emissions are calculated using emissions factors from the literature (Hospido et al., 2008; Lundin et al., 2000). Heavy metals and nutrients are gathered from experimental results (Morales-Amaral et al., 2015; Solé-Bundó et al., 2017). In electricity consume and heat production during the operation is also considered. Background data include data of construction materials, chemicals, energy production, avoided fertilizer, transportation, and sludge incineration process obtained from the Ecoinvent 3.1 database (Moreno-Ruiz et al., 2014; Weidema et al., 2013).

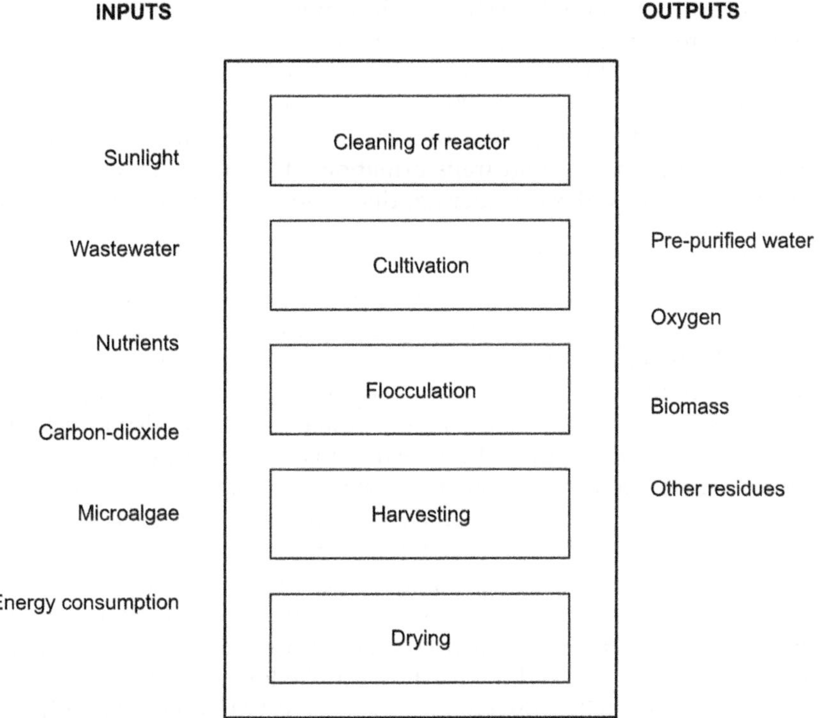

FIGURE 8.3   Life cycle inventory for microalgae biomass production.

As a general for any PBR system in LCI, the following aspects were measured and esti-mated, (1) weight of water consumed; (2) construction materials (steel, concrete, anticorro-sive paint used); (3) PVC materials (pipes, valves, connections, tubings, etc.); (4) packaging products; (5) chemicals used; and (6) energy consumption.

8.5.8.2 *Impacts assessment*

At this stage, environmental impact potentials are calculated using inventory data col-lected and compiled in the previous stage. The substages, including the classification, characterization, normalization of data, and weighting, are defined in this stage, (1) the *classification* includes the individual inventory item according to the relevant environmen-tal impact categories; (2) *characterization* includes inventory item contributing to the envi-ronmental impact category multiplied by a relevant coefficient and expressed in a common unit, and then the aggregated impact is calculated for each impact category; (3) in *normalization* the data collected were normalized, and (4) *weighting* indicates which envi-ronmental impact potential is more important. SimaPro (Pre-sustainability, 2014), CML2001 (Guinee, 2002), Cumulative Energy Demand (CED) by Ecoinvent version 2.0 (PRé Consultants) (Frischknecht et al., 2007), TRACI, GaBi, UMBERTO, Ecobilan, LCAiT, GEMIS, etc. are common software for commercial wastewater treatment and management in LCIA. The impact categories are environmental indicators, including global warming, climate change, ozone depletion, terrestrial acidification, freshwater eutrophication, marine eutrophication, photochemical oxidant formation, particulate matter formation, metal depletion, fossil depletion, human toxicity, and terrestrial ecotoxicity (Corominas, 2013; Fang, 2016; Gallego, 2008; Garfí et al., 2017; Goedkoop et al., 2013; Hospido et al., 2008), other categories include the energy resource, water demand indicator, and an indicator associated with land occupation and transformation, etc. The LCIA also implemented for the sludge management includes dewatering, thickening, and anaerobic digestion.

8.5.8.3 *Interpretation of the results*

The purpose of this step is to interpret the results of inventory and impact studies carried out for the purpose and scope of study undertaken. It also presents the recommendations and the products under consideration. The results may include the carbon footprint calculations, GHGs emitted; water footprints and soil emissions affecting the water quality may also be addressed. The sensitivity analysis (equation next) is also carried out to determine the inventory data which affected the result most significantly.

Some of the interpretations made on the wastewater treatment with microalgae based on LCA studies are as follows: (1) in a study carried out by Pérez-López et al. (2017) for comparative LCA between ORP, horizontal and vertical tubular PBRs in Netherland showed that the ORP system showed higher environmental impacts than both tubular PBRs, as the PBR systems compensate energy-consuming elements with higher productivity. (2) Rothermell et al. (2013) found that the energy use, global warming, eutrophication potential, and other impact categories were found to be lower when algae-based wastewater treatment is combined with conventional type of WWTPs. (3) The effluent (outflow) quality as studied by Craggs et al., 2014 was found to be better quality in terms of BOD, TSS, TN, TP, etc. in HRAP wastewater system than the conventional WWTPs in New Zealand. (4) In Kohlheb et al. (2020), LCA study showed that HRAP has great potential in municipal wastewater treatment wherever plenty of solar energy is available and it is more beneficial environmentally and economically also, contributing to CO_2 sequestration and eutrophication potentials of 146.27×10^{-3} kg CO_2 equiv m^{-3} and 126.14×10^{-6} kg PO_4 equiv m^{-3}, respectively. In this way, many studies were conducted in the present decade, which showed the better performance of algal-based wastewater treatment in terms of impact categories compared to the conventional WWTPs.

8.6 Conclusions

With the population explosion, urbanization, depletion of nonrenewable resources, overexploitation of freshwater, and other natural resources and the changing climatic conditions, the availability, quality (increase temperature, decrease dissolved oxygen, pathogen contamination, etc.) and the quantity of water will be affected in the coming years, posing threat to the biodiversity and the basic right to safe drinking water and sanitation of billions of people. The changing climatic conditions may also raise the frequency and severity of water scarcity and drought events, limiting the water resources. So, the best option is the recycling of wastewater to meet the demands. But the wastewater treatments also generate the residues and GHGs through energy consumption and biochemical processes accounting for 3%–7% emission, contributing to climate change; however, untreated wastewater on the other hand is a major source of methane which is the most potent for global warming. At present, LCA is the quantitative approach for addressing these potential environmental impacts for the different types of WWTPs throughout its life cycle, that is, from cradle to grave. These studies showed the advantages and disadvantages of various types of WWTPs and help guide in choosing the best processes out of many. These LCA studies have shown that microalgae-based wastewater treatment is the best option due to the optimal management of water resources, with reduced emissions and the ability to extract the potent GHGs to generate energy. This is also a highly productive wastewater treatment system that avoids the competition for the land requirement for food crops, as the wastewater is a valuable resource for its cultivation with the generation of high-value products like bioplastics, biofuel, biofertilizers, and clean water for smart irrigation. The uses of microalgae also had the advantages of lesser emission through the extraction of potent GHGs (carbon dioxide and methane) and converting them into sustainable energy applications. Besides the conventional wastewater treatment compared to the microalgae wastewater treatment has disadvantages, namely, variable nutrient removal efficiency, costly to operate, processes, leading to secondary pollutions and loss of potential nutrients like nitrogen and phosphorus.

References

Abhang, R., et al., 2013. Nanofiltration for recovery of heavy metal ions from waste water—a review. International Journal of Environmental Science and Technology 3 (1), 29–34.

Abou-Elela, S.I., Kamel, M.M., Fawzy, M.E., 2010. Biological treatment of saline wastewater using a salt-tolerant microorganism. Desalination. 250 (1), 1–5. Available from: https://doi.org/10.1016/j.desal.2009.03.022. Egypt.

Abou Shanab, R.A.I., et al., 2013. Microalgal species growing on piggery wastewater as a valuable candidate for nutrient removal and biodiesel production. Journal of Environmental Management. 115, 257–264. Available from: https://doi.org/10.1016/j.jenvman.2012.11.022. South Korea.

Absar, A., 2005. Water and wastewater properties and characteristics. Water Encyclopedia: Domestic, Municipal and Industrial Water Supply and Waste Disposal. John Wiley and Sons, Inc. Available from: https://doi.org/10.1002/047147844X.ww346.

Acero, A., Rodríguez, C., Ciroth, A., 2016. GreenDelta: 2016, LCIA methods. Impact Assessment Methods in Life Cycle Assessment and Their Impact Categories. Available from: https://www.openlca.org/wp-content/uploads/2016/08/LCIA-METHODS-v.1.5.5.pdf.

Aim Ben, R., Liu, M.G., Vigneswaran, S., 1993. Recent development of membrane processes for water and waste water treatment. Water Science and Technology 27 (10), 141–149. Available from: https://doi.org/10.2166/wst.1993.0221. IWA Publishing.

Akpor, O.B., Muchie, M., 2011. Environmental and public health implications of wastewater quality. African Journal of Biotechnology 10 (13), 2379–2387. South Africa. Available from: http://www.academicjournals.org/AJB/PDF/pdf2011/28Mar/Akpor%20and%20Muchie.pdf.

Alm, E., 2003. Implication of microbial heavy metal tolerance in the environment. Reviews in Undergraduate Research 2, 1–5.

Amini, A., et al., 2015. Environmental and economic sustainability of ion exchange drinking water treatment for organics removal. Journal of Cleaner Production 104, 413–421. Available from: https://doi.org/10.1016/j.jclepro.2015.05.056. United States: Elsevier Ltd.

An, J.Y., et al., 2003. Hydrocarbon production from secondarily treated piggery wastewater by the green alga *Botryococcus braunii*. Journal of Applied Phycology. Available from: https://doi.org/10.1023/A:1023855710410. South Korea: Springer Netherlands.

Arashiro, L.T., et al., 2019. The effect of primary treatment of wastewater in high rate algal pond systems: biomass and bioenergy recovery. Bioresource Technology 280, 27–36. Available from: https://doi.org/10.1016/j.biortech.2019.01.096. Spain: Elsevier Ltd.

Aslan, S., Kapdan, I.K., 2006. Batch kinetics of nitrogen and phosphorus removal from synthetic wastewater by algae. Ecological Engineering 28 (1), 64–70. Available from: https://doi.org/10.1016/j.ecoleng.2006.04.003. Turkey.

Bahri, A., 1998. Wastewater reclamation and reuse in Tunisia. Wastewater Reclamation and Reuse 877–916. Asano T. CRC Press.

Bai, Y., et al., 2018. Exploring the relationship between urbanization and urban eco-efficiency: evidence from prefecture-level cities in China. Journal of Cleaner Production 195, 1487–1496. Available from: https://doi.org/10.1016/j.jclepro.2017.11.115. China: Elsevier Ltd.

Bare, J., 2002. The tool for assessment of chemical and other environmental impacts (TRACI). Journal of Industrial Ecology 6, 3–4.

Barjoveanu, G., et al., 2013. Evaluation of water services system through LCA. A case study for Iasi City, Romania. International Journal of Life Cycle Assessment. Romania 19 (2), 449–462. Available from: https://doi.org/10.1007/s11367-013-0635-8.

Barros, A.I., et al., 2015. Harvesting techniques applied to microalgae: a review. Renewable and Sustainable Energy Reviews 41, 1489–1500. Available from: https://doi.org/10.1016/j.rser.2014.09.037. Portugal: Elsevier Ltd.

Becker, 1994. Microalgal biotechnology and microbiology. Cambridge Studies in Biotechnology. Cambridge University Press.

Benemann, J., 2008. Opportunities and Challenges in Algae Biofuels Production. Algae World, Presented at.

Best, G., et al., 2001. Application of immersed ultrafiltration membranes for organic removal and disinfection by-product reduction. Water Supply 1 (5–6), 221–231. Available from: https://doi.org/10.2166/ws.2001.0118. IWA Publishing.

Boduroğlu, G., Kiliç, N.K., Dönmez, G., 2014. Bioremoval of reactive blue 220 by *Gonium* sp. biomass. Environmental Technology (United Kingdom) 35 (19), 2410–2415. Available from: https://doi.org/10.1080/09593330.2014.908240. Turkey: Taylor and Francis Ltd.

Boelee, N.C., et al., 2011. Nitrogen and phosphorus removal from municipal wastewater effluent using microalgal biofilms. Water Research 45 (18), 5925–5933. Available from: https://doi.org/10.1016/j.watres.2011.08.044. Netherlands: Elsevier Ltd.

Bond, W.J., 1998. Effluent irrigation—an environmental challenge for soil science. Australian Journal of Soil Research 36 (4), 543–555. Available from: https://doi.org/10.1071/S98017. Australia: CSIRO.

Borchardt, D., Statzner, B., 1990. Ecological impact of urban stormwater runoff studied in experimental flumes: population loss by drift and availability of refugial space. Aquatic Sciences 52 (4), 299–314. Available from: https://doi.org/10.1007/BF00879759. Germany: Birkhäuser-Verlag.

Bosma, R., et al., 2014. Design and construction of the microalgal pilot facility AlgaePARC. Algal Research 6, 160–169. Available from: https://doi.org/10.1016/j.algal.2014.10.006. Netherlands: Elsevier B.V.

Bowen, W.R., Mukhtar, H., 1996. Characterisation and prediction of separation performance of nanofiltration membranes. Journal of Membrane Science 112 (2), 263–274. Available from: https://doi.org/10.1016/0376-7388(95)00302-9. United Kingdom: Elsevier B.V.

Bowen, W.R., Welfoot, J.S., 2002. Modelling the performance of membrane nanofiltration-critical assessment and model development. Chemical Engineering Science. United Kingdom 57 (7), 1121–1137. Available from: https://doi.org/10.1016/S0009-2509(01)00413-4.

Brennan, L., Owende, P., 2010. Biofuels from microalgae—a review of technologies for production, processing, and extractions of biofuels and co-products. Renewable and Sustainable Energy Reviews 14 (2), 557–577. Available from: https://doi.org/10.1016/j.rser.2009.10.009. Ireland.

Buonocore, E., et al., 2018. Life cycle assessment indicators of urban wastewater and sewage sludge treatment. Ecological Indicators 94, 13–23. Available from: https://doi.org/10.1016/j.ecolind.2016.04.047. Italy: Elsevier B.V.

Cadena, E., et al., 2018. Including an odor impact potential in life cycle assessment of waste treatment plants. International Journal of Environmental Science and Technology 15 (10), 2193–2202. Available from: https://doi.org/10.1007/s13762-017-1613-7. Spain: Center for Environmental and Energy Research and Studies.

Casazza, A.A., et al., 2016. Microalgae growth using winery wastewater for energetic and environmental purposes. Chemical Engineering Transactions 49, 565–570. Available from: https://doi.org/10.3303/CET1649095. Italy: Italian Association of Chemical Engineering - AIDIC.

Chakinala, A.G., et al., 2009. Industrial wastewater treatment using hydrodynamic cavitation and heterogeneous advanced Fenton processing. Chemical Engineering Journal 152 (2–3), 498–502. Available from: https://doi.org/10.1016/j.cej.2009.05.018. United Kingdom.

Chambers and Mills, 1996. "Dissolved Oxygen Conditions and Fish Requirements in the Athabasca, Peace and Slave Rivers: Assessment of Present Conditions and Future Trends. Northern River Basins Study Synthesis Report No. 5." Environment Canada/Alberta Environmental Protection. Edmonton, AB.

Chang, A.C., et al., 1996. Developing human health-related chemical guidelines for reclaimed wastewater irrigation. Water Science and Technology. Available from: https://doi.org/10.1016/0273-1223(96)00449-0. United States: Pergamon Press Inc.

Chen, C.Y., et al., 2011. Cultivation, photobioreactor design and harvesting of microalgae for biodiesel production: a critical review. Bioresource Technology. Taiwan 102 (1), 71–81. Available from: https://doi.org/10.1016/j.biortech.2010.06.159.

Chen, H., et al., 2015. Macroalgae for biofuels production: progress and perspectives. Renewable and Sustainable Energy Reviews 47, 427–437. Available from: https://doi.org/10.1016/j.rser.2015.03.086. China: Elsevier Ltd.

Chew, K.W., et al., 2017. Microalgae biorefinery: high value products perspectives. Bioresource Technology 229, 53–62. Available from: https://doi.org/10.1016/j.biortech.2017.01.006. Malaysia: Elsevier Ltd.

Chigor, V.N., Sibanda, T., Okoh, A.I., 2013. Studies on the bacteriological qualities of the Buffalo River and three source water dams along its course in the Eastern Cape Province of South Africa. Environmental Science and Pollution Research 20 (6), 4125–4136. Available from: https://doi.org/10.1007/s11356-012-1348-4. South Africa.

Chisti, Y., 2007. Biodiesel from microalgae. Biotechnology Advances 25 (3), 294–306. Available from: https://doi.org/10.1016/j.biotechadv.2007.02.001. New Zealand.

Chou, C., et al., 2011. "The combination of ozonation and ion exchange processes for treatment of a municipal wastewater plant effluent." In: 2011 International Conference on Electric Technology and Civil Engineering, ICETCE 2011. Taiwan. doi: 10.1109/ICETCE.2011.5775819.

Christenson, L., Sims, R., 2011. Production and harvesting of microalgae for wastewater treatment, biofuels, and bioproducts. Biotechnology Advances 29 (6), 686–702. Available from: https://doi.org/10.1016/j.biotechadv.2011.05.015. United States.

Chu, S., 1974. The influence of the mineral composition of the medium on the growth of planktonic algae. Journal of Ecology 30, 284–325.

Cohen, 2006. Water-desalination vs recycling. Press release, Ian Cohen, The Greens. Member of the Legislative Council.

Collet, P., et al., 2011. Life-cycle assessment of microalgae culture coupled to biogas production. Bioresource Technology 102 (1), 207–214. Available from: https://doi.org/10.1016/j.biortech.2010.06.154. France.

Coppens, J., et al., 2016. The use of microalgae as a high-value organic slow-release fertilizer results in tomatoes with increased carotenoid and sugar levels. Journal of Applied Phycology 28 (4), 2367–2377. Available from: https://doi.org/10.1007/s10811-015-0775-2. Belgium: Springer Netherlands.

Corominas, L., et al., 2013. Life cycle assessment applied to wastewater treatment: state of the art. Water Research 47 (15), 5480–5492. Available from: https://doi.org/10.1016/j.watres.2013.06.049. Spain: Elsevier Ltd.

Craggs, R., et al., 2014. High rate algal pond systems for low-energy wastewater treatment, nutrient recovery and energy production. New Zealand Journal of Botany 52 (1), 60–73. Available from: https://doi.org/10.1080/0028825X.2013.861855. New Zealand: Taylor and Francis Ltd.

Dang, H., Lovell, C.R., 2016. Microbial surface colonization and biofilm development in marine environments. Microbiology and Molecular Biology Reviews 80 (1), 91–138. Available from: https://doi.org/10.1128/MMBR.00037-15. China: American Society for Microbiology.

Das, C., et al., 2017. Efficient bioremediation of tannery wastewater by monostrains and consortium of marine *Chlorella* sp. and *Phormidium* sp. International journal of phytoremediation 20 (3), 284–292. Available from: https://doi.org/10.1080/15226514.2017.1374338. India: Taylor and Francis Inc.

De Bhowmick, G., Sarmah, A.K., Sen, R., 2019. Performance evaluation of an outdoor algal biorefinery for sustainable production of biomass, lipid and lutein valorizing flue-gas carbon dioxide and wastewater cocktail. Bioresource Technology 283, 198–206. Available from: https://doi.org/10.1016/j.biortech.2019.03.075. New Zealand: Elsevier Ltd.

Deprá, M.C., et al., 2020. Environmental impacts on commercial microalgae-based products: sustainability metrics and indicators. Algal Research 51, 102056. Available from: https://doi.org/10.1016/j.algal.2020.102056. Elsevier BV.

Deviller, G., et al., 2004. High-rate algal pond treatment for water reuse in an integrated marine fish recirculating system: effect on water quality and sea bass growth. Aquaculture. France 235 (1–4), 331–344. Available from: https://doi.org/10.1016/j.aquaculture.2004.01.023.

Dumas, A., et al., 1998. Biotreatment of fish farm effluents using the cyanobacterium Phormidium bohneri. Aquacultural Engineering 17 (1), 57–68. Available from: https://doi.org/10.1016/S0144-8609(97)01013-3. Canada: Elsevier.

Dunn, K., Maart, B., Rose, P., 2013. Arthrospira (Spirulina) in tannery wastewaters. Part 2: evaluation of tannery wastewater as production media for the mass culture of Arthrospira biomass. Water SA 39 (2), 279–284. Available from: https://doi.org/10.4314/wsa.v39i2.12. South Africa.

Eloussaief, M., et al., 2011. Mineralogical identification, spectroscopic characterization, and potential environmental use of natural clay materials on chromate removal from aqueous solutions. Chemical Engineering Journal 168 (3), 1024–1031. Available from: https://doi.org/10.1016/j.cej.2011.01.077. Tunisia: Elsevier B.V.

Eloussaief, M., Jarraya, I., Benzina, M., 2009. Adsorption of copper ions on two clays from Tunisia: pH and temperature effects. Applied Clay Science 46 (4), 409–413. Available from: https://doi.org/10.1016/j.clay.2009.10.008. Tunisia.

Enamala, M.K., et al., 2018. Production of biofuels from microalgae—a review on cultivation, harvesting, lipid extraction, and numerous applications of microalgae. Renewable and Sustainable Energy Reviews 94, 49–68. Available from: https://doi.org/10.1016/j.rser.2018.05.012. India: Elsevier Ltd.

Environmental Canada, 1999. State of the Great Lakes. Environmental Canada and the US. Environmental Protection Agency (no date).

Erdem, E., Karapinar, N., Donat, R., 2004. The removal of heavy metal cations by natural zeolites. Journal of Colloid and Interface Science 280 (2), 309–314. Available from: https://doi.org/10.1016/j.jcis.2004.08.028. Turkey.

Fahad, A., et al., 2019. Wastewater and its treatment techniques: an ample review. Indian Journal of Science and Technology. Indian Society for Education and Environment 12 (25), 1–13. Available from: https://doi.org/10.17485/ijst/2019/v12i25/146059.

Fang, L.L., et al., 2016. Life cycle assessment as development and decision support tool for wastewater resource recovery technology. Water Research 88, 538–549. Available from: https://doi.org/10.1016/j.watres.2015.10.016. Denmark: Elsevier Ltd.

Fegen, N., Gardner, T., Blackwell, P., 1998. Health risks associated with the reuse of effluent for irrigation—a literature review. State of Queensland Department of Natural Resources and Department of Primary Industries.

Fent, K., Weston, A.A., Caminada, D., 2006. Ecotoxicology of human pharmaceuticals. Aquatic Toxicology 76 (2), 122–159. Available from: https://doi.org/10.1016/j.aquatox.2005.09.009. Switzerland: Elsevier.

Ferreira, L., Torregrosa-López, V. Capuz-Rizo, S., 2017. "The accounting system as complementary data source for organizational life cycle assessment of higher education institutions." In: 21th International Congress on Project Management and Engineering.

Flörke, M., et al., 2013. Domestic and industrial water uses of the past 60 years as a mirror of socio-economic development: a global simulation study. Global Environmental Change 23 (1), 144–156. Available from: https://doi.org/10.1016/j.gloenvcha.2012.10.018. Germany.

Fontes, A.G., et al., 1987. Factors affecting the production of biomass by a nitrogen-fixing blue-green alga in outdoor culture. Biomass 13 (1), 33–43. Available from: https://doi.org/10.1016/0144-4565(87)90070-9. Spain.

Frischknecht, et al., 2007. "Implementation of Life Cycle Impact Assessment Methods:" Ecoinvent Report No. 3. v2.0. Swiss Centre for Life Cycle Inventories, Bern.

Fuchs, V.J., Mihelcic, J.R., Gierke, J.S., 2011. Life cycle assessment of vertical and horizontal flow constructed wetlands for wastewater treatment considering nitrogen and carbon greenhouse gas emissions. Water Research 45 (5), 2073–2081. Available from: https://doi.org/10.1016/j.watres.2010.12.021. United States: Elsevier Ltd.

Gallego, A., et al., 2008. Environmental performance of wastewater treatment plants for small populations. Resources, Conservation and Recycling 52 (6), 931–940. Available from: https://doi.org/10.1016/j.resconrec.2008.02.001. Spain.

Gao, F., et al., 2014. Concentrated microalgae cultivation in treated sewage by membrane photobioreactor operated in batch flow mode. Bioresource Technology 167, 441–446. Available from: https://doi.org/10.1016/j.biotech.2014.06.042. China: Elsevier Ltd.

García, J., Mujeriego, R., Hernández-Mariné, M., 2000. High rate algal pond operating strategies for urban wastewater nitrogen removal. Journal of Applied Phycology. Available from: https://doi.org/10.1023/a:1008146421368. Spain: Springer Netherlands.

Garfí, M., Flores, L., Ferrer, I., 2017. Life Cycle Assessment of wastewater treatment systems for small communities: activated sludge, constructed wetlands and high rate algal ponds. Journal of Cleaner Production 161, 211–219. Available from: https://doi.org/10.1016/j.jclepro.2017.05.116. Spain: Elsevier Ltd.

Garrett, M.K., Allen, M.D.B., 1976. Photosynthetic purification of the liquid phase of animal slurry. Environmental Pollution (1970) 10 (2), 127–139. Available from: https://doi.org/10.1016/0013-9327(76)90102-6. United Kingdom.

Ghimire, N., Wang, S., 2019. Biological Treatment of Petrochemical Wastewater. IntechOpen. Available from: https://doi.org/10.5772/intechopen.79655.

Goedkoop, M., Heijungs, R., Huijbregts, M., De Schryver, A., Struje, J., van Zelm, R., et al., 2009. A life cycle impact assessment method which comprises harmonized category indicators at the midpoint and then end point level. Report I: Characterisation, 1st edition, Ministerie van Volkshuisvesting, Ruimtelijke Ordering en Milieubeheer, Netherlands. PRé Consultants The Netherlands: Ruimte en Milieu, Ministerie van Volkshuisvesting, Ruimtelijke Ordening en Milieubeheer.

Goedkoop, M., Heijunngs, R., Huijbregts, M., Schryver, A., Struijs, J., van Zelm, R., et al., 2013. ReCiPE 2008: a life cycle impact assessment method which comprises harmonised category indicators at the midpoint and then endpoint level. PRé Consultants, Amersfoort, The Netherlands. Report I: Characterisation, 1st edition, Ministerie van Volkshuisvesting, Ruimtelijke Ordering en Milieubeheer, Netherlands. The Netherlands: Ruimte en Milieu, Ministerie van Volkshuisvesting, Ruimtelijke Ordening en Milieubeheer.

Guest, J.S., et al., 2009. A new planning and design paradigm to achieve sustainable resource recovery from wastewater. Environmental Science and Technology 43 (16), 6126–6130. Available from: https://doi.org/10.1021/es9010515. United States.

Guinee, J.B., 2002. Handbook on life cycle assessment operational guide to the ISO standards. The International Journal of Life Cycle Assessment 7 (5). Available from: https://doi.org/10.1007/bf02978897. Springer Science and Business Media LLC.

Guinée, J.B., Gorrée, M., Heijungs, R., Huppes, G., Kleijn, R., de Koning, A., van Oers, L., Wegener Sleeswijk, A., Suh, S., Udo de Haes, H.A., de Bruijn, H., van Duin, R., Huijbregts, M.A.J., Lindeijer, E., Roorda, A.A.H., Weidema, B.P. (2001a) Life cycle assessment; An operational guide to the ISO standards; Parts 1 and 2. Ministry of Housing, Spatial Planning and Environment (VROM) and Centre of Environmental Science (CML), Den Haag and Leiden, The Netherlands, Retrieved from: http://www.leidenuniv.nl/cml/ssp/projects/lca2/lca2.html.

Guinée, J.B., Gorrée, M., Heijungs, R., Huppes, G., Kleijn, R., de Koning, A., et al. 2001b. Life cycle assessment; An operational guide to the ISO standards; Part 3: Scientific Background. Ministry of Housing, Spatial Planning and Environment (VROM) and Centre of Environmental Science (CML), Den Haag and Leiden, The Netherlands. http://www.leidenuniv.nl/cml/ssp/projects/lca2/lca2.html.

Gustavsson, D.J.I., Tumlin, S., 2013. Carbon footprints of Scandinavian wastewater treatment plants. Water Science and Technology 68 (4), 887–893. Available from: https://doi.org/10.2166/wst.2013.318. IWA Publishing.

Habitat, U.N., WHO, 2021. Progress on Wastewater Treatment – Global Status and Acceleration Needs for SDG Indicator 6.3.1. United Nations Human Settlements Programme (UN-Habitat) and World Health Organization (WHO).

Hanotu, J., Bandulasena, H.C.H., Zimmerman, W.B., 2012. Microflotation performance for algal separation. Biotechnology and Bioengineering 109 (7), 1663–1673. Available from: https://doi.org/10.1002/bit.24449. United Kingdom.

Hase, R., et al., 2000. Photosynthetic production of microalgal biomass in a raceway system under greenhouse conditions in Sendai City. Journal of Bioscience and Bioengineering 89 (2), 157–163. Available from: https://doi.org/10.1016/S1389-1723(00)88730-7. Japan: Society of Fermentation and Bioengineering.

Hernández, D., 2016. Microalgae cultivation in high rate algal ponds using slaughterhouse wastewater for biofuel applications. Chemical Engineering Journal 285, 449–458. Available from: https://doi.org/10.1016/j.cej.2015.09.072. Spain: Elsevier.

Holloway, R.W., et al., 2016. Life-cycle assessment of two potable water reuse technologies: MF/RO/UV-AOP treatment and hybrid osmotic membrane bioreactors. Journal of Membrane Science 507, 165–178. Available from: https://doi.org/10.1016/j.memsci.2016.01.045. United States: Elsevier B.V.

Hong, J., et al., 2009. Environmental and economic life cycle assessment for sewage sludge treatment processes in Japan. Waste Management 29 (2), 696–703. Available from: https://doi.org/10.1016/j.wasman.2008.03.026. United States.

Hospido, A., Moreira, M.T., Feijoo, G., 2008. A comparison of municipal wastewater treatment plants for big centres of population in Galicia (Spain),". International Journal of Life Cycle Assessment 13 (1), 57–64. Available from: https://doi.org/10.1065/lca2007.03.314. Spain.

Huo, S., et al., 2012. Cultivation of *Chlorella zofingiensis* in bench-scale outdoor ponds by regulation of pH using dairy wastewater in winter, South China. Bioresource Technology 121, 76–82. Available from: https://doi.org/10.1016/j.biortech.2012.07.012. China.

Hussain, et al., 2002. Wastewater Use in Agriculture: Review of Impacts and Methodological Issues in Valuing Impacts. International Water Management Institute.

Hu, Q., Guterman, H., Richmond, A., 1996. "A flat inclined modular photobioreactor for outdoor mass cultivation of photoautotrophs." Biotechnology and Bioengineering. Israel: John Wiley & Sons Inc, vol. 51(1), pp. 51–60. Available from: https://doi.org/10.1002/(sici)1097-0290(19960705)51:1 < 51::aid-bit6 > 3.0.co;2-%23.

Ingraham, J., Ingraham, 1995. Introduction to Microbiology. Wadworth Publishing Company.

Itsubo, N., Inaba, A. (2012) LIME2—chapter 2: characterization and damage evaluation methods (2.1 Ozone layer depletion, 2.2 Global warming, 2.3 Acidification)." JLCA Newsletter. https://lca-forum.org/english/pdf/No15_Chapter2.1-2.3.pdf.

ISO 14040 International Standard, 2006. Environmental management—life cycle assessment – principles and framework. In: International Organization for Standardization, Geneva, Switzerland.

Jayalekshmi, S., Biju, M., Somarajan, J., 2021. Wastewater treatment technologies: a review. International Journal of Engineering Research and Technology 9, 9–12.

Jiménez-Pérez, M.V., et al., 2004. Growth and nutrient removal in free and immobilized planktonic green algae isolated from pig manure. Enzyme and Microbial Technology 34 (5), 392–398. Available from: https://doi.org/10.1016/j.enzmictec.2003.07.010. Elsevier BV.

Jolliet, O., et al., 2003. IMPACT 2002 + : a new life cycle impact assessment methodology. International Journal of Life Cycle Assessment 8 (6), 324–330. Available from: https://doi.org/10.1007/BF02978505. Switzerland: Springer Verlag.

Jones, E.R., et al., 2021. Country-level and gridded estimates of wastewater production, collection, treatment and reuse. Earth System Science Data 13 (2), 237–254. Available from: https://doi.org/10.5194/essd-13-237-2021. Netherlands: Copernicus GmbH.

Kanamarlapudi, S.L.R.K., Chintalpudi, V.K., Muddada, S., 2018. Application of biosorption for removal of heavy metals from wastewater. In: Derco, J., Vrana, B. (Eds.), Biosorption, IntechOpen, London. Available from: https://doi.org/10.5772/intechopen.77315.

Kesari, K.K., Verma, H.N., Behari, J., 2011. Physical methods in wastewater treatment. International Journal of Environmental Technology and Management 14 (1/2/3/4), 43. Available from: https://doi.org/10.1504/ijetm.2011.039257. Inderscience Publishers.

Kohlheb, N., et al., 2020. Assessing the life-cycle sustainability of algae and bacteria-based wastewater treatment systems: high-rate algae pond and sequencing batch reactor. Journal of Environmental Management 264, 110459. Available from: https://doi.org/10.1016/j.jenvman.2020.110459. Elsevier BV.

Komolafe, O., et al., 2014. Biodiesel production from indigenous microalgae grown in wastewater. Bioresource Technology 154, 297–304. Available from: https://doi.org/10.1016/j.biortech.2013.12.048. United Kingdom: Elsevier Ltd.

Laamanen, C.A., Ross, G.M., Scott, J.A., 2016. Flotation harvesting of microalgae. Renewable and Sustainable Energy Reviews 58, 75–86. Available from: https://doi.org/10.1016/j.rser.2015.12.293. Canada: Elsevier Ltd.

Lam, L., Kurisu, K., Hanaki, K., 2015. Comparative environmental impacts of source-separation systems for domestic wastewater management in rural China. Journal of Cleaner Production 104, 185–198. Available from: https://doi.org/10.1016/j.jclepro.2015.04.126. Japan: Elsevier Ltd.

Lam, M.K., Lee, K.T., 2012. Microalgae biofuels: a critical review of issues, problems and the way forward. Biotechnology Advances 30 (3), 673–690. Available from: https://doi.org/10.1016/j.biotechadv.2011.11.008. Malaysia.

Lee, K., Lee, C.G., 2001. Effect of light/dark cycles on wastewater treatments by microalgae. Biotechnology and Bioprocess Engineering 6 (3), 194–199. Available from: https://doi.org/10.1007/BF02932550. South Korea: Korean Society for Biotechnology and Bioengineering.

Lehtinen, et al., 2011. A review of LCA methods and tools and their suitability for. SMEs. Biochemistry 1.

Lehtoranta, S., Vilpas, R., Mattila, T.J., 2014. Comparison of carbon footprints and eutrophication impacts of rural on-site wastewater treatment plants in Finland. Journal of Cleaner Production 65, 439–446. Available from: https://doi.org/10.1016/j.jclepro.2013.08.024. Finland.

Li, Y., et al., 2011. Characterization of a microalga *Chlorella* sp. well adapted to highly concentrated municipal wastewater for nutrient removal and biodiesel production. Bioresource Technology 102 (8), 5138–5144. Available from: https://doi.org/10.1016/j.biortech.2011.01.091. United States.

Limphitakphong, N., Pharino, C., Kanchanapiya, P., 2016. Environmental impact assessment of centralized municipal wastewater management in Thailand. International Journal of Life Cycle Assessment 21 (12), 1789–1798. Available from: https://doi.org/10.1007/s11367-016-1130-9. Thailand: Springer Verlag.

Lundin, M., Bengtsson, M., Molander, S., 2000. Life cycle assessment of wastewater systems: influence of system boundaries and scale on calculated environmental loads. Environmental Science and Technology 34 (1), 180–1186. Available from: https://doi.org/10.1021/es990003f.

Machado, A.P., et al., 2007. Life cycle assessment of wastewater treatment options for small and decentralized communities. Water Science and Technology . Available from: https://doi.org/10.2166/wst.2007.497. Portugal.

Machineni, L., et al., 2017. Influence of nutrient availability and quorum sensing on the formation of metabolically inactive microcolonies within structurally heterogeneous bacterial biofilms: an individual-based 3D cellular automata model. Bulletin of Mathematical Biology 79 (3), 594–618. Available from: https://doi.org/10.1007/s11538-017-0246-9. India: Springer New York LLC.

Maekelae, K., Auvinen, H., 2009. LIPASTO-Transport Emission Database: Report No. VTT-SYMP-262, VTT Technical Research Centre of Finland, Espoo (Finland).

Mahvi, A.H., Gholami, F., Nazmara, S., 2008. Cadmium biosorption from wastewater by Ulmus leaves and their ash. European Journal of Scientific Research 23 (2), 197–203. Iran: EuroJournals, Inc. Available from: http://www.eurojournals.com/ejsr_23_2_02.pdf.

Marella, T.K., et al., 2019. Biodiesel production through algal cultivation in urban wastewater using algal floway. Bioresource Technology 280, 222–228. Available from: https://doi.org/10.1016/j.biortech.2019.02.031. India: Elsevier Ltd.

Martínez, M.E., et al., 2000. Nitrogen and phosphorus removal from urban wastewater by the microalga *Scenedesmus obliquus*. Bioresource Technology 73 (3), 263–272. Available from: https://doi.org/10.1016/S0960-8524(99)00121-2. Spain: Elsevier Science Ltd.

Martínez-Blanco, J., Inaba, A., Finkbeiner, M., 2016. Life Cycle Assessment of Organizations. Springer Science and Business Media LLC, pp. 333–394. Available from: https://doi.org/10.1007/978-94-017-7610-3_8.

Mathimani, T., Mallick, N., 2018. A comprehensive review on harvesting of microalgae for biodiesel—key challenges and future directions. Renewable and Sustainable Energy Reviews 91, 1103–1120. Available from: https://doi.org/10.1016/j.rser.2018.04.083. India: Elsevier Ltd.

Maximous, N.N., Nakhla, G.F., Wan, W.K., 2010. Removal of heavy metals from wastewater by adsorption and membrane processes: a comparative study. World Academy of Science, Engineering and Technology. Canada 64, 594–599. Available from: http://www.waset.org/journals/waset/v64/v64-109.pdf.

Mbewele, 2006. Microbial phosphorus removal in wastewater stabilization pond. A Licentiate Thesis from the School of Biotechnology: A Royal Institute of Technology Albanova.

McCasland, et al., 2008. Nitrate: Health Effects in Drinking Water. Cornell Cooperative Extension.

Metcalf, X., Eddy, X., 2003. Wastewater engineering: treatment and reuse. Wastewater Engineering, Treatment, Disposal and Reuse. Tata McGraw-Hill Publishing Company Limited.

Miller-Robbie, L., Ramaswami, A., Amerasinghe, P., 2017. Wastewater treatment and reuse in urban agriculture: Exploring the food, energy, water, and health nexus in Hyderabad, India. Environmental Research Letters 12 (7). Available from: https://doi.org/10.1088/1748-9326/aa6bfe. United States: Institute of Physics Publishing.

Min, M., et al., 2011. Cultivating *Chlorella* sp. in a pilot-scale photobioreactor using centrate wastewater for microalgae biomass production and wastewater nutrient removal. Applied Biochemistry and Biotechnology. United States 165 (1), 123–137. Available from: https://doi.org/10.1007/s12010-011-9238-7.

Miranda, A.F., et al., 2017. Applications of microalgal biofilms for wastewater treatment and bioenergy production. Biotechnology for Biofuels 10 (1). Available from: https://doi.org/10.1186/s13068-017-0798-9. Australia: BioMed Central Ltd.

Mohamed, N.A., 1994. "Application of Algal Ponds for Wastewater Treatment and Algal Production (M.Sc. thesis)." Faculty of Science, Cairo University branch, Beni-Suef.

Molina Grima, E., et al., 2003. Recovery of microalgal biomass and metabolites: process options and economics. Biotechnology Advances 20 (7–8), 491–515. Available from: https://doi.org/10.1016/S0734-9750(02)00050-2. Spain: Elsevier Inc.

Molinos-Senante, M., et al., 2014. Assessing the sustainability of small wastewater treatment systems: a composite indicator approach. Science of the Total Environment 497–498, 607–617. Available from: https://doi.org/10.1016/j.scitotenv.2014.08.026. Spain: Elsevier B.V.

Moore, M.R, Chapman, H.F, 2003. Endocrine disruptors in western society—are there any health risks in effluent reuse? In Proceedings of Water Recycling Australia, Second National Conference, Brisbane, 1-3 September. Second National Conference.

Morales-Amaral, M.d.M., et al., 2015. Outdoor production of Scenedesmus sp. in thin-layer and raceway reactors using centrate from anaerobic digestion as the sole nutrient source. Algal Research 12, 99–108. Available from: https://doi.org/10.1016/j.algal.2015.08.020. Spain: Elsevier.

Moreno-Ruiz, E., et al., 2014. Documentation of changes implemented in ecoinvent Data 3.1. Zurich: ecoinvent.

Mostafa, D., 2015. Waste water treatment in textile industries the concept and current removal technologies,". Journal of Biodiversity and Environmental Sciences 7 (1), 501–525.

Mulbry, W., Kangas, P., Kondrad, S., 2010. Toward scrubbing the bay: nutrient removal using small algal turf scrubbers on Chesapeake Bay tributaries. Ecological Engineering. United States 36 (4), 536–541. Available from: https://doi.org/10.1016/j.ecoleng.2009.11.026.

Musmarra, D., et al., 2016. Degradation of ibuprofen by hydrodynamic cavitation: Reaction pathways and effect of operational parameters. Ultrasonics Sonochemistry 29, 76–83. Available from: https://doi.org/10.1016/j.ultsonch.2015.09.002. Italy: Elsevier B.V.

Niero, M., et al., 2014. Comparative life cycle assessment of wastewater treatment in Denmark including sensitivity and uncertainty analysis. Journal of Cleaner Production 68, 25–35. Available from: https://doi.org/10.1016/j.jclepro.2013.12.051. Italy: Elsevier Ltd.

Nzayisenga, J.C., Eriksson, K., Sellstedt, A., 2018. Mixotrophic and heterotrophic production of lipids and carbohydrates by a locally isolated microalga using wastewater as a growth medium. Bioresource Technology 257, 260–265. Available from: https://doi.org/10.1016/j.biortech.2018.02.085. Sweden: Elsevier Ltd.

Oberdorff, T., et al., 2002. Development and validation of a fish-based index for the assessment of 'river health' in France. Freshwater Biology 47 (9), 1720–1734. Available from: https://doi.org/10.1046/j.1365-2427.2002.00884.x. France.

Okoh, A.I., et al., 2007. Wastewater treatment plants as a source of microbial pathogens in receiving watersheds. African Journal of Biotechnology 6 (25), 2932–2944. Available from: https://doi.org/10.5897/ajb2007.000-2462. South Africa: Academic Journals.

Olgun, H., Ceviz, N., Özkan, B., 2003. A case of dilated cardiomyopathy due to nutritional vitamin D deficiency rickets. Turkish Journal of Pediatrics 45 (2), 152–154. Turkey.

Olguín, E.J., et al., 2003. Annual productivity of Spirulina (Arthrospira) and nutrient removal in a pig wastewater recycling process under tropical conditions. Journal of Applied Phycology. Available from: https://doi.org/10.1023/A:1023856702544. Mexico: Springer Netherlands.

Oppong, S.O.B., Opoku, F., Govender, P.P., 2019. Tuning the electronic and structural properties of Gd-TiO$_2$-GO nanocomposites for enhancing photodegradation of IC dye: The role of Gd3 + ion. Applied Catalysis B: Environmental 243, 106–120. Available from: https://doi.org/10.1016/j.apcatb.2018.10.031. South Africa: Elsevier B.V.

Ortiz, M., Raluy, R.G., Serra, L., 2007. Life cycle assessment of water treatment technologies: wastewater and water-reuse in a small town. Desalination 204 (1–3), 121–131. Available from: https://doi.org/10.1016/j.desal.2006.04.026. Spain.

O'Connor, M., Garnier, G., Batchelor, W., 2014. Life cycle assessment comparison of industrial effluent management strategies. Journal of Cleaner Production 79, 168–181. Available from: https://doi.org/10.1016/j.jclepro.2014.05.066. Australia: Elsevier Ltd.

Pandey, A., Srivastava, S., Kumar, S., 2019. Isolation, screening and comprehensive characterization of candidate microalgae for biofuel feedstock production and dairy effluent treatment: a sustainable approach. Bioresource Technology 293, 121998. Available from: https://doi.org/10.1016/j.biortech.2019.121998. Elsevier BV.

Papazi, A., Makridis, P., Divanach, P., 2010. Harvesting Chlorella minutissima using cell coagulants. Journal of Applied Phycology 22 (3), 349–355. Available from: https://doi.org/10.1007/s10811-009-9465-2. Greece.

Peng, X.W., et al., 2012. Highly effective adsorption of heavy metal ions from aqueous solutions by macroporous xylan-rich hemicelluloses-based hydrogel. Journal of Agricultural and Food Chemistry. China 60 (15), 3909–3916. Available from: https://doi.org/10.1021/jf300387q.

Piao, W., et al., 2016. Life cycle assessment and economic efficiency analysis of integrated management of wastewater treatment plants. Journal of Cleaner Production 113, 325–337. Available from: https://doi.org/10.1016/j.jclepro.2015.11.012. South Korea: Elsevier Ltd.

Posten, C., 2009. Design principles of photo-bioreactors for cultivation of microalgae. Engineering in Life Sciences 9 (3), 165–177. Available from: https://doi.org/10.1002/elsc.200900003. Germany.

PRé Consultants, 2010. SimaPro 7: Introduction into LCA. http://www.pre-sustainability.com/download/manuals/SimaPro7. Introduction to LCA.

Pragya, N., Pandey, K.K., Sahoo, P.K., 2013. A review on harvesting, oil extraction and biofuels production technologies from microalgae. Renewable and Sustainable Energy Reviews 24, 159–171. Available from: https://doi.org/10.1016/j.rser.2013.03.034. India: Elsevier Ltd.

Prandini, J.M., et al., 2016. Enhancement of nutrient removal from swine wastewater digestate coupled to biogas purification by microalgae Scenedesmus spp. Bioresource Technology 202, 67–75. Available from: https://doi.org/10.1016/j.biortech.2015.11.082. Brazil: Elsevier Ltd.

Pulz, O., 2001. Photobioreactors: production systems for phototrophic microorganisms. Applied Microbiology and Biotechnology 57 (3), 287–293. Available from: https://doi.org/10.1007/s002530100702. Germany.

Pérez-López, P., et al., 2017. Comparative life cycle assessment of real pilot reactors for microalgae cultivation in different seasons. Applied Energy 205, 1151–1164. Available from: https://doi.org/10.1016/j.apenergy.2017.08.102. Spain: Elsevier Ltd.

Pérez-López, P., et al., 2014. Life cycle assessment of the production of the red antioxidant carotenoid astaxanthin by microalgae: from lab to pilot scale. Journal of Cleaner Production. Spain 64, 332–344. Available from: https://doi.org/10.1016/j.jclepro.2013.07.011.

Qadir, M., et al., 2020. Global and regional potential of wastewater as a water, nutrient and energy source. Natural Resources Forum 44 (1), 40–51. Available from: https://doi.org/10.1111/1477-8947.12187. Canada: Blackwell Publishing Ltd.

Qiang, H., Richmond, A., 1996. Productivity and photosynthetic efficiency of Spirulina platensis as affected by light intensity, algal density and rate of mixing in a flat plate photobioreactor. Journal of Applied Phycology 8 (2), 139–145. Available from: https://doi.org/10.1007/BF02186317. Israel: Kluwer Academic Publishers.

Quiroz Arita, C.E., Peebles, C., Bradley, T.H., 2015. Scalability of combining microalgae-based biofuels with wastewater facilities: a review. Algal Research 9, 160–169. Available from: https://doi.org/10.1016/j.algal.2015.03.001. United States: Elsevier B.V.

Rafati, L., et al., 2010. Removal of chromium (VI) from aqueous solutions using lewatit fo36 nano ion exchange resin. International Journal of Environmental Science and Technology. Iran: Center for Environmental and Energy Research and Studies 7 (1), 147–156. Available from: https://doi.org/10.1007/BF03326126.

Rahman, S.M.M., Handler, R.M., Mayer, A.L., 2016. Life cycle assessment of steel in the ship recycling industry in Bangladesh. Journal of Cleaner Production 135, 963–971. Available from: https://doi.org/10.1016/j.jclepro.2016.07.014. United States: Elsevier Ltd.

Rawat, I., et al., 2011. Dual role of microalgae: phycoremediation of domestic wastewater and biomass production for sustainable biofuels production. Applied Energy 88 (10), 3411–3424. Available from: https://doi.org/10.1016/j.apenergy.2010.11.025. South Africa: Elsevier Ltd.

Ray, N.E., Terlizzi, D.E., Kangas, P.C., 2015. Nitrogen and phosphorus removal by the Algal Turf Scrubber at an oyster aquaculture facility. Ecological Engineering 78, 27–32. Available from: https://doi.org/10.1016/j.ecoleng.2014.04.028. United States: Elsevier B.V.

Riaño, B., Molinuevo, B., García-González, M.C., 2011. Treatment of fish processing wastewater with microalgae-containing microbiota. Bioresource Technology 102 (23), 10829–10833. Available from: https://doi.org/10.1016/j.biortech.2011.09.022. Spain.

Risch, E., et al., 2014. How environmentally significant is water consumption during wastewater treatment?: application of recent developments in LCA to WWT technologies used at 3 contrasted geographical locations. Water Research 57, 20–30. Available from: https://doi.org/10.1016/j.watres.2014.03.023. France: Elsevier Ltd.

Rivas, J., et al., 2010. Treatment of cheese whey wastewater: combined coagulation—flocculation and aerobic biodegradation. Journal of Agricultural and Food Chemistry 58 (13), 7871–7877. Available from: https://doi.org/10.1021/jf100602j. Spain.

Rodrigues, D.B., et al., 2014. Production of carotenoids from microalgae cultivated using agroindustrial wastes. Food Research International 65, 144–148. Available from: https://doi.org/10.1016/j.foodres.2014.06.037. Brazil: Elsevier Ltd.

Rose, J.B., Gerba, C.P., 1991. Assessing potential health risks from viruses and parasites in reclaimed water in Arizona and Florida, USA. Water Science and Technology. Available from: https://doi.org/10.2166/wst.1991.0665. United States: IWA Publishing.

Rothermel, M.C., et al., 2013. A life cycle assessment based evaluation of a coupled wastewater treatment and biofuel production paradigm. Journal of Environmental Protection. s 04 (09), 1018–1033. Available from: https://doi.org/10.4236/jep.2013.49118.

Rubio, F. C., et al., 1999. "Prediction of dissolved oxygen and carbon dioxide concentration profiles in tubular photobioreactors for microalgal culture." In: Biotechnology and Bioengineering. Wiley-Blackwell, vol. 62 (1), pp. 71–86. Available from: https://doi.org/10.1002/(SICI)1097-0290(19990105)62:1 < 71::AID-BIT9 > 3.3.CO;2-K.

Sakaguchi, T., Nakajima, A., Horikoshi, T., 1981. Studies on the accumulation of heavy metal elements in biological systems. European Journal of Applied Microbiology and Biotechnology 12 (2), 84–89. Available from: https://doi.org/10.1007/bf01970039. Springer Science and Business Media LLC.

Salama, E.S., et al., 2017. Recent progress in microalgal biomass production coupled with wastewater treatment for biofuel generation. Renewable and Sustainable Energy Reviews 79, 1189–1211. Available from: https://doi.org/10.1016/j.rser.2017.05.091. South Korea: Elsevier Ltd.

Samer, M., 2015. Biological and Chemical Wastewater Treatment Processes. InTech, 10.5772/61250.

Sandnes, J.M., et al., 2005. Combined influence of light and temperature on growth rates of Nannochloropsis oceanica: linking cellular responses to large-scale biomass production. Journal of Applied Phycology. Norway 17 (6), 515–525. Available from: https://doi.org/10.1007/s10811-005-9002-x.

Schlesinger, A., et al., 2012. Inexpensive non-toxic flocculation of microalgae contradicts theories; overcoming a major hurdle to bulk algal production. Biotechnology Advances 30 (5), 1023–1030. Available from: https://doi.org/10.1016/j.biotechadv.2012.01.011. Israel.

Semerjian, L., Ayoub, G.M., 2003. High-pH-magnesium coagulation-flocculation in wastewater treatment. Advances in Environmental Research 7 (2), 389–403. Available from: https://doi.org/10.1016/S1093-0191(02)00009-6. Lebanon: Elsevier Ltd.

Sfez, S., et al., 2015. Environmental sustainability assessment of a microalgae raceway pond treating aquaculture wastewater: From up-scaling to system integration. Bioresource Technology 190, 321–331. Available from: https://doi.org/10.1016/j.biortech.2015.04.088. Belgium: Elsevier Ltd.

Sforza, E., et al., 2010. Vegetal oil from microalgae: mixotrophic growth in view of large-scale production. Journal of Biotechnology 150. Available from: https://doi.org/10.1016/j.jbiotec.2010.08.057. pp. 16–16 Elsevier BV.

Show, K.Y., Lee, D.J., 2013. Algal biomass harvesting. Biofuels from Algae 85–110. Available from: https://doi.org/10.1016/B978-0-444-59558-4.00005-X. China: Elsevier Inc.

Sincero, A.P., Sincero, G.A., 2002. Physical-chemical treatment of water and wastewater. In: Physical-Chemical Treatment of Water and Wastewater. CRC Press, United States, pp. 1–832. Available at: https://www.taylorfrancis.com/books/e/9781420031904.

Singh, A.K., et al., 2017. Phycoremediation of municipal wastewater by microalgae to produce biofuel. International Journal of Phytoremediation 19 (9), 805–812. Available from: https://doi.org/10.1080/15226514.2017.1284758. India: Taylor and Francis Inc.

Singh, A., Nigam, P.S., Murphy, J.D., 2011. Mechanism and challenges in commercialisation of algal biofuels. Bioresource Technology 102 (1), 26–34. Available from: https://doi.org/10.1016/j.biortech.2010.06.057. Ireland.

Singh, G., Patidar, S.K., 2018. Microalgae harvesting techniques: a review. Journal of Environmental Management 217, 499–508. Available from: https://doi.org/10.1016/j.jenvman.2018.04.010. India: Academic Press.

Snow, V.O., et al., 1999. Groundwater contamination from effluent irrigation. Water 26 (2), 26–29. Australia.

Solé-Bundó, M., et al., 2017. Assessing the agricultural reuse of the digestate from microalgae anaerobic digestion and co-digestion with sewage sludge. Science of the Total Environment 586, 1–9. Available from: https://doi.org/10.1016/j.scitotenv.2017.02.006. Spain: Elsevier B.V.

Spolaore, P., et al., 2006. Commercial applications of microalgae. Journal of Bioscience and Bioengineering. France 101 (2), 87–96. Available from: https://doi.org/10.1263/jbb.101.87.

Stagnitti, F., 1999. A model of the effects of nonuniform soil-water distribution on the subsurface migration of bacteria: implications for land disposal of sewage. Mathematical and Computer Modelling 29 (4), 41–52. Available from: https://doi.org/10.1016/S0895-7177(99)00038-2. Australia.

Suali, E., Sarbatly, R., 2012. Conversion of microalgae to biofuel. Renewable and Sustainable Energy Reviews 16 (6), 4316–4342. Available from: https://doi.org/10.1016/j.rser.2012.03.047. Malaysia.

Teitzel, G.M., Parsek, M.R., 2003. Heavy metal resistance of biofilm and planktonic Pseudomonas aeruginosa. Applied and Environmental Microbiology 69 (4), 2313–2320. Available from: https://doi.org/10.1128/AEM.69.4.2313-2320.2003. United States.

Thibodeau, C., et al., 2014. Comparison of black water source-separation and conventional sanitation systems using life cycle assessment. Journal of Cleaner Production 67, 45–57. Available from: https://doi.org/10.1016/j.jclepro.2013.12.012. Canada.

Toze, S. (1997) "Microbial Pathogens in Wastewater." CSIRO Land and Water Technical Report No. 1/97. https://publications.csiro.au/rpr/download?pid = procite:3effd8f9-a923-4ca9-a42a-9e6ce63ec1a6&dsid = DS1.

Thinkstep, 2019. GaBi Database & Modelling Principles. Thinkstep, Germany. Available from: http://www.gabi-software.com/fileadmin/gabi/Modeling_Principles_2019.pdf.

Toze, S., 2006. Reuse of effluent water—Benefits and risks. Agricultural Water Management. Available from: https://doi.org/10.1016/j.agwat.2005.07.010. Australia.

Uduman, N., et al., 2010. Dewatering of microalgal cultures—a major bottleneck to algae-based fuels. Renewable and Sustainable Energy Reviews 2. Available from: https://doi.org/10.1063/1.3294480.

Ugwu, C.U., Aoyagi, H., Uchiyama, H., 2008. Photobioreactors for mass cultivation of algae. Bioresource Technology 99 (10), 4021–4028. Available from: https://doi.org/10.1016/j.biortech.2007.01.046. Japan.

Van Den Hende, S., Vervaeren, H., Boon, N., 2012. Flue gas compounds and microalgae: (Bio-)chemical interactions leading to biotechnological opportunities. Biotechnology Advances 30 (6), 1405–1424. Available from: https://doi.org/10.1016/j.biotechadv.2012.02.015. Belgium.

Vree, et al., 2015. Comparison of four outdoor pilot-scale photobioreactors. Biotechnology for Biofuels 8. Available from: https://biotechnologyforbiofuels.biomedcentral.com/track/pdf/10.1186/s13068-015-0400-2.pdf.

Wang, H., et al., 2012. Mixotrophic cultivation of *Chlorella pyrenoidosa* with diluted primary piggery wastewater to produce lipids. Bioresource Technology 104, 215–220. Available from: https://doi.org/10.1016/j.biortech.2011.11.020. China.

Wang, J.H., et al., 2017. Microalgae-based advanced municipal wastewater treatment for reuse in water bodies. Applied Microbiology and Biotechnology 101 (7), 2659–2675. Available from: https://doi.org/10.1007/s00253-017-8184-x. China: Springer Verlag.

Wang, L., et al., 2010. Anaerobic digested dairy manure as a nutrient supplement for cultivation of oil-rich green microalgae Chlorella sp. Bioresource Technology 101 (8), 2623–2628. Available from: https://doi.org/10.1016/j.biortech.2009.10.062. United States.

Wang, Q., 2019. Growth enhancement of biodiesel-promising microalga Chlorella pyrenoidosa in municipal wastewater by polyphosphate-accumulating organisms. Journal of Cleaner Production. 240, 118148. Available from: https://doi.org/10.1016/j.jclepro.2019.118148. Elsevier BV.

Weidema, B., et al., 2013. Overview and methodology. Data quality guideline for the ecoinvent database version 3. Ecoinvent Report 1 (v3), The ecoinvent Centre.

Welch, E., 1992. Ecological effects of wastewater. Applied Limnology and Pollutant Effects. Chapman and Hall.

Yates, M., Gerba, 1997. Microbial considerations in wastewater reclamation and reuse,". Wastewater Reclamation and Reuse 1163–1192. Asano T. CRC Press.

Yin, H., et al., 2019. Textile wastewater treatment for water reuse: a case study. Processes 7 (1), 34. Available from: https://doi.org/10.3390/pr7010034. MDPI AG.

Yin, Z., et al., 2020. A comprehensive review on cultivation and harvesting of microalgae for biodiesel production: environmental pollution control and future directions. Bioresource Technology 301, 122804. Available from: https://doi.org/10.1016/j.biortech.2020.122804. Elsevier BV.

Yuan, Z., Pratt, S., Batstone, D.J., 2012. Phosphorus recovery from wastewater through microbial processes. Current Opinion in Biotechnology 23 (6), 878–883. Available from: https://doi.org/10.1016/j.copbio.2012.08.001. Australia.

Zang, Y., et al., 2015. Towards more accurate life cycle assessment of biological wastewater treatment plants: a review. Journal of Cleaner Production 107, 676–692. Available from: https://doi.org/10.1016/j.jclepro.2015.05.060. China: Elsevier Ltd.

Zhang, S., et al., 2014. Urban nutrient recovery from fresh human urine through cultivation of *Chlorella sorokiniana*. Journal of Environmental Management 145, 129–136. Available from: https://doi.org/10.1016/j.jenvman.2014.06.013. China: Academic Press.

Zhou, W., et al., 2012. A hetero-photoautotrophic two-stage cultivation process to improve wastewater nutrient removal and enhance algal lipid accumulation. Bioresource Technology 110, 448–455. Available from: https://doi.org/10.1016/j.biortech.2012.01.063. United States.

Zhou, Y., et al., 2013. A synergistic combination of algal wastewater treatment and hydrothermal biofuel production maximized by nutrient and carbon recycling. Energy and Environmental Science 6 (12), 3765–3779. Available from: https://doi.org/10.1039/c3ee24241b. United States.

Zhu, C., et al., 2018. Ultrasound enhanced electrochemical oxidation of Alizarin Red S on boron doped diamond (BDD) anode: effect of degradation process parameters. Chemosphere 209, 685–695. Available from: https://doi.org/10.1016/j.chemosphere.2018.06.137. China: Elsevier Ltd.

Life cycle assessment of microalgal biomass for valorization

Maria Lúcia Calijuri, Iara Barbosa Magalhães, Jessica Ferreira, Jackeline de Siqueira Castro and Bianca Barros Marangon

Post-Graduate Program in Civil Engineering, Department of Civil Engineering, Campus Universitário, Federal University of Viçosa (Universidade Federal de Viçosa/UFV), Viçosa, Minas Gerais, Brazil

9.1 Microalgae-based biorefinery and different routes

Researchers have been striving for products to become more sustainable than others with the same function that is already widespread and consolidated in the market. Products from fossil fuels are one of the leading examples of products to be replaced. In addition to representing a negative environmental impact due to the products and processes involved in their production, their future availability is compromised by the possibility of feedstock depletion (Londoño-Pulgarin et al., 2021).

Microalgae grown coupled to wastewater treatment can be harvested and used to produce biofuels and other nonfood bioproducts, providing an alternative to other potentially impacting sources. In this sense, they represent an opportunity for a paradigm shift in wastewater treatment for resource recovery, adding value to products obtained during treatment while mitigating the adverse effects of human activities (Arashiro et al., 2018; Catone et al., 2021).

The different characteristics of wastewater and, consequently, of the biomass produced in this cultivation medium provides a range of possibilities for valorization routes while at the same time limiting its use in other routes. Technical studies carried out with biomass from microalgae produced in wastewater show that not every route to use the biomass produced in this culture medium is economically or energetically viable (Assemany et al., 2016; Jankowska et al., 2017). Therefore the recovery route must be chosen to optimize the bearing of a given product, which is technically, economically, and environmentally viable.

Valorisation of Microalgal Biomass and Wastewater Treatment
DOI: https://doi.org/10.1016/B978-0-323-91869-5.00004-1

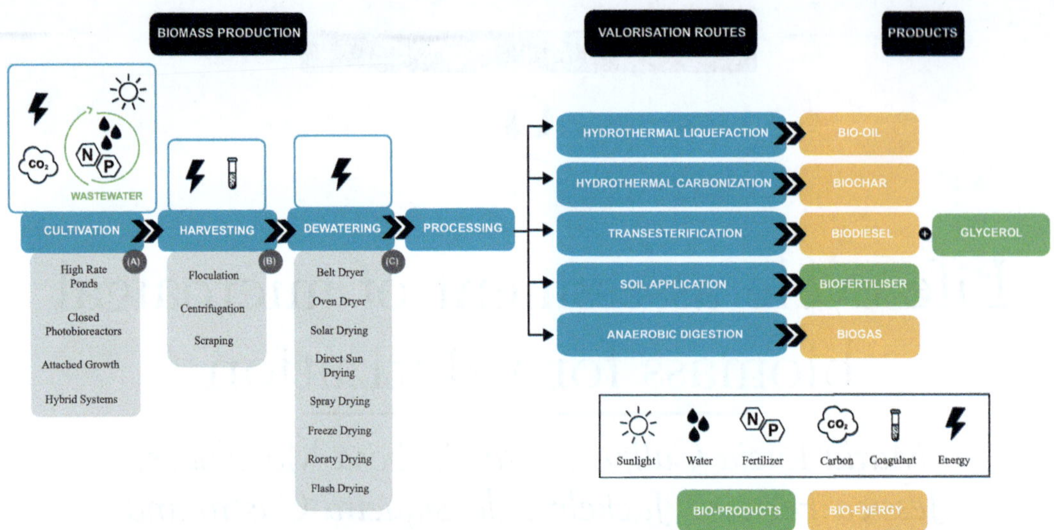

FIGURE 9.1 Microalgae biorefineries and valorization routes. Source: *Based on (a) Shahid et al. (2020); (b) de Assis (2020); (c) Agbede et al. (2020); Aliyu et al. (2021).*

In this sense, microalgae grown in wastewater are located at the intersection between the circular economy and the bioeconomy. Thus, the concept of "Circular Bioeconomy" has recently emerged, representing the convergence of these two approaches (Santagata et al., 2021; Hadley Kershaw et al., 2021). To facilitate the use of waste and the combined production of value-added products, biorefineries are considered a fundamental part of the circular bioeconomy (Stegmann et al., 2020) (Fig. 9.1). The main research objectives of biorefineries focus on identifying critical factors for large-scale development, increased microalgal biomass productivity, and composition and conversion efficiencies. Additionally, the environmental impacts associated with the valuation of microalgal biomass have been the focus of recent research through life cycle assessment (LCA) (Arashiro et al., 2018; Wu et al., 2018; Colzi Lopes et al., 2018; Hossain et al., 2019; Xiao et al., 2020).

In this chapter the focus will be given on LCAs of biomass recovery routes from microalgae cultivated in wastewater, namely, (1) routes that are technically widespread, like transesterification and anaerobic digestion (AD), and (2) routes with potential from a technical point of view (with technical research still in the initial phase) like hydrothermal carbonization (HTC) and hydrothermal liquefaction (HTL), bioethanol, and biofertilizers.

9.2 Life cycle assessment as a strategic tool for environmental feasibility analysis

The LCA has proven to be an essential tool to quantify the potential environmental impacts of a product or service. The methodology is based on tracking material, resources, energy requirements, and environmental outputs and emissions (Hijazi, 2016;

Yuan et al., 2015). Its application helps identify environmental bottlenecks of technologies within an existing production chain (Togarcheti et al., 2017) or in its early development stages (DeRose et al., 2019). In this way, its results can guide policymakers, industry, engineers, and other interested parties to outline promising strategies to pursue greater environmental viability.

The life cycle approach, known as cradle-to-grave, includes the analysis of environmental flows from the extraction of raw material to the final disposal of the product, going through all stages of its life (manufacturing, transport, distribution, and waste treatment). However, many studies are performed with cradle-to-gate, gate-to-gate, and gate-to-grave approaches depending on the objective. Also, LCA can be performed from two different perspectives: attributional and consequential (Hijazi et al., 2016).

Many documents have been prepared to guide and standardize LCAs (Frischknecht and Jolliet, 2019; ISO, 2006; EC-JRC 2011; Goedkoop et al., 2010). The standards developed by the International Organization for Standardization (ISO)—ISO 14040 and ISO 14044—are the most commonly used documents. The LCA has been standardized with a basic structure of four main steps: definition of objective and scope, life cycle inventory (LCI) analysis, life cycle impact assessment (LCIA), and interpretation of results (ISO, 2006).

The objective defines the study's application, motivation, and target audience (ISO, 2006). The scope of the study must clearly and transparently present the product system studied and its boundaries and functions (primary and secondary). Also, the functional unit (FU), allocation procedures, data quality requirements, all the assumptions and limitations are adopted, and, finally, which types of impacts and the methodology for LCIA will be used. The FU is a reference measure to assess the performance of the product system, endowed with function and value. Later, the LCI elaboration stage guides the reference for carrying out mass and energy balance calculations. Finally, boundaries clarify the full geographic, technological, and temporal coverage of the product's life cycle (ISO, 2006), delimiting which processes will be considered in the study.

The elaboration of the LCI occurs through the data collection and procedure for calculating the inputs (raw materials and products) and outputs (product, by-products, and liquid, solid, and gaseous emissions) of each process (ISO, 2006). It is essential to pay attention to all data quality requirements and predefined allocations during this step and even define them again if necessary. Allocation consists of allocating material or energy flows when a process generates one or more products. In this way, each product assumes part of the environmental responsibility. ISO 14044/2006 recommends avoiding allocations, as it is challenging to allocate impacts between by-products properly. In this attempt, more assumptions are applied, which can significantly interfere with the final result of the study. For this the system's expansion can be carried out, in which it is considered that, for by-products, their respective conventional products are considered avoided products (Arashiro et al., 2018). If this is not possible, the allocation can be made considering the physical characteristics (mass, energy, or nutritional content) or economic value of the products, according to the objective and products studied.

LCIA consists of correlating inputs, outputs, and emissions into quantitative results of environmental impact, which are commonly expressed in different categories. For this, factors and indicators are used, elaborated based on scientific evidence (Hijazi et al., 2016). The categories express the impacts associated with resources, human health, and ecological consequences (ISO, 2006). This step can be supported using specific software. The methodological and scientific frameworks for LCIA are constantly developed and improved (ISO, 2006), and the methodologies and databases available in the software are being updated over the years. Details about the software and database can be found in Campolina et al. (2015).

The results of impacts, by category, are expressed in characterization (final value of the environmental impact expressed, generally, by a substance that represents each category), normalization (magnitude on a standard scale for all categories), and weighting (values expressed considering the relevance between impact categories). Data quality analysis (contribution, sensitivity, and uncertainty) can be additionally performed to improve the understanding of the results' reliability (ISO, 2006), increasing the study's robustness. The interpretation step must be carried out in parallel with all the steps to guarantee their integrity. Finally, one should conclude the results of the stages of elaboration of the LCI and LCIA (Hijazi et al., 2016) based on a critical analysis that considers all processes and their technologies, limitations, and assumptions adopted (ISO, 2006).

9.3 Life cycle assessment in the context of microalgae biorefinery

9.3.1 Route 1—thermochemical processes: hydrothermal carbonization and hydrothermal liquefaction

HTC and HTL are thermochemical treatments for the conversion of wet biomass into bioenergy and bioproducts. When applied to microalgae, these processes promote the degradation of different biomass constituents, such as lipids, proteins, and carbohydrates. In addition, they prevent the drying of biomass, one of the bottlenecks for microalgal biorefineries (Choudhary et al., 2020). The water present in the biomass participates in the reactions that occur through the supply of thermal energy, using chemical catalysts being optional (Mathimani and Mallick 2019). HTC occurs in a temperature range of 180°C–260°C and pressure between 2 and 10 MPa. As a result, algal biomass is converted into hydrochar, which can be used as an energy source or as a nutrient source for the soil, in addition to gases and aqueous phases (AP) (Mishra et al., 2019; Castro et al., 2021). The operating temperature of the HTL ranges from 250°C to 380°C and the applied pressure from 2 to 25 MPa. Thus bio-oil is obtained, latter available for refining and use as fuel, solid waste, AP, and gases (Kazemi Shariat Panahi et al., 2019; Couto et al., 2020). Of the coproducts of HTC and HTL, AP represents a challenge to overcome for the application of hydrothermal technologies on an industrial scale due to their chemical complexity and toxicity (Watson et al., 2020; Leng et al., 2020).

LCA surveys of these valorization routes applied to biomass grown in wastewater are still developing. Therefore HTC and HTL LCAs applied to biomass cultivated in the synthetic medium were used as a source of information in this section. The research, in general, aims to quantify the potential environmental impacts of converting biomass to hydrochar and bio-oil (Fig. 9.2), as well as to identify process bottlenecks and propose technological improvements to optimize the environmental performance of microalgal biofuels. Relevant FUs are the generation of 1 kWh of electricity (Sztancs et al., 2021) and 1 MJ of fuel (Bennion et al., 2015; Sun et al., 2019; DeRose et al., 2019; Chen and Quinn 2021). The system boundaries applicable to these routes are cradle-to-gate (Sztancs et al., 2021), well-to-pump (Bennion et al., 2015;

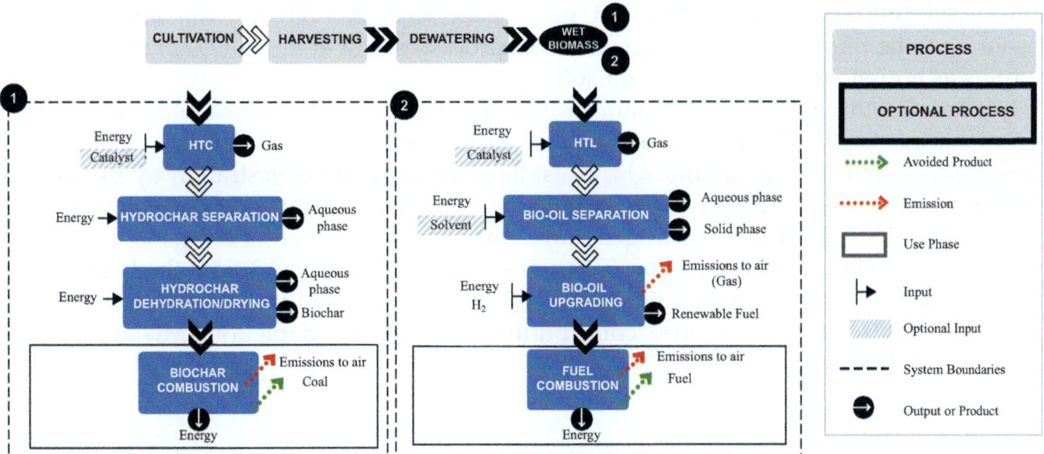

FIGURE 9.2 Main processes involved in the production of hydrochar and bio-oil included in the borders of LCAs. *LCAs*, Life cycle assessments.

DeRose et al., 2019), and well-to-wheels (Chen and Quinn 2021). It is noteworthy that both the FU and the system boundary are not limited to those mentioned earlier.

According to the proposed biorefinery configuration, cultivation, harvesting, and biomass dewatering (concentration) can be included within the system boundaries (Sun et al., 2019; Chen and Quinn 2021; Sztancs et al., 2021). However, there is the possibility of not considering these steps, as they are part of the wastewater treatment. Castro et al. (2020a,b) considered that, since wastewater treatment is a mandatory activity, it should not be considered in the microalgae product system, as it promotes the mitigation of environmental liability. In LCA modeling with wastewater, the water and nutrients can be considered avoided products, regardless of the recovery route (Marangon, 2021; de Souza et al., 2019).

In addition to HTC, the necessary steps for the production of hydrochar are its separation (e.g., in a centrifuge) and its dehydration/drying (e.g., in an oven) (Sztancs et al., 2021). Besides HTL, separation (e.g., in a centrifuge) and upgrading must be included in the production of bio-oil. According to the target fuel, the bio-oil upgrading step can vary, composed of one or more processes, but the entry of H_2 must be foreseen (Sun et al., 2019; DeRose et al., 2019; Chen and Quinn 2021).

Water and nutrient inputs such as BG-11 substrate, urea, and diammonium phosphate (Bennion et al., 2015; Sztancs et al., 2021) were included in the LCAs discussed in this section. These inputs can be saved with microalgae growth in wastewater. Another input can be included in carbon dioxide (CO_2) to supplement the needs of microalgae. CO_2 can be modeled as a pure (commercial) gas (Connelly et al., 2015) or a flue gas (Sun et al., 2019). However, the capture of CO_2 can also be considered from the atmosphere through biofixation by microalgae (Mu et al., 2017; Sztancs et al., 2021). Energy inputs, thermal or electrical, are included (Sun et al., 2019; Chen and Quinn 2021; Sztancs et al., 2021). Transport of biomass to the processing unit (Sun et al., 2019; Sztancs et al., 2021) of bio-oil to the upgrading plant (Mu et al., 2017) and from biofuel to the point of distribution or use (Chen and Quinn 2021) can be considered. DeRose et al. (2019) and Sun et al. (2019)

considered that solid waste, AP, and gases generated in HTL are not emissions but resources to be reused as by-products in other processes. Mu et al. (2017) considered these outputs as emissions to air, water, and soil.

It is noteworthy that, from the main results found in these researches, the negative CO_2 emissions in climate change/global warming (the most discussed category in the LCAs presented), found by the authors who considered the carbon sequestration by microalgae during their growth (Bennion et al., 2015; Sun et al., 2019; DeRose et al., 2019). Another result to be highlighted is the nitrogen emissions from the release of AP generated in the HTL, causing marine eutrophication (Chen and Quinn 2021). In addition, the high energy demand for HTL and the production of H_2 are highlighted as inputs that cause emissions in various impact categories, also contributing to fossil fuel depletion (Mu et al., 2017; Chen and Quinn 2021).

9.3.2 Route 2—transesterification

Microalgae can accumulate lipids during their growth, configuring a viable feedstock for biodiesel production (Karpagam et al., 2021). Among the transformation processes, transesterification is the most commonly applied. In this process, algal biomass (mainly esters) reacts with alcohol in the presence of a catalyst (Saranya and Ramachandra 2020; Karpagam et al., 2021). As a result, the lipids and alcohols are transformed into methyl or ethyl esters and glycerol (Collotta et al., 2016). The biodiesel derived from this process is called FAME—fatty acid methyl esters.

Transesterification can be reactive (in situ/direct) or extractive (indirect), depending on whether the process starts directly from the microalgal biomass or the lipids after the extraction process, respectively. In the extractive process, solvents such as chloroform, methanol, hexane, or isopropanol are used individually or as a mixture to extract lipids from microalgae biomass (Ghosh et al., 2017). Then, in a second step, the extracted triglycerides are transformed into FAME in the presence of monohydroxy alcohol (such as methanol) and a catalyst (an alkali or acid) with glycerol as a by-product (Ghosh et al., 2017; Branco-Vieira et al., 2020). Finally, in the in situ process, the wet microalgal biomass is used directly for transesterification (Chopra et al., 2020), mixed intermittently with methanol in the presence of a suitable catalyst for a certain period at a specific temperature (Ghosh et al., 2017). Therefore, in all forms of the process, it is essential to pay attention to adequate conditions that ensure high efficiencies of FAME formation, such as the molar ratio of alcohol to oil, moisture, agitation rate, reaction time, temperature, and catalyst type (Karpagam et al., 2021).

In terms of LCA, transesterification is usually evaluated as a process included in the boundaries of a microalgal biorefinery. The main objectives are to assess the environmental impacts of biodiesel production from microalgae biomass (Raghuvanshi et al., 2018; Branco-Vieira et al., 2020). Some works compare the results with the impact of similar energy production by fossil fuels (Adesanya et al., 2014) or other terrestrial crops (Jez et al., 2017). In this case, the compatibility is done by choosing the appropriate FU. Commonly, studies use FUs such as (1) energy production, for example, 1 MJ of energy produced by biodiesel (Raghuvanshi et al., 2018); (2) volumetric, for example, 300 L of biodiesel (Chopra et al., 2020); or (3) mass, such as 1 ton of biodiesel (Adesanya et al., 2014).

Typical boundaries include cultivation, harvesting, drying, lipid extraction, and transesterification (Collotta et al., 2016) for two-step transesterification (Fig. 9.3). LCA that considers in situ transesterification processes is still incipient, with only one work found so far (Chopra et al., 2020).

The reactor where the transformation occurs uses water, heat, and energy as its main inputs, necessary for temperature control ($\sim 70 °C$) and adequate agitation. In addition, the necessary chemical reagents such as solvents, catalysts, and alcohols are standard inputs. Depending on the process used, inventories contain methanol, sulfuric acid, and hexane (Chopra et al., 2020). The main outputs are the residues of the chemical reagents used and the associated CO_2 emissions. The glycerol resulting from the process can be considered an avoided product (Branco-Vieira et al., 2020) or reinserted at the cultivation stages (Chopra et al., 2020). The recovery of solvents throughout the process is one of the current bottlenecks, with hexane and methanol being responsible for the negative impacts of the lipid extraction and transesterification steps (Raghuvanshi et al., 2018; Chopra et al., 2020).

Among the evaluated impact categories, the most considered is greenhouse gas (GHG) emission potential. Despite being identified as relevant in microalgae energy transformation systems, categories such as land use and water consumption do not present such explicit data (Collotta et al., 2018). On the other hand, the transformation process impacts categories such as photochemical oxidation, referring to emissions of reactive substances harmful to human health and ecosystems (Lardon et al., 2009). Comparative studies with other biofuel production methods, such as pyrolysis and HTL for bio-oil production and anaerobic digestion for biogas production, indicate transesterification with a more favorable energy balance than pyrolysis and lower GHG emissions than all processes of biofuel production evaluated (Sun et al., 2019). These comparisons, however, do not present compatible results and differ mainly in terms of the LCA boundaries and considerations adopted. Furthermore, few works explain the allocation procedures used.

Studies using wastewater as a culture medium are already proposed and pointed to impact reduction compared to freshwater cultivation (Raghuvanshi et al., 2018). The main disadvantage identified in this configuration is the competition with the microorganisms present in the effluent,

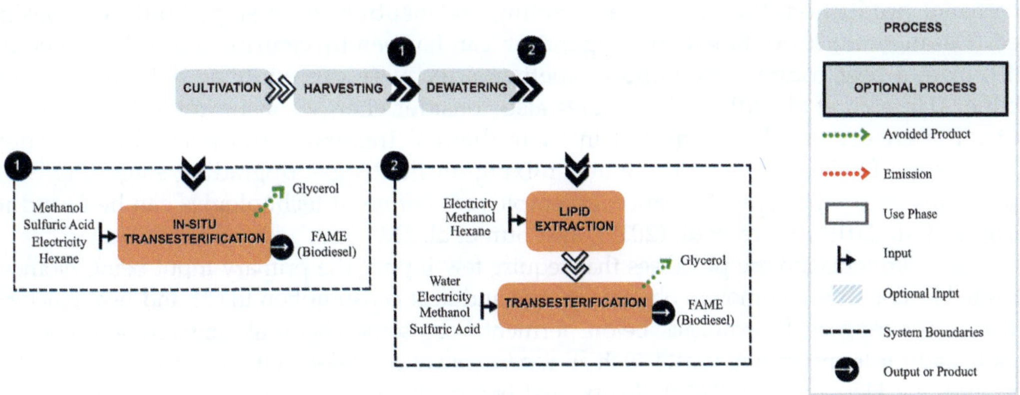

FIGURE 9.3   Main processes involved in the production of biodiesel included in the borders of LCAs. *LCAs,* Life cycle assessments.

limiting the growth and lipid accumulation in the biomass (Saranya and Shanthakumar 2019). Despite this, studies with strains more adapted to cultivation in wastewater from different sources showed satisfactory biodiesel yields (Fernández-Linares et al., 2017; Umamaheswari et al., 2020). The main advantage of this approach is the reduction of environmental impacts compared to freshwater cultivation (Raghuvanshi et al., 2018) and other biofuels (Nawkarkar et al., 2019), with lower atmospheric emissions, land use, eutrophication, and ecotoxicity.

9.3.3 Biological conversions of biomass: anaerobic digestion and fermentation

Given that biomass is submitted to a biotechnological process in the biochemical conversion pathways, the production yield depends on the microorganisms' performance. For biogas, anaerobic sludge composed of methanogenic bacteria and archaea is used, while for bioethanol, yeasts are used.

The LCA studies on biogas/biomethane and bioethanol aim to quantify the impacts of this bioenergy production. Several studies were performed to evaluate the energy and CO_2 emission balances. Some studies consider biodiesel production (main product) with the coproduction of biogas (Ventura et al., 2013; Kern, 2017; Yuan et al., 2015; Togarcheti et al., 2017) or bioethanol (Dasan et al., 2019) in their boundaries. A variety of FUs are used, such as treating 1 m^3 of water and 1 kg of biomass, 1 kg of biomethane and 1 MJ of energy, biofuel, or biomethane (Shimako et al., 2016; Jin et al., 2017; Arashiro et al., 2018; Dasan et al., 2019; Sun et al., 2019; Tasca et al., 2019; Xiao et al., 2020; Gholkar et al., 2021). As for bioethanol, the FUs used are 1 and 100,000 kg of biomass (Dasan et al., 2019; Hossain et al., 2019).

The main stages included in the boundaries of the studies (Fig. 9.4) are cultivation (mainly in ponds), harvesting performed by settling, flocculation/coagulation followed by AD or fermentation (after acid hydrolysis) steps. Centrifugation is observed in few cases, while biomass drying is not required. The biogas update step was performed by Ferreira et al. (2019), Sun et al. (2019), and Tasca et al. (2019). The later includes different technologies for removing H_2S, NH_3, and HCl. For biogas, there is also the possibility of biomass hydrothermal pretreatment (HTP) before AD. Due to the acid hydrolysis step during bioethanol production, the hydrolysate cooling and neutralization step should be considered (Dasan et al., 2019). Bioethanol separation can happen by centrifugation (Dasan et al., 2019) or in a beer column, while pure bioethanol recovery can occur by dehydration in a rectifier (Hossain et al., 2019). The studies also consider energy consumption to power the units. The energy and heat required in hydrothermal treatment and anaerobic digestion processes are for heating, tempering and mixing. In the biogas upgrading stage (Fig. 9.4), inputs depend on the type of technology employed. Different technologies can be found in Ferreira et al. (2019), Tasca et al. (2019), and Sun et al. (2019).

AD and fermentation are processes that require few inputs, the primary input being biomass (defatted or not). It is common to observe energy and heat consumption to stir and heat reactors. When considering acid hydrolysis before fermentation, some chemical consumption, such as H_2SO_4, and for fermentation, yeasts such as *Saccharomyces cerevisiae* are used (Dasan et al., 2019). The study by Hossain et al. (2019) also proved energy-intensive when considering microwave-assisted extraction at its frontier. From the biogas production scenarios' outputs, biogas is commonly directed to a combined heat and power unit (Shimako et al., 2016; Arashiro et al., 2018).

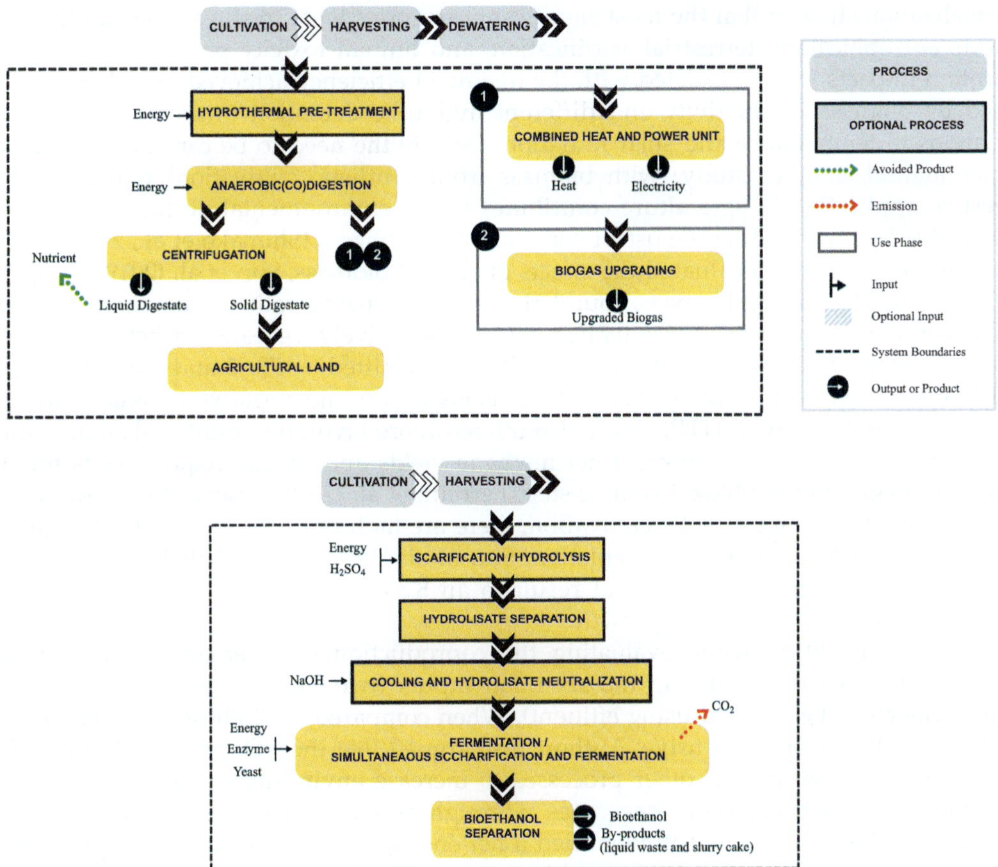

FIGURE 9.4 Main processes involved in the production of biogas included in the borders of LCAs. *LCAs*, Life cycle assessments.

The digestate and effluent can be used as biofertilizers; the latter can also be recirculated for biomass cultivation (Jin et al., 2017; Tasca et al., 2019).

In studies involving multifunctional systems (biodiesel and biogas), allocation by mass was performed (Shimako et al., 2016). System expansion was used for the digestate, considering that it replaces conventional fertilizer (Arashiro et al., 2018). Furthermore, during bioethanol production, the system's expansion can occur considering the energy credits of the by-products (liquid waste and slurry cake) (Hossain et al., 2019).

Considering the production of biogas with biomass grown in domestic sewage, Arashiro et al. (2018) demonstrated that the compensation (for the evaluated system) of the energy generated by the biogas and the avoided fertilizer, due to the application of digestate in the soil, were important factors that offered a reduction in environmental impacts, in some categories. On the other hand, the emission of NH_3 in the cultivation stage and heavy metals, from the digestate, in the soil stood out in the impact generated when the studied scenario was compared to the treatment in an activated sludge system.

Normalization showed that the most significant categories were freshwater eutrophication, marine eutrophication, terrestrial acidification, and human toxicity potentials. The eutrophication impacts were associated with the treatment efficiency achieved, which was influenced by biomass productivity and different hydraulic detention times due to seasonal variations in temperature and solar radiation. Despite the need to be careful to avoid soil contamination, another study (with biomass from synthetic cultivation) reinforced that digestate application in agriculture contributes to the environmental performance of biogas production by avoiding the use of conventional fertilizers (Shimako et al., 2016).

From the works that evaluated the effects of HTP of biomass, Sun et al. (2019), adopting FU of 100,000 kg of dry biomass, found that the net energy ratio of anaerobi digestion (AD) with and without HTP was 0.71 and 1.27, respectively, revealing a better result for the first system. The systems without and with HTP emitted -173.02 and -60.84 g CO_2eq (MJ biogas)$^{-1}$. Xiao et al. (2020) verified net energy gain and negative emissions for AD systems without and with HTP, and it also offered more favorable results. Also, the difference in results was due to different net methane yields and energy requirements for AD and HTP. Regarding the biogas update step, Ferreira et al. (2019) observed that, when carried out through photosynthetic cultivation, there are high impacts on climate change and low impacts on fossil and water depletion. Sun et al. (2019) found that the reuse of CO_2 lost during the upgrade steps would result in an 8.53% reduction in atmospheric emissions and less demand for algal growth.

Dasan et al. (2019), while evaluating the coproduction of bioethanol and biodiesel, observed that yeasts are among the six main inputs (two of these were fertilizers and water, which can be avoided using effluent). When compared to biodiesel, bioethanol production was less energy-intensive. Authors also argued that the CO_2 produced during fermentation could be used in other processes to increase environmental sustainability, for example, in the carbonation of beverages. Through an energy balance of bioethanol-only production, Hossain et al. (2019) identified a net energy ratio of 0.45 and a net production of 2749.6 GJ year^{-1} (for 100,000 kg of biomass per year). Considering the consumption (photosynthesis) and emission (use of bioethanol) of CO_2, the authors also demonstrated a positive net emission. However, comparing the CO_2 production for 1 gallon of pure ethanol with bioethanol, there was a 55% reduction. The authors highlighted that the application of effluent contributed to the environment.

9.3.4 Route 4—biofertilizer

Microalgal biomass cultivated in wastewater as a source of nutrients and soil conditioning in agriculture is an appealing practice. From a technical point of view, it is a product rich in nutrients and organic matter. Moreover, it can replace or amortize nutrients from fossil sources and industrial processes that cause severe environmental impacts. For years, studies evaluating the application of wet biomass to the soil and, more recently, organomineral fertilizers have been carried out from the perspective of agronomic benefits, such as plant growth and building soil fertility (Castro et al., 2017, Castro et al., 2020a,b; Marks et al., 2019; Lorentz et al., 2020; Pereira et al., 2021). Furthermore, research on this topic evaluated the emission to air, water, and soil, of substances that could potentially cause environmental impacts (Castro et al., 2017; Pereira et al., 2021). However, studies that assess the environmental impact of using this material as a source of nutrients are still small quantities (Arashiro, 2018; de Souza et al., 2019; Castro et al., 2020a,b) and have gaps to be explored.

In this sense, it is necessary to establish that, after harvesting, biomass can be applied moist to the soil as a source of nutrients (Fig. 9.5). Over time, it can provide the formation of a microalgal biofilm in the soil (Castro et al., 2017; Arashiro et al., 2018), promoting an increase in organic matter and consequently microbial activity. Additionally (Castro et al., 2020a,b; Pereira et al., 2021), microalgal biomass can be used to compose the mass of a granulated or pastilleted fertilizer to be bagged and marketed in molds, similar to commercial fertilizer nowadays. The main difference between these two biomass utilization routes is that, in the first one, the objective is to apply the biomass to the soil in regions close to where it was produced due to biomass moisture, thus optimizing the cost and impact on transport. On the other hand, in the second route, biofertilizer, the aim is to obtain a product with characteristics closer to the commercial one. That can be transported wherever needed, the main difference being the need for biomass drying and transport after production.

With distinctions made, the authors highlight that the main objectives of the LCAs were to analyze the impact of wastewater treatment through technologies that involve the use of microalgae for nutrient recovery. Therefore the volume of treated effluent or a unit of recovered nutrient was adopted as FU (Arashiro, 2018; de Souza et al., 2019). After applying biomass to the soil, the impacts related to emissions to water, air, and soil can be accounted for and incorporated into the modeling through the expansion of the system boundary. Additionally, Castro et al. (2020a,b) proposed drying and granulation within the product system boundary. The objective of this study was to evaluate the environmental impacts of the production of a biofertilizer composed of a mixture between triple superphosphate (commercial chemical fertilizer) and microalgal biomass, in a percentage defined in a previous experiment by Castro et al. (2020a,b) and to

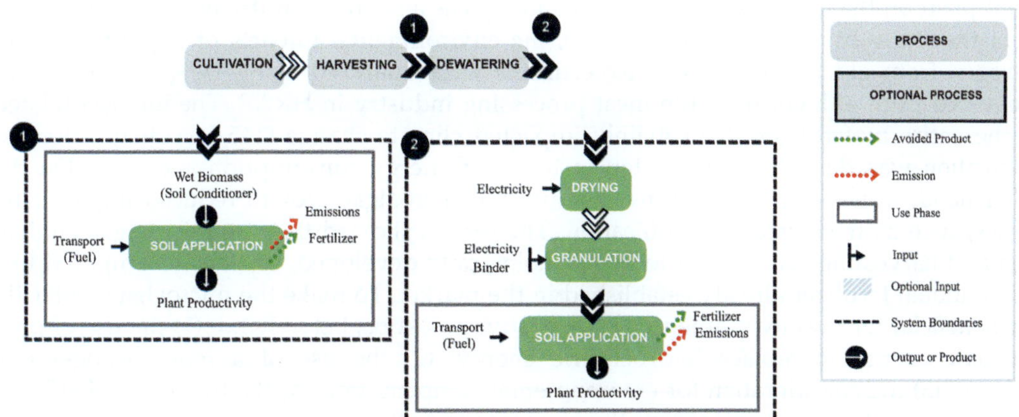

FIGURE 9.5 Main processes involved in the production of biofertilizer included in the borders of LCAs. *LCAs*, Life cycle assessments.

compare its impacts with those of commercially available conventional fertilizer production (100% triple superphosphate).

Phosphorus was recovered from a meat processing industry effluent in a high-rate algal pond (HRAP). The entire biofertilizer chain is mainly affected by climate changes (3.17 kg CO_2eq). Microalgal biofertilizer had a higher environmental impact than conventional fertilizer in all impact categories, highlighting climate change and terrestrial ecotoxicity. An ideal scenario was created considering that all energy used comes from photovoltaic panels; in the separation step, a physical method will be used, without energy expenditure (i.e., gravimetric sedimentation) and biomass will be dried in a drying bed instead of the thermal drying. In this scenario the impact of biofertilizer approaches those of triple superphosphate. When impacts of biomass cultivation and concentration stages were disregarded, the drying step was of great relevance, contributing to increasing biofertilizer impacts. More research is needed to optimize the algae production chain and determine the possibility of more environmentally attractive higher value-added products (Castro et al., 2020a,b).

As in most of the works already carried out with LCA, the authors see the need to compare the results obtained with other sources of nutrients already used, highlighting the importance of always considering that the final products must have the same functions. Thus, for example, fertilizer sources considered in the Ecoinvent database do not consider the granulation process (the processes that make up a product can be confirmed in the documentation of the product used). So the steps inherent to the systems must be equal as far as possible to avoid possible discrepancies.

In general, in works involving LCA of microalgae biomass biofertilizer, biomass production and effluent treatment are the first steps to be considered. At this stage, in general, the HRAPs are considered a production system. In addition to the material that makes up the HRAP, the electrical energy to supply the demand of the blade movement system is considered. As the structures that make up the system have a useful life of more than 25 years, the construction of this system can be disregarded in the model (Castro et al., 2020a,b; de Souza et al., 2019; Magalhães et al., 2021). Furthermore, Arashiro et al. (2018) considered the impact of the construction of component structures of the wastewater treatment plant on the life cycle of microalgae biomass as a source of nutrients for the soil.

In the study by de Souza et al. (2019), the environmental impacts of applying 1 kg of nitrogen from algal biomass (wet biofertilizer) were analyzed using LCA. Nitrogen was recovered from effluent from the meat processing industry in HRAP. The impacts related to the entire biofertilizer chain mainly impacted climate change (115 kgCO_2eq). Biomass cultivation was the most critical step in terms of energy consumption and time. On the other hand, wastewater as a culture medium for microalgae growth reduced impact categories, such as freshwater eutrophication. The results showed that the microalgae cultivation and harvesting stages need to be technologically developed, especially compared to a conventional fertilizer already established in the market. To make the microalgae biofertilizer beneficial for the environment, the alternatives must be, beforehand, (1) the use of photovoltaic energy to replace hydroelectric energy; (2) the use of a more nitrogen-rich effluent; (3) the consideration for environmental compensation for the treatment of effluent may be accounted for, disregarding the biomass production stage.

Arashiro et al. (2018) carried out an LCA of an HRAP system for wastewater treatment where microalgae biomass was reused for nutrient recovery (biofertilizer production) and bioenergy production (biogas). Furthermore, both alternatives were compared to a typical small activated sludge system. The results showed that the HRAP system coupled to biogas production is more environmentally viable than the HRAP system compared to biofertilizer production in climate change, ozone layer depletion, photochemical oxidant formation, and fossil destruction. Furthermore, different climatic conditions strongly influenced the results obtained in the eutrophication and metal depletion impact categories. The HRAP system where tropical temperatures and high solar radiation predominate (HRAP system coupled with biofertilizer production) had less impact on these categories. In addition, the characteristics (e.g., nutrient and heavy metal concentration) of microalgal biomass recovered from wastewater were crucial when assessing potential environmental impacts in the categories of terrestrial acidification, particulate material formation, and toxicity.

9.4 Uncertainty and sensitivity analysis

Uncertainty and sensitivity analysis are tools to test possibilities for reducing impacts and technical modifications in the microalgae valorization routes. The possibilities for carrying out these analyzes are diverse. Sun et al. (2019) performed sensitivity analysis on different bioenergy conversion processes, such as HTL, and found that microalgae cultivation had a significant impact on GHG emissions. This emission was due to the energy consumption in fertilizer production, microalgae productivity, and paddle rotation efficiency. Sztancs et al. (2021) examined the effects of the composition of the renewable energy mix on the sustainability indicator of the GHG footprint in hydrochar production by HTC. In biodiesel production the resulting impacts were more sensitive due to the electricity consumption, even surpassing the use of solvents (Liu et al., 2021). Also, in biodiesel production, higher yields tend to be the main path to reducing the impact on the route (Yang et al., 2011; Adesanya et al., 2014; Chopra et al., 2020). Arashiro et al. (2018) performed a sensitivity analysis considering a variation in NH_3 emissions and N_2O due to the application of digestate and biofertilizer in soils and the transport distance of these products to the application site. Yuan et al. (2015) evaluated the effects on GHG emissions by varying the CH_4 yield and Xiao et al. (2020) of biogas yield and nitrogen recovery from the digestate. Chen and Quinn (2021) performed a scenario analysis to verify the potential environmental impacts of HTL if other values found in the literature had been used. The authors varied the values of some operational parameters between conservative and optimistic limits. Also, de Souza et al. (2019) and Castro et al. (2020a,b) performed scenario analysis, in which the main inputs causing negative impacts were replaced by others considered to have less impact. Thus it is possible to investigate scenarios and identify aspects that need to be improved. However, concerning the routes for valuing biomass grown in wastewater, there is still a long path to go in this direction.

9.5 Challenges and opportunities in the light of life cycle assessment

Several studies have stated that wastewater can reduce impacts due to the use of water and nutrients. However, there are still few pieces of research that use primary data with the biomass cultivated in wastewater. This biomass's productivity and biochemical composition are different from those grown in media that ensure optimal nutritional conditions for microorganisms, which may require different inputs. Furthermore, as the biomass comprises other microorganisms, the same cultivation volume will lower the microalgae content. Therefore wastewater from different sources also offers different productivity and biochemical composition. Thus, presenting the wastewater parameters is interesting given they help explain microalgae' productivity, especially those referring to carbon, nitrogen, and phosphorus.

The main difficulties are mainly attributed to the lack of compatibility for comparing the results of microalgae-to-energy systems (Collotta et al., 2016). The studies must advance to present, clearly and objectively, the allocation procedures, scenario analysis, uncertainty, and sensitivity. The limitations in some studies are not stated or are not very evident. Thus there is a lack of clarity for readers to understand the interpretation of the results and even for future studies to carry out different models and seek to overcome the limitations found so far.

Advances in studies of conversion pathways should be encouraged. However, the cultivation of microalgae also needs to be improved. Ferreira et al. (2020) and Magalhães et al. (2021) highlighted the possibility of the concentration factor obtained in a hybrid system (composed of a pond and biofilm reactor) to provide better results for the production routes. The LCA from the recirculation of CO_2 can also be included produced during digestion or fermentation to the cultivation stages to identify its potential for reducing environmental impact and increasing productivity (Valente et al., 2019). In addition, there is the potential of cogeneration of products, helping to make the application of this biomass viable. The contribution of LCA to increase the use of water, nutrients, and energy, contributing to the circular economy in microalgae biorefineries, is also noteworthy.

References

Adesanya, V.O., et al., 2014. Life cycle assessment on microalgal biodiesel production using a hybrid cultivation system, Bioresource Technology, 163. Elsevier Ltd, United Kingdom, pp. 343–355. Available from: http://doi.org/10.1016/j.biortech.2014.04.051.

Agbede, O.O., et al., 2020. Thin layer drying of green microalgae (*Chlorella* sp.) paste biomass: drying characteristics, energy requirement and mathematical modeling, Bioresource Technology Reports, 11. Elsevier BV, p. 100467. Available from: http://doi.org/10.1016/j.biteb.2020.100467.

Aliyu, A., Lee, J.G.M., Harvey, A.P., 2021. Microalgae for biofuels via thermochemical conversion processes: a review of cultivation, harvesting and drying processes, and the associated opportunities for integrated production, Bioresource Technology Reports, 14. Elsevier BV, p. 100676. Available from: http://doi.org/10.1016/j.biteb.2021.100676.

Arashiro, L.T., et al., 2018. Life cycle assessment of high rate algal ponds for wastewater treatment and resource recovery, Science of the Total Environment, 622–623. Elsevier B.V, Spain, pp. 1118–1130. Available from: http://doi.org/10.1016/j.scitotenv.2017.12.051.

Assemany, P.P., et al., 2016. Energy potential of algal biomass cultivated in a photobioreactor using effluent from a meat processing plant, Algal Research, 17. Elsevier B.V, Brazil, pp. 53–60. Available from: http://doi.org/10.1016/j.algal.2016.04.018.

Bennion, E.P., et al., 2015. Lifecycle assessment of microalgae to biofuel: comparison of thermochemical processing pathways, Applied Energy, 154. Elsevier Ltd, United States, pp. 1062–1071. Available from: http://doi.org/10.1016/j.apenergy.2014.12.009.

Branco-Vieira, M., et al., 2020. Environmental assessment of industrial production of microalgal biodiesel in central-south Chile. Journal of Cleaner Production. Elsevier Ltd, Portugal, p. 266. Available from: http://doi.org/10.1016/j.jclepro.2020.121756.

Campolina, J., et al., 2015. A REVIEW OF LITERATURE ON SOFTWARE USED IN STUDIES OF LIFE CYCLE. Revista Eletrônica Em Gestão, Educação E Tecnologia Ambiental 19, 735–750. Available from: https://doi.org/10.5902/2236117015494.

Castro, J.d.S., Calijuri, M.L., Mattiello, E.M., et al., 2020a. Algal biomass from wastewater: soil phosphorus bioavailability and plants productivity. Science of the Total Environment. Elsevier B.V, Brazil, p. 711. Available from: http://doi.org/10.1016/j.scitotenv.2019.135088.

Castro, J.d.S., Calijuri, M.L., Ferreira, J., et al., 2020b. Microalgae based biofertilizer: a life cycle approach. Science of the Total Environment. Elsevier B.V, Brazil, p. 724. Available from: http://doi.org/10.1016/j.scitotenv.2020.138138.

Castro, J.d.S., et al., 2017. Microalgae biofilm in soil: greenhouse gas emissions, ammonia volatilization and plant growth, Science of the Total Environment, 574. Elsevier B.V, Brazil, pp. 1640–1648. Available from: http://doi.org/10.1016/j.scitotenv.2016.08.205.

Castro, J.d.S., et al., 2021. Hydrothermal carbonization of microalgae biomass produced in agro-industrial effluent: products, characterization and applications, Science of the Total Environment, 768. Elsevier BV, p. 144480. Available from: http://doi.org/10.1016/j.scitotenv.2020.144480.

Catone, C.M., et al., 2021. Bio-products from algae-based biorefinery on wastewater: a review, Journal of Environmental Management, 293. Elsevier BV, p. 112792. Available from: http://doi.org/10.1016/j.jenvman.2021.112792.

Chen, P.H., Quinn, J.C., 2021. Microalgae to biofuels through hydrothermal liquefaction: Open-source techno-economic analysis and life cycle assessment, Applied Energy, 289. Elsevier BV, p. 116613. Available from: http://doi.org/10.1016/j.apenergy.2021.116613.

Chopra, J., et al., 2020. Environmental impact analysis of oleaginous yeast based biodiesel and bio-crude production by life cycle assessment, Journal of Cleaner Production, 271. Elsevier BV, p. 122349. Available from: http://doi.org/10.1016/j.jclepro.2020.122349.

Choudhary, P., et al., 2020. A review of biochemical and thermochemical energy conversion routes of wastewater grown algal biomass, Science of the Total Environment, 726. Elsevier BV, p. 137961. Available from: http://doi.org/10.1016/j.scitotenv.2020.137961.

Collotta, M., et al., 2016. Evaluating microalgae-to-energy-systems: different approaches to life cycle assessment (LCA) studies. Biofuels, Bioproducts and Biorefining 10 (6), 883–895. Available from: https://doi.org/10.1002/bbb.1713. Italy: John Wiley and Sons Ltd.

Collotta, M., et al., 2018. Wastewater and waste CO_2 for sustainable biofuels from microalgae, Algal Research, 29. Elsevier B.V, Italy, pp. 12–21. Available from: http://doi.org/10.1016/j.algal.2017.11.013.

Colzi Lopes, A., et al., 2018. Energy balance and life cycle assessment of a microalgae based wastewater treatment plant: a focus on alternative biogas uses, Bioresource Technology, 270. Elsevier Ltd, Spain, pp. 138–146. Available from: http://doi.org/10.1016/j.biortech.2018.09.005.

Connelly, E.B., et al., 2015. Life cycle assessment of biofuels from algae hydrothermal liquefaction: the upstream and downstream factors affecting regulatory compliance. Energy and Fuels 29 (3), 1653–1661. Available from: https://doi.org/10.1021/ef502100f. United States: American Chemical Society.

Couto, E., Calijuri, M.L., Assemany, P., 2020. Biomass production in high rate ponds and hydrothermal liquefaction: wastewater treatment and bioenergy integration. Science of the Total Environment. Elsevier B.V, Brazil, p. 724. Available from: http://doi.org/10.1016/j.scitotenv.2020.138104.

Dasan, Y.K., et al., 2019. Life cycle evaluation of microalgae biofuels production: effect of cultivation system on energy, carbon emission and cost balance analysis, Science of the Total Environment, 688. Elsevier B.V, Malaysia, pp. 112–128. Available from: http://doi.org/10.1016/j.scitotenv.2019.06.181.

de Assis, L.R., et al., 2020. Innovative hybrid system for wastewater treatment: high-rate algal ponds for effluent treatment and biofilm reactor for biomass production and harvesting. Journal of Environmental Management 274, Brazil: Academic Press. Available from: https://doi.org/10.1016/j.jenvman.2020.111183.

DeRose, K., et al., 2019. Integrated techno economic and life cycle assessment of the conversion of high productivity, low lipid algae to renewable fuels, Algal Research, 38. Elsevier BV, p. 101412. Available from: http://doi.org/10.1016/j.algal.2019.101412.

de Souza M.H.B., Calijuri M.L., Assemany P.P., et al (2019) Soil application of microalgae for nitrogen recovery: a life-cycle approach. Journal of Cleaner Production 211:342–349. https://doi.org/10.1016/j.jclepro.2018.11.097

EC-JRC (Commission of the European Union. Joint Research Centre. Institute for Environment and Sustainability). (2011). International reference life cycle data system (ILCD) handbook:general guide for life cycle assessment:

provisions and action steps. Publications Office. https://nam11.safelinks.protection.outlook.com/?url=https%3A%2F%2Fdoi.org%2F10.2788%2F33030&data=05%7C01%7Ck.anbazhagan%40elsevier.com%7C8c44a7440cd74738bac908da7fa183dd%7C9274ee3f94254109a27f9fb15c10675d%7C0%7C0%7C637962628926965569%7CUnknown%7CTWFpbGZsb3d8eyJWIjoiMC4wLjAwMDAiLCJQIjoiV2luMzIiLCJBTiI6Ik1haWwiLCJXVCI6Mn0%3D%7C3000%7C%7C%7C&sdata=teg6sDcQBA8DLBSH3PRBWvs%2FMutR0pn4FAu7BKTDU58%3D&reserved=0; Available from: https://doi.org/10.2788/33030.

Fernández-Linares, L.C., et al., 2017. Assessment of *Chlorella vulgaris* and indigenous microalgae biomass with treated wastewater as growth culture medium, Bioresource Technology, 244. Elsevier Ltd, Mexico, pp. 400–406. Available from: http://doi.org/10.1016/j.biortech.2017.07.141.

Ferreira, A.F., et al., 2019. Life cycle assessment of pilot and real scale photosynthetic biogas upgrading units, Algal Research, 44. Elsevier BV, p. 101668. Available from: http://doi.org/10.1016/j.algal.2019.101668.

Ferreira, J., et al., 2020. Innovative microalgae biomass harvesting methods: technical feasibility and life cycle analysis, Science of The Total Environment, 746. Elsevier BV, p. 140939. Available from: http://doi.org/10.1016/j.scitotenv.2020.140939.

Frischknecht, R., Jolliet, O., 2019. Global Guidance on Environmental Life Cycle Impact Assessment Indicators, 2. UNEP/SETAC Life Cycle Initiative, Paris, France, pp. 80–103.

Gholkar, P., Shastri, Y., Tanksale, A., 2021. Renewable hydrogen and methane production from microalgae: a techno-economic and life cycle assessment study, Journal of Cleaner Production, 279. Elsevier BV, p. 123726. Available from: http://doi.org/10.1016/j.jclepro.2020.123726.

Ghosh, S., Banerjee, S., Das, D., 2017. Process intensification of biodiesel production from *Chlorella* sp. MJ 11/11 by single step transesterification, Algal Research, 27. Elsevier B.V, India, pp. 12–20. Available from: http://doi.org/10.1016/j.algal.2017.08.021.

Goedkoop, M., de Schryver, A., Oele, M., Durksz, S., de Roest, D. 2010. Introduction to LCA with SimaPro 7; Pré Consultants. Retrieved from http://www.pre-sustainability.com/download/manuals/SimaPro7Introduction ToLCA.pdf.

Hadley Kershaw, E., et al., 2021. The sustainable path to a circular bioeconomy. Trends in Biotechnology 39 (6), 542–545. Available from: https://doi.org/10.1016/j.tibtech.2020.10.015. United Kingdom: Elsevier Ltd.

Hijazi, O., et al., 2016. Review of life cycle assessment for biogas production in Europe, Renewable and Sustainable Energy Reviews, 54. Elsevier Ltd, Germany, pp. 1291–1300. Available from: http://doi.org/10.1016/j.rser.2015.10.013.

Hossain, N., Zaini, J., Indra Mahlia, T.M., 2019. Life cycle assessment, energy balance and sensitivity analysis of bioethanol production from microalgae in a tropical country. Renewable and Sustainable Energy Reviews. Elsevier Ltd, Australia, p. 115. Available from: http://doi.org/10.1016/j.rser.2019.109371.

International Organization for Standardization, 2006. Environmental management: life cycle assessment; requirements and guidelines, 14044. ISO, Geneva, Switzerland.

Jankowska, E., Sahu, A.K., Oleskowicz-Popiel, P., 2017. Biogas from microalgae: review on microalgae's cultivation, harvesting and pretreatment for anaerobic digestion, Renewable and Sustainable Energy Reviews, 75. Elsevier Ltd, Poland, pp. 692–709. Available from: http://doi.org/10.1016/j.rser.2016.11.045.

Jez, S., et al., 2017. Comparative life cycle assessment study on environmental impact of oil production from micro-algae and terrestrial oilseed crops, Bioresource Technology, 239. Elsevier Ltd, Italy, pp. 266–275. Available from: http://doi.org/10.1016/j.biortech.2017.05.027.

Jin, Q., et al., 2017. Comparison of biogas production from an advanced micro-bio-loop and conventional system, Journal of Cleaner Production, 148. Elsevier Ltd, China, pp. 245–253. Available from: http://doi.org/10.1016/j.jclepro.2017.02.021.

Karpagam, R., Jawaharraj, K., Gnanam, R., 2021. Review on integrated biofuel production from microalgal biomass through the outset of transesterification route: a cascade approach for sustainable bioenergy, Science of the Total Environment, 766. Elsevier BV, p. 144236. Available from: http://doi.org/10.1016/j.scitotenv.2020.144236.

Kazemi Shariat Panahi, H., et al., 2019. Recent updates on the production and upgrading of bio-crude oil from microalgae, Bioresource Technology Reports, 7. Elsevier BV, p. 100216. Available from: http://doi.org/10.1016/j.biteb.2019.100216.

Kern, J.D., et al., 2017. Using life cycle assessment and techno-economic analysis in a real options framework to inform the design of algal biofuel production facilities, Bioresource Technology, 225. Elsevier Ltd, United States, pp. 418–428. Available from: http://doi.org/10.1016/j.biortech.2016.11.116.

Lardon, L., et al., 2009. Life-cycle assessment of biodiesel production from microalgae. Environmental Science and Technology 43 (17), 6475–6481. Available from: https://doi.org/10.1021/es900705j. France.

Leng, L., et al., 2020. Bioenergy recovery from wastewater produced by hydrothermal processing biomass: progress, challenges, and opportunities, Science of The Total Environment, 748. Elsevier BV, p. 142383. Available from: http://doi.org/10.1016/j.scitotenv.2020.142383.

Liu, Y., et al., 2021. Review of waste biorefinery development towards a circular economy: from the perspective of a life cycle assessment, Renewable and Sustainable Energy Reviews, 139. Elsevier BV, p. 110716. Available from: http://doi.org/10.1016/j.rser.2021.110716.

Londoño-Pulgarin, D., et al., 2021. Fossil or bioenergy? Global fuel market trends, Renewable and Sustainable Energy Reviews, 143. Elsevier BV, p. 110905. Available from: http://doi.org/10.1016/j.rser.2021.110905.

Lorentz, J.F., et al., 2020. Microalgal biomass as a biofertilizer for pasture cultivation: plant productivity and chemical composition. Journal of Cleaner Production. Elsevier Ltd, Brazil, p. 276. Available from: http://doi.org/10.1016/j.jclepro.2020.124130.

Magalhães, I.B., et al., 2021. Technologies for improving microalgae biomass production coupled to effluent treatment: a life cycle approach, Algal Research, 57. Elsevier BV, p. 102346. Available from: http://doi.org/10.1016/j.algal.2021.102346.

Marangon, B.B., et al., 2021. A life cycle assessment of energy recovery using briquette from wastewater grown microalgae biomass. Journal of Environmental Management. Academic Press, Brazil, p. 285. Available from: http://doi.org/10.1016/j.jenvman.2021.112171.

Marks, E.A.N., Montero, O., Rad, C., 2019. The biostimulating effects of viable microalgal cells applied to a calcareous soil: increases in bacterial biomass, phosphorus scavenging, and precipitation of carbonates, Science of the Total Environment, 692. Elsevier B.V, Spain, pp. 784−790. Available from: http://doi.org/10.1016/j.scitotenv.2019.07.289.

Mathimani, T., Mallick, N., 2019. A review on the hydrothermal processing of microalgal biomass to bio-oil − knowledge gaps and recent advances, Journal of Cleaner Production, 217. Elsevier Ltd, India, pp. 69−84. Available from: http://doi.org/10.1016/j.jclepro.2019.01.129.

Mishra, S., Roy, M., Mohanty, K., 2019. Microalgal bioenergy production under zero-waste biorefinery approach: recent advances and future perspectives, Bioresource Technology, 292. Elsevier BV, p. 122008. Available from: http://doi.org/10.1016/j.biortech.2019.122008.

Mu, D., et al., 2017. Life cycle assessment and nutrient analysis of various processing pathways in algal biofuel production, Bioresource Technology, 230. Elsevier Ltd, United States, pp. 33−42. Available from: http://doi.org/10.1016/j.biortech.2016.12.108.

Nawkarkar, P., et al., 2019. Life cycle assessment of *Chlorella* species producing biodiesel and remediating wastewater. Journal of Biosciences 44 (4), Available from: https://doi.org/10.1007/s12038-019-9896-0. India: Springer.

Pereira, A.S.A.d.P., et al., 2021. Organomineral fertilizers pastilles from microalgae grown in wastewater: ammonia volatilization and plant growth. Science of the Total Environment. Elsevier B.V, Brazil, p. 779. Available from: http://doi.org/10.1016/j.scitotenv.2021.146205.

Raghuvanshi, S., et al., 2018. Comparative study using life cycle approach for the biodiesel production from microalgae grown in wastewater and fresh water. Procedia CIRP. Elsevier B.V, India. Available from: http://doi.org/10.1016/j.procir.2017.11.030.

Santagata, R., et al., 2021. Food waste recovery pathways: challenges and opportunities for an emerging bio-based circular economy. A systematic review and an assessment, Journal of Cleaner Production, 286. Elsevier BV, p. 125490. Available from: http://doi.org/10.1016/j.jclepro.2020.125490.

Saranya, D., Shanthakumar, S., 2019. Green microalgae for combined sewage and tannery effluent treatment: performance and lipid accumulation potential. Journal of Environmental Management. India: Academic Press 241, 167−178. Available from: http://doi.org/10.1016/j.jenvman.2019.04.031.

Saranya, G., Ramachandra, T.V., 2020. Life cycle assessment of biodiesel from estuarine microalgae, Energy Conversion and Management: X, 8. Elsevier BV, p. 100065. Available from: http://doi.org/10.1016/j.ecmx.2020.100065.

Shahid, A., et al., 2020. Cultivating microalgae in wastewater for biomass production, pollutant removal, and atmospheric carbon mitigation; a review, Science of the Total Environment, 704. Elsevier BV, p. 135303. Available from: http://doi.org/10.1016/j.scitotenv.2019.135303.

Shimako, A.H., et al., 2016. Environmental assessment of bioenergy production from microalgae based systems, Journal of Cleaner Production, 139. Elsevier Ltd, France, pp. 51−60. Available from: http://doi.org/10.1016/j.jclepro.2016.08.003.

Stegmann, P., Londo, M., Junginger, M., 2020. The circular bioeconomy: its elements and role in European bioeconomy clusters, Resources, Conservation & Recycling: X, 6. Elsevier BV, p. 100029. Available from: http://doi.org/10.1016/j.rcrx.2019.100029.

Sun, C.H., et al., 2019. Life-cycle assessment of biofuel production from microalgae via various bioenergy conversion systems, Energy, 171. Elsevier Ltd, China, pp. 1033–1045. Available from: http://doi.org/10.1016/j.energy.2019.01.074.

Sztancs, G., et al., 2021. Catalytic hydrothermal carbonization of microalgae biomass for low-carbon emission power generation: the environmental impacts of hydrochar co-firing, Fuel, 300. Elsevier BV, p. 120927. Available from: http://doi.org/10.1016/j.fuel.2021.120927.

Tasca, A.L., et al., 2019. Biomethane from short rotation forestry and microalgal open ponds: system modeling and life cycle assessment, Bioresource Technology, 273. Elsevier Ltd, Italy, pp. 468–477. Available from: http://doi.org/10.1016/j.biortech.2018.11.038.

Togarcheti, S.C., et al., 2017. Life cycle assessment of microalgae based biodiesel production to evaluate the impact of biomass productivity and energy source, Resources, Conservation and Recycling, 122. Elsevier B.V, India, pp. 286–294. Available from: http://doi.org/10.1016/j.resconrec.2017.01.008.

Umamaheswari, J., Kavitha, M.S., Shanthakumar, S., 2020. Outdoor cultivation of *Chlorella pyrenoidosa* in paddy-soaked wastewater and a feasibility study on biodiesel production from wet algal biomass through in-situ transesterification, Biomass and Bioenergy, 143. Elsevier BV, p. 105853. Available from: http://doi.org/10.1016/j.biombioe.2020.105853.

Valente, A., Iribarren, D., Dufour, J., 2019. How do methodological choices affect the carbon footprint of microalgal biodiesel? A harmonised life cycle assessment, Journal of Cleaner Production, 207. Elsevier Ltd, Spain, pp. 560–568. Available from: http://doi.org/10.1016/j.jclepro.2018.10.020.

Ventura, J.R.S., et al., 2013. Life cycle analyses of CO_2, energy, and cost for four different routes of microalgal bioenergy conversion, Bioresource Technology, 137. Elsevier Ltd, South Korea, pp. 302–310. Available from: http://doi.org/10.1016/j.biortech.2013.02.104.

Watson, J., et al., 2020. Valorization of hydrothermal liquefaction aqueous phase: pathways towards commercial viability, Progress in Energy and Combustion Science, 77. Elsevier BV, p. 100819. Available from: http://doi.org/10.1016/j.pecs.2019.100819.

Wu, W., Lin, K.H., Chang, J.S., 2018. Economic and life-cycle greenhouse gas optimization of microalgae-to-biofuels chains, Bioresource Technology, 267. Elsevier Ltd, Taiwan, pp. 550–559. Available from: http://doi.org/10.1016/j.biortech.2018.07.083.

Xiao, C., et al., 2020. Life cycle and economic assessments of biogas production from microalgae biomass with hydrothermal pretreatment via anaerobic digestion, Renewable Energy, 151. Elsevier Ltd, China, pp. 70–78. Available from: http://doi.org/10.1016/j.renene.2019.10.145.

Yang, J., et al., 2011. Life-cycle analysis on biodiesel production from microalgae: water footprint and nutrients balance. Bioresource Technology. United States 102 (1), 159–165. Available from: http://doi.org/10.1016/j.biortech.2010.07.017.

Yuan, J., Kendall, A., Zhang, Y., 2015. Mass balance and life cycle assessment of biodiesel from microalgae incorporated with nutrient recycling options and technology uncertainties. GCB Bioenergy 7, 1245–1259. Available from: https://doi.org/10.1111/gcbb.12229.

Biorefinery and bioremediation potential of microalgae

Eleni Koutra[1], Sameh Samir Ali[2,3], Myrsini Sakarika[1,4] and Michael Kornaros[1]

[1]Laboratory of Biochemical Engineering and Environmental Technology (LBEET), Department of Chemical Engineering, University of Patras, Patras, Greece [2]Biofuels Institute, School of the Environment and Safety Engineering, Jiangsu University, Zhenjiang, P.R. China [3]Botany Department, Faculty of Science, Tanta University, Tanta, Egypt [4]Center for Microbial Ecology and Technology (CMET), Ghent University, Gent, Belgium

10.1 Introduction

Microalgae represent a highly diverse group of microorganisms, with impressive potential for advanced biotechnological applications, including CO_2 sequestration and climate change mitigation, wastewater treatment, biofuel production, and value-added product formation through the utilization of an unlimited variety of valuable compounds, including proteins, carbohydrates, lipids, pigments, vitamins, minerals, and plenty of secondary metabolites (Fig. 10.1) (Koller et al., 2014a). However, effective upstream and downstream processing of microalgae is necessary for marketable products and microalgal applications at an industrial scale. Upstream processing results in microalgal biomass production through appropriate cultivation, while downstream processing includes biomass harvesting, drying, cell disruption, extraction, and fractionation, thus producing multiple microalgal products (Vanthoor-Koopmans et al., 2013). Currently, commercial applications of microalgae are limited to only a few microalgal species, mainly intended for food, feed, and added-value compounds. At the same time, several economic and technical obstacles hinder the extensive industrialization of microalgal technology ('t Lam et al., 2018).

Like a petroleum refinery, microalgal biomass can be appropriately processed and valorized within a biorefinery, thus significantly improving process sustainability (Chew et al., 2017). Under this scope, effective fractionation of *Chlorella vulgaris* biomass toward lipids,

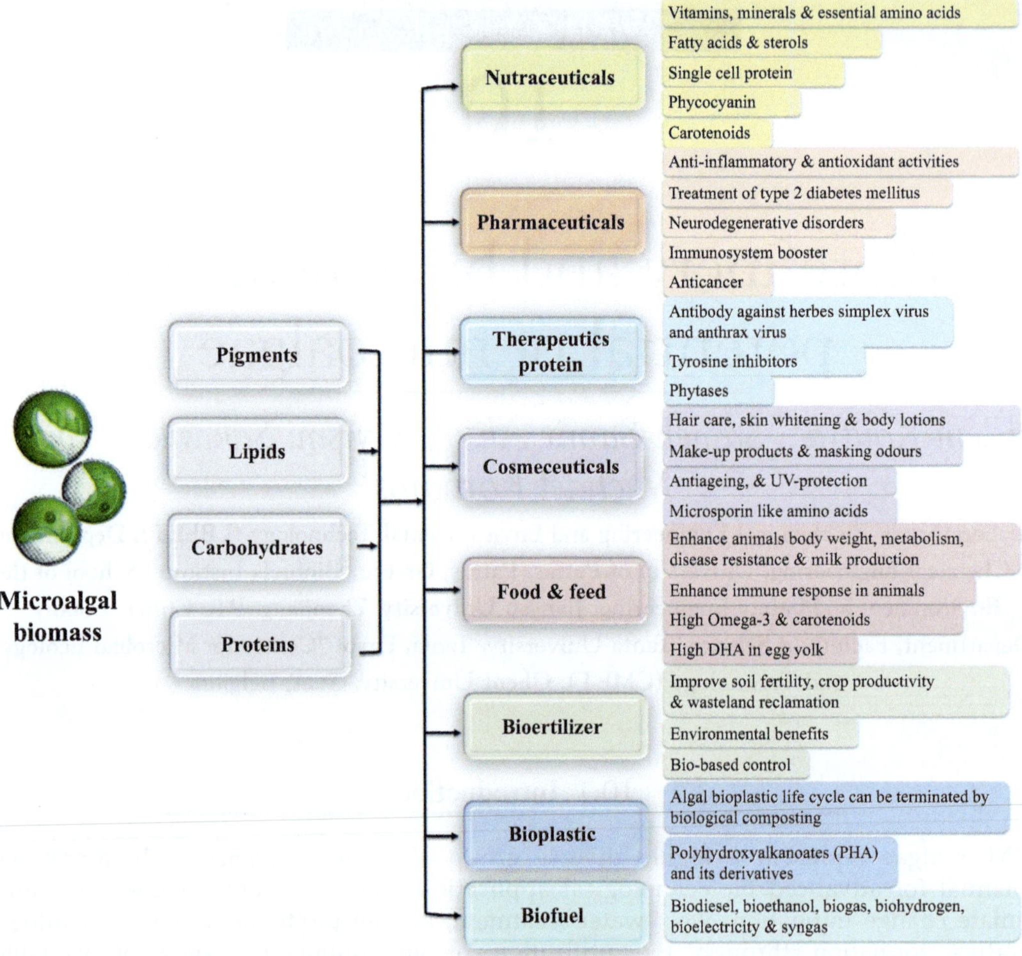

FIGURE 10.1　Microalgae-based products, including biofuels, food, feed, nutraceuticals, pharmaceuticals, biofertilizers, bioplastics, etc.

proteins, and carbohydrates was achieved upon bead milling and lipase treatment, resulting in 88%, 68%, and 74% recovery, respectively (Alavijeh et al., 2020). Furthermore, 11% lipid yield was obtained by *Tetraselmis* sp., while the defatted biomass was further used for biogas production through anaerobic digestion, resulting in 236 mL $CH_4\ g^{-1}\ VS_{added}$ (Hernández et al., 2014). Undoubtedly, the overall profit of the process can further increase through the utilization of wastewater and effluents as a substrate for microalgal cultivation (Bhatia et al., 2021), thus contributing to wastewater treatment and decrease in cost associated with resource utilization. Under this scope, industrial wastewater supplemented with 5% flue gas CO_2 was used for *Chlorella* sp. and *Chlorococcum* sp. cultivation, resulting in higher biomass productivity of the former, 209 compared to 105.4 mg L^{-1} day^{-1} and more efficient nutrient removal. However, the biomass of *Chlorococcum* sp. was characterized by high lipid (34.1%) and carbohydrate (18.7%) content, potentially applied for biofuel production (Yadav, Dash and Sen, 2019). Lastly, a microalgal consortium composed of the phycobiliprotein-rich *Nostoc*, *Phormidium*, and *Geitlerinema* sp. was grown on mixed secondary effluent from anaerobic digestion of microalgae, resulting in total phosphorus and 86% ammonium nitrogen removal. At the same time, after phycocyanin and phycoerythrin recovery, the residual biomass was used for methane production, up to $199\,\mathrm{mL\,g^{-1}}$ VS (Arashiro et al., 2020).

10.2 Microalgae-based biofuels and green energy production

Microalgae represent one of the most promising sources of biofuels, and extensive research has been carried out in this field to increase bioenergy production and decrease dependence on fossil-based fuels. Due to their intrinsic characteristics, including high growth rate and biomass productivity, high concentration of valuable compounds, no need for arable land and fresh water, as well as wastewater treatment, and CO_2 sequestration, microalgae effectively address the challenges of first- and second-generation biofuels, setting high potential for environment-friendly bioenergy applications (Lee et al., 2021). However, currently, large-scale microalgal biofuels production is not feasible, owing to high production cost and lack of the appropriate infrastructure for industrial application ('t Lam et al., 2018). Under this scope, efforts have focused on robust strains, either natural isolates or upon genetic and metabolic engineering, effective cultivation systems, and downstream processing, thus resulting in the improved conversion of microalgal biomass to biofuels (Peng et al., 2020). In addition, concomitant treatment of different types of wastewater and biofuel production has been proven as an advantageous integrated process, resulting in organic and inorganic nutrient removal and remediation of persistent contaminants, such as heavy metals (Hussain et al., 2021). Among the various types of microalgal biofuels, biodiesel, biogas, bioalcohols, biocrude oil, and biohydrogen are the most common. At the same time, electricity can directly be produced through a microalgae-based microbial fuel cell (mMFC).

Biodiesel is by far the most well-studied case of microalgal fuel, and it can be produced through transesterification of the intracellular lipids, mainly neutral lipids, while the ratio of saturated and unsaturated fatty acids is critical for high biodiesel quality, depending on the climate conditions and biodiesel use (Asadi, Rad and Qaderi, 2020). Besides neutral lipids serving as energy reserves, the lipid fraction of microalgal biomass also includes structural lipids, mainly polar lipids and sterols, as well as waxes and hydrocarbons (Sharma, Schuhmann and Schenk, 2012). Concerning lipid composition, fatty acid chain length varies between C10 and C24. However, C16–C18 is considered the most suitable for high-quality biodiesel, as determined by the International Standards ASTM D6751 in the United States and EN 14214 in Europe. Lipid biosynthesis is a multistep reaction initiating from atmospheric CO_2 fixation in microalgal chloroplasts, which results in acetyl-CoA and finally triacylglycerols (TAGs) production (Fig. 10.2). Lipid content varies greatly between different species. However, several strategies can significantly enhance lipid accumulation and fatty acid composition, including culture illumination,

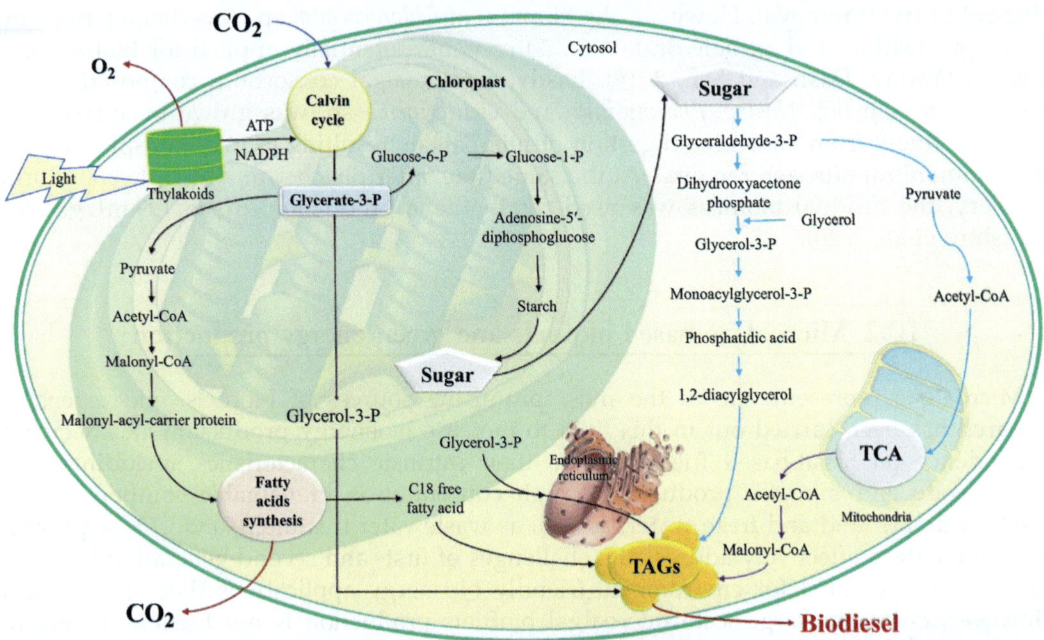

FIGURE 10.2   Biochemical pathways for CO_2 fixation and production of TAGs in microalgae. *TAGs*, Triacylglycerols.

temperature, nutrient limitation, pH, and salinity (Gouveia et al., 2017). To date, numerous microalgal species and cultivation conditions have shown high potential for biodiesel production, while effective processing, in terms of harvesting, drying, and lipid extraction, has been investigated to effectively surpass the existing obstacles that hinder microalgae biodiesel scale-up (Deshmukh, Bala and Kumar, 2019; Rahul et al., 2020; Tandon and Jin, 2017).

In addition, bioethanol, which currently represents the most widely used type of biofuel, can be produced through saccharification of intracellular carbohydrates, followed by alcoholic fermentation (Cheah et al., 2016), performed by different microorganisms, including bacteria and yeasts (Phwan et al., 2018). Third-generation bioethanol is a nontoxic and sustainable biofuel, potentially produced by various microalgal species with a carbohydrate content between 5.2% and 64%, as observed in the case of *Isochrysis* sp. and *Spirogyra* sp., respectively, after the appropriate pretreatment and monosaccharides release (Phwan et al., 2018). However, several advancements are still needed before bioethanol commercialization, aiming to optimize biomass productivity and composition, cost-effective processing steps, and simultaneous production of value-added products (Maia et al., 2020). Microalgal carbohydrates can also be valorized toward producing biobutanol, a similar biofuel to gasoline, which can be produced upon the appropriate biomass pretreatment and fermentation by saccharolytic microorganisms that produce butyric acid, such as most *Clostridium* species (Abomohra and Elshobary, 2019). Furthermore, *Chlorella, Chlamydomonas,* and *Scenedesmus* sp. represent promising candidates for biobutanol production, owing to their high carbohydrate content, surpassing 50%, while engineered microalgae can also directly produce biobutanol through photosynthesis

(Shanmugam et al., 2021). Furthermore, microalgal biomass can be converted to biocrude oil through hydrothermal liquefaction and pyrolysis. At the same time, in the case of the former, specific advantages contributing to technology transfer at an industrial scale include application to wet biomass and dilute cultures (Kazemi Shariat Panahi et al., 2019), which are mainly observed in open systems commonly hindering downstream processing due to high cost of harvesting.

Concerning gaseous biofuels, biogas production through anaerobic digestion represents a widely investigated option for microalgal biomass, especially in cultivating in wastewater (Acién et al., 2016). There are numerous advantages concerning the application of anaerobic digestion for biomass valorization. However, several pretreatment methods and strategies, such as codigestion, have been previously performed to surpass process obstacles, mainly associated with biomass biodegradability, low conversion rate, and ammonia inhibition (Solé-Bundó et al., 2019). Furthermore, microalgal technology can be applied to the production of bio-H_2, which will probably represent the dominant biofuel in the future, characterized by high energy yield and environmental friendliness. The underlying processes include photosynthesis or microbial conversion under dark or photofermentation (Anwar et al., 2019). Lastly, mMFC constitutes an integrated electricity generation process, while effective wastewater treatment, CO_2 sequestration, and valuable biomass production can be accomplished (Kusmayadi et al., 2020). This hybrid technology makes use of the oxygen produced through microalgal photosynthesis for cathode reduction (Fig. 10.3). In contrast, the CO_2 produced by the bacteria inhabiting the anode is used for carbon fixation by microalgae. In addition, mMFC can be further exploited for biofuels production through microalgal biomass processing, as in the case of an integrated system of a

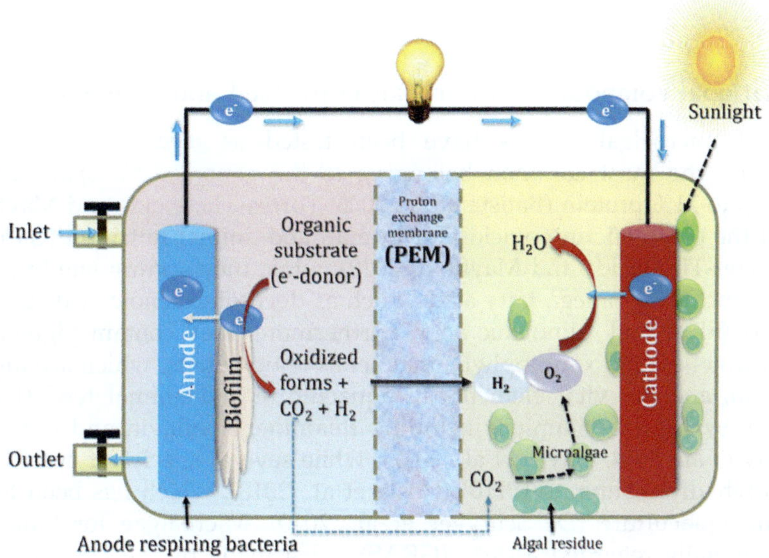

FIGURE 10.3 Schematic diagram of the microbial fuel cell with microalgal biocathodes and anode respiring bacteria.

mMFC treating industrial wastewater, concurrently producing power of $838.68 \, \text{mW m}^{-2}$, along with the lipid-accumulating (up to 44%) microalga *Scenedesmus abundans*, which can be further used for biodiesel production (Nayak and Ghosh, 2019).

10.3 Microalgal food and feed applications

One of the future challenges is meeting the nutritional requirements of the ever-growing world population while natural resources become scarcer. Microalgae are a promising alternative protein source with high nutritional value that can be produced more sustainably than conventional food or feed sources (Torres-Tiji, Fields and Mayfield, 2020). Apart from their increased sustainability, microalgae have been proven to be highly nutritious to humans (Torres-Tiji, Fields and Mayfield, 2020), as well as to monogastric and ruminant animals, aquaculture, or even pets (Madeira et al., 2017). Due to their high nutritional value, they can (partly) substitute conventional protein ingredients in human nutrition, such as meat (Torres-Tiji, Fields and Mayfield, 2020). They have a high positive environmental impact, directly contributing to a more sustainable food chain (Pikaar et al., 2017). Even when they are not used as food, they can replace conventional feed supplements such as soybean (Pikaar et al., 2018), fish meal (Tocher, 2015), and fish oil (Haas et al., 2016) and contribute to the increased sustainability of food production. This section discusses the nutritional compounds contained in microalgal cells, the health benefits arising from the consumption of microalgae, and the increased quality of animal products due to the inclusion of microalgae in animal feed. Finally, the potential limitations for the commercial success of microalgae as food and feed ingredients are briefly discussed.

10.3.1 Nutritional compounds of microalgae for feed and food applications

A variety of microalgal species have been tested as feed and food ingredients (Table 10.1), with the most common belonging to the genera *Arthrospira* and *Chlorella*. They contain up to 71% protein (Batista et al., 2006; Torres-Tiji, Fields and Mayfield, 2020), comprising all the essential amino acids for human and animal nutrition (García-Garibay et al., 2014; Torres-Tiji, Fields and Mayfield, 2020), while they accumulate lipids at a content up to 75%, including omega fatty acids, such as docosahexaenoic acid (DHA), eicosapentaenoic acid (EPA) and γ-linolenic acid. Furthermore, they contain pigments such as chlorophylls, carotenes, and xanthophylls that act as antioxidants, which are important for human nutrition, and are vital elements in aquaculture and animal feed (Hayes et al., 2017). They can synthesize vitamins, including thiamine, riboflavin and cyanocobalamin (García-Garibay et al., 2014; Hayes et al., 2017), while several species of cyanobacteria can accumulate polyhydroxybutyrate (PHB) (Costa et al., 2018), which has been found to act as prebiotic in aquaculture (De Schryver et al., 2010). Microalgae for food should be approved as generally regarded as safe (GRAS), a list made by the US Food and Drugs Administration (FDA) (Hayes et al., 2017; Torres-Tiji, Fields and Mayfield, 2020), while when targeting animal nutrition, the regulations are less strict, but harmful compounds

TABLE 10.1 Microalgal species that have been tested for their use as human food and as feed for aquaculture and shrimps, monogastric and ruminant animals.

	Humans	Aquaculture and shrimp	Monogastric animals	Ruminant animals
Species	*Arthrospira* sp., *Chlamydomonas reinhardtii, Chlorella protothecoides, Chlorella pyrenoidesa, Chlorella regularis, Chlorella vulgaris, Chlorella* sp., *Dunaliella salina, Dunaliella* sp., *Euglena gracilis, Haematococcus pluvialis, Haematococcus* sp., *Schizochytrium* sp.	*Chlorella* sp., *Cryptonemia crenulata, Haematococcus* sp., *H. pluvialis, Hypnea cervicornis, Isochrysis* sp., *Nannochloropsis salina, Nannochloropsis* sp., *Navicula* sp., *Pavlova* sp., *Pavlova viridis, Phaeodactylum* sp., *Schizochytrium* sp., *Skeletonema* sp. *Tetraselmis* sp., *Thalassiosira* sp., *Thalassiosira pseudonana, Tisochrysis lutea*	*Amphora coffeaeformis, Arthrospira* sp., *Arthrospira platensis, Chlorella* sp., *Aurantiochytrium limacinum, Desmodesmus* sp., *Dunaliella* sp., *Nannochloropsis oceanica, Nannochloropsis* sp., *Oocystis* sp., *Porphyridium* sp., *Scenedesmus* sp., *Schizochytrium* sp.	*Arthrospira* sp., *A. platensis Chlorella* sp., *Nannochloropsis* sp., *Schizochytrium* sp.
References	Hayes et al. (2017); Kusmayadi et al. (2021); Torres-Tiji, Fields and Mayfield (2020)	Chauton et al. (2015); FDA (2021); Ghosh et al. (2016); Graham et al. (2009); Haas et al. (2016); Harun et al. (2010); de Medeiros et al. (2021); Patterson and Gatlin (2013)	Ekmay et al. (2014); El-bahr et al. (2020); Ginzberg et al. (2000); Kusmayadi et al. (2021); de Medeiros et al. (2021)	Boeckaert et al. (2008); Hayes et al. (2017); Holman (2012); Lodge-Ivey et al. (2014); Madeira et al. (2017); de Medeiros et al. (2021); Yaakob et al. (2014)

like heavy metals and pathogens should be monitored and maintained below certain thresholds (van der Spiegel, Noordam and van der Fels-Klerx, 2013).

10.3.2 Health benefits arising from the production of microalgal products

There are a large number of health benefits arising from the consumption of compounds contained in microalgal biomass. For instance, bioactive compounds produced by microalgae, including flavonoids, terpenoids, and phenolics, are associated with increased cellular defense mechanisms (Sudhakar et al., 2019); can have antiallergenic, antidiabetic, and antiinflammatory actions (De Jesus Raposo, De Morais and De Morais, 2013); and reduce the cholesterol and triglycerides (Navarro et al., 2016). Furthermore, microalgae have antibacterial and anticancer effects, show potential in cancer treatment, and act as cytotoxic, antiapoptotic, and antimetastatic agents (Hayes et al., 2017). Astaxanthin has been shown to increase the survival and improve the growth performance, reproductivity, and disease resistance of fish (Lim et al., 2018), while astaxanthin from *Haematococcus* sp. protects human skin from sun damage (Hayes et al., 2017). When incorporated in poultry feed, microalgae decreased cholesterol and increased disease resistance (Hayes et al., 2017). The addition of *Arthrospira platensis* in pig feed resulted in 1.3 times increased average weight gain (Simkus et al., 2013), the use of *Chlorella* sp. in pig feed resulted in higher weight gain (Yan, Lim and Kim, 2012), while *Schizochytrium* JB5 inclusion in pork feed led to increased lymphocyte counts, therefore, contributing to an enforced immune system (Kibria and Kim, 2019). Cattle fed with *Arthrospira* sp. presented increased fertility, an enhanced immune system, and improved feed intake, increased weight gain, and milk production (Hayes et al., 2017). Similarly, *A. platensis* inclusion in Australian sheep feed resulted in increased final weight and weight gain (Holman, 2012), while the addition of *Arthrospira* sp. to sheep feed resulted in reduced cholesterol, TAG, and body fat while improving immunity parameters, such as serum, and red and white blood cell counts (Liang et al., 2020).

10.3.3 Microalgae as feed ingredients result in increased animal product quality

Apart from the nutritional characteristics, microalgal incorporation in feed results in increased quality of the animal products aimed at human consumption. Specifically, astaxanthin, contained in microalgae, has been generally found to increase the color of fish (Lim et al., 2018). *H. pluvialis*—a microalgal species with high astaxanthin content (Lim et al., 2018)—is approved by the US FDA as a salmonoid feed resulting in enhanced color (FDA, 2021). Similarly, the incorporation of microalgae in poultry feed has been shown to enhance the color of the skin (Yaakob et al., 2014), meat (Altmann et al., 2018), and egg yolk (Hayes et al., 2017; Yaakob et al., 2014). Supplementation of *A. platensis* in broiler feed resulted in increased EPA and DHA content in thigh meat (De Jesus Raposo, De Morais and De Morais, 2013), resulted in softer and more tender meat and decreased metallic flavor (Altmann et al., 2018), and gave enriched umami and chicken flavor in the meat (Altmann et al., 2020). The use of *Schizochytrium* sp. resulted in higher contents of $n-3$ polyunsaturated fatty acids (PUFA) in chicken breast meat (Madeira et al., 2017). Inclusion of *Porphyridium* sp. apart from enhancing the color of egg yolk reduced its cholesterol content (Ginzberg et al., 2000), while *Amphora coffeaeformis* resulted in higher body weight gain and final weight, increased PUFA and amino acid content, and lower microbial counts (El-Bahr et al., 2020). Beneficial effects of the supplementation of microalgae in the feed have also been reported for pork, cattle, and cows. The inclusion of *Schizochytrium* sp. in the feed increased the DHA content of pork (Vossen et al., 2017), *A. platensis* addition in pork feed resulted in increased PUFA in the meat and improved aroma (Altmann et al., 2019). Feeding trials with *Arthrospira* sp. showed an increased milk protein content in cattle (Hayes et al., 2017). Other reports also showed an increased EPA and DHA content in cattle meat arising from the inclusion of microalgae in their feed (Madeira et al., 2017). Regarding other ruminants, including *Schizochytrium* sp. in dairy cow feed, it results in milk with reduced fat content and increased linolenic acid (Boeckaert et al., 2008).

10.3.4 Parameters affecting the commercial success of microalgae as food and feed ingredients

Even though microalgae present promising features in food and feed, there are still some constraints that limit its overall commercial success. First and foremost, the production costs of microalgal biomass are still high, and therefore microalgae currently do not favorably compete with conventional animal feed (Kusmayadi et al., 2021). To tackle this issue, research is dedicated to finding ways to minimize production costs by increasing biomass productivity (Kusmayadi et al., 2021). Two main concerns using microalgal proteins as food or feed in the context of nutrition are related to (1) their low digestibility due to the rigid cell wall (Hayes et al., 2017; Madeira et al., 2017), which can be problematic for monogastric animals (Madeira et al., 2017), and (2) the relatively high nucleic acid content compared to conventional protein sources that can lead to gout or kidney stones due to the accumulation of uric acid (García-Garibay et al., 2014; Navarro et al., 2016). The latter requires either a carefully selected number of microalgae for inclusion in food or feed, or treatment methods to decrease the nucleic acid content, therefore increasing the production costs. The biggest constraints affecting the commercial success of microalgae as animal feed are the variability in the nutritional value and the adoption of a new product from the customers (Rajesh Banu et al., 2020). Cultivation conditions such as nutrient availability, salinity, pH, growth phase, light intensity, and illumination period have been found to affect the nutritional profile of microalgae (Sui et al., 2019) and, therefore, should be carefully selected according to the product requirements. Furthermore, it was recently demonstrated that the olfactory properties (e.g., the aroma) are affected by the cultivation conditions, such as the choice of substrate and microorganism (Sakarika et al., 2020). This is an important aspect to consider since one of the main criteria defining the acceptability, and therefore the commercial success of a food/feed source is the flavor (i.e., aroma and taste) (Lawless, 1991).

10.4 Biofertilizers

Agriculture is largely based on fertilizers for effective macro- and micronutrient supply and improved primary production, which is anticipated to further increase in the future, owing to the ever-increasing population. Undoubtedly, chemical fertilizers' use is not sustainable, resulting in pollution of natural ecosystems, soil acidification, and loss of microbial biodiversity (Kang et al., 2021). Besides synthetic fertilizers, various types of wastewater have been used for fertilization purposes. However, nutrient runoff from the agricultural fields can adversely affect adjacent aquifers. Under this scope, alternative fertilizers with improved properties have been investigated in modern agriculture, including utilization of microorganisms in different formulations, which can be applied either on soil or various plant parts, significantly improving plant characteristics and yield (Renuka et al., 2016). Between them, microalgae offer significant advantages, including the slow-releasing effect of the intracellular nutrients, along with various bioactive compounds, including pigments, phytohormones, and amino acids, thus promoting and protecting plants while diminishing potential risk of eutrophication and deterioration of natural ecosystems (Kang et al., 2021; Lorentz et al., 2020). Microbial consortia and microalgae-based fertilizers can significantly contribute to atmospheric nitrogen fixation, soil nutrient enrichment, enhanced plant yield, and effective defense against various pathogens by releasing secondary metabolites. However, concerns may arise from the application of biofertilizers derived from microalgae cultivated in wastewater due to the presence of pathogens or other contaminants, which can potentially affect the food chain in the long run (Koutra et al., 2018).

Currently, several studies have proved the beneficial effect of microalgal biofertilizers on crops, as in the case of maize (*Zea mays* L.) seeds treated with cow dung manure, along with 3-g dry biomass of *C. vulgaris* and *Spirulina platensis*, respectively, per kg of soil, resulting in enhanced seed germination. More than 50% growth increase compared to the control condition within 2 months (Dineshkumar et al., 2019). Furthermore, enhanced flowering and growth were observed in tomato plants upon application of dry biomass or extracts of the microalga *Acutodesmus dimorphus*, while the time of application is crucial until biomass efficiently decomposes and nutrients become available (Garcia-Gonzalez and Sommerfeld, 2016). In addition, comparable productivity of the grass species *Uruchloa brizantha* was observed upon fertilization with a conventional fertilizer rich in nitrogen (N) and potassium (K), and the biomass produced by a high rate algal pond treating dairy wastewater (Lorentz et al., 2020). Similarly, plant tissues of the protein-rich wheat (*Triticum aestivum* L.) were significantly increased with N, P, and K, while growth was promoted in the case of the application of two dry microalgal consortia comprising unicellular and filamentous species, respectively, upon cultivation in municipal wastewater (Renuka et al., 2016).

Among the most common nutrients, phosphorus (P) plays a crucial role since it represents a limited resource that necessitates the establishment of sustainable agricultural practices. The life cycle assessment (LCA) of a microalgae-based P fertilizer revealed a higher environmental impact of biofertilizer compared to conventional triple superphosphate in all categories, especially concerning climate change and soil toxicity, while several improvements in upstream and downstream processes are anticipated to enhance biofertilizers' impact (Castro et al., 2020). In general, the production of multiple products within a microalgal biorefinery approach has been proposed as an effective way to enhance the sustainability of microalgal technology (Eppink et al., 2019). To this end, microalgae-based fertilization of rice (*Oryza sativa*) using residual *Scenedesmus* sp. biomass after lipid extraction intended for biodiesel production resulted in increased grain yield, thus decreasing dependence on chemical fertilizers, currently estimated as 13.7% of global use (Nayak, Swain and Sen, 2019). Interestingly, multiple utilization routes of *Scenedesmus obliquus* biomass derived from cultivation in brewery wastewater revealed its high potential for biofuel production, including H_2, biogas, biochar, and biocrude oil, extraction of value-added metabolites (phenols and flavonoids), while germination and growth of barley were significantly stimulated in case of microalgal pellet application (Ferreira et al., 2019). Furthermore, the application of microalgae-based fertilizers plays a significant role in regions with permanent or low water scarcity, resulting in efficient plant growth promoting, as in the case of *Phoenix dactylifera* L. supplementation with 0.5-g *Tetraselmis* sp., concurrently enhancing soil properties and contributing to environmental protection (Saadaoui et al., 2019).

10.5 Pharmaceuticals, cosmetics, and microalgal bioplastics

A wide range of value-added products can also be produced by microalgae, making the cultivation of these photosynthetic microorganisms desirable for multiple industrial products and applications. In addition to biofuels, biofertilizers, feed and nutraceuticals, highly promising opportunities for marketable products also arise in the case of pharmaceuticals, cosmetics, and bioplastics (Rumin et al., 2020). Concerning pharmaceuticals, the therapeutic properties of several bioactive compounds derived from microalgal cells, either intracellularly or secreted in the cultivation medium, have long been investigated, while chemical synthesis of such effective analogs seems currently unfeasible. Microalgal metabolites are characterized by antimicrobial activity against pathogenic bacteria and fungi, antiviral, antioxidant, antiinflammatory, and neuroprotective properties, while recombinant proteins can also be expressed in microalgal cells with multiple therapeutic effects (Jha et al., 2017). Furthermore, microalgal products retain brain functionality, act preventively or therapeutically against cancer, and stimulate an immune response (Mehariya et al., 2021). The most commonly applied pharmaceutics compounds naturally produced by microalgae include antimicrobial agents, such as alcohols, phenols, terpenoids, tannins, polysaccharides, fatty acids, toxins, and algicides (Rizwan et al., 2018), while production of more advanced pharmaceuticals, including growth factors, hormones, enzymes, antibodies, and vaccines, is currently possible through genetic engineering of microalgae (Yan et al., 2016). Nevertheless, it is worth noting that the significance of screening studies since high microalgal biodiversity represents an almost unexplored source of therapeutic molecules, with free radical scavenging, cancer-preventing, and targeted bioactivity (Senousy, Abd Ellatif and Ali, 2020). A galactose-composed heteropolysaccharide extracted by *Tribonema* sp. could effectively stimulate RAW264.7 macrophages, while at a concentration of 250 μg mL^{-1}, significant anticancer activity was observed against the hepatic cancer cell line (HepG2), with up to 66.8% inhibition rate (Chen et al., 2019).

Furthermore, the neuroprotective effect of *Spirulina* sp. compounds has been well documented, also highlighting the significance of its uptake for brain development in case of malnutrition (Sorrenti et al., 2021). However, specifically designed carriers are usually needed for bioactive compounds' encapsulation to maintain their stability and activity, design a marketable product, and have a targeted effect (Vieira, Pastrana and Fuciños, 2020). The strong antioxidant astaxanthin is a common example, while encapsulation in poly (L-lactic acid), calcium-alginate, and numerous other coating materials can preserve its pharmaceutical properties and enhance its stability and administration (Liu et al., 2019; Vieira, Pastrana and Fuciños, 2020).

Furthermore, eukaryotic microalgae and cyanobacteria have been used as valuable sources of compounds applied in cosmetics to maintain and improve skin appearance and functionality over the last few decades. Such a commercial application derives from multiple beneficial properties, including sun and UV protection, whitening, healing, and antiinflammatory activity, as well as antiaging properties mainly associated with the moisturizing effect of microalgal ingredients, such as hydroxy acids and extracellular polymeric substances (Yarkent, Gürlek and Oncel, 2020). In addition, microalgal polysaccharides have gelling, thickening and moisturizing effects. Intracellular pigments and vitamins such as A, B, C, and E can be applied as colorants and natural antioxidants. Phytohormones play a significant role against aging, while valuable poly-unsaturated fatty acids or total microalgal extracts constitute highly attractive, environment-friendly, and healthy bio-based products, potentially applicable in thalassotherapy (Mourelle, Gómez and Legido, 2017). Microalgal pigments, including carotenoids, chlorophylls, and phycobiliproteins, represent one of the primary targeted bioactive compounds, as in the case of astaxanthin with over 500 times higher antioxidant activity than vitamin E, and a cost exceeding 6000 € kg^{-1} (Marino et al., 2020). Similar to pharmaceuticals, delivery systems also determine cosmetics' effectiveness. Therefore the appropriate formulations have to be designed. For example, a highly homogeneous and stable nanoemulsion was prepared to contain biomass extract from the marine microalga *Tetraselmis tetrathele*, with high phenolic content, comparable to commercial products (Farahin et al., 2019). In addition, *Scenedesmus rubescens* extract proved effective in tackling UV-induced aging through beneficial effects on skin collagen and fibroblast, as well as through protection from discoloration, and formation of sunburn cells, with higher efficiency than a commercial sunscreen of high protection (Campiche et al., 2018). However, safety issues should be carefully considered, and toxicological assessment must accompany evaluation of microalgal extracts for health-affecting applications since several toxin-producing microalgal species, mainly diatoms and dinoflagellates, can have adverse effects upon use (Morocho-Jácome et al., 2020).

Concerning plastic production, a highly promising alternative to conventional plastics includes microbial-based biodegradable polymers, which release CO_2 and water upon decomposition, closing the carbon cycle (Vanessa et al., 2015). Concerning the global market, production of bioplastics is estimated to reach 5800 million $ in 2021, and the global demand is anticipated to increase up to threefold within the century (Costa et al., 2018; Kavitha et al., 2016). Bioplastics can be categorized into photodegradable, semibiodegradable, and chemically synthesized biopolymers, as well as polyhydroxyalkanoates (PHAs), which can be divided into short-, medium-, and long-chain length PHAs, while the most common PHA accumulated in microbial cells is the short-chain length polymer PHB. PHAs are recyclable and biocompatible microbial polyesters, with similar mechanical and thermal properties to polypropylene, which can be used in several medical, agricultural, and energy applications and food packaging and paints (Costa et al., 2018; Martins et al., 2014). To date, several bacterial species have been used as potential bioplastics producers, accumulating PHAs as energy storage materials, usually under stress conditions. However, bacterial PHAs are usually characterized by high cost, thus hindering industrial production, mainly due to the high cost of organic carbon sources, oxygen requirements and downstream processing (Samantaray and Mallick, 2015). To this end, microalgae represent highly promising microorganisms for PHAs production, concurrently utilizing atmospheric CO_2 through photosynthesis. PHAs production has been investigated mostly in cyanobacteria, mainly *Synechococcus* sp., *Synechocystis* sp., *Spirulina* sp., etc. However, microalgal biopolymers are still too expensive compared to conventional plastics, while, at the same time, competition with food and feed applications should be considered (Abdo and Ali, 2019). So far, rather low PHA accumulation, mostly a PHB content below 10% on a dry weight basis, has been documented in only a few hundred microalgae, mainly cyanobacteria, under photoautotrophic conditions (Ansari, Fatma and Appanna, 2016). In contrast, under specific culture conditions, such as nutrient limitation and carbon supplementation, PHB or copolymer, such as 3-hydroxybutyric acid-co-3-hydroxyvaleric acid, content can significantly increase depending on the microbial host, reaching up to or exceeding 50% (w/w) (Bhati and Mallick, 2016; Samantaray and Mallick, 2015). Lastly, several interesting options for microalgae-based bioplastics production also exist, including the processing of starch-rich biomass, as in the case of *Chlamydomonas reinhardtii* characterized by up to 49% starch, thus contributing to effective plasticization in glycerol (Mathiot et al., 2019), as well as utilization of *Scenedesmus* sp., *Desmodesmus* sp., and *Spirulina* sp. toward bioplastics formation (López Rocha et al., 2020; Zhang et al., 2020).

10.6 Bioremediation potential of microalgae

Huge amounts of municipal, agricultural, and industrial wastewater are generated worldwide, adversely affecting natural ecosystems, polluting soil, air, and aquifers, and threatening public health. Therefore effective wastewater treatment is imperative for environmental protection and sustainability, human well-being, and socioeconomic development. Currently, conventional wastewater treatment systems include various physical, chemical, and biological processes. However, the complexity of wastewater types and the presence of persistent contaminants and economic and technical obstacles call for advanced processes and multifaceted approaches (Li et al., 2019). Besides common contaminants, emerging organic pollutants, such as pharmaceuticals, hormones, polyaromatic hydrocarbons, pesticides, and heavy inorganic metals, can cause toxic symptoms to receive organisms, endocrine disruption, and pathogens' resistance. Therefore emphasis has been given to the effective treatment of the entire range of available pollutants (Nguyen et al., 2021; Nie et al., 2020). Under this scope, microalgal technology can be effectively applied in wastewater treatment, thus diminishing the levels of organic and inorganic nutrients present in discharged effluents, mitigating CO_2 from flue gases, and concurrently producing valuable biomass at low cost, which can be further processed for multiple bioproducts (Koutra et al., 2018; Zhang et al., 2016). More specifically, microalgae can effectively remove inorganic compounds, mainly nitrogen and phosphorus, from wastewater, degrade organic substances resulting in chemical and biological oxygen demand decrease, respectively, and also eliminate heavy metals, pathogens, dyes, pesticides, and other contaminants through different mechanisms (Chai et al., 2021). Removal of contaminants by microalgal cells can be performed through biosorption, bioabsorption, bioaccumulation, and biodegradation (Fig. 10.4).

Through biosorption, microalgae can passively remove organic contaminants, based on specified interactions with functional groups of cell wall compounds, while active removal includes microbial uptake of compounds present in wastewater, which are metabolized before (biodegradation) or after uptake in the cell and are subsequently accumulated or further metabolized depending on cellular needs (Mustafa et al., 2021). Bioremediation potential of several species has been evaluated to date, highlighting the effective use of robust strains of *Chlorella*, *Scenedesmus*, *Nannochloropsis*, *Dunaliella*, *Spirulina* sp., and others, while many factors contribute to process efficiency, including C, N, P levels, cultivation mode, illumination, and CO_2 supply (Li et al., 2019). *Chlorella* sp. and *Scenedesmus* sp. were effectively cultivated in raw municipal wastewater, supplemented with CO_2 or flue gas, thus exceeding 95% nitrogen removal, while the biofertilizing effect of the produced biomass was validated in wheat plants (Das et al., 2019). Furthermore, *Chlorella* sp. could effectively fix 0.12 g of CO_2 L^{-1} day^{-1} of flue gas, and 81% total nitrogen from palm oil mill effluent, under optimized conditions (Hariz et al., 2018). In addition, up to 80% Fe and 60% Mn removal were achieved by the microalgal species *Desmodesmus* sp. and *Heterochlorella* sp. at acidic conditions, while further biomass processing enhanced biodiesel production. However, despite the high potential of microalgae for effective wastewater treatment, several limitations still have to be met, including the application of robust, acclimatized or improved strains, co-cultivation with bacteria, and other microalgal species,

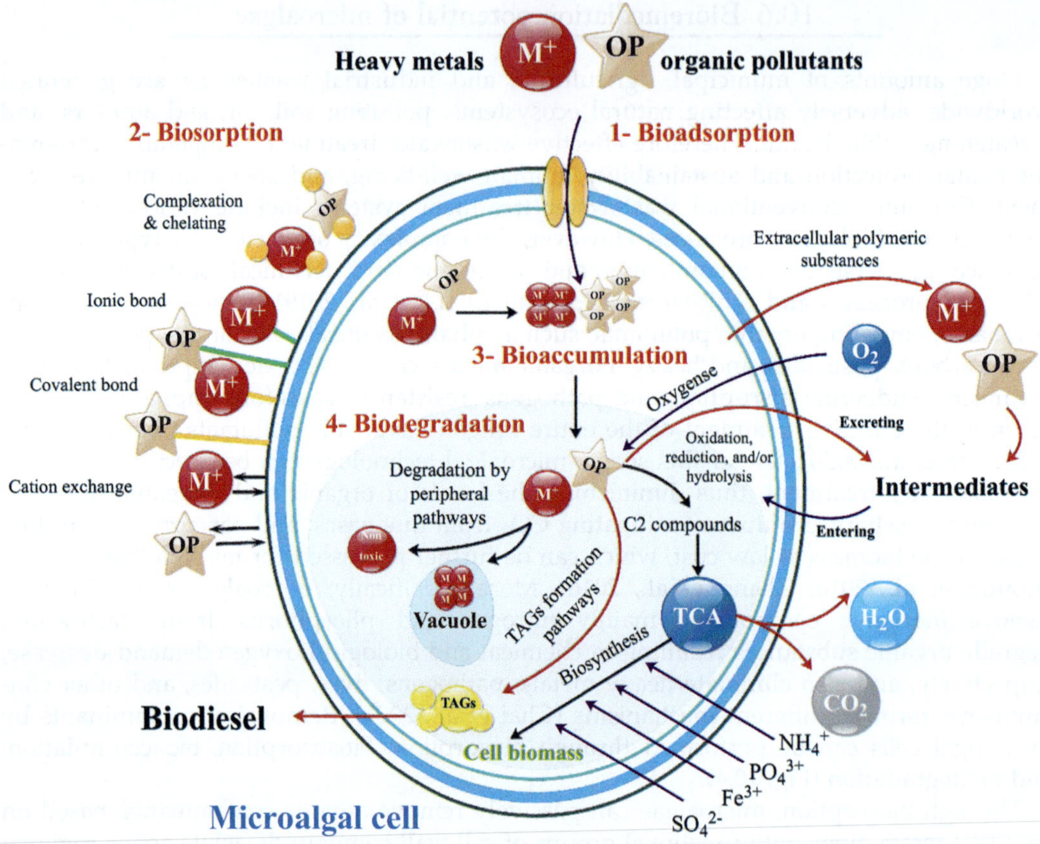

FIGURE 10.4   Bioremediation mechanisms for contaminant removal by microalgal cells.

thus taking advantage of consortial properties, appropriate pretreatment of wastewater, cost-effective downstream processing, while safety assessment of the produced biomass is necessary to highlight potential risks of subsequent applications.

10.7 Conclusions

Microalgal technology offers unique opportunities for effective wastewater treatment, energy applications, and production of food, feed, fertilizers, pharmaceuticals, cosmetics, and bioplastics. However, the industrial application needs further technological and economic improvement, which can derive from scientific advancements, and integrated processes, in the framework of a microalgal biorefinery, where complete valorization of microalgal biomass is performed. In addition, wastewater can be effectively used for microalgal cultivation, concurrently reducing production costs and addressing current environmental issues toward a more circular economy.

References

Abdo, S.M., Ali, G.H., 2019. Analysis of polyhydroxybutrate and bioplastic production from microalgae. Bulletin of the National Research Centre 43 (1). Available from: https://doi.org/10.1186/s42269-019-0135-5. Springer Science and Business Media LLC.

Abomohra, A.E.F., Elshobary, M., 2019. Biodiesel, bioethanol, and biobutanol production from microalgae. Microalgae Biotechnology for Development of Biofuel and Wastewater Treatment. Springer Singapore, China, pp. 293–321. Available from: http://doi.org/10.1007/978-981-13-2264-8_13.

Acién, F.G., et al., 2016. Wastewater treatment using microalgae: how realistic a contribution might it be to significant urban wastewater treatment? Applied Microbiology and Biotechnology 100 (21), 9013–9022. Available from: https://doi.org/10.1007/s00253-016-7835-7. Spain: Springer Verlag.

Alavijeh, R.S., et al., 2020. Combined bead milling and enzymatic hydrolysis for efficient fractionation of lipids, proteins, and carbohydrates of *Chlorella vulgaris* microalgae. Bioresource Technology 309. Available from: https://doi.org/10.1016/j.biortech.2020.123321. Iran: Elsevier Ltd.

Altmann, B.A., et al., 2018. Meat quality derived from high inclusion of a micro-alga or insect meal as an alternative protein source in poultry diets: a pilot study. Foods 7 (3). Available from: https://doi.org/10.3390/foods7030034. Germany: MDPI Multidisciplinary Digital Publishing Institute.

Altmann, B.A., et al., 2019. Do dietary soy alternatives lead to pork quality improvements or drawbacks? A look into micro-alga and insect protein in swine diets. Meat Science 153, 26–34. Available from: https://doi.org/10.1016/j.meatsci.2019.03.001. Germany: Elsevier Ltd.

Altmann, B.A., et al., 2020. The effect of insect or microalga alternative protein feeds on broiler meat quality. Journal of the Science of Food and Agriculture 100 (11), 4292–4302. Available from: https://doi.org/10.1002/jsfa.10473. Germany: John Wiley and Sons Ltd.

Ansari, S., Fatma, T., Appanna, V.D., 2016. Cyanobacterial polyhydroxybutyrate (PHB): screening, optimization and characterization. PLOS ONE. Public Library of Science (PLoS) e0158168. Available from: https://doi.org/10.1371/journal.pone.0158168.

Anwar, M., et al., 2019. Recent advancement and strategy on bio-hydrogen production from photosynthetic microalgae. Bioresource Technology 292, 121972. Available from: https://doi.org/10.1016/j.biortech.2019.121972. Elsevier BV.

Arashiro, L.T., et al., 2020. Natural pigments and biogas recovery from microalgae grown in wastewater. ACS Sustainable Chemistry and Engineering 8 (29), 10691–10701. Available from: https://doi.org/10.1021/acssuschemeng.0c01106. Spain: American Chemical Society.

Asadi, P., Rad, H.A., Qaderi, F., 2020. Lipid and biodiesel production by cultivation isolated strain *Chlorella sorokiniana* pa.91 and *Chlorella vulgaris* in dairy wastewater treatment plant effluents. Journal of Environmental Health Science and Engineering 18 (2), 573–585. Available from: https://doi.org/10.1007/s40201-020-00483-y. Iran: Springer Science and Business Media Deutschland GmbH.

Batista, A.P., et al., 2006. Rheological characterization of coloured oil-in-water food emulsions with lutein and phycocyanin added to the oil and aqueous phases. Food Hydrocolloids 20 (1), 44–52. Available from: https://doi.org/10.1016/j.foodhyd.2005.02.009. Portugal.

Bhati, R., Mallick, N., 2016. Carbon dioxide and poultry waste utilization for production of polyhydroxyalkanoate biopolymers by Nostoc muscorum Agardh: a sustainable approach. Journal of Applied Phycology 28 (1), 161–168. Available from: https://doi.org/10.1007/s10811-015-0573-x. India: Springer Netherlands.

Bhatia, S.K., et al., 2021. Wastewater based microalgal biorefinery for bioenergy production: progress and challenges. Science of the Total Environment 751. Available from: https://doi.org/10.1016/j.scitotenv.2020.141599. South Korea: Elsevier B.V.

Boeckaert, C., et al., 2008. Effect of dietary starch or micro algae supplementation on rumen fermentation and milk fatty acid composition of dairy cows. Journal of Dairy Science 91 (12), 4714–4727. Available from: https://doi.org/10.3168/jds.2008-1178. Belgium: American Dairy Science Association.

Campiche, R., et al., 2018. Protective effects of an extract of the freshwater microalga *Scenedesmus rubescens* on UV-irradiated skin cells. International Journal of Cosmetic Science 40 (2), 187–192. Available from: https://doi.org/10.1111/ics.12450. Wiley.

Castro, J.d S., et al., 2020. Microalgae based biofertilizer: a life cycle approach. Science of the Total Environment 724. Available from: https://doi.org/10.1016/j.scitotenv.2020.138138. Brazil: Elsevier B.V.

Chai, W.S., et al., 2021. Multifaceted roles of microalgae in the application of wastewater biotreatment: a review. Environmental Pollution 269. Available from: https://doi.org/10.1016/j.envpol.2020.116236. Malaysia: Elsevier Ltd.

Chauton, M.S., et al., 2015. A techno-economic analysis of industrial production of marine microalgae as a source of EPA and DHA-rich raw material for aquafeed: research challenges and possibilities. Aquaculture 436, 95−103. Available from: https://doi.org/10.1016/j.aquaculture.2014.10.038. Norway: Elsevier.

Cheah, W.Y., et al., 2016. Cultivation in wastewaters for energy: a microalgae platform. Applied Energy 179, 609−625. Available from: https://doi.org/10.1016/j.apenergy.2016.07.015. Malaysia: Elsevier Ltd.

Chen, X., et al., 2019. Partial characterization, the immune modulation and anticancer activities of sulfated polysaccharides from filamentous microalgae *Tribonema* sp. Molecules 24 (2). Available from: https://doi.org/10.3390/molecules24020322. China: MDPI AG.

Chew, K.W., et al., 2017. Microalgae biorefinery: high value products perspectives. Bioresource Technology 229, 53−62. Available from: https://doi.org/10.1016/j.biortech.2017.01.006. Malaysia: Elsevier Ltd.

Costa, J.A.V., et al., 2018. Recent advances and future perspectives of PHB production by cyanobacteria. Industrial Biotechnology 14 (5), 249−256. Available from: https://doi.org/10.1089/ind.2018.0017. Brazil: Mary Ann Liebert Inc.

Das, P., et al., 2019. Microalgal nutrients recycling from the primary effluent of municipal wastewater and use of the produced biomass as bio-fertilizer. International Journal of Environmental Science and Technology. Qatar: Center for Environmental and Energy Research and Studies 16 (7), 3355−3364. Available from: https://doi.org/10.1007/s13762-018-1867-8.

De Jesus Raposo, M.F., De Morais, R.M.S.C., De Morais, A.M.M.B., 2013. Health applications of bioactive compounds from marine microalgae. Life Sciences 93 (15), 479−486. Available from: https://doi.org/10.1016/j.lfs.2013.08.002. Portugal.

de Medeiros, V.P.B., et al., 2021. Microalgae in the meat processing chain: feed for animal production or source of techno-functional ingredients. Current Opinion in Food Science 37, 125−134. Available from: https://doi.org/10.1016/j.cofs.2020.10.014. Brazil: Elsevier Ltd.

De Schryver, P., et al., 2010. Poly-β-hydroxybutyrate (PHB) increases growth performance and intestinal bacterial range-weighted richness in juvenile European sea bass, *Dicentrarchus labrax*. Applied Microbiology and Biotechnology 86 (5), 1535−1541. Available from: https://doi.org/10.1007/s00253-009-2414-9. Belgium.

Deshmukh, S., Bala, K., Kumar, R., 2019. Selection of microalgae species based on their lipid content, fatty acid profile and apparent fuel properties for biodiesel production. Environmental Science and Pollution Research 26 (24), 24462−24473. Available from: https://doi.org/10.1007/s11356-019-05692-z. India: Springer Verlag.

Dineshkumar, R., et al., 2019. The impact of using microalgae as biofertilizer in maize (*Zea mays* L.). Waste and Biomass Valorization 10 (5), 1101−1110. Available from: https://doi.org/10.1007/s12649-017-0123-7. India: Springer Netherlands.

Ekmay, R., et al., 2014. Nutritional and metabolic impacts of a defatted green marine microalgal (*Desmodesmus* sp.) biomass in diets for weanling pigs and broiler chickens. Journal of Agricultural and Food Chemistry 62 (40), 9783−9791. Available from: https://doi.org/10.1021/jf501155n. United States: American Chemical Society.

El-Bahr, S., et al., 2020. Effect of dietary microalgae on growth performance, profiles of amino and fatty acids, antioxidant status, and meat quality of broiler chickens. Animals 10 (5). Available from: https://doi.org/10.3390/ani10050761. Saudi Arabia: MDPI AG.

Eppink, M.H.M., et al., 2019. From current algae products to future biorefinery practices: A review. Advances in Biochemical Engineering/Biotechnology 99−123. Available from: https://doi.org/10.1007/10_2016_64. Netherlands: Springer Science and Business Media Deutschland GmbH.

Farahin, A.W., et al., 2019. Use of microalgae: *Tetraselmis tetrathele* extract in formulation of nanoemulsions for cosmeceutical application. Journal of Applied Phycology 31 (3), 1743−1752. Available from: https://doi.org/10.1007/s10811-018-1694-9. Malaysia: Springer Netherlands.

FDA, 2021. Code of Federal Regulations (CFR) − Title 21. <https://www.fda.gov/medical-devices/medical-device-databases/code-federal-regulations-title-21-food-and-drugs> (accessed 17.08.21).

Ferreira, A., et al., 2019. *Scenedesmus obliquus* microalga-based biorefinery − from brewery effluent to bioactive compounds, biofuels and biofertilizers − aiming at a circular bioeconomy. Biofuels, Bioproducts and Biorefining 13 (5), 1169−1186. Available from: https://doi.org/10.1002/bbb.2032. Portugal: John Wiley and Sons Ltd.

García-Garibay, M., et al., 2014. Single cell protein: the algae, Encyclopedia of Food Microbiology, Second Edition Elsevier Inc, Mexico, pp. 425–430. Available from: http://doi.org/10.1016/B978-0-12-384730-0.00309-8.

Garcia-Gonzalez, J., Sommerfeld, M., 2016. Biofertilizer and biostimulant properties of the microalga *Acutodesmus dimorphus*. Journal of Applied Phycology 28 (2), 1051–1061. Available from: https://doi.org/10.1007/s10811-015-0625-2. United States: Springer Netherlands.

Ghosh, S., et al., 2016. Live feed for marine finfish and shellfish culture, training manual on live feed for marine finfish and shellfish cultureIn: Biji, Xavier, et al., (Eds.), Visakhapatnam Regional Centre Central Marine Fisheries Research Institute, Andhra Pradesh, India.

Ginzberg, A., et al., 2000. Chickens fed with biomass of the red microalga *Porphyridium* sp. have reduced blood cholesterol level and modified fatty acid composition in egg yolk. Journal of Applied Phycology. Available from: https://doi.org/10.1023/a:1008102622276. Israel: Springer Netherlands.

Gouveia, L., et al., 2017. Biodiesel from microalgae. Microalgae-Based Biofuels and Bioproducts: From Feedstock Cultivation to End-Products. Elsevier Inc., Portugal, pp. 235–258. Available from: http://doi.org/10.1016/B978-0-08-101023-5.00010-8.

Graham, L.E., Graham, J.M., Wilcox, L.W., 2009. Algae. Benjamin Cummings, San Francisco, USA.

Haas, S., et al., 2016. Marine microalgae Pavlova viridis and *Nannochloropsis* sp. as n-3 PUFA source in diets for juvenile European sea bass (*Dicentrarchus labrax* L.). Journal of Applied Phycology 28 (2), 1011–1021. Available from: https://doi.org/10.1007/s10811-015-0622-5. Germany: Springer Netherlands.

Hariz, H.B., et al., 2018. CO_2 fixation capability of Chlorella sp. and its use in treating agricultural wastewater. Journal of Applied Phycology 30 (6), 3017–3027. Available from: https://doi.org/10.1007/s10811-018-1488-0. Malaysia: Springer Netherlands.

Harun, R., et al., 2010. Bioprocess engineering of microalgae to produce a variety of consumer products. Renewable and Sustainable Energy Reviews 14 (3), 1037–1047. Available from: https://doi.org/10.1016/j.rser.2009.11.004. Australia.

Hayes, M., et al., 2017. Microalgal proteins for feed, food and health. Microalgae-Based Biofuels and Bioproducts: From Feedstock Cultivation to End-Products. Elsevier Inc, Ireland, pp. 347–368. Available from: http://doi.org/10.1016/B978-0-08-101023-5.00015-7.

Hernández, D., et al., 2014. Biofuels from microalgae: lipid extraction and methane production from the residual biomass in a biorefinery approach. Bioresource Technology 170, 370–378. Available from: https://doi.org/10.1016/j.biortech.2014.07.109. Spain: Elsevier Ltd.

Holman, B., 2012. Growth and body conformation responses of genetically divergent Australian sheep to Spirulina (*Arthrospira platensis*) supplementation. American Journal of Experimental Agriculture. Sciencedomain International 160–173. Available from: https://doi.org/10.9734/ajea/2012/992.

Hussain, F., et al., 2021. Microalgae an ecofriendly and sustainable wastewater treatment option: biomass application in biofuel and bio-fertilizer production. A review. Renewable and Sustainable Energy Reviews 137. Available from: https://doi.org/10.1016/j.rser.2020.110603. Pakistan: Elsevier Ltd.

Jha, D., et al., 2017. Microalgae-based pharmaceuticals and nutraceuticals: an emerging field with immense market potential. ChemBioEng Reviews 4 (4), 257–272. Available from: https://doi.org/10.1002/cben.201600023. India: Wiley-Blackwell.

Kang, Y., et al., 2021. Potential of algae–bacteria synergistic effects on vegetable production. Frontiers in Plant Science 12. Available from: https://doi.org/10.3389/fpls.2021.656662. South Korea: Frontiers Media S.A. (656662).

Kavitha, G., et al., 2016. Optimization of polyhydroxybutyrate production utilizing waste water as nutrient source by *Botryococcus braunii* Kütz using response surface methodology. International Journal of Biological Macromolecules 93, 534–542. Available from: https://doi.org/10.1016/j.ijbiomac.2016.09.019. India: Elsevier B.V.

Kazemi Shariat Panahi, H., et al., 2019. Recent updates on the production and upgrading of bio-crude oil from microalgae. Bioresource Technology Reports 7. Available from: https://doi.org/10.1016/j.biteb.2019.100216. Iran: Elsevier Ltd.

Kibria, S., Kim, I.H., 2019. Impacts of dietary microalgae (*Schizochytrium* JB5) on growth performance, blood profiles, apparent total tract digestibility, and ileal nutrient digestibility in weaning pigs. Journal of the Science of Food and Agriculture 99 (13), 6084–6088. Available from: https://doi.org/10.1002/jsfa.9886. South Korea: John Wiley and Sons Ltd.

Koller, M., Muhr, A., Braunegg, G., 2014a. Microalgae as versatile cellular factories for valued products. Algal Research 6, 52–63. Available from: https://doi.org/10.1016/j.algal.2014.09.002. Austria: Elsevier.

Koutra, E., et al., 2018. Bio-based products from microalgae cultivated in digestates. Trends in Biotechnology 36 (8), 819–833. Available from: https://doi.org/10.1016/j.tibtech.2018.02.015. Greece: Elsevier Ltd.

Kusmayadi, A., et al., 2020. Microalgae-microbial fuel cell (mMFC): an integrated process for electricity generation, wastewater treatment, CO_2 sequestration and biomass production. International Journal of Energy Research 44 (12), 9254–9265. Available from: https://doi.org/10.1002/er.5531. Taiwan: John Wiley and Sons Ltd.

Kusmayadi, A., et al., 2021. Microalgae as sustainable food and feed sources for animals and humans – biotechnological and environmental aspects. Chemosphere 271, 129800. Available from: https://doi.org/10.1016/j.chemosphere.2021.129800. Elsevier BV.

Lawless, H., 1991. The sense of smell in food quality and sensory evaluation. Journal of Food Quality. United States 14 (1), 33–60. Available from: https://doi.org/10.1111/j.1745-4557.1991.tb00046.x.

Lee, S.Y., et al., 2021. Techniques of lipid extraction from microalgae for biofuel production: a review. Environmental Chemistry Letters 19 (1), 231–251. Available from: https://doi.org/10.1007/s10311-020-01088-5. Malaysia: Springer Science and Business Media Deutschland GmbH.

Li, K., et al., 2019. Microalgae-based wastewater treatment for nutrients recovery: a review. Bioresource Technology 291, 121934. Available from: https://doi.org/10.1016/j.biortech.2019.121934. Elsevier BV.

Liang, Y., et al., 2020. Effects of *Spirulina* supplementation on lipid metabolism disorder, oxidative stress caused by high-energy dietary in Hu sheep. Meat Science 164, 108094. Available from: https://doi.org/10.1016/j.meatsci.2020.108094. Elsevier BV.

Lim, K.C., et al., 2018. Astaxanthin as feed supplement in aquatic animals. Reviews in Aquaculture 10 (3), 738–773. Available from: https://doi.org/10.1111/raq.12200. Malaysia: Wiley-Blackwell.

Liu, G., et al., 2019. Enhancing the stability of astaxanthin by encapsulation in poly (L-lactic acid) microspheres using a supercritical anti-solvent process. Particuology 44, 54–62. Available from: https://doi.org/10.1016/j.partic.2018.04.006. China: Elsevier B.V.

Lodge-Ivey, S.L., Tracey, L.N., Salazar, A., 2014. Ruminant nutrition symposium: the utility of lipid extracted algae as a protein source in forage or starch-based ruminant diets. Journal of Animal Science 92 (4), 1331–1342. Available from: https://doi.org/10.2527/jas.2013-7027. United States: American Society of Animal Science.

López Rocha, C.J., et al., 2020. Development of bioplastics from a microalgae consortium from wastewater. Journal of Environmental Management 263. Available from: https://doi.org/10.1016/j.jenvman.2020.110353. Mexico: Academic Press.

Lorentz, J.F., et al., 2020. Microalgal biomass as a biofertilizer for pasture cultivation: plant productivity and chemical composition. Journal of Cleaner Production 276. Available from: https://doi.org/10.1016/j.jclepro.2020.124130. Brazil: Elsevier Ltd.

Madeira, M.S., et al., 2017. Microalgae as feed ingredients for livestock production and meat quality: a review. Livestock Science 205, 111–121. Available from: https://doi.org/10.1016/j.livsci.2017.09.020. Portugal: Elsevier B.V.

Maia, J.L.D., et al., 2020. Microalgae starch: a promising raw material for the bioethanol production. International Journal of Biological Macromolecules 165, 2739–2749. Available from: https://doi.org/10.1016/j.ijbiomac.2020.10.159. Brazil: Elsevier B.V.

Marino, T., et al., 2020. From *Haematococcus pluvialis* microalgae a powerful antioxidant for cosmetic applications. Chemical Engineering Transactions 79, 271–276. Available from: https://doi.org/10.3303/CET2079046. Italy: Italian Association of Chemical Engineering - AIDIC.

Martins, R.G., et al., 2014. Bioprocess engineering aspects of biopolymer production by the cyanobacterium *Spirulina* strain LEB 18. International Journal of Polymer Science 2014. Available from: https://doi.org/10.1155/2014/895237. Brazil: Hindawi Publishing Corporation.

Mathiot, C., et al., 2019. Microalgae starch-based bioplastics: screening of ten strains and plasticization of unfractionated microalgae by extrusion. Carbohydrate Polymers 208, 142–151. Available from: https://doi.org/10.1016/j.carbpol.2018.12.057. France: Elsevier Ltd.

Mehariya, S., et al., 2021. Microalgae for high-value products: a way towards green nutraceutical and pharmaceutical compounds. Chemosphere 280. Available from: https://doi.org/10.1016/j.chemosphere.2021.130553. Italy: Elsevier Ltd.

Morocho-Jácome, A.L., et al., 2020. (Bio)technological aspects of microalgae pigments for cosmetics. Applied Microbiology and Biotechnology 104 (22), 9513–9522. Available from: https://doi.org/10.1007/s00253-020-10936-x. Brazil: Springer Science and Business Media Deutschland GmbH.

Mourelle, M.L., Gómez, C.P., Legido, J.L., 2017. The potential use of marine microalgae and cyanobacteria in cosmetics and thalassotherapy. Cosmetics 4 (4). Available from: https://doi.org/10.3390/cosmetics4040046. Spain: MDPI AG.

Mustafa, S., et al., 2021. Microalgae biosorption, bioaccumulation and biodegradation efficiency for the remediation of wastewater and carbon dioxide mitigation: Prospects, challenges and opportunities. Journal of Water Process Engineering. Elsevier BV 41, 102009. Available from: https://doi.org/10.1016/j.jwpe.2021.102009.

Navarro, F., et al., 2016. Microalgae as a safe food source for animals: nutritional characteristics of the acidophilic microalga *Coccomyxa onubensis*. Food and Nutrition Research 60 (1). Available from: https://doi.org/10.3402/fnr.v60.30472. Spain: Taylor and Francis Ltd.

Nayak, J.K., Ghosh, U.K., 2019. Post treatment of microalgae treated pharmaceutical wastewater in photosynthetic microbial fuel cell (PMFC) and biodiesel production. Biomass and Bioenergy 131. Available from: https://doi.org/10.1016/j.biombioe.2019.105415. India: Elsevier Ltd.

Nayak, M., Swain, D.K., Sen, R., 2019. Strategic valorization of de-oiled microalgal biomass waste as biofertilizer for sustainable and improved agriculture of rice (Oryza sativa L.)crop. Science of the Total Environment 682, 475−484. Available from: https://doi.org/10.1016/j.scitotenv.2019.05.123. India: Elsevier B.V.

Nguyen, H.T., et al., 2021. The application of microalgae in removing organic micropollutants in wastewater. Critical Reviews in Environmental Science and Technology 51 (12), 1187−1220. Available from: https://doi.org/10.1080/10643389.2020.1753633. Informa UK Limited.

Nie, J., et al., 2020. Bioremediation of water containing pesticides by microalgae: mechanisms, methods, and prospects for future research. Science of the Total Environment 707. Available from: https://doi.org/10.1016/j.scitotenv.2019.136080. China: Elsevier B.V.

Patterson, D., Gatlin, D.M., 2013. Evaluation of whole and lipid-extracted algae meals in the diets of juvenile red drum (*Sciaenops ocellatus*). Aquaculture (Amsterdam, Netherlands) 416−417, 92−98. Available from: https://doi.org/10.1016/j.aquaculture.2013.08.033. United States.

Peng, L., et al., 2020. Biofuel production from microalgae: a review. Environmental Chemistry Letters 18 (2), 285−297. Available from: https://doi.org/10.1007/s10311-019-00939-0. China: Springer.

Phwan, C.K., et al., 2018. Overview: comparison of pretreatment technologies and fermentation processes of bioethanol from microalgae. Energy Conversion and Management 173, 81−94. Available from: https://doi.org/10.1016/j.enconman.2018.07.054. Malaysia: Elsevier Ltd.

Pikaar, I., et al., 2017. Microbes and the next Nitrogen Revolution. Environmental Science and Technology 51 (13), 7297−7303. Available from: https://doi.org/10.1021/acs.est.7b00916. Australia: American Chemical Society.

Pikaar, I., et al., 2018. Decoupling livestock from land use through industrial feed production pathways. Environmental Science and Technology 52 (13), 7351−7359 Available from: https://doi.org/10.1021/acs.est.8b00216. Australia: American Chemical Society.

Rahul, S.M., et al., 2020. Insights about sustainable biodiesel production from microalgae biomass: a review. International Journal of Energy Research. Available from: https://doi.org/10.1002/er.6138. India: John Wiley and Sons Ltd.

Rajesh Banu, J., et al., 2020. Microalgae based biorefinery promoting circular bioeconomy-techno economic and life-cycle analysis. Bioresource Technology 302, 122822. Available from: https://doi.org/10.1016/j.biortech.2020.122822. Elsevier BV.

Renuka, N., et al., 2016. Exploring the efficacy of wastewater-grown microalgal biomass as a biofertilizer for wheat. Environmental Science and Pollution Research 23 (7), 6608−6620. Available from: https://doi.org/10.1007/s11356-015-5884-6. India: Springer Verlag.

Rizwan, M., et al., 2018. Exploring the potential of microalgae for new biotechnology applications and beyond: a review. Renewable and Sustainable Energy Reviews 92, 394−404. Available from: https://doi.org/10.1016/j.rser.2018.04.034. Pakistan: Elsevier Ltd.

Rumin, J., et al., 2020. Analysis of scientific research driving microalgae market opportunities in Europe. Marine Drugs 18 (5). Available from: https://doi.org/10.3390/md18050264. France: MDPI AG.

Saadaoui, I., et al., 2019. Assessment of the algae-based biofertilizer influence on date palm (*Phoenix dactylifera* L.) cultivation. Journal of Applied Phycology 31 (1), 457−463. Available from: https://doi.org/10.1007/s10811-018-1539-6. Qatar: Springer Netherlands.

Sakarika, M., et al., 2020. The type of microorganism and substrate determines the odor fingerprint of dried bacteria targeting microbial protein production. FEMS Microbiology Letters 367 (18). Available from: https://doi.org/10.1093/femsle/fnaa138. Belgium: Oxford University Press.

Samantaray, S., Mallick, N., 2015. Impact of various stress conditions on poly-β-hydroxybutyrate (PHB) Accumulation in *Aulosira fertilissima* CCC 444. Current Biotechnology 366–372. Available from: https://doi.org/10.2174/2211550104666150806000642. Bentham Science Publishers Ltd.

Senousy, H.H., Abd Ellatif, S., Ali, S., 2020. Assessment of the antioxidant and anticancer potential of different isolated strains of cyanobacteria and microalgae from soil and agriculture drain water. Environmental Science and Pollution Research 27 (15), 18463–18474. Available from: https://doi.org/10.1007/s11356-020-08332-z. Egypt: Springer.

Shanmugam, S., et al., 2021. Recent developments and strategies in genome engineering and integrated fermentation approaches for biobutanol production from microalgae. Fuel 285, 119052. Available from: https://doi.org/10.1016/j.fuel.2020.119052. Elsevier BV.

Sharma, K.K., Schuhmann, H., Schenk, P.M., 2012. High lipid induction in microalgae for biodiesel production. Energies 5 (5), 1532–1553. Available from: https://doi.org/10.3390/en5051532. Australia: MDPI AG.

Simkus, A., et al., 2013. The effect of blue algae *Spirulina platensis* on pig growth; Melsvadumblio *Spirulina platensis*. Veterinarija ir Zootechnika 61, 70–74.

Solé-Bundó, M., et al., 2019. Co-digestion strategies to enhance microalgae anaerobic digestion: a review. Renewable and Sustainable Energy Reviews 112, 471–482. Available from: https://doi.org/10.1016/j.rser.2019.05.036. Spain: Elsevier Ltd.

Sorrenti, V., et al., 2021. Spirulina microalgae and brain health: a scoping review of experimental and clinical evidence. Marine Drugs 19 (6), . Available from: https://doi.org/10.3390/md19060293Italy: MDPI AG.

Sudhakar, M.P., et al., 2019. A review on bioenergy and bioactive compounds from microalgae and macroalgae-sustainable energy perspective. Journal of Cleaner Production 228, 1320–1333. Available from: https://doi.org/10.1016/j.jclepro.2019.04.287. India: Elsevier Ltd.

Sui, Y., et al., 2019. Enhancement of co-production of nutritional protein and carotenoids in *Dunaliella salina* using a two-phase cultivation assisted by nitrogen level and light intensity. Bioresource Technology 287. Available from: https://doi.org/10.1016/j.biortech.2019.121398. Belgium: Elsevier Ltd.

Tandon, P., Jin, Q., 2017. Microalgae culture enhancement through key microbial approaches. Renewable and Sustainable Energy Reviews 80, 1089–1099. Available from: https://doi.org/10.1016/j.rser.2017.05.260. China: Elsevier Ltd.

't Lam, G.P., et al., 2018. Multi-product microalgae biorefineries: from concept towards reality. Trends in Biotechnology 36 (2), 216–227. Available from: https://doi.org/10.1016/j.tibtech.2017.10.011. Netherlands: Elsevier Ltd.

Tocher, D.R., 2015. Omega-3 long-chain polyunsaturated fatty acids and aquaculture in perspective. Aquaculture 449, 94–107. Available from: https://doi.org/10.1016/j.aquaculture.2015.01.010. United Kingdom: Elsevier B.V.

Torres-Tiji, Y., Fields, F.J., Mayfield, S.P., 2020. Microalgae as a future food source. Biotechnology Advances 41. Available from: https://doi.org/10.1016/j.biotechadv.2020.107536. United States: Elsevier Inc.

van der Spiegel, M., Noordam, M.Y., van der Fels-Klerx, H.J., 2013. Safety of novel protein sources (insects, microalgae, seaweed, duckweed, and rapeseed) and legislative aspects for their application in food and feed production. Comprehensive Reviews in Food Science and Food Safety 12 (6), 662–678. Available from: https://doi.org/10.1111/1541-4337.12032. Netherlands.

Vanessa, C.C., et al., 2015. Polyhydroxybutyrate production by *Spirulina* sp. LEB 18 grown under different nutrient concentrations. African Journal of Microbiology Research. Academic Journals 1586–1594. Available from: https://doi.org/10.5897/AJMR2015.7530.

Vanthoor-Koopmans, M., et al., 2013. Biorefinery of microalgae for food and fuel. Bioresource Technology 135, 142–149. Available from: https://doi.org/10.1016/j.biortech.2012.10.135. Netherlands: Elsevier Ltd.

Vieira, M.V., Pastrana, L.M., Fuciños, P., 2020. Microalgae encapsulation systems for food, pharmaceutical and cosmetics applications. Marine drugs 18 (12). Available from: https://doi.org/10.3390/md18120644. Portugal: NLM (Medline).

Vossen, E., et al., 2017. Production of docosahexaenoic acid (DHA) enriched loin and dry cured ham from pigs fed algae: nutritional and sensory quality. European Journal of Lipid Science and Technology 119 (5). Available from: https://doi.org/10.1002/ejlt.201600144. Belgium: Wiley-VCH Verlag.

Yaakob, Z., et al., 2014. An overview: biomolecules from microalgae for animal feed and aquaculture. Journal of Biological Research (Greece) 21 (1). Available from: https://doi.org/10.1186/2241-5793-21-6. Malaysia: BioMed Central Ltd.

Yadav, G., Dash, S.K., Sen, R., 2019. A biorefinery for valorization of industrial waste-water and flue gas by microalgae for waste mitigation, carbon-dioxide sequestration and algal biomass production. Science of the

Total Environment 688, 129–135. Available from: https://doi.org/10.1016/j.scitotenv.2019.06.024. India: Elsevier B.V.

Yan, L., Lim, S.U., Kim, I.H., 2012. Effect of fermented chlorella supplementation on growth performance, nutrient digestibility, blood characteristics, fecal microbial and fecal noxious gas content in growing pigs. Asian-Australasian Journal of Animal Sciences 25 (12), 1742–1747. Available from: https://doi.org/10.5713/ajas.2012.12352. South Korea.

Yan, N., et al., 2016. The potential for microalgae as bioreactors to produce pharmaceuticals. International Journal of Molecular Sciences 17 (6). Available from: https://doi.org/10.3390/ijms17060962. China: MDPI AG.

Yarkent, Ç., Gürlek, C., Oncel, S.S., 2020. Potential of microalgal compounds in trending natural cosmetics: a review. Sustainable Chemistry and Pharmacy 17. Available from: https://doi.org/10.1016/j.scp.2020.100304. Turkey: Elsevier B.V.

Zhang, T.Y., et al., 2016. Promising solutions to solve the bottlenecks in the large-scale cultivation of microalgae for biomass/bioenergy production. Renewable and Sustainable Energy Reviews 60, 1602–1614. Available from: https://doi.org/10.1016/j.rser.2016.02.008. China: Elsevier Ltd.

Zhang, C., et al., 2020. A sustainable solution to plastics pollution: an eco-friendly bioplastic film production from high-salt contained *Spirulina* sp. residues. Journal of Hazardous Materials 388. Available from: https://doi.org/10.1016/j.jhazmat.2019.121773. China: Elsevier B.V.

Recent developments and challenges: a prospectus of microalgal biomass valorization

Maria Lúcia Calijuri[1], Paula Assemany[2], Eduardo Couto[3], Adriana Paulo de Sousa Oliveira[1], Juliana F. Lorentz[1] and Letícia Rodrigues de Assis[1]

[1]Post-Graduate Program in Civil Engineering, Department of Civil Engineering, Campus Universitário, Federal University of Viçosa (Universidade Federal de Viçosa/UFV), Viçosa, Minas Gerais, Brazil [2]Department of Environmental Engineering, Campus Universitário, Federal University of Lavras (Universidade Federal de Lavras/UFLA), Lavras, Minas Gerais, Brazil [3]Institute of Applied and Pure Sciences, Rua Irmã Ivone Drumond, Federal University of Itajubá, Campus Itabira (Universidade Federal de Itajubá, Campus Itabira/Unifei), Itabira, Minas Gerais, Brazil

11.1 Historical perspective of microalgal biomass utilization

For a long time, microalgae have been explored as a source of human and animal food. They have been primarily cultivated throughout history for commercial production of high nutritional value products for both humans and animals. Microalgal antioxidants and their immunological and anticancer effects are widely known, being a source of high added-value products such as carotenoids and omega-3.

In the 1950s, Burlew (1953) proposed using microalgae as an alternative source of protein in the face of world food demand. In 1960 Japan began constructing the first industrial plant to produce *Chlorella* for human consumption (Vigani et al., 2015). In the 1980s, 46 large-scale factories in Asia had monthly productions of over 100 kg, mainly of the *Chlorella* genus (Fig. 11.1). The large-scale production of cyanobacteria also started in the Asian continent, more precisely in India, in the 1980s. In addition, several industrial plants

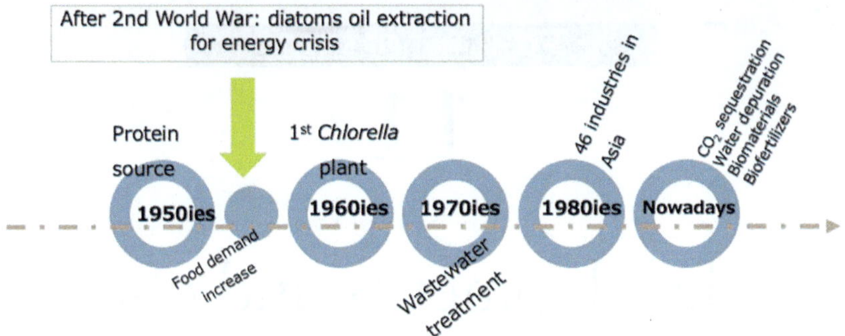

FIGURE 11.1   Timeline of research and commercial applications of microalgae biomass.

producing *Haematococcus pluvialis* as a source of astaxanthin were established in the United States and India. Thus, over 30 years, the microalgae biotechnology industry acquired an increasingly significant diversity (Olaizola, 2003).

In the context of renewable energies, studies with liquid fuels from microalgae intensified in the mid-1980s (Huang et al., 2010). During World War II, German scientists extracted lipids from diatoms to solve the energy crisis (Cohen et al., 1995). Soon after, the research continued in the United States at the Carnegie Institute, with the results published in the book *Algal Culture from Laboratory to Pilot Plant* (Burlew, 1953).

In the 2000s, the main cultivated genera were *Spirulina*, *Chlorella*, *Dunaliella*, and *Haematococcus*. Almost half of microalgae production was located in mainland China, with the remainder in Japan, Taiwan, the United States, India, and Australia, with small producers in other countries (Benemann, 2009). *Chlorella*, *Nannochloropsis* spp., *H. pluvialis*, and *Spirulina* stand out as the most exploited species for biotechnological purposes world over for several years (Mobin and Alam, 2017). In Europe, Araújo et al. (2021) reported that these are currently the most cultivated genera concerning the number of companies. *Chlorella* and *Spirulina* are the most produced species in dry algae volume (Vigani et al., 2015).

Approximately 7000 t of dry algae are produced worldwide annually (Brasil et al., 2017). The global market of microalgae-based products was valued at US$1.6 billions in 2021 and it may reach US$2.8 billions by 2028 (The Insight Partners, 2021). Most countries produce biomass for human consumption (dry or paste) in addition to its production for the extraction of high added-value products (pigments, antioxidants, long-chain polyunsaturated fatty acids, phycobiliproteins, and polysaccharides) and as food for aquaculture.

Nowadays, the expansion of production is highlighted, mainly in the European, North American, and Asian continents, with a significant contribution of investments from India and Israel (Vieira de Mendonça et al., 2021). The European continent is responsible for the production of approximately 142 t of dry biomass. Of this total, food supplements (such as nutraceuticals) and cosmetics, each corresponding to 24% of the total number of companies, are the main applications of the produced algal biomass. Next, emphasis is given to the use of biomass for animal feed, such as aquaculture (19%) (Araújo et al., 2021). In addition to the applications already highlighted, other uses for microalgae more recently explored on a global scale are solutions for bioremediation of water and effluents, biofertilizers, CO_2 sequestration from industrial systems, and production of biomaterials (Vieira de Mendonça et al., 2021).

11.2 Wastewater as culture medium

The use of different wastewaters, such as domestic, industrial, and agricultural, as a culture medium for microalgae cultivation can significantly reduce the costs related to their production since wastewater can be a source of water and nutrients (Fernández et al., 2021). The use of microalgae-based systems and the subsequent biomass recovery represents an opportunity for a paradigm shift in the wastewater treatment sector based on resource recovery.

In general, wastewater treatment occurs from establishing a symbiotic relationship between microalgae and other microorganisms (Rashid et al., 2018). For example, microalgae produce oxygen used by bacteria in organic matter degradation. In contrast, the CO_2 produced in heterotrophic metabolism is used by microalgae in photosynthesis. In addition, nutrients are recovered from biomass assimilation, which can be used for different purposes. Therefore various microalgae strains are used in wastewater treatment, such as *Chlorella vulgaris* (Wollmann et al., 2019), *C. sorokiniana, C. minutíssima, Scenedesmus acutus, S. quadricauda, Botryococcus braunii, Oscillatoria, Nostoc, Chlorococcum* (Emparan et al., 2019), *Coelastrum* sp. (Villar-Navarro et al., 2018), among others.

The high rate ponds (HRPs) can be highlighted regarding the systems used to cultivate microalgae in wastewater. HRPs are open systems with continuous mixing of the wastewater by rotating paddlewheels, which gives HRPs a potential advantage over conventional wastewater treatment ponds. It provides intermittent exposure of microalgae to solar radiations, which reduces the photoinhibition effects and eliminates thermal and nutrient stratification (Couto et al., 2020).

Concerning cultivation with domestic sewage, carbon is the limiting nutrient for microalgal growth (Couto et al., 2018a,b). Thus CO_2 addition can optimize biomass growth. In addition to balancing the C/N/P ratio in the culture medium, the CO_2 addition also controls pH values, which increase the consumption of inorganic carbon in photosynthesis. This may be related to the availability of nutrients for microalgae since phosphorus is removed by chemical precipitation at high pH values, and ammonia nitrogen is volatilized (Young et al., 2017; Robles et al., 2020). Moreover, NH_3 is toxic to many microalgae species. Thus its presence can impair biomass productivity. Therefore the CO_2 addition and the consequent pH control optimize nutrient recovery through biomass uptake since they prevent nutrient loss.

However, the CO_2 addition may represent an increase in operating costs. Thus CO_2 from atmospheric emissions is an alternative to optimize biomass production without increasing operating costs (Assis et al., 2019; Vieira de Mendonça et al., 2021) and may minimize negative environmental impacts from pure CO_2 production. Regarding the HRP depth, it is known that reduced values allow radiation penetration in the entire water column. On the other hand, it may represent greater area demand and thermal instability due to the greater variation in temperature (Couto et al., 2020). The typical values reported in the literature vary between 0.15 and 0.50 m (Arbib et al., 2017; Sutherland et al., 2017).

In addition to productivity, the characteristics of wastewater can interfere with the biomass biochemical composition. Commonly, without nutrient limitation, such as N and P, microalgae can produce great amounts of biomass without a considerable accumulation of lipids. This is because photosynthetic products are used for microalgal growth and not for the production and storage of lipids (Bélanger-Lépine et al., 2018). Moreover, the biomass produced in wastewater with an abundant presence of N has high protein concentrations (Ljubic et al., 2018). For example, carbon limitation provides a biomass with lower carbohydrate and lipid contents compared to protein when concerning domestic sewage.

Contrary, when using carbon-rich wastewater, such as landfill leachates, agroindustrial, or brewery effluents, the concentration of lipids and carbohydrates in biomass tends to increase (Gifuni et al., 2018; Hernández-García et al., 2019). Assemany et al. (2020) found greater potential for methane generation in biomass produced from brewery effluent than in biomass produced in domestic sewage, mainly due to the higher carbohydrate content. The authors stated that a suitable culture medium could replace the need for biomass pretreatment for anaerobic digestion (AD). The shortage of nutrients is identified as one of the main alternatives for increasing the lipid content and other carbon-rich biomolecules, such as polyhydroxyalkanoates (Costa et al., 2018a,b).

The characteristics of the effluent can also influence the ash content in the biomass. For example, the biomass produced in domestic sewage may have ash concentrations above 40% due to inorganic solids, such as sand (Chen et al., 2014; Couto et al., 2018a,b). This content may come from trace metals, which can cause human health and environmental problems even in low concentrations (Leong and Chang, 2020). In addition, the content and composition of the ash can interfere with the use of biomass. The presence of metals can make the application in the soil unfeasible, and the inorganic content, in general, can harm energy use (Couto et al., 2018a,b).

11.3 Main valorization routes of wastewater-grown microalgal biomass: recent developments

11.3.1 Animal food

As discussed in the proceeding section, microalgal biomass is mainly composed of proteins ($\sim$50%), carbohydrates (20%–30%), lipids (20%–30%), and other compounds in smaller fractions ($\sim$5%) (Alam et al., 2020). This biochemical composition makes them attractive animal feed supplements (Dineshbabu et al., 2019; Kusmayadi et al., 2021). The importance is also highlighted given the expectation that by 2050, there will be an increase in the demand for animal products, more specifically of proteins, as a result of the population increase (FAO, 2018; de Medeiros et al., 2021). It is estimated that in the next 30 years, the world population will reach 9.7 billion inhabitants (UNESCO, 2020), making evident the greater demand for resources, including food (de Medeiros et al., 2021).

Consequently, conventional sources of biomass (soybeans, corn, rice, among others) that support the agricultural and animal feed market may suffer limitations due to overexploitation of raw materials, in addition to competition for food (Dineshbabu et al., 2019; de Medeiros et al., 2021). Thus, to meet future protein demands, the adoption of unconventional sources, such as microalgae, emerged as a potential alternative to meet animals' nutritional and physiological demands (Madeira et al., 2017; Kusmayadi et al., 2021; Vieira de Mendonça et al., 2021).

However, microalgae cultivation in wastewater to produce biomass rich in nutritional compounds of interest for animal supplementation still faces challenges. Few studies have inserted biomass grown in wastewater into the animal diet. But, some have found promising results of protein sources (between 27% and 43%), cultivating microalgae in different municipal wastewater (Assis et al., 2019; Rodrigues de Assis et al., 2020), swine wastewater (Moheimani et al., 2018), livestock (Choudhary et al., 2017), and meat processing wastewater (Pereira et al., 2021). It is noteworthy that before processing biomass grown in wastewater, removing pathogens or any other potentially toxic compounds is recommended (Choudhary et al., 2017). Furthermore, strategies, such as genetic and metabolic engineering, can assist in species selection and modification techniques to promote the accumulation of the specific nutritional compound (Balamurugan et al., 2021; Sproles et al., 2021). In this context, it is believed that the feasibility of wastewater-grown biomass will soon become a reality.

Unlike the cultivation in wastewater, microalgae cultivation in synthetic media with interest in animal feed supplementation is increasing. The primary purpose of inserting microalgae in animal feed is to use them as an ingredient due to their nutritional properties. They can also be used as a coloring agent and improve the organoleptic properties of the final product (Lafarga, 2019; Levasseur et al., 2020). In addition, the insertion of microalgae in food products can be carried out in dry (powder) or wet (extruded or extracts) form (de Medeiros et al., 2021). The application may contain all the biomass or just a few isolated compounds as the ingredients of the formulations (Lafarga, 2019). The total insertion of biomass must be carefully evaluated, as this practice can impart undesirable characteristics such as color, aroma, and flavor to the product (de Medeiros et al., 2021). For example, this possibility should be investigated in ruminants that can digest the total biomass without processing (Fernández et al., 2021).

Regarding the benefits of microalgal for animal feed, the most attractive isolated compounds are proteins, carotenoids, omega-3 fatty acids, vitamins, and antioxidants (Dineshbabu et al., 2019; Kusmayadi et al., 2021). Moreover, given these nutritional components of microalgae, low percentages of their components are recommended in young animals (Fernández et al., 2021). The protein quality from microalgae such as *C. vulgaris* is similar to soybean meal and milk protein (Madeira et al., 2017). Thus replacing conventional protein sources by microalgae can occur in up to 10% and 33%, respectively, for poultry and swine (Fernández et al., 2021).

Among the researches that added microalgae (or isolated compounds) to the animal diet, a variety of animals have already been evaluated, such as fish (Cardinaletti et al., 2018; Qiao et al., 2019; Roohani et al., 2019), cattle (Carvalho et al., 2018; Lamminen et al., 2019), sheep (Fan et al., 2019; Kholif et al., 2017; Liang et al., 2020), swine (Altmann, 2019; de Tonnac et al., 2018; Moheimani, 2018; Kibria and Kim, 2019; Manor et al., 2017), and poultry (Park et al., 2018; Wu et al., 2019). These studies, as well as other literature (Dineshbabu et al., 2019; Fernández et al., 2021; de Medeiros et al., 2021), highlighted the positive effects of microalgae on digestibility, intestinal function, immune system; in addition to greater growth of animals; increased appetite, weight, milk production, improving the nutritional quality of meat and eggs. Adverse effects, such as changes in the palatability of the products (Altmann et al., 2019; Lamminen et al., 2019), are also reported. From these recently published papers, the applicability of microalgae in animal nutrition was verified, reinforcing the search for strategies to accelerate the production of commodities to, finally, materialize their insertion in the feed industries (Balamurugan et al., 2021; Dineshbabu et al., 2019).

11.3.2 Energy routes

Fossil energy sources are in a phase of sharp decline and are recognized for their devastating effect on climate change. One attractive solution to overcome these problems is the production of bioenergy using microalgal biomass cultivated in wastewater. Several conversion routes are suggested, but those using wet biomass are the most promising, as they do not require a biomass drying step (Couto et al., 2020; Bhatia et al., 2021). The main processes of converting wet microalgal biomass into bioenergy are biochemical and hydrothermal (Table 11.1).

11.3.2.1 Biochemical processes

AD allows obtaining biogas (a mixture of methane, CO_2, and other gases) from the microbial decomposition of biomass under anaerobic conditions. It is a simple process with low energy demand. However, the AD of microalgal biomass presents a negative aspect due to the high protein content in the biomass, contributing to an unbalanced carbon/nitrogen (C/N) ratio, resulting in an excess of ammonia nitrogen, and a reduction in CH_4 production. Another characteristic refers to the rigidity of the microalgal cell wall, which confers low biodegradability to the biomass. Therefore codigestion with other substrates and pretreatments (mechanical, thermal, chemical, or biological) to solubilize cell components are suggested (Ganesh Saratale et al., 2018). Dębowski et al. (2017) reported $0.183-0.267$ L CH_4 g^{-1} VS using *Chlorella* sp. cultivated in different wastewaters. The best result was observed for the biomass with the highest C/N ratio. Assemany et al. (2020) used biomass cultivated in domestic sewage as substrate in AD and obtained 62 mL CH_4 g^{-1} VS. This performance increased to 130 mL CH4 g^{-1} VS with substrate complementarity using 10% olive mill wastewater.

Dark fermentation (DF) results in the production of biohydrogen ($bioH_2$) through anaerobic fermentation. The process is similar to AD, so similar challenges occur in DF. However, obtaining $bioH_2$ is highlighted in some recent researches, as it has the advantage of being a carbon-free energy route. Additionally, hydrogen has a high energy yield of 141.65 MJ kg^{-1} (Bhatia et al., 2021). Three different cultures (*C. vulgaris*, *Scenedesmus obliquus*, and a consortium composed of *Chlorella*, *Scenedesmus*, *Chaetophora*, and *Navicula*) were grown in domestic sewage and used to obtain $bioH_2$ (Batista et al., 2015) with yield ranging from 40.8 mL H_2 g^{-1} VS for *C. vulgaris* to 56.8 mL H_2 g^{-1} VS for *S. obliquus*. Variations were proportionally related to sugar content in biomass (Batista et al., 2015). In another study, the biomass grown in swine wastewater was pretreated by ultrasonication followed by enzymatic saccharification with subsequent fermentation to obtain $bioH_2$, and

TABLE 11.1 Performance of different conversion routes of wastewater-grown microalgae biomass into bioenergy.

Effluent used as a growth medium	Microalgae specie	Methane yield (L CH$_4$ g^{-1} VS)	References
Brewery industry	The most abundant species were *Scenedesmus obliquus* and *Chlorella vulgaris*	160	Assemany et al. (2020)
Swine	Consortium with the dominance of *Scenedesmus* sp.	44	Perazzoli et al. (2016)
Chicken manure-based liquid digestate (CM)	*Chlorella* sp.	Algae biomass = 0.19[a] Codigestion of CM: biomass, 8:2 = 0.23[a]	Li et al. (2017)
Domestic sewage	Consortium with the dominance of *Chlorella* sp.	Raw biomass = 0.14 thermally pretreated biomass = 0.23	Solé-Bundó et al. (2020)

Effluent used as a growth medium	Microalgae specie	bioH$_2$ yield (mL H$_2$ g^{-1} VS)	References
Different effluents: poultry, swine and cattle breeding, brewery and dairy industries, and urban	*Scenedesmus obliquus*	Swine = 390 Poultry = 378 Brewery = 67 Urban = 57 Dairy = 56 Cattle = 50	Ferreira et al. (2018)
Domestic sewage	Green microalgae	18.2−22.1	Wang et al. (2019)

Effluent used as a growth medium	Microalgae specie	Energy in hydrochar (MJ kg^{-1})	References
Swine	−	16.7−18.6	Marin-Batista et al. (2019)
Meat processing industry	58.5% of *Chlorella vulgaris* and 37.4% of *Navicula* sp.	∼17	Castro et al. (2021)

Effluent used as a growth medium	Microalgae specie	Energy in bio-oil (MJ kg^{-1})	References
Domestic sewage	Consortium of *Chlorella pyrenoidosa* and *Phormidium*	24.8	Naaz et al. (2019)
Domestic sewage	*Galdieria sulphuraria* monoculture and *Galdieria sulphuraria* polyculture	38.2−38.9	Cheng et al. (2019)

[a]*Maximum potential of methane production.*
VS, Volatile solids.

here the production ranged from 103 to 116 mL H_2 g^{-1} VS, emphasizing the pretreatments that favored the solubilization of carbohydrates and proteins, and improved the hydrogen production (Kumar et al., 2018).

11.3.2.2 Hydrothermal process

Hydrothermal carbonization (HTC) is a process in which wet biomass is heated under moderate pressure and temperature (>2 MPa and 180°C–220°C, respectively), resulting in the production of hydrochar, a carbon-rich material that can be used as solid fuel. HTC is a quick process as the reaction time is usually less than 1 hour. In addition, the process requires less aggressive conditions compared to hydrothermal liquefaction (HTL), providing less energy demand. However, as highlighted in Section 11.2, the high levels of ash and proteins in wastewater-grown microalgal biomass may present a challenge. It can reduce the calorific value of hydrochar and present a risk of equipment damage due to incrustation. At the same time, the N from proteins can integrate the hydrochar and contribute to the emission of nitrogen oxides during burning (Marin-Batista et al., 2019; Hess et al., 2019; Castro et al., 2021).

The meat processing industry effluent grown microalgal biomass was converted into hydrochar by providing 170°C temperature and 10-minute reaction time, and the process yielded 77.7% solids and 78.2% energy. Despite the promising results, the calorific value of hydrochar was not influenced by the treatments, attributed to the high ash content of the biomass ($\sim$33%) (Castro et al., 2021). In another study, the biomass grown in swine wastewater presented a C concentration of 38.4% and a calorific value of 16.9 MJ kg^{-1}. However, after the HTC, hydrochar did not significantly increase the concentration of C (38.8%–41.8%) and the calorific value (16.7–18.6 MJ kg^{-1}). In an attempt to improve the performance of HTC, the hydrochar was treated with a solution of HCl which allowed the removal of ash and a $1.37x$ increase in the calorific value (Marin-Batista et al., 2019).

HTL allows obtaining bio-oil by subjecting the biomass to high pressure and temperature conditions (5–30 MPa and 240°C–380°C, respectively) (Leng et al., 2020). Biomass grown in wastewater is recognized for having a reduced concentration of lipids (Section 11.2). However, due to extreme HTL conditions, the conversion of the whole biomass into bio-oil occurs, improving the process yield. The addition of catalysts also increases the performance of HTL. As the high levels of ash and proteins present in the biomass are among the main factors that limit the application of HTL, a denitrogenation step may be necessary before using bio-oil as fuel (Goswami et al., 2019).

Microalgal biomass produced during domestic sewage treatment was used as a substrate for the HTL process. The best result was achieved with a reaction time of 15 minutes at 300°C and a biomass/water ratio of 1/10, which provided a bio-oil yield of 44.4% in ash-free dry biomass. Thus almost half of the energy available in biomass was converted into bio-oil. However, the N content in the bio-oil was significant and varied between 50% and 80% (Couto et al., 2018a,b). The microalgal biomass from a swine wastewater treatment plant was subjected to a previous demineralization with the formic acid step. Energy recovery using raw biomass was 24.3%, and it varied from 32.5% to 45.2% for demineralized biomass. Demineralization reduced ash content by 35.7% and possibly changed the biomass structure, improving thermal decomposition (Carpio et al., 2021).

11.3.3 Fertilizers

Food security requires high productivity agricultural practices. This productivity has been achieved through fertilization (organic and inorganic), which provides a proper nutrient input to influence the development and yield of crops (Suleiman et al., 2020). Research indicates that the type and amount of fertilizer influences the physicochemical properties of the soil, leading to significant long-term impacts on its health and productive capacity (Cassman et al., 2016; Zhu et al., 2019; Chen et al., 2021).

N, P, and K are the essential nutrients in agricultural production (Huang et al., 2019). They are most from non-renewable sources and have widespread use in world agriculture (Silambarasan et al., 2021). N is the most limiting, followed by P (Alvarez et al., 2021), with the former demanded in larger quantities and at a higher cost. It presents huge environmental losses (Oita et al., 2020), leading to water pollution (van der Kooij et al., 2020). Moreover, in the case of N, it is naturally transformed through the reduction of atmospheric N_2 to NH_3, by diazotrophic microorganisms (bacteria, cyanobacteria, and some species of microalgae), known as biological nitrogen fixation (Kour et al., 2020; Chen et al., 2021). Phosphorus is usually found in soils in its nonavailable forms (for plant absorption) due to its tendency to complexation with organic molecules or insoluble inorganic salts, such as calcium (Ca), iron (Fe), or aluminum (Al) (Alvarez et al., 2021).

To minimize the negative impacts caused by conventional chemical fertilizers, the use of organic fertilizers (biofertilizers) has been increasingly researched (Tahami et al., 2017). According to Naher et al. (2019), besides providing one or more essential macro or micronutrients, biofertilizers also provide phytohormones for the growth and development of plants. In this context, microalgal biomass appears to be a natural material that contains vitamins, minerals, amino acids, IAA (indole-3-acetic acid), and other compounds beneficial for plant nutrition (Döring et al., 2015) and the improvement of soil' physical properties (Kumar and Singh, 2020). Fig. 11.2 shows various uses of microalgal biomass for agricultural purposes as an eco-friendly and sustainable alternative.

In recent years, microalgal strains have become a promising source of biostimulants and biofertilizers in agriculture to produce healthy plants and increase crop productivity (Suleiman et al., 2020; Supraja et al., 2020). Microalgae are known as fixatives of carbon dioxide (CO_2) and atmospheric nitrogen (N_2) (Renuka et al., 2018). In addition, they also solubilize immobilized phosphorus (Liu et al., 2019). The composition of microalgal biomass contains phytohormones, proteins, lipids, and polysaccharides (Chiaiese et al., 2018), vitamins, and amino acids (Uthirapandi et al., 2018), which promotes its use as a potential source of nutrients to improve agriculture production (Supraja et al., 2020; Silambarasan et al., 2021).

As a fertilizer source, the use of wastewater-grown microalgal biomass has been studied, and positive effects of its use have been reported (Coppens et al., 2016; Castro et al., 2017; Renuka et al., 2017; Das et al., 2019). Some studies have tested microalgal biomass as a P source, assimilated from wastewater, bringing it back into the cycle as a valuable nutrient (Siebers et al., 2019). It aims to release inorganic P, contributing to its availability to plants (mineralization of P) (Siebers et al., 2019; Alvarez et al., 2021). Moreover, microalgal biomass due to the cell structure has been considered a slow-release fertilizer (Coppens et al., 2016; Kumar et al., 2017; Volf and Rosolem, 2021), which can delay the nutrient release, resulting in less environmental pollution and a higher rate of nutrient utilization (Siebers et al., 2019).

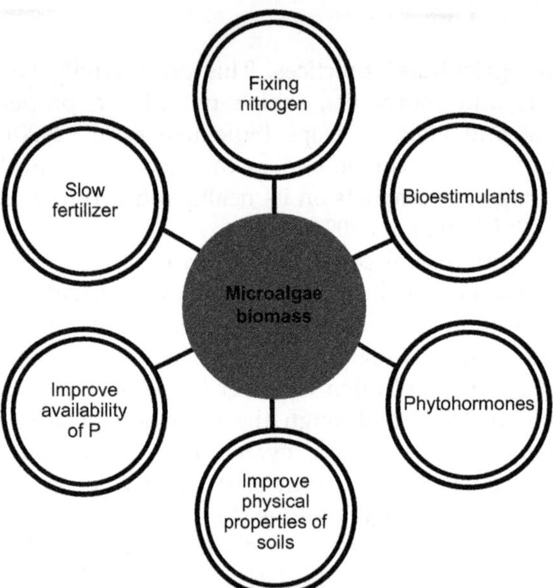

FIGURE 11.2   Microalgae biomass for agricultural purposes as an eco-friendly and sustainable alternative.

Microalgal cell properties protect against stress conditions, such as drought, salinity, light intensity variation, frost, and colonization by bacteria or fungi (Michalak et al., 2016a,b), making them ideal for agricultural use. Recent studies have shown that polysaccharides contribute to a plant's response process in its defense against salt stress and pathogen stress (Michalak et al., 2016a,b; Costa et al., 2018a,b) and act as a source and sink of C, in addition to improving soil aggregation and stabilization (Costa et al., 2018a,b). Chiaiese et al. (2018) reported the use of microalgal biomass as an effective plant growth promoter. Cellular extracts from microalgae and cyanobacteria have also been postulated to release growth-promoting substances and inorganic elements that improve microbial consortia in the soil and provide resistance against diseases (Koffi et al., 2018). Ronga et al. (2019) stated that microalgal proteins could act as metabolic precursors of phytohormones that play an essential role in embryogenesis, organogenesis, and protection against osmotic stress.

It can be seen that microalgal valorization as a biofertilizer has aroused increasing interest. Its use should be highlighted as a promising way of using biomass from enriched organic waste streams, closing the cycle in the context of the circular economy and minimizing the adverse effects of synthetic fertilizers.

11.3.4 Other valorization routes

In addition to products such as those mentioned in previous sections, algal biomass produced in wastewaters can be used to obtain higher value-added products, such as pigments and biopolymers.

Microalgae have different metabolic routes that can be explored to produce biopolymers. Some main routes are (1) use of entire algal cell; (2) use of biopolymers synthesized by algae, such as polysaccharides; (3) obtaining of low-molecular-weight molecules that can be used as a substrate for polymerization, such as ethanol and lactic acid (Lutzu et al., 2021).

Microalgal biomass with high starch content (up to 50%) (Gifuni et al., 2017) and/or proteins (up to 60%) (Hayes et al., 2017) can be converted directly into thermoplastic materials via biomass extrusion after harvesting and drying and, in some cases, after pretreatment for cell disruption (Fabra et al., 2017). López Rocha et al. (2020) evaluated the production of bioplastics from blends of glycerol and biomass produced during wastewater treatment, composed of *Scenedesmus obliquus*, *Desmodesmus communis*, *Nannochloropsis gaditana*, and *Arthrospira platensis*. Bioplastics showed superior viscoelastic properties than pure biomass (*Arthrospira* species), especially at higher temperatures, implying greater thermal resistance. Furthermore, the water absorption of bioplastics from the consortium was lower due to their lower deformability, inhibiting the swelling of the samples (López Rocha et al., 2020).

Regarding biopolymer synthesis, microalgae can accumulate polysaccharides, such as starch, cellulose, and hemicellulose (Chen et al., 2013), in addition to proteins. Typical characteristics of starch produced from microalgae are the crystal structure and small sizes, such as 1.5 μm for starch from *C. sorokiniana* (Gifuni et al., 2017), suitable for thin-film applications, and flavor carriers in the food industry (Lutzu et al., 2021). Furthermore, starch accumulation by algal cells is usually a metabolic response to nutrient deprivation and alternation between light and dark cycles (León-Saiki et al., 2017). However, few studies explore the microalgal starch accumulation potential to produce bioplastics (Lutzu et al., 2021). Other examples of polysaccharides from microalgae, but most commonly found in cyanobacteria, are glycogen and amylopectin-like glucans, cell wall polysaccharides (Bertocchi et al., 1990), and extracellular polysaccharides.

As highlighted in Section 11.2, wastewater-grown microalgal biomass may be rich in protein content, which can also be used to obtain thermoplastic biopolymers. To avoid the protein purification stage, studies are focusing on using the whole protein-rich biomass to explore the thermoplastic effects of proteins (Torres et al., 2015).

Regarding pigments, carotenoids, which are natural pigments recognized for their many beneficial effects on human health and several industrial applications, stand out (Kalra et al., 2021). Pigments, like lipids, are generally produced by algal cells as natural responses to stressful conditions. Recent studies demonstrate the potential of algal biomass produced in effluents to obtain pigments. For example, the accumulation of 805 μg g^{-1} of carotenoids (chl-*a* and chl-*b*) via heterotrophic cultivation of *Chlamydomonas biconvexa* in sugarcane vinasse (Santana et al., 2017). With effluent from the paper mill industry, 11.88 μg mL^{-1} of carotenoid was reported for *Chlamydomonas debaryana* in heterotrophic culture (Arora et al., 2016). After cultivation in municipal wastewater, a consortium of microalgae showed 72.3 μg g^{-1} of β carotene, 36.4 μg g^{-1} of lutein, and 14.7 μg g^{-1} of lycopene (Basily et al., 2018). Phycobiliproteins are auxiliary pigments with a high commercial value and antioxidative, anticarcinogenic, antiinflammatory, antiangiogenic, and neuro and hepatoprotective functions (Cuellar-Bermudeza et al., 2014). Arashiro et al. (2020) studied phycobiliproteins accumulation in *Nostoc* sp., *A. platensis*, and *Porphyridium purpureum* cultivated in a food industry wastewater and reported concentrations in the final biomass up to 103, 57, and 30 mg g^{-1} of phycocyanin, allophycocyanin, and phycoerythrin, respectively.

11.4 Challenges and opportunities for wastewater-grown microalgal biomass valorization

The search for the inclusion of microalgae as a supplement to animal feed has been growing in recent years. However, challenges are still recurrent in the cultivation stages, which still do not match the biomass productivity capable of meeting commercial demands (Kusmayadi et al., 2021). Synthetic crops have higher biomass yields, but they also increase costs due to nutrients and water demands. New cultivation systems that reconcile resource recovery and productivity need to be investigated as strategies to reduce biomass production costs and, consequently, allow microalgae-supplemented diets to have a competitive price concerning normal ones (Kusmayadi et al., 2021). However, the importance of ensuring that the biomass cultivated in wastewater does not contain harmful substances is reinforced (Dineshbabu et al., 2019).

The conversion of biomass to bioenergy requires the adoption of strategies to increase the performance of conversion routes, which include the reduction of ash and nitrogen in the biomass. In addition, the wastewater treatment system and the biomass processing into bioenergy must be designed in an integrated and optimized way. The choice of the biomass-to-bioenergy conversion route must consider the characteristics of the biomass, which is directly related to the characteristics of the effluent used as a culture medium (Couto et al., 2020).

Aiming at the need for changes in agriculture fertilization and the circular economy concept, the use of wastewater for microalgal fertilizers has been considered promising. As it is a recent focus area for research, studies and evaluations are needed considering production on a commercial scale (Renuka et al., 2018; Siebers et al., 2019), in addition to research on the accumulation of bioactive molecules and their effects on the soil and plant health (Chiaiese et al., 2018; Renuka et al., 2018; Supraja et al., 2020). Aiming to mitigate N_2O emissions from microalgae biofertilizers, Suleiman et al. (2020) suggested future research to investigate nitrification inhibition. Moreover, further assessments of the optimal amount of biofertilizer to be applied are needed to avoid an inhibitory effect (Chiaiese et al., 2018). In addition, given the vast diversity of microalgae, investigations aimed at their selection and genetic improvement for enhancing their use in agriculture is highly needed. Evaluation of the duration of the effect of this biofertilizer after application and the impact of its application on climatic factors are also stated (Chiaiese et al., 2018) as the prospects of microalgal valorization.

Besides the most traditional routes, it is urgent to focus on the valorization routes of high-value products (biopolymers and pigments) from wastewater-grown microalgal biomass so that the full potential of the biomass is explored and environmental and economic benefits are maximized throughout the production chain. Further researches are needed to integrate wastewater treatment and biopolymers accumulating algae strains (Lutzu et al., 2021). The development of cultivation systems focusing on higher biomass productivity and carotenoid yields should be highlighted, particularly screening microalgae species naturally present in wastewater (Kalra et al., 2021).

References

Alam, M.A., Xu, J.L., Wang, Z., 2020. Microalgae Biotechnology for Food, Health and High Value Products. Springer Singapore, China, pp. 1–483. Available from: https://doi.org/10.1007/978-981-15-0169-2.

Altmann, B.A., et al., 2019. Do dietary soy alternatives lead to pork quality improvements or drawbacks? A look into micro-alga and insect protein in swine diets. Meat Science 153, 26–34. Available from: https://doi.org/10.1016/j.meatsci.2019.03.001. Germany: Elsevier Ltd.

Alvarez, A.L., et al., 2021. Microalgae, soil and plants: a critical review of microalgae as renewable resources for agriculture. Algal Research 54, 102200. Available from: https://doi.org/10.1016/j.algal.2021.102200. Elsevier BV.

Arashiro, L.T., et al., 2020. Natural pigments from microalgae grown in industrial wastewater. Bioresource Technology 303. Available from: https://doi.org/10.1016/j.biortech.2020.122894. Spain: Elsevier Ltd.

Araújo, R., et al., 2021. Current status of the algae production industry in Europe: an emerging sector of the blue bioeconomy. Frontiers in Marine Science 7. Available from: https://doi.org/10.3389/fmars.2020.626389. Italy: Frontiers Media S.A.

Arbib, Z., et al., 2017. Optimization of pilot high rate algal ponds for simultaneous nutrient removal and lipids production. Science of the Total Environment 589, 66–72. Available from: https://doi.org/10.1016/j.scitotenv.2017.02.206. Spain: Elsevier B.V.

Arora, N., et al., 2016. Bioremediation of domestic and industrial wastewaters integrated with enhanced biodiesel production using novel oleaginous microalgae. Environmental Science and Pollution Research 23 (20), 20997–21007. Available from: https://doi.org/10.1007/s11356-016-7320-y. India: Springer Verlag.

Assemany, P., et al., 2020. Complementarity of substrates in anaerobic digestion of wastewater grown algal biomass. Waste and Biomass Valorization 11 (11), 5759–5770. Available from: https://doi.org/10.1007/s12649-019-00875-8. Brazil: Springer Science and Business Media B.V.

Assis, T.C.d, et al., 2019. Using atmospheric emissions as CO_2 source in the cultivation of microalgae: productivity and economic viability. Journal of Cleaner Production 215, 1160–1169. Available from: https://doi.org/10.1016/j.jclepro.2019.01.093. Brazil: Elsevier Ltd.

Balamurugan, S., Sathishkumar, R., Li, H.-Y., 2021. Biotechnological perspectives to augment the synthesis of valuable biomolecules from microalgae by employing wastewater. Journal of Water Process Engineering 39, 101713. Available from: https://doi.org/10.1016/j.jwpe.2020.101713. Elsevier BV.

Basily, H., et al., 2018. Exploration of using the algal bioactive compounds for cosmeceuticals and pharmaceutical applications. Egyptian Pharmaceutical Journal 109. Available from: https://doi.org/10.4103/epj.epj_6_18. Medknow.

Batista, A.P., et al., 2015. Combining urban wastewater treatment with biohydrogen production – an integrated microalgae-based approach. Bioresource Technology 184, 230–235. Available from: https://doi.org/10.1016/j.biortech.2014.10.064. Portugal: Elsevier Ltd.

Bélanger-Lépine, F., et al., 2018. Cultivation of an algae-bacteria consortium in wastewater from an industrial park: effect of environmental stress and nutrient deficiency on lipid production. Bioresource Technology 267, 657–665. Available from: https://doi.org/10.1016/j.biortech.2018.07.099. Canada: Elsevier Ltd.

Benemann, J., 2009. Microalgae biofuels: a brief introduction. Benemann Associates and Microbio Engineering 1.

Bertocchi, C., et al., 1990. Polysaccharides from cyanobacteria. Carbohydrate Polymers 12 (2), 127–153. Available from: https://doi.org/10.1016/0144-8617(90)90015-K. Italy.

Bhatia, S.K., et al., 2021. Wastewater based microalgal biorefinery for bioenergy production: Progress and challenges. Science of the Total Environment 751, 141599. Available from: https://doi.org/10.1016/j.scitotenv.2020.141599. Elsevier BV.

Brasil, B.d.S.A.F., et al., 2017. Microalgae and cyanobacteria as enzyme biofactories. Algal Research 25, 76–89. Available from: https://doi.org/10.1016/j.algal.2017.04.035. Brazil: Elsevier B.V.

Burlew, J., 1953. Algal Culture From Laboratory to Pilot Plant, 1st edn Publication, p. 600.

Cardinaletti, G., et al., 2018. Effects of graded levels of a blend of *Tisochrysis lutea* and *Tetraselmis suecica* dried biomass on growth and muscle tissue composition of European sea bass (Dicentrarchus labrax) fed diets low in fish meal and oil. Aquaculture (Amsterdam, Netherlands) 485, 173–182. Available from: https://doi.org/10.1016/j.aquaculture.2017.11.049. Italy: Elsevier B.V.

Carpio, R.B., et al., 2021. Effects of reaction temperature and reaction time on the hydrothermal liquefaction of demineralized wastewater algal biomass. Bioresource Technology Reports 14, 100679. Available from: https://doi.org/10.1016/j.biteb.2021.100679. Elsevier BV.

Carvalho, J.R.R., et al., 2018. Performance, insulin sensitivity, carcass characteristics, and fatty acid profile of beef from steers fed microalgae. Journal of Animal Science 96 (8), 3433–3445. Available from: https://doi.org/10.1093/jas/sky210. Brazil: Oxford University Press.

Cassman, N.A., et al., 2016. Plant and soil fungal but not soil bacterial communities are linked in long-term fertilized grassland. Scientific Reports 6. Available from: https://doi.org/10.1038/srep23680. Netherlands: Nature Publishing Group.

Castro, J.d.S., et al., 2017. Microalgae biofilm in soil: greenhouse gas emissions, ammonia volatilization and plant growth. Science of the Total Environment 574, 1640–1648. Available from: https://doi.org/10.1016/j.scitotenv.2016.08.205. Brazil: Elsevier B.V.

Castro, J.d.S., et al., 2021. Hydrothermal carbonization of microalgae biomass produced in agro-industrial effluent: products, characterization and applications. Science of the Total Environment 768, 144480. Available from: https://doi.org/10.1016/j.scitotenv.2020.144480. Elsevier BV.

Chen, C.Y., et al., 2013. Microalgae-based carbohydrates for biofuel production. Biochemical Engineering Journal 78, 1–10. Available from: https://doi.org/10.1016/j.bej.2013.03.006. Taiwan.

Chen, W.T., et al., 2014. Hydrothermal liquefaction of mixed-culture algal biomass from wastewater treatment system into bio-crude oil. Bioresource Technology 152, 130–139. Available from: https://doi.org/10.1016/j.biortech.2013.10.111. United States: Elsevier Ltd.

Chen, H., et al., 2021. Long-term organic and inorganic fertilization alters the diazotrophic abundance, community structure, and co-occurrence patterns in a vertisol. Science of the Total Environment 766, 142441. Available from: https://doi.org/10.1016/j.scitotenv.2020.142441. Elsevier BV.

Cheng, F., et al., 2019. Bio-crude oil from hydrothermal liquefaction of wastewater microalgae in a pilot-scale continuous flow reactor. Bioresource Technology 294, 122184. Available from: https://doi.org/10.1016/j.biortech.2019.122184. Elsevier BV.

Chiaiese, P., et al., 2018. Renewable sources of plant biostimulation: microalgae as a sustainable means to improve crop performance. Frontiers in Plant Science 871. Available from: https://doi.org/10.3389/fpls.2018.01782. Italy: Frontiers Media S.A.

Choudhary, P., et al., 2017. Development and performance evaluation of an algal biofilm reactor for treatment of multiple wastewaters and characterization of biomass for diverse applications. Bioresource Technology 224, 276–284. Available from: https://doi.org/10.1016/j.biortech.2016.10.078. India: Elsevier Ltd.

Cohen, Z., Norman, H.A., Heimer, Y.M., 1995. Microalgae as a source of omega 3 fatty acids. World Review of Nutrition and Dietetics 77, 1–31. undefined.

Coppens, J., et al., 2016. The use of microalgae as a high-value organic slow-release fertilizer results in tomatoes with increased carotenoid and sugar levels. Journal of Applied Phycology 28 (4), 2367–2377. Available from: https://doi.org/10.1007/s10811-015-0775-2. Belgium: Springer Netherlands.

Costa, O.Y.A., Raaijmakers, J.M., Kuramae, E.E., 2018a. Microbial extracellular polymeric substances: ecological function and impact on soil aggregation. Frontiers in Microbiology 9. Available from: https://doi.org/10.3389/fmicb.2018.01636. Netherlands: Frontiers Media S.A.

Costa, S.S., et al., 2018b. Influence of nitrogen on growth, biomass composition, production, and properties of polyhydroxyalkanoates (PHAs) by microalgae. International Journal of Biological Macromolecules 116, 552–562. Available from: https://doi.org/10.1016/j.ijbiomac.2018.05.064. Brazil: Elsevier B.V.

Couto, E.A., Calijuri, M.L., et al., 2018a. Effect of depth of high-rate ponds on the assimilation of CO_2 by microalgae cultivated in domestic sewage. Environmental Technology (United Kingdom) 39 (20), 2653–2661. Available from: https://doi.org/10.1080/09593330.2017.1364302. Brazil: Taylor and Francis Ltd.

Couto, E.A., Pinto, F., et al., 2018b. Hydrothermal liquefaction of biomass produced from domestic sewage treatment in high-rate ponds. Renewable Energy 118, 644–653. Available from: https://doi.org/10.1016/j.renene.2017.11.041. Brazil: Elsevier Ltd.

Couto, E., Calijuri, M.L., Assemany, P., 2020. Biomass production in high rate ponds and hydrothermal liquefaction: wastewater treatment and bioenergy integration. Science of the Total Environment 724. Available from: https://doi.org/10.1016/j.scitotenv.2020.138104. Brazil: Elsevier B.V.

Cuellar-Bermudeza, S.P., Romero-Ogawaa, M.A., Rittmannb, B.E., Parra-Saldivara, R., 2014. Algae biofuels production processes, carbon dioxide fixation and biorefinery concept. Journal of Petroleum & Environmental Biotechnology 5, 2157–7463. Available from: https://doi.org/10.4172/2157-7463.1000185.

Das, P., et al., 2019. Microalgal nutrients recycling from the primary effluent of municipal wastewater and use of the produced biomass as bio-fertilizer. International Journal of Environmental Science and Technology 3355–3364. Available from: https://doi.org/10.1007/s13762-018-1867-8. Springer Science and Business Media LLC.

de Medeiros, V.P.B., et al., 2021. Microalgae in the meat processing chain: feed for animal production or source of techno-functional ingredients. Current Opinion in Food Science 37, 125–134. Available from: https://doi.org/10.1016/j.cofs.2020.10.014. Brazil: Elsevier Ltd.

de Tonnac, A., Guillevic, M., Mourot, J., 2018. Fatty acid composition of several muscles and adipose tissues of pigs fed n-3 PUFA rich diets. Meat Science 140, 1–8. Available from: https://doi.org/10.1016/j.meatsci.2017.11.023.

Dębowski, M., et al., 2017. The Influence of anaerobic digestion effluents (ADEs) used as the nutrient sources for *Chlorella* sp. cultivation on fermentative biogas production. Waste and Biomass Valorization 8 (4), 1153–1161. Available from: https://doi.org/10.1007/s12649-016-9667-1. Poland: Springer Science and Business Media B.V.

Dineshbabu, G., et al., 2019. Microalgae—nutritious, sustainable aqua- and animal feed source. Journal of Functional Foods 62, 103545. Available from: https://doi.org/10.1016/j.jff.2019.103545. Elsevier BV.

Döring, J., et al., 2015. Growth, yield and fruit quality of grapevines under organic and biodynamic management. PLoS One 10 (10). Available from: https://doi.org/10.1371/journal.pone.0138445. Germany: Public Library of Science.

Emparan, Q., Harun, R., Danquah, M.K., 2019. Role of phycoremediation for nutrient removal from wastewaters: a review. Applied Ecology and Environmental Research 17 (1), 889–915. Available from: https://doi.org/10.15666/aeer/1701_889915. Malaysia: Corvinus University of Budapest.

Fabra, M.J., et al., 2017. Development and characterization of hybrid corn starch-microalgae films: effect of ultrasound pre-treatment on structural, barrier and mechanical performance. Algal Research 28, 80–87. Available from: https://doi.org/10.1016/j.algal.2017.10.010. Spain: Elsevier B.V.

Fan, Y., et al., 2019. Effects of algae supplementation in high-energy dietary on fatty acid composition and the expression of genes involved in lipid metabolism in Hu sheep managed under intensive finishing system. Meat Science 157, 107872. Available from: https://doi.org/10.1016/j.meatsci.2019.06.008. Elsevier BV.

FAO, 2018. The Future of Food and Agriculture – Alternative Pathways to 2050. FAO, Rome, Italy.

Fernández, F.G.A., et al., 2021. The role of microalgae in the bioeconomy. New Biotechnology 61, 99–107. Available from: https://doi.org/10.1016/j.nbt.2020.11.011. Spain: Elsevier B.V.

Ferreira, A., et al., 2018. Combining biotechnology with circular bioeconomy: from poultry, swine, cattle, brewery, dairy and urban wastewaters to biohydrogen. Environmental Research 164, 32–38. Available from: https://doi.org/10.1016/j.envres.2018.02.007. Portugal: Academic Press Inc.

Ganesh Saratale, R., et al., 2018. A critical review on anaerobic digestion of microalgae and macroalgae and co-digestion of biomass for enhanced methane generation. Bioresource Technology 262, 319–332. Available from: https://doi.org/10.1016/j.biortech.2018.03.030. South Korea: Elsevier Ltd.

Gifuni, I., et al., 2017. Microalgae as new sources of starch: isolation and characterization of microalgal starch granules. Chemical Engineering Transactions 57, 1423–1428. Available from: https://doi.org/10.3303/CET1757238. Italy: Italian Association of Chemical Engineering – AIDIC.

Gifuni, I., et al., 2018. Identification of an industrial microalgal strain for starch production in biorefinery context: the effect of nitrogen and carbon concentration on starch accumulation. New Biotechnology 41, 46–54. Available from: https://doi.org/10.1016/j.nbt.2017.12.003. Italy: Elsevier B.V.

Goswami, G., Makut, B.B., Das, D., 2019. Sustainable production of bio-crude oil via hydrothermal liquefaction of symbiotically grown biomass of microalgae-bacteria coupled with effective wastewater treatment. Scientific Reports 9 (1). Available from: https://doi.org/10.1038/s41598-019-51315-5. India: Nature Publishing Group.

Hayes, M., et al., 2017. Microalgal proteins for feed, food and health. Microalgae-Based Biofuels and Bioproducts: From Feedstock Cultivation to End-Products. Elsevier Inc, Ireland, pp. 347–368. Available from: https://doi.org/10.1016/B978-0-08-101023-5.00015-7.

Hernández-García, A., et al., 2019. Wastewater-leachate treatment by microalgae: biomass, carbohydrate and lipid production. Ecotoxicology and Environmental Safety 174, 435–444. Available from: https://doi.org/10.1016/j.ecoenv.2019.02.052. Mexico: Academic Press.

Hess, D., et al., 2019. Techno-economic analysis of ash removal in biomass harvested from algal turf scrubbers. Biomass and Bioenergy 123, 149–158. Available from: https://doi.org/10.1016/j.biombioe.2019.02.010. United States: Elsevier Ltd.

Huang, G.H., et al., 2010. Biodiesel production by microalgal biotechnology. Applied Energy 87 (1), 38−46. Available from: https://doi.org/10.1016/j.apenergy.2009.06.016. China: Elsevier Ltd.

Huang, Q., et al., 2019. The 19-years inorganic fertilization increased bacterial diversity and altered bacterial community composition and potential functions in a paddy soil. Applied Soil Ecology 144, 60−67. Available from: https://doi.org/10.1016/j.apsoil.2019.07.009. China: Elsevier B.V.

Kalra, R., Gaur, S., Goel, M., 2021. Microalgae bioremediation: a perspective towards wastewater treatment along with industrial carotenoids production. Journal of Water Process Engineering 40. Available from: https://doi.org/10.1016/j.jwpe.2020.101794. India: Elsevier Ltd.

Kholif, A.E., et al., 2017. Dietary *Chlorella vulgaris* microalgae improves feed utilization, milk production and concentrations of conjugated linoleic acids in the milk of Damascus goats. Journal of Agricultural Science 155 (3), 508−518. Available from: https://doi.org/10.1017/S0021859616000824. Egypt: Cambridge University Press.

Kibria, S., Kim, I.H., 2019. Impacts of dietary microalgae (*Schizochytrium* JB5) on growth performance, blood profiles, apparent total tract digestibility, and ileal nutrient digestibility in weaning pigs. Journal of the Science of Food and Agriculture 99 (13), 6084−6088. Available from: https://doi.org/10.1002/jsfa.9886. Wiley.

Koffi, K.T., Kumar, S., Sur, D.H., 2018. Extraction of plant nutrients from freshwater algae and their role in sustainable agriculture. International Journal of Current Biotechnology 6, 1−8. Available from: http://ijcb.mainspringer.com/6_4/cb604001.pdf.

Kour, D., et al., 2020. Microbial biofertilizers: bioresources and eco-friendly technologies for agricultural and environmental sustainability. Biocatalysis and Agricultural Biotechnology 23. Available from: https://doi.org/10.1016/j.bcab.2019.101487. India: Elsevier Ltd.

Kumar, A., Singh, J.S., 2020. Microalgal Bio-Fertilizers. Elsevier BV, pp. 445−463. Available from: https://doi.org/10.1016/b978-0-12-818536-0.00017-8.

Kumar, A., et al., 2017. Cyanobacterial biotechnology: an opportunity for sustainable industrial production. Climate Change and Environmental Sustainability 97. Available from: https://doi.org/10.5958/2320-642X.2017.00011.4. Diva Enterprises Private Limited.

Kumar, G., et al., 2018. Cultivation of microalgal biomass using swine manure for biohydrogen production: impact of dilution ratio and pretreatment. Bioresource Technology 260, 16−22. Available from: https://doi.org/10.1016/j.biortech.2018.03.029. South Korea: Elsevier Ltd.

Kusmayadi, A., et al., 2021. Microalgae as sustainable food and feed sources for animals and humans − biotechnological and environmental aspects. Chemosphere 271, 129800. Available from: https://doi.org/10.1016/j.chemosphere.2021.129800. Elsevier BV.

Lafarga, T., 2019. Effect of microalgal biomass incorporation into foods: nutritional and sensorial attributes of the end products. Algal Research 41, 101566. Available from: https://doi.org/10.1016/j.algal.2019.101566. Elsevier BV.

Lamminen, M., et al., 2019. Different microalgae species as a substitutive protein feed for soya bean meal in grass silage based dairy cow diets. Animal Feed Science and Technology 247, 112−126. Available from: https://doi.org/10.1016/j.anifeedsci.2018.11.005. Finland: Elsevier B.V.

Leng, L., et al., 2020. Nitrogen in bio-oil produced from hydrothermal liquefaction of biomass: a review. Chemical Engineering Journal 401, 126030. Available from: https://doi.org/10.1016/j.cej.2020.126030. Elsevier BV.

Leong, Y.K., Chang, J.-S., 2020. "Bioremediation of heavy metals using microalgae: recent advances and mechanisms. Bioresource Technology 303, 122886. Available from: https://doi.org/10.1016/j.biortech.2020.122886. Elsevier BV.

León-Saiki, G.M., et al., 2017. The role of starch as transient energy buffer in synchronized microalgal growth in *Acutodesmus obliquus*. Algal Research 25, 160−167. Available from: https://doi.org/10.1016/j.algal.2017.05.018. Netherlands: Elsevier B.V.

Levasseur, W., Perré, P., Pozzobon, V., 2020. A review of high value-added molecules production by microalgae in light of the classification. Biotechnology Advances 41, 107545. Available from: https://doi.org/10.1016/j.biotechadv.2020.107545. Elsevier BV.

Li, R., et al., 2017. Co-digestion of chicken manure and microalgae *Chlorella* 1067 grown in the recycled digestate: nutrients reuse and biogas enhancement. Waste Management 70, 247−254. Available from: https://doi.org/10.1016/j.wasman.2017.09.016. China: Elsevier Ltd.

Liang, Y., et al., 2020. Effects of *Spirulina* supplementation on lipid metabolism disorder, oxidative stress caused by high-energy dietary in Hu sheep. Meat Science 164, 108094. Available from: https://doi.org/10.1016/j.meatsci.2020.108094. Elsevier BV.

Liu, J., et al., 2019. Carbon-nutrient stoichiometry drives phosphorus immobilization in phototrophic biofilms at the soil-water interface in paddy fields. Water Research 167. Available from: https://doi.org/10.1016/j.watres.2019.115129. China: Elsevier Ltd.

Ljubic, A., et al., 2018. Biomass composition of *Arthrospira platensis* during cultivation on industrial process water and harvesting. Journal of Applied Phycology 30 (2), 943–954. Available from: https://doi.org/10.1007/s10811-017-1332-y. Denmark: Springer Netherlands.

López Rocha, C.J., et al., 2020. Development of bioplastics from a microalgae consortium from wastewater. Journal of Environmental Management 263. Available from: https://doi.org/10.1016/j.jenvman.2020.110353. Mexico: Academic Press.

Lutzu, G.A., et al., 2021. Latest developments in wastewater treatment and biopolymer production by microalgae. Journal of Environmental Chemical Engineering 9 (1), 104926. Available from: https://doi.org/10.1016/j.jece.2020.104926. Elsevier BV.

Madeira, M.S., et al., 2017. Microalgae as feed ingredients for livestock production and meat quality: a review. Livestock Science 205, 111–121. Available from: https://doi.org/10.1016/j.livsci.2017.09.020. Portugal: Elsevier B.V.

Manor, M.L., et al., 2017. Defatted microalgae serve as a dual dietary source of highly bioavailable iron and protein in an anemic pig model. Algal Research 26, 409–414. Available from: https://doi.org/10.1016/j.algal.2017.07.018. United States: Elsevier B.V.

Marin-Batista, J.D., et al., 2019. Valorization of microalgal biomass by hydrothermal carbonization and anaerobic digestion. Bioresource Technology 274, 395–402. Available from: https://doi.org/10.1016/j.biortech.2018.11.103. Spain: Elsevier Ltd.

Michalak, I., Chojnacka, K., et al., 2016a. Evaluation of supercritical extracts of algae as biostimulants of plant growth in field trials. Frontiers in Plant Science 7 (2016). Available from: https://doi.org/10.3389/fpls.2016.01591. Poland: Frontiers Research Foundation.

Michalak, I., Górka, B., et al., 2016b. Supercritical fluid extraction of algae enhances levels of biologically active compounds promoting plant growth. European Journal of Phycology 51 (3), 243–252. Available from: https://doi.org/10.1080/09670262.2015.1134813. Poland: Taylor and Francis Ltd.

Mobin, S., Alam, F., 2017. Some promising microalgal species for commercial applications: a review. Energy Procedia. Available from: https://doi.org/10.1016/j.egypro.2017.03.177. Australia: Elsevier Ltd.

Moheimani, N.R., et al., 2018. Nutritional profile and in vitro digestibility of microalgae grown in anaerobically digested piggery effluent. Algal Research 35, 362–369. Available from: https://doi.org/10.1016/j.algal.2018.09.007. Australia: Elsevier B.V.

Naaz, F., et al., 2019. Investigations on energy efficiency of biomethane/biocrude production from pilot scale wastewater grown algal biomass. Applied Energy 254. Available from: https://doi.org/10.1016/j.apenergy.2019.113656. India: Elsevier Ltd.

Naher, U.A., et al., 2019. Fertilizer Management Strategies for Sustainable Rice Production. Elsevier BV, pp. 251–267. Available from: https://doi.org/10.1016/b978-0-12-813272-2.00009-4.

Oita, A., et al., 2020. Trends in the food nitrogen and phosphorus footprints for Asia's giants: China, India, and Japan. Resources, Conservation and Recycling 157, 104752. Available from: https://doi.org/10.1016/j.resconrec.2020.104752. Elsevier BV.

Olaizola, M., 2003. Commercial development of microalgal biotechnology: from the test tube to the marketplace. Biomolecular Engineering. Available from: https://doi.org/10.1016/S1389-0344(03)00076-5. United States: Elsevier.

Park, J.H., Lee, S.I., Kim, I.H., 2018. Effect of dietary *Spirulina* (Arthrospira) platensis on the growth performance, antioxidant enzyme activity, nutrient digestibility, cecal microflora, excreta noxious gas emission, and breast meat quality of broiler chickens. Poultry Science 97 (7), 2451–2459. Available from: https://doi.org/10.3382/ps/pey093. South Korea: Oxford University Press.

Perazzoli, S., et al., 2016. Optimizing biomethane production from anaerobic degradation of Scenedesmus spp. biomass harvested from algae-based swine digestate treatment. International Biodeterioration and Biodegradation 109, 23–28. Available from: https://doi.org/10.1016/j.ibiod.2015.12.027. Brazil: Elsevier Ltd.

Pereira, A.S.A.d.P., et al., 2021. Organomineral fertilizers pastilles from microalgae grown in wastewater: ammonia volatilization and plant growth. Science of the Total Environment 779. Available from: https://doi.org/10.1016/j.scitotenv.2021.146205. Brazil: Elsevier B.V.

Qiao, H., et al., 2019. Feeding effects of the microalga *Nannochloropsis* sp. on juvenile turbot (*Scophthalmus maximus* L.). Algal Research 41. Available from: https://doi.org/10.1016/j.algal.2019.101540. China: Elsevier B.V.

Rashid, N., Park, W.K., Selvaratnam, T., 2018. Binary culture of microalgae as an integrated approach for enhanced biomass and metabolites productivity, wastewater treatment, and bioflocculation. Chemosphere 194, 67–75. Available from: https://doi.org/10.1016/j.chemosphere.2017.11.108. United States: Elsevier Ltd.

Renuka, N., et al., 2017. Wastewater grown microalgal biomass as inoculants for improving micronutrient availability in wheat. Rhizosphere 3, 150–159. Available from: https://doi.org/10.1016/j.rhisph.2017.04.005. India: Elsevier B.V.

Renuka, N., et al., 2018. Microalgae as multi-functional options in modern agriculture: current trends, prospects and challenges. Biotechnology Advances 36 (4), 1255–1273. Available from: https://doi.org/10.1016/j.biotechadv.2018.04.004. South Africa: Elsevier Inc.

Robles, Á., et al., 2020. Microalgae-bacteria consortia in high-rate ponds for treating urban wastewater: elucidating the key state indicators under dynamic conditions. Journal of Environmental Management 261. Available from: https://doi.org/10.1016/j.jenvman.2020.110244. Spain: Academic Press.

Rodrigues de Assis, L., et al., 2020. Innovative hybrid system for wastewater treatment: high-rate algal ponds for effluent treatment and biofilm reactor for biomass production and harvesting. Journal of Environmental Management 274. Available from: https://doi.org/10.1016/j.jenvman.2020.111183. Brazil: Academic Press.

Ronga, D., et al., 2019. Microalgal biostimulants and biofertilisers in crop productions. Agronomy 9 (4). Available from: https://doi.org/10.3390/agronomy9040192. Italy: MDPI AG.

Roohani, A.M., et al., 2019. Effect of spirulina *Spirulina platensis* as a complementary ingredient to reduce dietary fish meal on the growth performance, whole-body composition, fatty acid and amino acid profiles, and pigmentation of Caspian brown trout (Salmo trutta caspius) juveniles. Aquaculture Nutrition 25 (3), 633–645. Available from: https://doi.org/10.1111/anu.12885. Iran: Blackwell Publishing Ltd.

Santana, H., et al., 2017. Microalgae cultivation in sugarcane vinasse: selection, growth and biochemical characterization. Bioresource Technology 228, 133–140. Available from: https://doi.org/10.1016/j.biortech.2016.12.075. Brazil: Elsevier Ltd.

Siebers, N., et al., 2019. Towards phosphorus recycling for agriculture by algae: soil incubation and rhizotron studies using 33P-labeled microalgal biomass. Algal Research 43. Available from: https://doi.org/10.1016/j.algal.2019.101634. Germany: Elsevier B.V.

Silambarasan, S., et al., 2021. Removal of nutrients from domestic wastewater by microalgae coupled to lipid augmentation for biodiesel production and influence of deoiled algal biomass as biofertilizer for *Solanum lycopersicum* cultivation. Chemosphere 268, 129323. Available from: https://doi.org/10.1016/j.chemosphere.2020.129323. Elsevier BV.

Solé-Bundó, M., Garfí, M., Ferrer, I., 2020. Pretreatment and co-digestion of microalgae, sludge and fat oil and grease (FOG) from microalgae-based wastewater treatment plants. Bioresource Technology 298, 122563. Available from: https://doi.org/10.1016/j.biortech.2019.122563. Elsevier BV.

Sproles, A.E., et al., 2021. Recent advancements in the genetic engineering of microalgae. Algal Research 53, 102158. Available from: https://doi.org/10.1016/j.algal.2020.102158. Elsevier BV.

Suleiman, A.K.A., et al., 2020. From toilet to agriculture: fertilization with microalgal biomass from wastewater impacts the soil and rhizosphere active microbiomes, greenhouse gas emissions and plant growth. Resources, Conservation and Recycling 161, 104924. Available from: https://doi.org/10.1016/j.resconrec.2020.104924. Elsevier BV.

Supraja, K.V., Behera, B., Balasubramanian, P., 2020. Efficacy of microalgal extracts as biostimulants through seed treatment and foliar spray for tomato cultivation. Industrial Crops and Products 151, 112453. Available from: https://doi.org/10.1016/j.indcrop.2020.112453. Elsevier BV.

Sutherland, D.L., Turnbull, M.H., Craggs, R.J., 2017. Environmental drivers that influence microalgal species in fullscale wastewater treatment high rate algal ponds. Water Research 124, 504–512. Available from: https://doi.org/10.1016/j.watres.2017.08.012. New Zealand: Elsevier Ltd.

Tahami, M.K., et al., 2017. Plant growth promoting rhizobacteria in an ecological cropping system: a study on basil (*Ocimum basilicum* L.) essential oil production. Industrial Crops and Products 107, 97–104. Available from: https://doi.org/10.1016/j.indcrop.2017.05.020. Iran: Elsevier B.V.

Torres, S., et al., 2015. Green Composites from residual microalgae biomass and poly(butylene adipate- co -terephthalate): processing and plasticization. ACS Sustainable Chemistry & Engineering 3 (4), 614–624. Available from: https://doi.org/10.1021/sc500753h. American Chemical Society (ACS).

The Insight Partners (2021). Microalgae-Based Products Market Forecast to 2028 - COVID-19 Impact and Global Analysis by Type (Spirulina, Chlorella, Astaxanthin, Beta Carotene, and Others) and Application (Food & Beverages, Animal Feed, Pharmaceuticals and Nutraceuticals, Personal Care, and Others), Report. Available at: https://www.theinsightpartners.com/reports/microalgae-based-products-market/. (acessed 12 July 2022).

UNESCO, 2020. The United Nations World Water Development Report 2020—Water and Climate Change. https://en.unesco.org/themes/water-security/wwap/wwdr/2020.

Uthirapandi, V., Suriya, S., Boomibalagan, P., 2018. Bio-fertilizer potential of seaweed liquid extracts of marine macro algae on growth and biochemical parameters of *Ocimum sanctum*. Journal of Pharmacognosy and Phytochemistry 7, 3528–3532.

van der Kooij, S., et al., 2020. Phosphorus recovered from human excreta: a socio-ecological-technical approach to phosphorus recycling. Resources, Conservation and Recycling 157. Available from: https://doi.org/10.1016/j.resconrec.2020.104744. Netherlands: Elsevier B.V.

Vieira de Mendonça, H., et al., 2021. Microalgae in a global world: New solutions for old problems? Renewable Energy 165, 842–862. Available from: https://doi.org/10.1016/j.renene.2020.11.014. Brazil: Elsevier Ltd.

Vigani, M., et al., 2015. Food and feed products from micro-algae: market opportunities and challenges for the EU. Trends in Food Science and Technology 42 (1), 81–92. Available from: https://doi.org/10.1016/j.tifs.2014.12.004. Spain: Elsevier Ltd.

Villar-Navarro, E., et al., 2018. Removal of pharmaceuticals in urban wastewater: high rate algae pond (HRAP) based technologies as an alternative to activated sludge based processes. Water Research 139, 19–29. Available from: https://doi.org/10.1016/j.watres.2018.03.072. Spain: Elsevier Ltd.

Volf, M.R., Rosolem, C.A., 2021. Soil P diffusion and availability modified by controlled-release P fertilizers. Journal of Soil Science and Plant Nutrition 21 (1), 162–172. Available from: https://doi.org/10.1007/s42729-020-00350-7. Brazil: Springer Science and Business Media Deutschland GmbH.

Wang, Q., et al., 2019. Free ammonia pretreatment to improve bio-hydrogen production from anaerobic dark fermentation of microalgae. ACS Sustainable Chemistry and Engineering 7 (1), 1642–1647. Available from: https://doi.org/10.1021/acssuschemeng.8b05405. Australia: American Chemical Society.

Wollmann, F., et al., 2019. Microalgae wastewater treatment: biological and technological approaches. Engineering in Life Sciences 19 (12), 860–871. Available from: https://doi.org/10.1002/elsc.201900071. Germany: Wiley-VCH Verlag.

Wu, Y.B., et al., 2019. Dual functions of eicosapentaenoic acid-rich microalgae: enrichment of yolk with n-3 polyunsaturated fatty acids and partial replacement for soybean meal in diet of laying hens. Poultry Science 98 (1), 350–357. Available from: https://doi.org/10.3382/ps/pey372. China: Oxford University Press.

Young, P., Taylor, M., Fallowfield, H.J., 2017. Mini-review: high rate algal ponds, flexible systems for sustainable wastewater treatment. World Journal of Microbiology and Biotechnology 33 (6). Available from: https://doi.org/10.1007/s11274-017-2282-x. Australia: Springer Netherlands.

Zhu, J., et al., 2019. Effects of reduced inorganic fertilization and rice straw recovery on soil enzyme activities and bacterial community in double-rice paddy soils. European Journal of Soil Biology 94, 103116. Available from: https://doi.org/10.1016/j.ejsobi.2019.103116. Elsevier BV.

Nonconventional treatments of agro-industrial wastes and wastewaters by heterotrophic/mixotrophic cultivations of microalgae and Cyanobacteria

Rihab Hachicha[1,2], Fatma Elleuch[3], Hajer Ben Hlima[3], Pascal Dubessay[2], Helene de Baynast[2], Cedric Delattre[2,4], Guillaume Pierre[2], Ridha Hachicha[3], Slim Abdelkafi[3], Imen Fendri[1] and Philippe Michaud[2]

[1]Laboratoroire de Biotechnologies Végétales Appliquées à l'Amélioration des Cultures, Faculté des Sciences de Sfax, Université de Sfax, Sfax, Tunisia [2]Université Clermont Auvergne, CNRS, SIGMA Clermont, Institut Pascal, Clermont-Ferrand, France [3]Laboratoire de Génie Enzymatique et Microbiologie, Equipe de Biotechnologie des Algues, Ecole Nationale d'Ingénieurs de Sfax, Université de Sfax, Sfax, Tunisia [4]Institut Universitaire de France (IUF), 1 rue Descartes Paris, France

12.1 Introduction

The continuous discharge of agro-industrial wastes and wastewaters (W&WW) has raised several environmental problems, affecting soil, air, and aquatic ecosystems (Gonçalves et al., 2016; Mo et al., 2018). Conventional physicochemical and biological processes for wastewater treatments remove some nutrients from wastewaters and reduce biological oxygen demand (Crini and Lichtfouse, 2019). They eliminate solid particles before the active intervention of microorganisms for the bioremediation of organic compounds (Hodaifa et al., 2020).

Valorisation of Microalgal Biomass and Wastewater Treatment
DOI: https://doi.org/10.1016/B978-0-323-91869-5.00002-8

However, they are expensive and limited for the elimination of inorganic nitrogen and phosphorous, heavy metals, and xenobiotics, in addition to neglecting the resourceful utilization of these effluents (Markou and Georgakakis, 2011; Wollmann et al., 2019; Elleuch et al., 2021). W&WW composition is characterized by a high organic and/or inorganic content, including high concentrations of phenolic compounds, lipids, and organic acids as reported for olive mill and winery wastewaters (Dermeche et al., 2013; Markou and Georgakakis, 2011; Rizzo et al., 2010). As early as the 1950s, microalgae cultivation was proposed as nonconventional wastewater treatment, but despite its potential, it had not been developed enough (Fabris et al., 2020; Wang et al., 2014).

Microalgae are single-celled plants of growing interest in many areas (Lauersen, 2019) because of their ability to produce biofuel, foods, or feeds and also high-value compounds such as pigments, lipids, peptides, proteins, carbohydrates, and vitamins (Fendri et al., 2013; Pignolet et al., 2013). Despite their maximum theoric photosynthetic efficiency in terms of conversion of light energy into the biomass of up to 3 and 10%, only a few genera (100 genera) of microalgae are cultivated at the laboratory scale, and no more than 10 were exploited in an industrial context (Gaignard et al., 2019). Similarly, Cyanobacteria are prokaryotic microorganisms that use CO_2 as the sole carbon source in photoautotrophy and are known for producing a wide variety of compounds with high potential for industrial applications (Markou and Georgakakis, 2011; Bhargava, 2017). There is a disagreement among taxonomists in classifying this polyphyletic group of organisms as microalgae or bacteria (Chapman and Chapman, 1973; Palinska and Surosz, 2014). Cyanobacteria are neither considered nor referred to as microalgae, and the term microalgae will only concern eukaryotes. The cultivation of Cyanobacteria and microalgae in wastewaters allows the reduction of their organic and inorganic contents and their cleaning up.

Moreover, the obtained biomass could be used as a commercial value-added product. The global consumption of water for domestic and industrial purposes can reach 450 billion $m^3 year^{-1}$, where the domestic sector alone could contribute to 70% (315 billion $m^3 year^{-1}$). The use of these 70% domestic wastewaters for microalgae cultivation could generate 23.5 billion t of oil (Abinandan and Shanthakumar, 2015). The major disadvantage for the bioconversion of W&WWs by microalgae is correlated to the cost required by the photoautotrophic treatments of large volumes of discharges (De-Bashan and Bashan, 2010). Also, most W&WWs are opaque with a high load of suspended solids, thus limiting the penetration of light resulting in poor biomass production (Bhatnagar et al., 2011; Wang et al., 2015). The alternative adopted is the use of heterotrophic/mixotrophic cultures in which W&WWs rich in carbon, nitrogen, and other minerals are used as a substrate by microalgae or Cyanobacteria (Li et al., 2011a,b). Unlike photoautotrophy, heterotrophy is a mode in which the absence of light, organic carbon dissolved in the culture media are used as substrates for energy production and growth in aerobic using respiration. Mixotrophy is another process of a culture where microalgae combine the two modes (photoautotrophy and heterotrophy). Indeed, during mixotrophy, inorganic carbon (CO_2) is fixed by photosynthesis in the presence of light. In contrast, organic carbon is assimilated and metabolized to generate energy and metabolites by aerobic respiration (Lowrey et al., 2015). The purpose of this chapter is to describe the ability of photosynthetic microorganisms to metabolize carbon heterotrophically and mixotrophically in agro-industrial W&WWs and the potential benefit which presents this combination to the overall economy of production while generating value-added biomolecules with significant industrial interests.

12.2 Heterotrophic and mixotrophic culture of microalgae and Cyanobacteria versus autotrophy

Currently, photoautotrophic culture is the most common and the lowest-cost technology used (open ponds) to cultivate microalgae and Cyanobacteria. The carbon of microalgal biomass (50% DW) is from CO_2 when these microorganisms are cultivated in photoautotrophy. Likewise, to produce 100 t of algal biomass, approximately 183 t of CO_2, generally supplied by the atmosphere, are fixed (Chisti, 2007). It can also be added to the culture medium in the form of HCO_3^- or by gas bubbling with pH-dependent proportions (Espinosa-Carreón and Escobedo-Urías, 2017).

Carbonic anhydrases (CAs) catalyze the reversible hydration of CO_2 in Chlorophyceae (Amoroso et al., 1998), Bacillariophyceae (Iglesias-Rodriguez and Merrett, 1997), Haptophyceae (Nimer et al., 1996), and Rhodophyceae (Aizawa and Miyachi, 1986). Among the four families of CAs (α-, β-, γ-, and ε-CA) identified in nature, only α- and β-CA(s) were described in microalgae (Mitsuhashi et al., 2000; Moroney et al., 2001). These enzymes can be either extracellular, ensuring the diffusion of CO_2 through the cell membrane or can be located into the cytoplasm, mitochondria and mainly in the chloroplast (Moroney and Ynalvez, 2007). In the cytoplasm, the CA converts CO_2 to HCO_3^- to prevent its avoidance from the cell. HCO_3^- is converted to CO_2 in the chloroplast by another CA, which concentrates it around the ribulose-1,5-bisphosphate carboxylase/oxygenase (RuBisCo) (Rigobello-Masini et al., 2003). This protein uses ribulose-1,5 bisphosphate and CO_2 as substrates yielding two molecules of 3-phosphoglycerate (3-PGA). adenosine triphosphate (ATP) and nicotinamide adenine dinucleotide (NADH) H, H^+ formed during the photochemical phase are necessary for carbon fixation as 3-PGA is subsequently phosphorylated into 1,3-bisphosphoglycerate and reduced to glyceraldehyde-3-phosphate (G3P) (Mohsenpour et al., 2021) (Fig. 12.1).

Open ponds are the widest reactors used because of their low cost, simplicity, and standardization. However, they are more confronted with microbial contaminations and dependence on the natural environment. The water used to supply these systems sometimes comes from wastewaters. Photobioreactors (PBRs), more complex to set up and much more expensive, have higher productivity avoiding biological and/or chemical contaminations and have better control of culture parameters (Perez-Garcia et al., 2011). Microalgae production cost can reach 55.45 USD dollars kg^{-1} with tubular PBRsrs against 5.56 USD dollars kg^{-1} for open ponds (Tang et al., 2020). Although most microalgae are photoautotrophic organisms, some species are able to grow in heterotrophy or mixotrophy (Lowrey et al., 2015).

During mixotrophic cultures, the specific growth rate of microalgae can be measured as the sum of those in photoautotrophic and heterotrophic modes (analyzing specific growth rate or lipid content and dry weight). Other studies revealed that the two metabolisms (photosynthesis and aerobic respiration) are mutually influenced, contributing to a synergistic effect and improving biomass productivity (Wang et al., 2014). There are no specific growing conditions

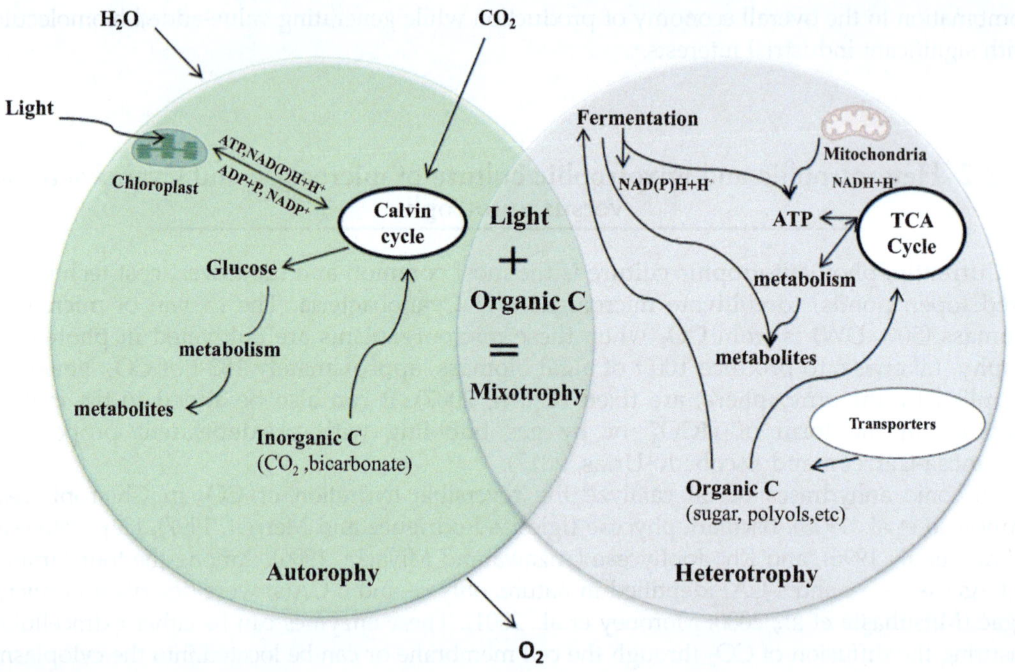

FIGURE 12.1   Autotrophic, heterotrophic, and mixotrohpic metabolisms in microalgae.

for mixotrophy. It is a variant of the autotrophic mode with adequate light conditions in a culture medium supplemented with organic carbon (Chen and Zhang, 1997; Lowrey et al., 2015).

In comparison to autotrophic mode, heterotrophic and mixotrophic growths offer some advantages such as (1) lower cost production and better control of microalgae growth, due to the absence of light and the capacity of holding higher algal biomass; (2) greater quality and quantity of biosynthesis products such as pigments and lipids; (3) higher cell density, growth rate, and energetic cell metabolism; (4) potential of using a carbon source derived from agro-industrial by-products or wastewaters; (5) low contamination risk; (6) cheap and simple bioreactor design (Perez-Garcia et al., 2015; Gaignard et al., 2019). Yet, production at an industrial scale with these culture regimes is not profitable until now (Kröger and Müller-Langer, 2011; Liang, 2013; Tabernero et al., 2012; Wang et al., 2014), taking into account (1) the bacterial and fungal proliferation in competition with that of nonaxenic monoalgal culture; (2) increased energy expenditure using organic carbon sources such as glucose; (3) restricted number of microalgae able to heterotrophically and mixotrophically growth; and (5) ineptitude to produce light-induced metabolites such as photosynthetic pigments into heterotrophic cultures. To compromise between the problem of cost and the desire to carry out economic microalgae cultures, the use of a low-value carbon source such as wastewater, lignocellulosic material, and agro-industrial process wastes seems to be a solution. Likewise, it is necessary to design economical bioreactors suitable for heterotrophic and/or mixotrophic industrial production using cheaper materials and to implement alternative mixing strategies using a renewable energy source as well as cheap sterilization strategies. Finally, reducing downstream processing costs (harvesting and processing biomass) is becoming crucial (Kröger and Müller-Langer, 2011; Tabernero et al., 2012; Liang, 2013). Perez-Garcia et al. (2015) opined that heterotrophic and mixotrophic cultivations of microalgae could be used in biorefinery systems, the global process of which involves the preparation of growth medium (filtration, mixing, dilution, sterilization, and hydrolysis of the carbon sources) and the choice of the adapted strains. Indeed, the appropriate microalgae should be selected according to their genetic and metabolic capacities. Microalgal culture can be carried out according to the four operating modes: batch, fed-batch, continuous, or semicontinuous; laboratory, pilot, or commercial scales. All previous steps (cultivation and growth medium preparation) are owned by the "upstream" part of the process. "Downstream" part of the culture will be carried out to obtain the desired products from biomass, which consists of cell harvesting using various methods (filtration, flocculation, flotation centrifugation, and sedimentation) and addition to drying and cell disruption. The biomass obtained subsequently passes through cascaded extraction to convert raw materials into useful products and finally purification for commercial products.

The carbon sources metabolized by the microalgae are mainly pyruvate, acetate, glycerol, hexoses (glucose or fructose) and pentoses (e.g., xylose or arabinose). Lactate, ethanol, and disaccharides such as lactose, sucrose, and cellobiose have been unsuccessfully tested and known to have negative results in heterotrophic growth and metabolite production (Perez-Garcia et al., 2011). Ratledge et al. (2001) showed that *Crypthecodinium cohnii* was able to grow in cultures containing 8 g L^{-1} acetate. In another research, 10 g L^{-1} of sodium acetate added to municipal wastewater inoculated with *Chlorella vulgaris* for tertiary treatment was metabolized in heterotrophy. This was not possible when calcium acetate replaced sodium acetate (Perez-Garcia et al., 2011). Acetate could be toxic for many microorganisms at high concentrations, even if it was mentioned by Perez-Garcia et al. (2011) that several species of microalgae can assimilate it at low concentrations. Despite this variety of carbon sources, glucose remains the most widely used carbon substrate for heterotrophic and mixotrophic cultures (Perez-Garcia et al., 2015). Understanding how and where the different metabolisms happen in mixotrophic and heterotrophic growths is essential for scale-up and fermentation process design (Fig. 12.1). Glucose is typically metabolized in the cytoplasm of microalgae via the oxidative pathways of pentosephosphate and Embden—Meyerhof—Parnas (EMP) (Neilson and Lewin, 1974; Perez-Garcia et al., 2015). Acetate is first thioesterified with coenzyme A by acetyl coenzyme A (acetyl-CoA) synthetase (EC 6.2.1.1) to generate acetyl-CoA using a single ATP molecule. Then acetyl-CoA is oxidized in the (1) glyoxylate cycle leading to malate in the glyoxysome after the action of malate synthase and isocitrate lyase and the (2) *tricarboxylic acid* (TCA) cycle providing carbon skeletons, energy in the form of ATP and reducing molecules (NADH) in mitochondria (Perez-Garcia et al., 2011; Perez-Garcia et al., 2015). Glycerol is metabolized after being phosphorylated to glyceraldehyde-3-phosphate and glycerate via several enzymes such as glycerol kinase (EC 2.7.1.30), sn-glycerol-3-phosphate NAD oxidoreductase (EC 1.1.1.8) and triose-phosphate (EC 5.3.1.1) to form (via the EMP pathway) pyruvate, which subsequently enters the TCA cycle (Perez-Garcia et al., 2015).

In microalgae, carbon and nitrogen metabolisms are closely linked since they share carbon sources supplied in the autotrophic mode and assimilated in the heterotrophic one. In the presence of nitrogen the carbon flow, resulting from the binding of CO_2, is directed toward the Calvin cycle to allow the production of precursors of amino acids and lipids (Mohsenpour et al., 2021). Under photoautotrophic conditions, nitrogen starvation may lead to the decrease of microalgae growth and discolouration of the biomass marked by a reduction in pigment contents. This phenomenon causes an early enter in stationary phase and an accumulation of lipids and/or carbohydrates polymers at the expense of proteins and pigments (da Silva et al., 2009; Li et al., 2011a,b), causing an increase in cell volume (da Silva et al., 2009; Lamers et al., 2012). For example, in response to nitrogen starvation, some *Chlorella* accumulates neutral lipids (Li et al., 2011a,b). Strains of the genus *Scenedesmus* increase their intracellular lipid content (from 11.5% to 22.4%) and change their fatty acid composition (Vidyashankar et al., 2013). An increase in exopolysaccharide (EPS) content has also been observed in *Rhodella violacea*, *Porphyridium marinum*, or *Flintiella sanguinaria* in the case of nitrogen starvation (Villay et al., 2013; Soanen et al., 2016; Gaignard et al., 2018). In addition, a modification of the constituent fatty acid profile was observed in *Dunaliella salina* where unsaturated molecules such as C16: 1, C16: 3, C16: 4, and C18: 3 decreased in the absence of nitrogen at the expense of other fatty acids such as C16: 0, C16: 2, C18: 0, C18: 1, and C18: 2 with content increased. The concentration of some pigments increased also during a nitrogen deficiency. In the same way, intracellular carotenoids of *D. salina*e evolved from 1.28 to 14.4 g L^{-1} of biomass in the case of nitrogen deficiency (Lamers et al., 2012).

The most used nitrogen sources by microalgae are minerals (NH_4^+, NO_3^-, and NO_2), but there are also organic sources such as urea or amino acids. Microalgae principally assimilate nitrogen via the glutamine synthetase (GS)—glutamate synthase (GOGAT) cycle. Likewise, it is also assimilated via the NADP-glutamate dehydrogenase (GDH) pathway in some green algae under certain conditions (Hellebust and Ahmad, 1989). Several strains of microalgae better assimilate ammonium instead of nitrate. For example, the red microalgae *Galdieria sulphuraria* uptakes only ammonium (Reeb and Bhattacharya, 2010). The GS/GOGAT pathway is the primary pathway for ammonium assimilation under either autotrophic or heterotrophic conditions. It needs carbon skeletons in the form of keto acids (2 oxoglutarate and oxaloacetate) and energy in the form of ATP and NADPH (Perez-Garcia et al., 2011) (Fig. 12.1). Millbank (1957) highlighted that a higher level of keto acids was found in *Chlorella* spp. cultivated under heterotrophic conditions and nitrogen starvation compared to the low quantities obtained when the microalga was grown autotrophically.

Nevertheless, ammonium incorporation into glutamate can also be performed by GDH (EC 1.4.1.2). Ammonium transport across the membranes and under autotrophic and heterotrophic conditions is carried out by means of AMT-type transporters (Wilhelm et al., 2006). Eight putative ammonium- and thirteen putative nitrate/nitrite transporters have been identified in *Chlamydomonas* sp. and other ammonium transporters belonging to the AMT family in diatoms (Allen et al., 2005; Fernandez and Galvan, 2007). Before the GS reaction, assimilation of nitrate requires its transport across the membrane consuming energy, carbon, and protons before its reduction to ammonia. In higher plants and microalgae, this reaction is catalyzed by nitrate- (NR; EC 1.6.6.1—3) and nitrite reductase (NiR; EC 1.7.7.1), which work sequentially (Perez-Garcia et al., 2011). The reduction of nitrate to nitrite is performed through NAD(P)H-dependent nitrate reductase (NR) in the cytosol. NiR catalyzes in plastids nitrite reduction to ammonium using ferredoxin as electron donor (Fernandez and Galvan, 2007). Some microalgae are also able to use urea and organic nitrogen under autotrophic and heterotrophic conditions.

Despite the dependence of the growth yield on the organic nitrogen source, the carbon source, and the strain/species, it has been shown that comparable growth yields have been reported with organic nitrogen compounds compared to those obtained with nitrate or ammonia (Neilson and Lewin, 1974). Urea is the most extensive organic nitrogen source that helps in the growth of algae. It is metabolized by urease (EC: 3.5.1.5) and urea amidolyase (also called urea carboxylase, UALse, EC: 6.3.4.6) (Perez-Garcia et al., 2011).

During dark aerobic respiration, microalgae use organic compounds as electron donors while oxygen is consumed as the final electron acceptor. Throughout dark conditions, respiration provides energy (ATP and NADH) for maintenance and biosynthesis and essential carbon skeletons for biosynthesis. Moreover, dark respiration rates (mol O_2 mol^{-1} carbon d) increase with growth rates and reach under optimal conditions, about 20%–30% of growth rates (Perez-Garcia et al., 2011; Perez-Garcia et al., 2015). During heterotrophic or mixotrophic growth, microalgae biomass may significantly improve and reach over 40 g L^{-1} (Alkhamis, 2015). Lipids and storage polysaccharides (starch and glycogen) are generally accumulated under heterotrophic and mixotrophic conditions contrary to chlorophyll content that decreases considerably (Choix et al., 2012; Wang et al., 2014). Nevertheless, the study carried out by Marquardt (1998) has shown that the red microalgae *G. sulphuraria* treated with the inhibitor of the carotenoid biosynthetic pathway, norflurazon, retained its photosynthetic apparatus when grown in the dark. Unlike autotrophic growth, where some strains were not able to grow with a partially decomposed chloroplast structure, heterotrophic growth shows a reduction of the chlorophyll content, leading to the accumulation of a pheophorbide a-like pigment, an almost total loss of carotenoids in addition to intact algal ultrastructure. Authors suggest that the accumulation of pheophorbide a-like is probably not a direct result of NF treatment but of culturing the algae in the dark. In another study, Gross and Schnarrenberger (1995) cultivated two strains of *G. sulphuraria* under mixo- and heterotrophic modes using 27 different sugars as substrates. They found that one of the strains keeps in higher pigment content with a long lag period when grown on the sugar alcohol D-mannitol, suggesting that these results consist of an ideal system to study the metabolism of rare sugars and sugar alcohols. Moreover, Chen et al. (2017) determined the optimal concentration of acetate (≈ 5 g L^{-1}) used to increase lutein production by *Chlorella sorokiniana* cultivated under mixotrophy. Under these conditions, lutein productivity was improved by 42%.

Chlorella is the oldest and the most used genus for lipid production due to its similar composition to vegetable oils rich in saturated and unsaturated C18s fatty acids. Supplemented with 1% (w/v) glucose, *C. vulgaris* showed lipid productivity under mixotrophic growth 1.5 times and 13.5 times higher than heterotrophic and photoautotrophic cultivation, respectively (Liang et al., 2009). Under heterotrophy, with hydrolyzate of Jerusalem artichoke tuber as carbon source, *Chlorella protothecoides* accumulated lipids up to $\sim 50\%$ of its dry weight in comparison to 15% reached under autotrophic mode. These lipids have been known for their potential to be 100% converted into biodiesel. More than 82% of the total biodiesel composes of linoleic and oleic acid methyl esters (Cheng et al., 2009). Perez-Garcia et al. (2015) pointed out that in mixotrophic growth and lipid production, *C. sorokiniana* can hold out 51%. A real-time polymerase chain reaction study carried out on genes acc D (heteromeric acetyl-CoA carboxylase beta subunit), acc 1 (homomeric acetyl-CoA carboxylase), and rbc L (ribulose-1,5-bisphosphate carboxylase/oxygenase large subunit) showed elevated expression levels of the acc D reflecting the significant lipid content of this strain during the stationary phase along with low-gene expression of acc1 and rbc L. Compared with autotrophic cultivation and with the addition of glucose, a high biomass and lipid production of mixotrophically cultivated *Nannochloropsis oculata*, *D. salina*, *C. sorokiniana*, and *Scenedesmus obliquus* have been achieved with an increase of 1.1−1.6 times, 1.8−2.4 times, 4.1−8.0, and 50 times, respectively. Cecchin et al. (2018) highlighted the molecular basis of acetate assimilation by *C. sorokiniana* in autotrophic versus mixotrophic growths by transcriptome de novo assembly and analyzing by the mean of RNA-sequencing. In the presence of a reduced carbon source as acetate, an improvement of biomass, proteins, lipids, and starch productivities has been shown in addition to a decrease of chlorophyll content. Moreover, they reported that acetate assimilation caused upregulation of the phosphoenolpyruvate carboxylase enzyme, enabling the potential recovery of carbon atoms lost by acetate oxidation. As a result, the increase in productivity observed in mixotrophic cultivation of *C. sorokiniana* may be associated with a different genetic regulation leading to fine regulation of cell metabolism. In another study, under mixotrophic conditions, an increase of growth and lipid productivity of *Chlorella pyrenoidosa* was observed when it was grown in the presence of acetate (Liu et al., 2018). Morais et al. (2017) showed that the mixotrophic growth of *Phaeodactylum tricornutum* with continuous light and glycerol (0.1 M) in the culture medium provided 338.97 mg L^{-1} of lipids, which was ∼80% higher than the concentration obtained with autotrophic mode. Furthermore, the best salinity value allowing the highest microalgae growth was 15‰.

The red microalgae *Porphyridium cruentum* accumulated lipids around 11% in heterotrophic conditions (Oh et al., 2009). The latter, mixotrophically cultivated on a boiled fraction of *Solanum tuberosum* flour, also showed an increase in production of phycoerythrin (PE) and EPS under 12 hours: 12 hours light/dark with a light intensity of 234 μmol photons m^{-2} s^{-1} to reach concentrations of the order of 10 and 330 μg mL^{-1} instead of 7 and 129 μg mL^{-1}, respectively, in autotrophic conditions (Fabregas et al., 1999). Likewise, storage biomolecules may be accumulated during mixotrophic and heterotrophic conditions. For example, glycogen amount in *G. sulphuraria* mixotrophically and heterotrophically cultivated was significantly increased when grown on glucose, while polyunsaturated fatty acids content increased only in heterotrophic conditions (Sakurai et al., 2016). The more studied microalgae cultivated in complete darkness were *Amphora*, *Ankistrodesmus*, *Chlamydomonas*, *Chlorella*, *Chlorococcum*, *Crypthecodinium*, *Cyclotella*, *Dunaliella*, *Euglena*, *Nannochloropsis*, *Nitzschia*, *Ochromonas*, and *Tetraselmis* (Xu et al., 2006; Liang, 2013; Perez-Garcia et al., 2015; Ye et al., 2018). It has been shown that *C. prototothecoides* can grow on various carbon sources in darkness, but it is mysterious whether it can grow on organic carbon ones in the presence of light (Xu et al., 2006). For mixotrophy, literature reports the ability of *Brachiomonas submarina*, *Chlorella spp.*, *Chlorococcum sp.*, *Cyclotella cryptica*, *Euglena gracilis*, *Haematococcus pluvialis*, *Nannochloropsis spp.*, *Naviculasaprophila*, *Nitzschia sp.*, *Ochromonas minima*, *P. tricornutum*, *Rhodomonas reticulate*, *P. cruentum*, *G. sulphuraria*, *Cyanidium caldarium*, *Porphyridium purpureum*, *Rhodella*, and *S. obliquus* (Bassi et al., 2014;

Liang et al., 2009; Fabregas et al., 1999; Gaignard et al., 2019). It was noticed that some microalgae are not truly mixotrophs but have the capacity to grow under different trophic modes and especially the possibility of switching from one to another depending on environmental conditions (Perez-Garcia et al., 2011).

Cyanobacterial genera such as *Anabaena, Arthrospira*, and *Synechococcus* can grow heterotrophically and mixotrophically (Perez-Garcia et al., 2015).

Despite the ability of several species of microalgae to grow on several carbon sources, only a few organic nutrients can be successfully fed for biomass production. Perez-Garcia et al. (2011) reported that the culture of *C. vulgaris* on different carbon sources led to variabilities in biomass productivity. Consequently, it is necessary to choose the carbon source fitting with the specific microalgae. Among red microalgae, *P. cruentum* grew using glucose and acetate to achieve high cell densities (Oh et al., 2009). The Cyanidiaceae family, particularly the genera *Cyanidium, Cyanidioschyzon*, and *Galdieria*, also exhibits a great multiplication capacity on various carbon sources. These microalgae are very sensitive to desiccation, which is why cells grow endolithically and are subjected to strong light limitations that hinder autotrophic growth. *G. sulphuraria* can metabolize at least 27 different sugars or polyols: disaccharides, hexoses, pentoses, deoxy sugars, hexitols, pentiols, tetriols and tiols, amino acids and certain organic acids (Gross and Schnarrenberger, 1995; Oesterhelt et al., 1999; Schilling and Oesterhelt, 2007).

All the advantages of heterotrophic/mixotrophic cultures of microalgae can jump new research for more exploitation. Genomics, bioinformatics analyses, genetic and metabolic engineering currently lead to new approaches in microalgae biotechnology (Boyle and Morgan, 2009). As a result, it is now possible to transform some obligate photoautotrophs to heterotrophs by introducing genes of sugar transporters in their genome. For example, *Volvox carteri* was the first microalgae successfully transformed with the hexose/H^+ symporter gene from *Chlorella* sp. (Hallmann and Sumper, 1996). In the same way, other trophic conversion by simple genetic transformations has been reported in *Chlamydomonas reinhardtii*. In fact, via the heterologously expressed HUP1 hexose (hexose uptake protein) symporter derived from *Chlorella kessleri*, this strain has become able to use glucose for heterotrophic growth in complete darkness. Furthermore, an increment in H_2 production capacity was performed by this newly developed strain opening new perspectives on future strategies for enhancing bio-H_2 production efficiency under natural day/night regimes. In another study, by inserting a single gene encoding a glucose transporter (glut1 or hup1), *P. tricornutum* has become able to grow heterotrophically using the organic carbon substrate glucose (Doebbe et al., 2007; Zaslavskaia et al., 2001).

12.3 Heterotrophic/mixotrophic cultivations of microalgae and Cyanobacteria in agro-industrial wastes and wastewaters

Agro-industrial W&WW treatment by microalgae dates for a long time since this combination provides a biomass production with high economic value (Table 1; Leite et al., 2019). These treatments are still less developed despite their potential. The major disadvantage is correlated to the cost required in the autotrophic treatment of very large volumes of these discharges promptly (de-Bashan and Bashan, 2010). Another issue represented by the autotrophic mode is that most W&WWs are opaque with a high load of suspended solids, thus limiting light penetration (Bhatnagar et al., 2011). In addition, most of the treatment processes of these effluents (physicochemical, biological, thermal, radiative, etc.) seem to be expensive and ineffective in removing inorganic pollutants (Markou and Georgakakis, 2011). For those reasons, adopting heterotrophy and mixotrophy has become a novel solution to overcome the problems caused by autotrophy (light, cost, etc.) as well as removing at most of the organic and inorganic (mainly nitrogen and phosphorous) pollutants (Wang et al., 2012). Research dealing with the mixotrophic/heterotrophic cultivations of microalgae using various W&WWs as a cultivation medium or as an organic substrate is briefly discussed in this section.

12.3.1 Pork wastes and wastewaters

Pork constitutes the most widely consumed meat in the world, resulting in large quantities of waste production. Feces, urine, and wash water are the essential compounds of pork wastes that are responsible for the constitution of a slurry material rich in organic and inorganic pollutants. The potential for converting the nutrients from swine W&WWs into valuable biomass through microalgal cultivation has attracted recently much attention.

As reviewed by Leite et al. (2019), the pilot treatment of wastewater mixture (municipal and piggery wastewaters) in an upflow anaerobic sludge blanket reactor followed by *C. sorokiniana* cultivation in three flat panel PBRs was monitored for 4 weeks. This allowed more than 90% of organic matter removal in addition to the microalgae final dry weight of around 1 g L^{-1}. The results showed an average removal of dissolved inorganic carbon and PO_4^{3-} from 46% to 56% and from 40% to 60%, respectively, as well as total removal of NH_3. They also reported that this relevant removal reduced the N: P ratio during the cultivation and, consequently, affected the biomass productivity and nutrient uptake. Montero et al. (2018) cultivated under controlled and noncontrolled mixotrophies of *Chlorococcum* sp. using a digestate from pig manure. With nitrogen starvation and under controlled conditions (irradiance of $46 \pm 3 \text{ μmol photons m}^{-2} \text{ s}^{-1}$ and temperature of 30°C), tested dilutions (2, 5, 6, and 8% v/v) of water promoted higher biomass density compared to the control medium. In addition, they found that intracellular lipids remained constant throughout tested dilutions and control cultures. Carbohydrates increased from 20% to 45% and dry cell weight from 20% to 42% in tested dilutions compared to the control medium. Cheng et al. (2013) investigated the effect of the synergism of algal-bacterial systems to treat piggery effluent. By adopting air-stripping as pretreatment to reduce the high concentration of ammonical nitrogen in the effluent and the mixotrophic cultivation mode, an enhancement of biomass and lipid productivity of *Desmodesmus sp.* CHX1 were obtained to reach 0.869 and $0.1182 \text{ g}^{-1} \text{ day}^{-1}$, respectively. Moreover, an efficient reduction of nutrient content was reported. The nitrogen, ammonical nitrogen, and total phosphorus removal rates reached 87.3%, over 95%, and 93.1%, respectively. Wang et al. (2015) used diluted nitrogen-abundant swine wastewater to cultivate *C. vulgaris* JSC-6 for simultaneous nutrient/chemical oxygen deman (COD) removal and carbohydrate production. Under mixotrophic and heterotrophic conditions and depending on the dilution ratio of the wastewater (20-fold diluted), 76% of COD and 91% of nitrogen removal were achieved.

In addition, using fivefold dilution, a maximal biomass concentration of 3.96 g L^{-1} was attained. Regarding carbohydrate content, the latter reached up to 58% (dry weight). The authors also demonstrated that the growth of this carbohydrate-rich strain was accompanied by the formation of extracellular organic matter (EOM), which may be responsible for increased total organic carbon level in the water leading to adverse effects on microalgal wastewater treatment. They also suggested that it is crucial to reach optimal conditions of the best timing for harvesting carbohydrate-rich microalgal biomass to avoid increases in the COD level due to the formation of EOM. Diluted piggery wastewater was used for mixotrophic cultivation of *C. pyrenoidosa* by Wang et al. (2012). The objective was to combine primary piggery wastewater treatment with oil production using technologies, economically acceptable. *C. pyrenoidosa* biomass increased as the initial COD value passed from 250 to 1000 mg L^{-1}. In addition, an efficient removal of ammonium which was over 90% in all diluted samples was noted. The maximum lipid productivity $(6.3 \text{ mg L}^{-1} \text{ day}^{-1})$ was achieved when the initial COD was set to 1000 mg L^{-1}. Fatty acid profiles of extracted lipids were different to those obtained with the microalgae cultivated in the control medium. Acidogenically fermented liquid swine manure was used by Hu et al. (2012a,b) as nutrient for heterotrophic/mixotrophic cultivation of *Chlorella sp.* UMN271 for nutrient removal and conversion into biomass. Furthermore, exogenous acids, namely, acetic, propionic, and butyric acids added at 0.1% (v/v) during 7-day batch cultivation, support *Chlorella* sp's growth. The latter displayed a high growth rate and lipid productivities. They also reported that acidogenic fermentation of raw LSM might promote the volatile fatty acids accumulation, suggesting that the profile obtained may be considered feedstock for biodiesel and other biofuel production. Luo et al. (2016) demonstrated that the use of 40% diluted supernatant of anaerobically and aerobically treated swine wastewater (AnATSW) was ideal for mixotrophic cultivation of *Coelastrella* sp. QY01. The latter was screened and isolated for its potential of growth in AnATSW. The maximum growth rate of this strain was 0.325 day^{-1} with biomass productivity of $57.46 \text{ mg L}^{-1} \text{ day}^{-1}$. Ammonia nitrogen and total phosphorus removal in AnATSW reached 100%, while inorganic carbon removal attained 78%. The highest biomass and lipid productivities were 57.46 and $13.42 \text{ mg L}^{-1} \text{ day}^{-1}$, respectively, after 10 days of cultivation. The combined central composite design (CCD) approach and response surface analysis were used for semicontinuously growth optimization of *Chlorella* sp. on acidogenically digested swine wastewater in bench-scale multilayer PBR (Hu et al., 2013). Adopting the optimal conditions of the two key parameters, wastewater dilution rate = 8.00 times and hydraulic retention time = 2.26 days estimated from the significant 2^2 CCD. The resulting mass productivity and nutrient removal rates matched with the predicted values. Moreover, the high algal biomass contents obtained (protein 58.78% and lipid 26.09% of the dry weight) can be considered an ideal alternative for biofuel and feed production.

12.3.2 Poultry wastes and wastewaters

In the United States, and especially in Georgia, annual poultry litter production can reach up to 1.5 million, equal to 108 t of urea and 85 t of diammonium phosphate (Bhatnagar et al., 2011). Litter, manner, and slaughterhouse are considered to be the most important wastes from poultry.

Poultry manure is characterized by high total nitrogen ($\sim 46\,\mathrm{g\,kg^{-1}}$ manure) and ammonical nitrogen ($\sim 14.4\,\mathrm{g\,kg^{-1}}$ manure). Poultry litter contains $\sim 3.3\%$ nitrogen and $\sim 2.6\%$ phosphorus (Mukhtar, 2005; Bhatnagar et al., 2011; Markou and Georgakakis, 2011).

Bhatnagar et al. (2011) evaluated mixotrophic cultivation of *Chlamydomonas globosa*, *Chlorella minutissima*, and *Scenedesmus bijuga* in the control medium BG11 supplemented with different organic carbon substrates (glucose, sucrose, and acetate) and poultry litter extract (PLE) wastewaters. The authors observed, first, higher biomass production for the three strains growing mixotrophically compared to photoautotrophic conditions. Second, the addition of glucose and nitrogen in PLE resulted in an increase of algal biomass by 180% compared to the BG11 medium. Diluted supernatant poultry litter leachate (PLL) obtained from acid extraction of nutrients contained in poultry litter was used by Markou et al. (2016), as a medium for the cultivation of *Spirulina platensis* and *C. vulgaris*. *Spirulina* was not capable of growing in $15\times$ and $10\times$ diluted leachates.

Nonetheless, in $20\times$ and $25\times$ dilutions, the biomass production was half of that in a Zarrouk's medium. This biomass was characterized by high carbohydrate content (37%−44%) due to the stress caused by nutrient limitation. Regarding growth and biomass production of *C. vulgaris* in PLL are better than those obtained using the control medium (BG11). Biomass composition had a lower protein and higher carbohydrate and lipid contents compared to the control medium. In another study, *C. vulgaris* and *Tetraselmis chuii* were heterotrophically cultivated in the poultry manure extract (Lowrey and Yildiz, 2014). Poor growth of both strains was observed, meaning that the physicochemical composition of this specific wastewater is not adapted to support their heterotrophic growth. It was concluded that these microalgae were either unsuitable for heterotrophy (*T. chuii*) or unable to utilize available organic carbon (*C. vulgaris*).

12.3.3 Cattle wastes and wastewater

The discharge of cattle wastewaters causes major environmental problems because of its huge production, reaching around 1.3 billion head of cattle worldwide (Markou and Georgakakis, 2011). In China, cattle production is estimated to be 53 million year^{-1}. Cattle manure contains an appreciable amount of nitrogen and organic pollutant (Markou and Georgakakis, 2011). Data about the use of cattle manure for microalgae cultivation are scarce. In 2019 a study by Luo et al. explored the cultivation of *Scenedesmus* sp. in anaerobic digestion effluent from cattle manure (ADEC) combined with municipal wastewater for nutrient removal and lipid accumulation (Luo et al., 2019). Considering wastewater as a diluent, this strain showed rapid growth in 10% ADEC. With the following optimized culture condition; pH of 7.0 and light intensity of 5000 lx. *Scenedesmus* sp. achieved, during a 7-day culture, a biomass concentration of $4.65\,\mathrm{g\,L^{-1}}$ and lipid productivity of $81.9\,\mathrm{mg\,L^{-1}\,day^{-1}}$. Regarding nutrient removal, they attain about 90% of COD, NO_3^--N, NH_4^--N, and 79%−88% of PO_4^{3-}-P. The profile of fatty acid methyl esters (FAMEs) in ADEC growth medium was analyzed and showed saturated (39.48%) and monounsaturated (60.52%) fatty acids with a higher proportion of oleic and palmitic acids. In addition to this latter, 16- to 18-chain-length fatty acids were estimated for more than 98% of total FAMEs. This diverse lipid content may be a viable biodiesel feedstock.

12.3.4 Dairy wastes and wastewaters

The dairy industry is considered the largest food sector, characterized by huge amounts of water mainly coming from cleaning, sanitization, heat exchange (heating/cooling), and floor washing (Tocchi et al., 2012). The generated effluent contains high biological oxygen deman (BOD), COD, and macro-and micronutrient concentrations and high total nitrogen and phosphorus. In addition, dairy wastewaters (DWWs) accommodate huge compounds such as wasted milk, minerals, proteins, lactose, and fats (Markou and Georgakakis, 2011; Chokshi et al., 2016). Therefore lactose, the main constituent of cheese whey, can be used as a valuable carbon source to heterotrophic/mixotrophic cultivations of microalgae. In this context, Girard et al. (2014) studied mixotrophic and heterotrophic growths of *S. obliquus* using cheese whey permeate (WP) as a novel alternative to produce biodiesel. Compared to photoautotrophic and heterotrophic modes, the partial substitution of the Bold's Basal Medium (BBM) by 40% (v/v) of WP has stimulated specific growth rates and biomass yields under mixotrophy. A reduction of the lactose concentration (54.4%) concomitantly to increased galactose and glucose concentrations was attributed to extracellular lactose hydrolysis by the microalgae. Nitrogen (nitrate and ammonium) removal was almost complete, and neutral lipids were found after intracellular lipid analyses, particularly under conditions of high pH (>9.5). Daneshvar et al. (2019) reported the growth of *Scenedesmus quadricauda* and *Tetraselmis suecica* in the reused DWW. Dry biomass, pigment contents of these microalgae and pollutant exhaustion have been analyzed under mixotrophic and heterotrophic cultivation in the first step, then under a second mixotrophic mode as well. Dry weights and Chl a content of both strains were found to be higher in the first cycle (mixotrophy and heterotrophy) than the second one (mixotrophy). The best nutrient removal was obtained with *S. quadricauda* after two cycles of cultivation (76.77% of total organic carbon, 92.15% of total nitrogen, 100% of phosphate, and 100% of sulfate removals). Fatty acids analysis of *S. quadricauda* and *T. suecica* lipids during the two consecutive cycles revealed a dominance of C18:1 and C18:3n-3 fatty acids, respectively. Biomass and lipid productions were enhanced for *C. protothecoides* growing heterotrophically in dairy coproduct. Espinosa-Gonzalez et al. (2014) used glucose and galactose from prehydrolyzed WP using encapsulated β-galactosidase, for batch and fed-batch cultures. Biomass production and total lipid accumulation reached $9.1 \pm 0.2 \,\mathrm{g\,L^{-1}}$, $17.2 \pm 1.3 \,\mathrm{g\,L^{-1}}$, and $42.0\% \pm 6.6\%$, $20.5\% \pm 0.3\%$ (dry weight basis), in batch and fed-batch cultures, respectively. Direct utilization of WP by simultaneous saccharification and fermentation has led to $7.3 \pm 1.3 \,\mathrm{g\,L^{-1}}$ of biomass with $49.9\% \pm 3.3\%$ of lipid (dry weight basis) in batch mode. Anaerobic digestion of dairy manure wastewaters is often used to reduce the concentration of nutrients, but also for its advantage of converting organic carbon into methane. Therefore Wang et al. (2010) tried among the dilutions tested, $25 \times$ anaerobically diluted digested dairy manure wastewater for *Chlorella* sp. cultivation. Within this dilution, they found a higher average specific growth rate ($0.409 \,\mathrm{d^{-1}}$) and fatty acid content (13.7%), besides the efficiently removing of COD and nutrients, mainly ammonia, total nitrogen, and total phosphorus (27.9%, 100%, 77.6%, 74.7%, respectively). A variety of fatty acids were obtained from this strain, mainly octadecadienoic (C18:2) and hexadecanoic (C16:0) acids derived from triacylglyceride phospholipid and free fatty acids. A nonhydrolyzed cheese whey powder solution (nhCW), a mixture of pure glucose and galactose, and a hydrolyzed cheese whey powder solution (hCW) were tested as organic carbon sources for *C. vulgaris*

P12 cultivated in mixotrophy (Abreu et al., 2012). hCW induced a significant increase in final biomass concentration (3.58 ± 0.12 g L^{-1}), specific growth rate (0.43 ± 00 day^{-1}), lipid productivities, and starch/proteins contents, compared to photoautotrophic conditions or mixotrophy with others carbon sources. The highest pigment content (0.74%) was achieved in the photoautotrophic growth.

12.3.5 Olive oil mill wastes and wastewaters

The Mediterranean countries producing olive oil, such as Spain, Italy, Greece, Tunisia, and Turkey, are faced with the problem of eliminating the agro-industrial residues resulting from this production, namely, olive-oil mill wastewaters (OMWWs), which constitutes about $1-1.6$ m^3 per ton of olive fruits processed (Markou et al., 2012). During the olive oil extraction process, the worldwide estimated volume of OMWW is $7-30$ million m^3 year^{-1}, where Mediterranean areas generate 30 hm^3. This affluent, characterized by a brown aqueous liquid with a cloudy appearance, contains a very high saline charge and is acidic. Its polluting organic load composed of sugars, tannins, polyalcohols, pectins, lipids, and mainly phenolic compounds makes it hardly biodegradable essentially by traditional biological processes (Dareioti et al., 2010; Dermeche et al., 2013; Hodaifa et al., 2020). Hodaifa et al. (2012) investigated the growth of *S. obliquus* in diluted OMWWs without fat matter (WF) and pretreated with active carbon (WC). Two other comparative cultures were also carried out using a synthetic mineral medium (RL) and urban wastewater (UW). Maximum biomass productivity and specific growth rates were achieved in the presence of $5\%-15\%$ (v/v) OMWWs. They also resulted, in a maximum growth rate of 0.052 ± 0.012, 0.048 ± 0.0, 0.049 ± 0.001, 0.04 ± 0.002, and 0.037 ± 0.003 hour^{-1} in UW, RL, WF, WC, and 10% OMWW, respectively, suggesting that the dark color and the absence of essential nutrients were at the origin of limited growth of this strain. Sánchez et al. (2001) investigated the effect of OMWWs concentration and the type of lightning (continuous or intermittent) on the mixotrophic growth of *C. pyrenoidosa*. The best experiment leading to biomass productivity of $1.4 \ 10^{-3}$ g L^{-1} hour^{-1} and a growth rate of 0.04 hour^{-1} was a 10% (v/v) initial concentration of OMWW, a continuous illumination (55 W m^{-2}), and aeration rate (CO_2 supply) of 1 L (liter cell suspension)$^{-1}$ min^{-1} (v/v/min). Markou et al. (2012) have used sodium hypochlorite (NaOCl) to decrease the phenol concentration and turbidity of OMWWs, making them suitable for mixotrophically cultivated *S. platensis*. OMWWs were settled for 10 days and treated with 5% of sodium hypochlorite (NaOCl) and 5% calcium hypochlorite Ca (OCl)$_2$ (called tOMWW$_{10}$). With 10% t OMWW$_{10}$ complemented with a nitrogen source, maximum biomass production reached 1696 mg L^{-1}. COD and carbohydrate removals were found to be 73.18% and 91.19%, respectively, while phenols, phosphorus, and nitrates were completely removed in runs with 5% t OMWW$_{10}$. In another study, Di Caprio et al. (2018) proposed OMWWs fed-batch supply and two-stage supply strategies (photoautotrophic cultivation followed by heterotrophic cultivation) instead of batch supply to enhance biomass production of *Scenedesmus* sp. They found that the progressive addition of OMWWs allowed a higher biomass content of 0.86 g L^{-1}. During nitrogen starvation and in the two-stage strategy, biomass production reached 1.4 g L^{-1}, OMWW sugar removal was $60\%-70\%$, leading to 44% carbohydrate accumulation. In addition, phenol removal was about 55% in the two-stage strategy when the heterotrophic stage lasted longer than 8–10 days. Hodaifa et al. (2009) reported the influence of daily doses of light (in the range 0–36 E m^{-2} d^{-1}) on the growth of *S. obliquus* in diluted OMWWs. They emphasized an inhibition of mixotrophic cultures in the presence of high doses of OMWWs. At OMWW 5%, this strain exhibited a maximum specific growth rate of 0.026 hour^{-1} under mixotrophic and heterotrophic conditions. In addition, maximum removal of BOD was obtained in the heterotrophic cultures. Finally, the highest content in carbohydrate (65.8%) and lowest protein content (258 mg g^{-1}) were achieved, confirming the deficiency of OMWW in nitrogen.

12.3.6 Other agro-industrial wastes and wastewaters

Piasecka et al. (2017) used beet molasses as an alternative and sole carbon source for mixotrophic cultivation of *Parachlorella kessleri*, obtaining a maximum specific growth rate of 0.80 day^{-1}. Biomass, oil, and protein productivities as the calorific value of algal cell biomass reached 0.42 g L^{-1} days, 112.56 and 244.95 mg L^{-1} days, and 22.1 MJ kg^{-1}, respectively. Mitra et al. (2012) demonstrated inexpensive agro-industrial coproducts from the extraction process of corn oil, the dry-grind ethanol thin stillage, as nutrient raw materials for heterotrophic/mixotrophic growth of *C. vulgaris*. Microalgal biomass and oil content rich in linoleic acid reached 9.8 g L^{-1} and 43%, respectively. Enhanced lipids production by *C. sorokiniana* using carob pod extract and industrial glycerol and acetate-rich oxidized wine waste lees as carbon sources were reported by León-Vaz et al. (2019). The fed-batch mixotrophic culture in the presence of 100 mM of acetate coming from the oxidized wine waste lees added with 30 mM of ammonium resulted in a specific growth rate, a biomass concentration and a lipid content of 0.052 hour^{-1}, 11 g L^{-1} and 38% (w/w), respectively. To increase lipid yield and reduce biodiesel cost, Gao et al. (2010) used sweet sorghum juice as an alternative carbon source to glucose for heterotrophic cultivation of *C. protothecoides*. At an initial reducing sugar concentration of 10 g L^{-1} in the enzymatic hydrolyzsates of sweet sorghum juice, the dry cell yield and lipid content achieved 5.1 g L^{-1} and 52.5%, respectively, after 12-h culture in flasks. To reach maximum lipids accumulation, Giovanardi et al. (2013) investigated the use of carbon-rich wastes derived from the apple vinegar production for mixotrophic growth of *Neochloris oleoabundans*. They result in the highest final cell density besides lipid accumulation. Xu et al. (2006) investigated the use of corn powder hydrolysate to heterotrophically cultivated *C. protothecoides*. Maximum biomass and lipid content of 3.92 g L^{-1}, 55.3% were achieved in 6 days.

12.4 Conclusions and future prospects

Thanks to their characteristics, microalgae and Cyanobacteria can contribute significantly to waste management through their cultivation in heterotrophy and mixotrophy using W&WWs. Although most studies mainly focused on algae productivity and nutrient recovery efficiency, a few of them discuss the algal growth type (heterotrophy, mixotrophy, or autotrophy). These indicate that the alternate metabolic pathways of heterotrophic and mixotrophic modes of cultivation can yield a considerable reduction of inorganic and organic pollutants and improve production economics while providing concentrated and valuable biomass and bioproduct accumulation. Wastewater treatment using microalgae or Cyanobacteria is a well-established technology that offers new insights for the microalgae industry and the wastewater treatment industry. It presents several advantages compared to traditional wastewater treatments such as lower capital and operational costs, pollutants and pathogen decrease, nutrient recovery in the form of valuable biomass, energy savings, and CO_2 emissions reduction.

Moreover, the application of microalgae for wastewater treatment might be considered a prerequisite for them to enter the energy market through biofuels. However, growing microalgae on wastewater are facing challenges that induce the development of other alternatives. They include mainly the land requirement, environmental and operational condition influence, effect of wastewater characteristics, and biomass harvesting and valorization. In this chapter, more emphasis has been given to wastewater treatment using heterotrophic/mixotrophic cultivation of microalgae and Cyanobacteria upon considering its technical challenges and opportunities. Still, on an industrial scale, it was in the experimental stage. It should be noted that the number of companies of varying sizes specializing in the production of microalgae has increased significantly in recent years, which is an indication of the economic potential of this technology.

Nevertheless, the available data on the investment and operating costs involved in wastewater treatment technology are deficient. There are thus only a few analyses limited to small territories, but some of them provide exciting indications. To better benefit from the full potential of the combination of microalgae production and wastewater treatment, many research and development challenges have still to be focused on the development of robust, productive wastewater-adapted microalgal species and in the development and innovation of cultivation and downstream processing systems, which will allow for better growth, harvesting and conversion of the algal biomass. Further, some improvement levers may be provided by physicochemical and/or enzymatic pretreatments, diluting or introducing some growth enhancers to the wastewater to adapt microalgae to prevailing conditions and improve process efficiency. Likewise, there is undoubtedly a need for further studies to understand the biology and physiology of mixotrophic/heterotrophic microalgae metabolic pathways and optimize the cultivation process under these conditions for target performances on sustainability and cost. In addition, genomic and transcriptomic information and molecular and biochemical characterization of the key genes could be continued to maximize productivity and nutrient removal.

Acknowledgments

This work has been funded by the Franco-Tunisian program (PHC UTIQUE—46201NH).

References

Abinandan, S., Shanthakumar, S., 2015. Challenges and opportunities in application of microalgae (Chlorophyta) for wastewater treatment: a review. Renewable and Sustainable Energy Reviews 52, 123–132. Available from: https://doi.org/10.1016/j.rser.2015.07.086. India: Elsevier Ltd.

Abreu, A.P., et al., 2012. Mixotrophic cultivation of *Chlorella vulgaris* using industrial dairy waste as organic carbon source. Bioresource Technology 118, 61–66. Available from: https://doi.org/10.1016/j.biortech.2012.05.055. Portugal.

Aizawa, K., Miyachi, S., 1986. Carbonic anhydrase and CO_2 concentrating mechanisms in microalgae and cyanobacteria. FEMS Microbiology Letters 39 (3), 215–233. Available from: https://doi.org/10.1016/0378-1097(86)90447-7. undefined.

Alkhamis, Y.A., 2015. Cultivation of Microalgae in Phototrophic, Mixotrophic and Heterotrophic Conditions (PhD Thesis). Flinders University, School of Biological Sciences.

Allen, A.E., Ward, B.B., Song, B., 2005. Characterization of diatom (Bacillariophyceae) nitrate reductase genes and their detection in marine phytoplankton communities. Journal of Phycology 41 (1), 95–104. Available from: https://doi.org/10.1111/j.1529-8817.2005.04090.x. United States.

Amoroso, G., et al., 1998. Uptake of $HCO3^-$ and CO_2 in cells and chloroplasts from the microalgae *Chlamydomonas reinhardtii* and Dunaliella tertiolecta. Plant Physiology 116 (1), 193–201. Available from: https://doi.org/10.1104/pp.116.1.193. Germany: American Society of Plant Biologists.

Bassi, A., Saxena, P., Aguirre, A.M., 2014. Mixotrophic algae cultivation for energy production and other applications. Algal Biorefineries. Springer Netherlands, Canada, pp. 177–202. Available from: https://doi.org/10.1007/978-94-007-7494-0_7.

Bhargava, S., 2017. Cyanobacterial diversity: a potential source of bioactive compounds. Advances in Biotechnology & Microbiology 4.

Bhatnagar, A., et al., 2011. Renewable biomass production by mixotrophic algae in the presence of various carbon sources and wastewaters. Applied Energy 88 (10), 3425–3431. Available from: https://doi.org/10.1016/j.apenergy.2010.12.064. United States: Elsevier Ltd.

Boyle, N.R., Morgan, J.A., 2009. Flux balance analysis of primary metabolism in *Chlamydomonas reinhardtii*. BMC Systems Biology 3. Available from: https://doi.org/10.1186/1752-0509-3-4. United States.

Cecchin, M., et al., 2018. Molecular basis of autotrophic vs mixotrophic growth in *Chlorella sorokiniana*. Scientific Reports 8 (1),. Available from: https://doi.org/10.1038/s41598-018-24979-8. Italy: Nature Publishing Group.

Chen, F., Zhang, Y., 1997. High cell density mixotrophic culture of *Spirulina platensis* on glucose for phycocyanin production using a fed-batch system. Enzyme and Microbial Technology 20 (3), 221–224. Available from: https://doi.org/10.1016/S0141-0229(96)00116-0. Hong Kong.

Chapman, V.J., Chapman, D.J., 1973. The Algae. Springer.

Chen, C.Y., et al., 2017. Enhancing lutein production with *Chlorella sorokiniana* Mb-1 by optimizing acetate and nitrate concentrations under mixotrophic growth. Journal of the Taiwan Institute of Chemical Engineers 79, 88–96. Available from: https://doi.org/10.1016/j.jtice.2017.04.020. Taiwan: Taiwan Institute of Chemical Engineers.

Cheng, Y., et al., 2009. Biodiesel production from Jerusalem artichoke (*Helianthus Tuberosus* L.) tuber by heterotrophic microalgae *Chlorella protothecoides*. Journal of Chemical Technology & Biotechnology 84 (5), 777–781. Available from: https://doi.org/10.1002/jctb.2111. Wiley.

Cheng, H., Tian, G., Liu, J., 2013. Enhancement of biomass productivity and nutrients removal from pretreated piggery wastewater by mixotrophic cultivation of *Desmodesmus* sp. CHX1. Desalination and Water Treatment 51 (37–39), 7004–7011. Available from: https://doi.org/10.1080/19443994.2013.769917. China: Desalination Publications.

Chisti, Y., 2007. Biodiesel from microalgae. Biotechnology Advances 25 (3), 294–306. Available from: https://doi.org/10.1016/j.biotechadv.2007.02.001. New Zealand.

Choix, F.J., de-Bashan, L.E., Bashan, Y., 2012. Enhanced accumulation of starch and total carbohydrates in alginate-immobilized Chlorella spp. induced by *Azospirillum brasilense*: I. Autotrophic conditions. Enzyme and Microbial Technology 51 (5), 294–299. Available from: https://doi.org/10.1016/j.enzmictec.2012.07.013. Mexico.

Chokshi, K., et al., 2016. Microalgal biomass generation by phycoremediation of dairy industry wastewater: an integrated approach towards sustainable biofuel production. Bioresource Technology 221, 455–460. Available from: https://doi.org/10.1016/j.biortech.2016.09.070. India: Elsevier Ltd.

Crini, G., Lichtfouse, E., 2019. Advantages and disadvantages of techniques used for wastewater treatment. Environmental Chemistry Letters 17 (1), 145–155. Available from: https://doi.org/10.1007/s10311-018-0785-9. France: Springer Verlag.

da Silva, A.F., Lourenço, S.O., Chaloub, R.M., 2009. Effects of nitrogen starvation on the photosynthetic physiology of a tropical marine microalga *Rhodomonas* sp. (Cryptophyceae). Aquatic Botany 91 (4), 291–297. Available from: https://doi.org/10.1016/j.aquabot.2009.08.001. Brazil.

Daneshvar, E., et al., 2019. Sequential cultivation of microalgae in raw and recycled dairy wastewater: microalgal growth, wastewater treatment and biochemical composition. Bioresource Technology 273, 556–564. Available from: https://doi.org/10.1016/j.biortech.2018.11.059. Finland: Elsevier Ltd.

Dareioti, M.A., et al., 2010. Exploitation of olive mill wastewater and liquid cow manure for biogas production,". Waste Management 30 (10), 1841–1848. Available from: https://doi.org/10.1016/j.wasman.2010.02.035. Greece.

de-Bashan, L.E., Bashan, Y., 2010. Immobilized microalgae for removing pollutants: review of practical aspects. Bioresource Technology 101 (6), 1611–1627. Available from: https://doi.org/10.1016/j.biortech.2009.09.043. Mexico: Elsevier Ltd.

Dermeche, S., et al., 2013. Olive mill wastes: biochemical characterizations and valorization strategies. Process Biochemistry 48 (10), 1532–1552. Available from: https://doi.org/10.1016/j.procbio.2013.07.010. Algeria.

Di Caprio, F., Altimari, P., Pagnanelli, F., 2018. Integrated microalgae biomass production and olive mill wastewater biodegradation: optimization of the wastewater supply strategy. Chemical Engineering Journal 349, 539–546. Available from: https://doi.org/10.1016/j.cej.2018.05.084. Italy: Elsevier B.V.

Doebbe, A., et al., 2007. Functional integration of the HUP1 hexose symporter gene into the genome of *C. reinhardtii*: impacts on biological H_2 production. Journal of Biotechnology 131 (1), 27–33. Available from: https://doi.org/10.1016/j.jbiotec.2007.05.017. Germany.

Elleuch, J., et al., 2021. Zinc biosorption by *Dunaliella* sp. AL-1: mechanism and effects on cell metabolism. Science of the Total Environment 773, 145024. Available from: https://doi.org/10.1016/j.scitotenv.2021.145024. Elsevier BV.

Espinosa-Carreón, T.L., Escobedo-Urías, D., 2017. South region of the gulf of California large marine ecosystem upwelling, fluxes of CO_2 and nutrients. Environmental Development 22, 42–51. Available from: https://doi.org/10.1016/j.envdev.2017.03.005. Mexico: Elsevier B.V.

Espinosa-Gonzalez, I., Parashar, A., Bressler, D.C., 2014. Heterotrophic growth and lipid accumulation of *Chlorella protothecoides* in whey permeate, a dairy by-product stream, for biofuel production. Bioresource Technology 155, 170–176. Available from: https://doi.org/10.1016/j.biortech.2013.12.028. Canada: Elsevier Ltd.

Fabregas, J., et al., 1999. Mixotrophic production of phycoerythrin and exopolysaccharide by the microalga. Cryptogamie Algologie 89–94. Available from: https://doi.org/10.1016/s0181-1568(99)80009-9. Elsevier BV.

Fabris, M, Abbriano, R.M., Pernice, M., Sutherland, D.L., Commault, A.S., Hall, C.C., Labeeuw, L., McCauley, J.I., Kuzhiuparambil, U., Ray, P., Kahlke, T., Ralph, P.J., 2020. Emerging technologies in algal biotechnology: toward the establishment of a sustainable, algae-based bioeconomy. Frontiers in Plant Sciences 11. Available from: https://doi.org/10.3389/fpls.2020.00279.

Fendri, I., et al., 2013. Olive fermentation brine: Biotechnological potentialities and valorization. Environmental Technology (United Kingdom) 34 (2), 181–193. Available from: https://doi.org/10.1080/09593330.2012.689364. Tunisia: Taylor and Francis Ltd.

Fernandez, E., Galvan, A., 2007. Inorganic nitrogen assimilation in Chlamydomonas. Journal of Experimental Botany . Available from: https://doi.org/10.1093/jxb/erm106. Spain.

Gaignard, C., et al., 2018. The red microalga *Flintiella sanguinaria* as a new exopolysaccharide producer. Journal of Applied Phycology 30 (5), 2803–2814. Available from: https://doi.org/10.1007/s10811-018-1389-2. France: Springer Netherlands.

Gaignard, C., et al., 2019. New horizons in culture and valorization of red microalgae. Biotechnology Advances 37 (1), 193–222. Available from: https://doi.org/10.1016/j.biotechadv.2018.11.014. France: Elsevier Inc.

Gao, C., et al., 2010. Application of sweet sorghum for biodiesel production by heterotrophic microalga *Chlorella protothecoides*. Applied Energy 87 (3), 756–761. Available from: https://doi.org/10.1016/j.apenergy.2009.09.006. China: Elsevier Ltd.

Giovanardi, M., et al., 2013. Morphophysiological analyses of *Neochloris oleoabundans* (Chlorophyta) grown mixotrophically in a carbon-rich waste product. Protoplasma 250 (1), 161–174. Available from: https://doi.org/10.1007/s00709-012-0390-x. Italy: Springer-Verlag Wien.

Girard, J.M., et al., 2014. Mixotrophic cultivation of green microalgae *Scenedesmus obliquus* on cheese whey permeate for biodiesel production. Algal Research 5 (1), 241–248. Available from: https://doi.org/10.1016/j.algal.2014.03.002. Canada: Elsevier.

Gonçalves, A.L., et al., 2016. The effect of increasing CO_2 concentrations on its capture, biomass production and wastewater bioremediation by microalgae and cyanobacteria. Algal Research 14, 127–136. Available from: https://doi.org/10.1016/j.algal.2016.01.008. Portugal: Elsevier B.V.

Gross, W., Schnarrenberger, C., 1995. Heterotrophic growth of two strains of the acido-thermophilic red alga *Galdieria sulphuraria*. Plant and Cell Physiology 36 (4), 633–638. Germany.

Hallmann, A., Sumper, M., 1996. "The Chlorella hexose/H+ symporter is a useful selectable marker and biochemical reagent when expressed in Volvox," Proceedings of the National Academy of Sciences of the United States of America. Germany, 93(2), 669–673. https://www.pnas.org/doi/10.1073/pnas.93.2.669.

Hellebust, J.A., Ahmad, I., 1989. Regulation of nitrogen assimilation in green microalgae. Journal of Marine Biology & Oceanography 6, 241–255.

Hodaifa, G., Martínez, M.E., Sánchez, S., 2009. Daily doses of light in relation to the growth of *Scenedesmus obliquus* in diluted three-phase olive mill wastewater. Journal of Chemical Technology and Biotechnology 84 (10), 1550–1558. Available from: https://doi.org/10.1002/jctb.2219. Spain.

Hodaifa, G., et al., 2012. Inhibitory effects of industrial olive-oil mill wastewater on biomass production of *Scenedesmus obliquus*. Ecological Engineering 42, 30–34. Available from: https://doi.org/10.1016/j.ecoleng.2012.01.020. Spain.

Hodaifa, G., et al., 2020. Combination of physicochemical operations and algal culture as a new bioprocess for olive mill wastewater treatment. Biomass and Bioenergy 138, 105603. Available from: https://doi.org/10.1016/j.biombioe.2020.105603. Elsevier BV.

Hu, B., Min, M., Du, Z.Y., Chen, P., Ma, X., Liu, Y., Lei, H., Shi, J., Ruan, R., Liu, Y., Lei, H., Shi, J., Ruan, R., 2013. Development of an effective acidogenically digested swine manure-based algal system for improved wastewater treatment and biofuel and feed production. Applied Energy 107, 255–263. Available from: https://doi.org/10.1016/j.apenergy.2013.02.033.

Hu, B., Min, M., Zhou, W., Du, Z., et al., 2012a. Enhanced mixotrophic growth of microalga *Chlorella* sp. on pretreated swine manure for simultaneous biofuel feedstock production and nutrient removal. Bioresource Technology. Available from: https://doi.org/10.1016/j.biortech.2012.09.031. United States: Elsevier Ltd.

Hu, B., Min, M., Zhou, W., Li, Y., et al., 2012b. Influence of exogenous CO_2 on biomass and lipid accumulation of microalgae *Auxenochlorella protothecoides* cultivated in concentrated municipal wastewater. Applied Biochemistry and Biotechnology 166 (7), 1661–1673. Available from: https://doi.org/10.1007/s12010-012-9566-2. United States.

Iglesias-Rodriguez, M.D., Merrett, M.J., 1997. Dissolved inorganic carbon utilization and the development of extracellular carbonic anhydrase by the marine diatom *Phaeodactylum tricornutum*. New Phytologist 135 (1), 163–168. Available from: https://doi.org/10.1046/j.1469-8137.1997.00625.x. Wiley.

Kröger, M., Müller-Langer, F., 2011. Impact of heterotrophic and mixotrophic growth of microalgae on the production of future biofuels. Biofuels 2 (2), 145–151. Available from: https://doi.org/10.4155/bfs.11.5. Germany.

Lamers, P.P., Janssen, M., De Vos, R.C.H., Bino, R J., Wijffels, R.H., 2012. Carotenoid and fatty acid metabolism in nitrogen-starved *Dunaliella salina*, a unicellular green microalga. Journal of Biotechnoly 162, 21–27. Available from: https://doi.org/10.1016/j.jbiotec.2012.04.018.

Lauersen, K.J., 2019. Eukaryotic microalgae as hosts for light-driven heterologous isoprenoid production. Planta 249 (1), 155–180. Available from: https://doi.org/10.1007/s00425-018-3048-x. Germany: Springer Verlag.

Leite, L.d S., Hoffmann, M.T., Daniel, L.A., 2019. Microalgae cultivation for municipal and piggery wastewater treatment in Brazil. Journal of Water Process Engineering 31. Available from: https://doi.org/10.1016/j.jwpe.2019.100821. Brazil: Elsevier Ltd.

León-Vaz, A., et al., 2019. Using agro-industrial wastes for mixotrophic growth and lipids production by the green microalga *Chlorella sorokiniana*. New Biotechnology 51, 31–38. Available from: https://doi.org/10.1016/j.nbt.2019.02.001. Spain: Elsevier B.V.

Li, Y., Han, D., et al., 2011a. Photosynthetic carbon partitioning and lipid production in the oleaginous microalga *Pseudochlorococcum* sp. (Chlorophyceae) under nitrogen-limited conditions. Bioresource Technology 102 (1), 123–129. Available from: https://doi.org/10.1016/j.biortech.2010.06.036. United States.

Li, Y., Zhou, W., et al., 2011b. Integration of algae cultivation as biodiesel production feedstock with municipal wastewater treatment: Strains screening and significance evaluation of environmental factors. Bioresource Technology 102 (23), 10861–10867. Available from: https://doi.org/10.1016/j.biortech.2011.09.064. United States.

Liang, Y., 2013. Producing liquid transportation fuels from heterotrophic microalgae. Applied Energy 104, 860–868. Available from: https://doi.org/10.1016/j.apenergy.2012.10.067. United States: Elsevier Ltd.

Liang, Y., Sarkany, N., Cui, Y., 2009. Biomass and lipid productivities of *Chlorella vulgaris* under autotrophic, heterotrophic and mixotrophic growth conditions. Biotechnology Letters 31 (7), 1043–1049. Available from: https://doi.org/10.1007/s10529-009-9975-7. United States.

Liu, L., et al., 2018. Lipid accumulation of *Chlorella pyrenoidosa* under mixotrophic cultivation using acetate and ammonium. Bioresource Technology 262, 342–346. Available from: https://doi.org/10.1016/j.biortech.2018.04.092. China: Elsevier Ltd.

Lowrey, J., Yildiz, I., 2014. Investigation of heterotrophic cultivation potential of *Chlorella vulgaris* and *Tetraselmis chuii* in controlled environment wastewater growth media from dairy, poultry and aquaculture industries. Acta Horticulturae 1037, 1109–1114. Available from: https://doi.org/10.17660/ActaHortic.2014.1037.147. Canada: International Society for Horticultural Science.

Lowrey, J., Brooks, M.S., McGinn, P.J., 2015. Heterotrophic and mixotrophic cultivation of microalgae for biodiesel production in agricultural wastewaters and associated challenges—a critical review. Journal of Applied Phycology 27 (4), 1485–1498. Available from: https://doi.org/10.1007/s10811-014-0459-3. Canada: Kluwer Academic Publishers.

Luo, L., He, H., Yang, C., Wen, S., Zeng, G., Wu, M., Zhou, Z., Lou, W., 2016. Nutrient removal and lipid production by *Coelastrella* sp. in anaerobically and aerobically treated swine wastewater. Bioresource Technology 216, 135–141. Available from: https://doi.org/10.1016/j.biortech.2016.05.059.

Luo, L., et al., 2019. Simultaneous nutrition removal and high-efficiency biomass and lipid accumulation by microalgae using anaerobic digested effluent from cattle manure combined with municipal wastewater. Biotechnology for Biofuels 12 (1). Available from: https://doi.org/10.1186/s13068-019-1553-1. China: BioMed Central Ltd.

Markou, G., Georgakakis, D., 2011. Cultivation of filamentous cyanobacteria (blue-green algae) in agro-industrial wastes and wastewaters: a review. Applied Energy 88 (10), 3389–3401. Available from: https://doi.org/10.1016/j.apenergy.2010.12.042. Greece: Elsevier Ltd.

Markou, G., Chatzipavlidis, I., Georgakakis, D., 2012. Cultivation of *Arthrospira* (Spirulina) platensis in olive-oil mill wastewater treated with sodium hypochlorite. Bioresource Technology 112, 234–241. Available from: https://doi.org/10.1016/j.biortech.2012.02.098. Greece.

Markou, G., Iconomou, D., Muylaert, K., 2016. Applying raw poultry litter leachate for the cultivation of *Arthrospira platensis* and *Chlorella vulgaris*. Algal Research 13, 79–84. Available from: https://doi.org/10.1016/j.algal.2015.11.018. Greece: Elsevier B.V.

Marquardt, J., 1998. Effects of carotenoid-depletion on the photosynthetic apparatus of a *Galdieria sulphuraria* (Rhodophyta) strain that retains its photosynthetic apparatus in the dark. Journal of Plant Physiology 152 (4–5), 372–380. Available from: https://doi.org/10.1016/S0176-1617(98)80250-2. Germany: Elsevier GmbH.

Millbank, J.W., 1957. Microalgae – commercial potential for fuel, food and feed. Journal of Food Science and Technology 21, 28–30.

Mitra, D., van Leeuwen, J.(Hans), Lamsal, B., 2012. Heterotrophic/mixotrophic cultivation of oleaginous *Chlorella vulgaris* on industrial co-products. Algal Research 1 (1), 40–48. Available from: https://doi.org/10.1016/j.algal.2012.03.002. Elsevier BV.

Mitsuhashi, S., et al., 2000. X-ray structure of β-carbonic anhydrase from the red alga, *Porphyridium purpureum*, reveals a novel catalytic site for CO2 hydration. Journal of Biological Chemistry 275 (8), 5521–5526. Available from: https://doi.org/10.1074/jbc.275.8.5521. Japan.

Mo, J., et al., 2018. A review on agro-industrial waste (AIW) derived adsorbents for water and wastewater treatment. Journal of Environmental Management 227, 395–405. Available from: https://doi.org/10.1016/j.jenvman.2018.08.069. China: Academic Press.

Mohsenpour, S.F., et al., 2021. Integrating micro-algae into wastewater treatment: a review. Science of the Total Environment 752. Available from: https://doi.org/10.1016/j.scitotenv.2020.142168. United Kingdom: Elsevier B.V.

Montero, E., et al., 2018. Mixotrophic cultivation of *Chlorococcum* sp. under non-controlled conditions using a digestate from pig manure within a biorefinery. Journal of Applied Phycology 30 (5), 2847–2857. Available from: https://doi.org/10.1007/s10811-018-1467-5. Mexico: Springer Netherlands.

Morais, K.C.C. et al., 2017. "Sustainable energy via biodiesel production from autotrophic and mixotrophic growth of the microalga *Phaeodactylum tricornutum* in compact photobioreactors." In: 2016 IEEE Conference on Technologies for Sustainability, SusTech 2016. Brazil: Institute of Electrical and Electronics Engineers Inc. doi: 10.1109/SusTech.2016.7897177.

Moroney, J.V., Ynalvez, R.A., 2007. Proposed carbon dioxide concentrating mechanism in *Chlamydomonas reinhardtii*. Eukaryotic Cell 6 (8), 1251–1259. Available from: https://doi.org/10.1128/EC.00064-07. United States.

Moroney, J.V., Bartlett, S.G., Samuelsson, G., 2001. Carbonic anhydrases in plants and algae. Plant, Cell and Environment 24 (2), 141–153. Available from: https://doi.org/10.1046/j.1365-3040.2001.00669.x. Wiley.

Mukhtar, S., 2005. Poultry Production: Manure and Wastewater Management. Marcel Dekker.

Neilson, A.H., Lewin, R.A., 1974. The uptake and utilization of organic carbon by algae: an essay in comparative biochemistry. Phycologia 13, 227–264.

Nimer, N.A., Merrett, M.J., Brownlee, C., 1996. Inorganic carbon transport in relation to culture age and inorganic carbon concentration in a high-calcifying strain of *Emiliania huxleyi* (prymnesiophyceae). Journal of Phycology 32 (5), 813–818. Available from: https://doi.org/10.1111/j.0022-3646.1996.00813.x. United Kingdom: Blackwell Publishing Inc.

Oesterhelt, C., Schnarrenberger, C., Gross, W., 1999. Characterization of a sugar/polyol uptake system in the red alga *Galdieria sulphuraria*. European Journal of Phycology 34 (3), 271–277. Available from: https://doi.org/10.1017/S0967026299002176. Germany.

Oh, S.H., et al., 2009. Lipid production in *Porphyridium cruentum* grown under different culture conditions. Journal of Bioscience and Bioengineering 108 (5), 429–434. Available from: https://doi.org/10.1016/j.jbiosc.2009.05.020. South Korea.

Palinska, K.A., Surosz, W., 2014. Taxonomy of cyanobacteria: a contribution to consensus approach. Hydrobiologia 740 (1), 1–11. Available from: https://doi.org/10.1007/s10750-014-1971-9. Poland: Kluwer Academic Publishers.

Perez-Garcia, O., et al., 2011. Heterotrophic cultures of microalgae: metabolism and potential products. Water Research 45 (1), 11–36. Available from: https://doi.org/10.1016/j.watres.2010.08.037. Mexico: Elsevier Ltd.

Perez-Garcia, O., et al., 2015. Microalgal heterotrophic and mixotrophic culturing for bio-refining: from metabolic routes to techno-economics. Algal Biorefineries: Volume 2: Products and Refinery Design. Springer International Publishing, New Zealand, pp. 61–131. Available from: https://doi.org/10.1007/978-3-319-20200-6_3.

Perez-Garcia, O., Bashan, Y., Esther Puente, M., 2011. Organic carbon supplementation of sterilized municipal wastewater is essential for heterotrophic growth and removing ammonium by the microalga *chlorella vulgaris*1. Journal of Phycology 47 (1), 190–199. Available from: https://doi.org/10.1111/j.1529-8817.2010.00934.x. Wiley.

Piasecka, A., Krzemińska, I., Tys, J., 2017. Enrichment of *Parachlorella kessleri* biomass with bioproducts: oil and protein by utilization of beet molasses. Journal of Applied Phycology 29 (4), 1735–1743. Available from: https://doi.org/10.1007/s10811-017-1081-y. Poland: Springer Netherlands.

Pignolet, O., et al., 2013. Highly valuable microalgae: biochemical and topological aspects. Journal of Industrial Microbiology and Biotechnology 40 (8), 781–796. Available from: https://doi.org/10.1007/s10295-013-1281-7. France.

Ratledge, C., Kanagachandran, K., Anderson, A.J., Grantham, D.J., Stephenson, J.C., 2001. Production of docosahexaenoic acid by *Crypthecodinium cohnii* grown in a pH-auxostat culture with acetic acid as principal carbon source. Lipids 36, 1241–1246. Available from: https://doi.org/10.1007/s11745-001-0838-x.

Reeb, V., Bhattacharya, D., 2010. The thermo-acidophilic cyanidiophyceae (Cyanidiales). In: Seckbach, J., Chapman, D. (Eds.), Red Algae in the Genomic Age. Cellular Origin, Life in Extreme Habitats and Astrobiology, 13. Springer, Dordrecht, pp. 409–426. Available from: https://doi.org/10.1007/978-90-481-3795-4_22.

Rigobello-Masini, M., Aidar, E., Masini, J.C., 2003. Extra and intracelular activities of carbonic anhydrase of the marine microalga *Tetraselmis gracilis* (Chlorophyta). Brazilian Journal of Microbiology 34 (3), 267–272. Available from: https://doi.org/10.1590/S1517-83822003000300017. Brazil: Sociedade Brasileira de Microbiologia.

Rizzo, L., Lofrano, G., Belgiorno, V., 2010. Olive mill and winery wastewaters pre-treatment by coagulation with chitosan. Separation Science and Technology 45 (16), 2447–2452. Available from: https://doi.org/10.1080/01496395.2010.487845. Italy.

Sakurai, T., et al., 2016. Profiling of lipid and glycogen accumulations under different growth conditions in the sulfothermophilic red alga *Galdieria sulphuraria*. Bioresource Technology 200, 861–866. Available from: https://doi.org/10.1016/j.biortech.2015.11.014. Japan: Elsevier Ltd.

Sánchez, S., et al., 2001. Mixotrophic culture of *Chlorella pyrenoidosa* with olive-mill wastewater as the nutrient medium. Journal of Applied Phycology 13 (5), 443–449. Available from: https://doi.org/10.1023/A:1011929723586. Spain.

Schilling, S., Oesterhelt, C., 2007. Structurally reduced monosaccharide transporters in an evolutionary conserved red alga. Biochemical Journal 406 (2), 325–331. Available from: https://doi.org/10.1042/BJ20070448. Germany.

Soanen, N., et al., 2016. Improvement of exopolysaccharide production by *Porphyridium marinum*. Bioresource Technology 213, 231–238. Available from: https://doi.org/10.1016/j.biortech.2016.02.075. France: Elsevier Ltd.

Tabernero, A., Martín del Valle, E.M., Galán, M.A., 2012. Evaluating the industrial potential of biodiesel from a microalgae heterotrophic culture: Scale-up and economics. Biochemical Engineering Journal 63, 104−115. Available from: https://doi.org/10.1016/j.bej.2011.11.006. Spain.

Tang, D.Y.Y., et al., 2020. Potential utilization of bioproducts from microalgae for the quality enhancement of natural products. Bioresource Technology 304. Available from: https://doi.org/10.1016/j.biortech.2020.122997. Malaysia: Elsevier Ltd.

Tocchi, C., et al., 2012. Aerobic treatment of dairy wastewater in an industrial three-reactor plant: effect of aeration regime on performances and on protozoan and bacterial communities. Water Research 46 (10), 3334−3344. Available from: https://doi.org/10.1016/j.watres.2012.03.032. Italy: Elsevier Ltd.

Vidyashankar, S., et al., 2013. Selection and evaluation of CO_2 tolerant indigenous microalga *Scenedesmus dimorphus* for unsaturated fatty acid rich lipid production under different culture conditions. Bioresource Technology 144, 28−37. Available from: https://doi.org/10.1016/j.biortech.2013.06.054. India: Elsevier Ltd.

Villay, A., et al., 2013. Optimisation of culture parameters for exopolysaccharides production by the microalga *Rhodella violacea*. Bioresource Technology 146, 732−735. Available from: https://doi.org/10.1016/j.biortech.2013.07.030. France: Elsevier Ltd.

Wang, L., et al., 2010. Anaerobic digested dairy manure as a nutrient supplement for cultivation of oil-rich green microalgae *Chlorella* sp. Bioresource Technology 101 (8), 2623−2628. Available from: https://doi.org/10.1016/j.biortech.2009.10.062. United States.

Wang, H., et al., 2012. Mixotrophic cultivation of *Chlorella pyrenoidosa* with diluted primary piggery wastewater to produce lipids. Bioresource Technology 104, 215−220. Available from: https://doi.org/10.1016/j.biortech.2011.11.020. China.

Wang, J., Yang, H., Wang, F., 2014. Mixotrophic cultivation of microalgae for biodiesel production: Status and prospects. Applied Biochemistry and Biotechnology 172 (7), 3307−3329. Available from: https://doi.org/10.1007/s12010-014-0729-1. China: Humana Press Inc.

Wang, Y., et al., 2015. Cultivation of *Chlorella vulgaris* JSC-6 with swine wastewater for simultaneous nutrient/COD removal and carbohydrate production. Bioresource Technology 198, 619−625. Available from: https://doi.org/10.1016/j.biortech.2015.09.067. China: Elsevier Ltd.

Wilhelm, C., et al., 2006. The regulation of carbon and nutrient assimilation in diatoms is significantly different from green algae. Protist 157 (2), 91−124. Available from: https://doi.org/10.1016/j.protis.2006.02.003. Germany: Elsevier GmbH.

Wollmann, F., et al., 2019. Microalgae wastewater treatment: biological and technological approaches. Engineering in Life Sciences 19 (12), 860−871. Available from: https://doi.org/10.1002/elsc.201900071. Germany: Wiley-VCH Verlag.

Xu, H., Miao, X., Wu, Q., 2006. High quality biodiesel production from a microalga *Chlorella protothecoides* by heterotrophic growth in fermenters. Journal of Biotechnology 126 (4), 499−507. Available from: https://doi.org/10.1016/j.jbiotec.2006.05.002. China.

Ye, Y., et al., 2018. Optimizing culture conditions for heterotrophic-assisted photoautotrophic biofilm growth of *Chlorella vulgaris* to simultaneously improve microalgae biomass and lipid productivity. Bioresource Technology 270, 80−87. Available from: https://doi.org/10.1016/j.biortech.2018.08.116. China: Elsevier Ltd.

Zaslavskaia, L.A., et al., 2001. Trophic conversion of an obligate photoautotrophic organism through metabolic engineering. Science (New York, N.Y.) 292 (5524), 2073−2075. Available from: https://doi.org/10.1126/science.160015. United States.

Ecological and environmental services of microalgae

Archita Sharma and Shailendra Kumar Arya

Department of Biotechnology, University Institute of Engineering and Technology (UIET),
Panjab University (PU), Chandigarh, India

13.1 Introduction

A technologically advanced world where natural resources are getting scarce with each passing day is on the verge of developing a potential alternative to provide nourishment, chemicals, and energy for *Homo sapiens*. Microalgae—a unicellular microorganism of marine and freshwater habitat with single eukaryotic cell—have grabbed the attention of researchers because of unique characteristics such as rapid growth, quick acclimatization to the environment, synthesis of diverse bioactive substances (with a potential to exhibit varied properties such as antioxidant, antibacterial, antiviral, antitumor, antihypertensive, neuroprotective, and immune stimulant) (Christaki et al., 2015; Silva et al., 2020), increased yield of biomass per unit area, the potential to concentrate beneficial chemicals and occupy various nutrients economically (Dolganyuk et al., 2020). The biggest group constitutes green algae (*Chlorophyceae*), golden algae (*Chrysophyceae*), diatoms (*Bacillariophyceae*), cyanobacteria/blue-green algae (*Cyanophyceae*), which are majorly exploited considering their biotechnological applications (Olasehinde et al., 2017). The contribution of total biomass by microalgae is huge enough to regulate the global ecosystem and environment. According to reports, there are over 800,000 microalgal species with only 50,000 documented ones to date, where every species has been adapted to the specific environment like extreme climate zones, salt conditions, pH, and light, etc. (Usher et al., 2014; Mata et al., 2010; Chowdhury and Loganathan, 2019). The technologies exploited to produce biomass from microalgal species have negligible contribution toward polluting the environment by consuming carbon dioxide (CO_2) and generating oxygen (O_2). Additionally, it consumes very less water and can occupy barren land or land unsuitable for cultivating crops. Hence, microalgae have exhibited colossal potential as an alternative in the era of the resource-limited world (Demirel et al., 2018; Torres-Tiji et al., 2020). Microalgae-derived bioactive compounds (Fig. 13.1) have significant demand in diverse industries such as pharmacology, medicine, cosmetics, chemicals, fish farming, energy, and agriculture (Jacob-Lopes et al., 2019). Significantly, microalgae are considered the magnificent source of value-added products exhibiting benefits to health, ecosystem, and environment and thus representing a huge potential in a variety of markets such as proteins, pigments, polyunsaturated fatty acids (PUFA), lipids, vitamins, minerals, and polysaccharides (Khanra et al., 2018; Christaki et al., 2015) (Fig. 13.1). Topographically, the prominent producers of microalgal biomass are Taiwan, the United States, Japan, China, Spain, Brazil, Israel, Myanmar, and Germany, with an annual dry biomass production of 19,000 tons resulting in the generation of revenue of around USD 5.7 billion (Jacob-Lopes et al., 2019; Silva et al., 2020).

As a rule of thumb, large-scale production of sources has both positive and negative influences on the environment. Currently, a wide-ranging scope exists considering microalgal species such as domestic and industrial treatment of wastewaters (Christenson and Sims, 2011; Abdel-Raouf et al., 2012), fertilizers for producing food, health supplements for humans, mitigation of greenhouse gases (GHGs) (Fagerstone et al., 2011; Ferrón et al., 2012), improving marine and terrestrial habitats, etc. Various reviews discuss the significant impact on the environment by cultivation of microalgae, such as the appropriate design of pond-based cultivation systems through the exploitation of ecological principles intending to decrease the detrimental effects on the environment (Smith et al., 2010; Usher et al., 2014).

The purpose of this chapter is to scrutinize and render significant knowledge regarding potential applications of microalgae bolstering the environment and ecosystem. This chapter focuses on assessing the capabilities of microalgal species as a raw material for developing biologically significant products with applications in diverse spectra. The chapter is divided into three major sections: (1) microalgae—which discusses, in brief, the versatility of microalgae and different species, various cultivation methods and production processes

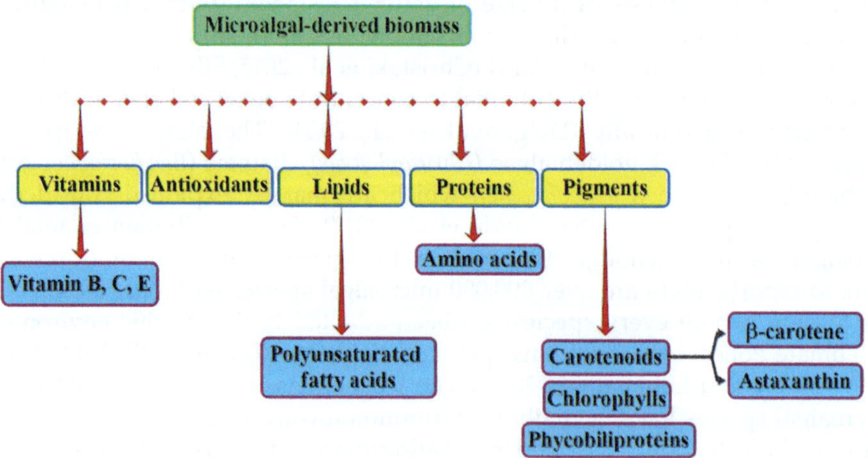

FIGURE 13.1　Microalgae and their respective bioactive compounds. Source: *From Koyande, A.K., Chew, K.W., Rambabu, K., Tao, Y., Chu, D.-T., Show, P.-L., 2019. Microalgae: a potential alternative to health supplementation for humans. Food Science and Human Wellness 8 (1), 16–24. https://doi.org/10.1016/j.fshw.2019.03.001.*

following the roadmap and their market potential respectively; (2) ecological services—which deals with broad range applications of microalgae benefiting human and social welfare, for example, health supplements, poultry feeding, antitumor activity, antioxidant properties, cosmetics, and therapeutic applications; (3) environment-based services—which illustrate the potential role of microalgae in treating wastewaters, pest control, aquaculture, mitigation of harmful gases, production of diverse biofuels like biodiesel, and bioethanol, remediation of heavy metals toxicity such as mercury and cadmium. Furthermore, this chapter gives a brief overview of different regulatory policies associated with the exploitation of microalgae for serving mankind and improving the condition of the environment, indenting to make the world a better place to live in.

13.2 Microalgae—a multifaceted microorganism

Microalgae are unicellular or multicellular organisms that constitute prokaryotic organisms like cyanobacteria/blue-green algae, eukaryotic algae, etc. Microalgae belong to a polyphyletic group of autotrophic and heterotrophic microorganisms generally found in fresh water and marine niches (Alam and Wang, 2019; Gangl et al., 2015). There are a variety of microorganisms with approximately 200,000–800,000 species where only nearly 35,000 have been classified and elucidated (Ebenezer et al., 2012). Microalgae being epitomized as a unicellular model for plants (Ball, 2005) exhibit higher growth and productivity (Lu et al., 2014). Additionally, marine algal species can show growth under an extreme saline environment, decreasing the risks associated with contamination (Bacellar Mendes and Vermelho, 2013). Fig. 10.1 shows the extreme versatility of the microalgal kingdom (Abidin et al., 2020). The onset of increasing interest in the microalgal studies has resulted in recognition of a variety of novel microalgal species, which ultimately is the focal point of a potential source of diverse compounds such as feed and food supplementations, nutrition, cosmetics industry, pharmaceutical industries, and bioenergy production (Gong et al., 2011; de Carvalho and Caramujo, 2017; Balia Yusof et al., 2018).

13.2.1 The diverse approaches of cultivation

Microalgal cultivation is the introductory step for producing bioactive compounds. Majorly, there are three approaches to cultivate microalgae: (1) batch culture—species cultivated in a closed vessel/environment with fixed specifications. Usually, there is no addition of fresh nutrients during the ongoing cultivation process, resulting in a decrease in the number of nutrients with time (Lim and Shin, 2013). This cultivation process is generally regarded as the easiest one. Still, the biggest disadvantage is the self-shading effect of microalgal cells and erratic irradiance because of the close environment and repetitive nutrient consumption, respectively (Zhu, 2015). (2) Fed-batch/semicontinuous culture—it provides a continuous supply of nutrients to microalgae with the simultaneous discharge of effluents/waste intending to maintain the concentration of culture medium in the respective system (Tan et al., 2018), rendering a more effective and relative process to cultivate microalgae. The disadvantages lie in difficulty controlling and monitoring contamination in the growth pattern with a fixed feed system and no feedback control (Tan et al., 2018). (3) Continuous culture—it is analogous to batch cultivation with minor modifications like adding fresh nutrients/medium up till an exponential growth phase is attained. This ensures the continuous reproduction of microalgae at a median rate following a normal growth cycle. The advantage relies upon the high yield of a specific product for long durations and smooth manipulation of the concentration and pH of nutrients (Hoskisson and Hobbs, 2005). Nevertheless, this technique is usually not favored at the industrial level because of a high chance of contamination and the difficulty and intricacies of the cultivation process (Egli, 2015).

The significant modes of microalgae cultivation are photoautotrophic, heterotrophic, mixotrophic, and photoheterotrophic, with photoautotrophic and heterotrophic as the most common modes (Chen et al., 2017). Photoautotrophic cultivation exploits carbon dioxide and is designated as a sole source of carbon and light. In contrast, heterotrophic cultivation utilizes organic carbon like sugars, wastewater, acetate, and organic acids as a source of carbon and energy-deprived of solar/light. Mixotrophic cultivation consumes carbon dioxide and organic compounds following concomitant respiratory and photosynthetic metabolic processes. On the other hand, in photoheterotrophic cultivation, light and carbon source are designated as carbon sources (Table 13.1).

Precisely, there are two major categories of systems for cultivating microalgae: (1) open systems such as open ponds or raceway ponds. Open ponds have paddlewheel-based raceway ponds (with 0.2×0.5 m depth), ensuring the flow of nutrients and water alongside microalgal species for constant exposure to the surroundings following the maintenance of the integrity of the pond. These ponds have a continuous mode of operation with the steady addition of carbon dioxide and nutrients intending to maintain the system's equilibrium. Simultaneously, microalgae are discharged from the open pond tank from the opposite end. For optimizing the performance, it is generally suggested to culture the microalgae with a maximal volume of 30 m^3(Santos-Sánchez et al., 2016). On the other hand, the closed systems, also called photobioreactors, utilize light energy to administer a photobiological reaction and are mostly exploited for producing microalgal biomass in bulk quantities under a controlled environment (Chisti, 2007; Mata et al., 2010). The three devised configurations of photobioreactors, spiral tubes, flat panels, and bubbles column, constitute plastic/glass (Santos-Sánchez et al., 2016). Economically, in comparison to the open raceway ponds, a closed system is expensive because of the high cost of capital, such as production cost associated with the cultivation system, which ranges between 5.56 USD kg^{-1} for open raceway

TABLE 13.1   Different types of cultivation methods of microalgae and their respective characteristics.

Method of cultivation	Source of energy	Source of carbon	Type of reactors	Characteristic features
Phototrophic	Light	Inorganic	Photobioreactors	Low culture cell density, evaporation of water
Heterotrophic	Organic matter	Organic	Bioreactors	Chances of contamination by microbes, costly components associated with the nutrient medium
Mixotrophic	Light and organic matter	Inorganic and organic	Closed photobioreactor	

and 55.45 USD kg^{-1} for tubular photobioreactors, respectively (Acién Fernández et al., 2019). Thus, because of the high cost of producing biomass, a team of researchers fabricated a unique concept of microalgal cultivation, which includes liquid form-bed photobioreactor intending to decrease the cost of harvestation and energy needs (Janoska et al., 2018a,b). Briefly, a closed system of cultivation is the most suitable cultivation system due to controlled culture conditions but is usually restricted via the high cost of production and less yield of biomass. Table 13.1 summarizes the advantages and disadvantages of both open- and closed systems (Tang et al., 2020).

13.3 Ecological services of microalgae

Microalgae can produce numerous biochemicals for their exploitation in food and medical research domains (Table 13.1) (Chowdhury and Loganathan, 2019; Demirel, 2018; Jacob-Lopes et al., 2008; Mata et al., 2010; Tan, 2018; Torres-Tiji et al., 2020; Usher, 2014). There are diverse ecological services of microalgae such as the production of biofuels, wastewater treatment (WWT), mitigation of GHGs, human health supplements, animal feed, cosmetic or therapeutic products, antioxidative property, anticancer properties, production of microalgal pigments such as astaxanthin, β-carotenes, and phycobiliproteins, production of biofertilizers, enhancing the marine biodiversity, etc. There are a variety of important microalgal species that are generally exploited for producing commercialized products, some of them are *Spirulina*, *Chaetoceros*, *Chlorella*, *Dunaliella*, *Isochrysis*, etc. Here, in this review, we have explained a few examples of the commercialized services or benefits rendered by microalgae that further ameliorate the ecological system (Fig. 13.2).

13.3.1 Microalgae-derived human health supplements

For over a thousand years, numerous microalgae species have been consumed or incorporated into the human diet. Microalgae are an underused crop intending to produce dietary food items. Their cultivation does not go up against contest with land and resources, which is generally required for the production of conventional crops. Additionally, they show remarkable yield in comparison to terrestrial crops. The high protein content has illustrated a significant potential intending to the requirements of dietary food items for the burgeoning population. Other than protein sources, there are various bioactive compounds present in microalgal species that render additional benefits to human health (Koyande et al., 2019).

There are certain developed countries where the local population consumes food products with high calorific values due to a busy and modern style of living. The implications are usually numerous health issues like obesity, high blood pressure, diabetes, and cardiovascular diseases. For a healthy habit a balanced diet is a must which constitutes the presence of antioxidants, vitamins, PUFAs, etc. In documented research work, it has been reported that microalgal species are a rich source of proteins, carbohydrates, lipids, PUFAs, bioactive compounds, etc. (Suganya et al., 2016; Sathasivam et al., 2019). Microalgae are an exquisite source of vitamins (vitamin A, B1, B2, B6, B12, C, and E) and minerals (potassium, iron, magnesium, calcium, and iodine) (Becker, 2013). In ancient times, people of Taiwan,

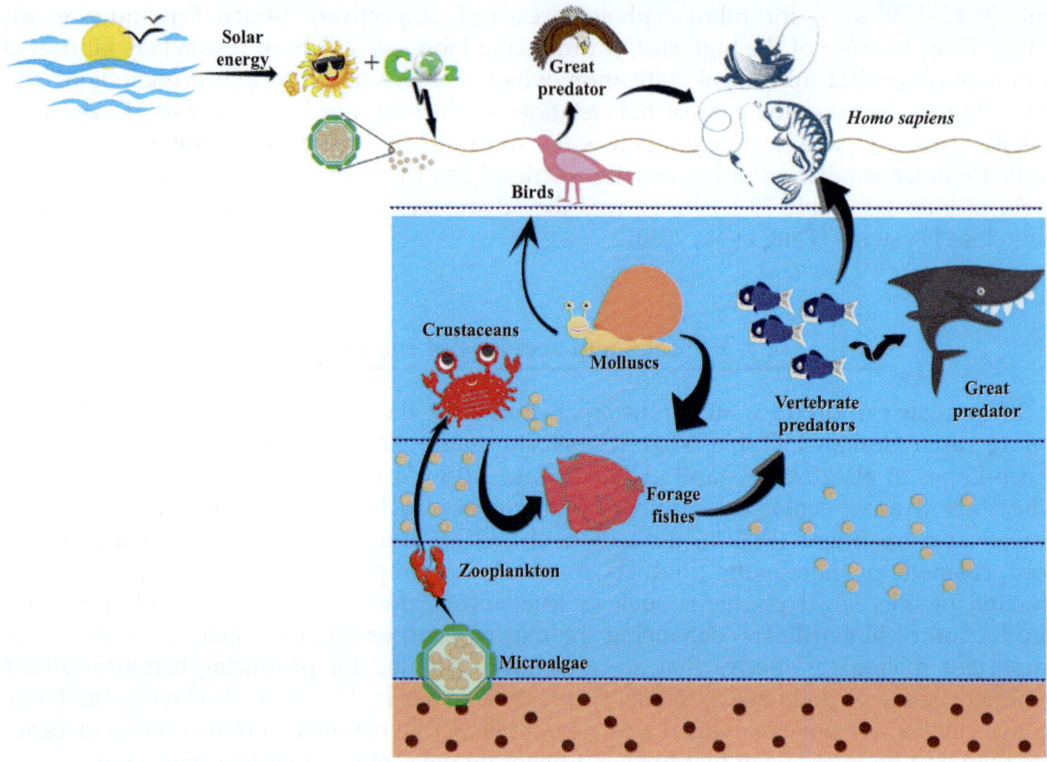

FIGURE 13.2   Diagrammatic representation of the significance of microalgal species in our ecosystem.

Japan, and Mexico had the habit of consuming *Nostoc*, *Chlorella*, and *Spirulina* species as a part of their dietary requirements (Sathasivam et al., 2019). In the present era, food derived from various microalgal strains has commercial market potential recognized as healthy foods, accessible in the form of capsules, tablets, powders, and liquids (Pulz and Gross, 2004). They are amalgamated with candies, gums, snacks, pastes, noodles, breakfast cereals, wine, and other beverages (Liang et al., 2004). The majorly exploited microalgal species are *Spirulina platensis*, *Chlorella* sp., *Dunaliella tertiolecta*, *Dunaliella salina*, and *Aphanizomenon flos-aquae* because of their high protein content and nutritional properties (Soletto et al., 2005).

Chlorella sp. is frequently marketed as a healthy and functional food due to its potential to curb or prevent common disorders such as Alzheimer's disease and cancer (Caporgno and Mathys, 2018). With the spiraling concern of a healthy and nutritious diet, the sales of *Chlorella*-derived food products skyrocketed in the global market with a worth of USD 138 million in 2016, reaching USD 164 million in 2021 (Koyande et al., 2019; Market Research Future, 2020). In the 1940s, Jorgenson and Convit conducted research work in Venezuela by feeding 80 leper patients with *Chlorella*-derived plankton soup, exhibiting better energy, health, and weight in them. This marked the first reported demonstration of microalgae as a potential supplement for improving human health (Rani et al., 2018). According to the documented reports, the extract of *Chlorella* sp. has shown various health benefits (Barrow and Shahidi, 2007), such as hepatoprotective agent and hypocholesterolemic agent in the case of malnutrition and intoxication by ethionine by lowering the blood sugar concentration and increasing the concentration of hemoglobin. *Chlorella* consists of an active immunostimulatory-β-1,3-glucan, the function of which is to reduce the levels of lipids in blood and behave as a free radical scavenger (Rizwan et al., 2018).

As labeled by World Health Organization (WHO), *Spirulina* sp. is another superfood and is a member of the blue-green photoautotrophic genus of unicellular microalgae. Their cells are a rich source of proteins with a 70% dry weight (Soletto et al., 2005; Barkia et al., 2019; Dolganyuk et al., 2020). It is an excellent source of vitamin A, B1, B2, and B12, essential fatty acids (FAs), and other beneficial pigments like xanthophylls and carotenoids. Humans fail to synthesize these effective bioactive compounds and hence are produced in bulk quantities (Sathasivam et al., 2019). A minimal amount (7 g) of dried biomass of *Spirulina* constitutes nearly 4-g protein, 1-g fat (PUFAs such as omega-3 and omega-6 FAs), and 11%, 15%, and 4% of Required Daily Allowance (RDA) of vitamin B1, B2, and B3, respectively. Additionally, it consists of 21% and 11% of RDA of copper (Cu) and iron (Fe), respectively. Other minerals like magnesium (Mg), manganese (Mn), and potassium (K) are also reported to be present in small amounts (Koyande et al., 2019).

Reports are suggesting the beneficial health properties of *Spirulina* like lowering low-desity lipoprotein (LDL) cholesterol and levels of triglycerides (Parikh et al., 2001), reducing blood pressure (Mazokopakis et al., 2014; Torres-Duran et al., 2007) and controlling blood sugar (Parikh et al., 2001), increasing hemoglobin levels, and improving the immune system of the aged population (Selmi et al., 2011). According to the reports, *Spirulina* consists of high (1) protein content as that of tofu, (2) calcium in comparison to milk, (3) iron as compared to spinach, and (4) β-carotene in comparison to carrots (Capelli and Cysewski, 2010). Table 13.2 describes certain examples of exploitation of microalgae in diverse food products.

13.3.2 Microalgae as a feed ingredient for livestock feed ingredient

The poultry industry is designated as the most significant source of income all over the globe. With the continuous and intensive escalation of the human population, there is a dire need for a secure and protein-rich feeding source (El-Ghany, 2020). The addition of antibiotics in the poultry feed has led to the (1) growth of drug-resistant bacteria (Manyi-Loh et al., 2018; SØrum and Sunde, 2001), (2) presence of residues in the poultry-derived products (Daeseleire et al., 2017; Tufa, 2015), and (3) reduction in the microbiota present in intestines (Ribeiro et al., 2020; Andremont, 2000). The extreme exploitation of hazardous antimicrobials in poultry industries has broadly affected the health of both birds and animals. The European Union (European Commision, 2001; Prestinaci et al., 2015) has decided to ban antibiotics as nutrition growth enhancers of poultry feed. So, numerous biological alternative approaches are under investigation for incorporating antimicrobe devoid of any adverse effects on productivity and health, respectively (Madeira et al., 2017; Swiatkiewicz et al., 2016).

Microalgae have grabbed international attention and are considered an apt source of important and effective nutrients (Shokri et al., 2014; Aly, 2015). Blue-green microalgae/cyanobacteria, namely, *Arthrospira*, which includes *S. platensis* and *Spirulina maxima*, are the most effective and broadly distributed edible additives of humans and animals and

TABLE 13.2 Commercial exploitation of microalgal species in diverse food products.

Product	Microalgal species	Advantages	References
Biscuits	*Arthrospira platensis, Chlorella vulgaris, Tetraselmis suecica, Isochrysis galbana, Aporosa fusiformis, Phaeodactylum tricornutum*	Nutritional and techno-functional properties like proteins, fibers, antioxidants, mineral content, omega-3 polyunsaturated fatty acids, etc.	Soletto et al. (2005); Bruton, 2009; García et al. (2017)
Bread	*A. platensis* (gluten free bread), *Arthrospira sp., Dunaliella sp., I. galbana, Nannochloropsis gaditana, Scenedesmus almeriensis, T. suecica*		Cai et al. (2013); Caporgno and Mathys (2018)
Cookies	*C. vulgaris, Haematococcus pluvialis*	Coloring agent, antioxidant activity	Rani et al. (2018)
Extruded snacks	*Arthrospira sp.*	Nutritional characteristics like high protein content	Barrow and Shahidi (2007)
Biomass (phycocyanin)	*Spirulina platensis*	Healthy food supplement	Chu et al. (2002)
Capsules, crystals powder	*Aphanizomenon flos-aquae*	Human nutrition	Yamaguchi (1996)
β-Carotene-rich powder	*Dunaliella salina*		Ben-Amotz and Avron (1992)
Arachidonic acid (AA)	*Parietochloris incise*	Food additives, nutraceuticals	Solana et al. (2014); Cohen and Cohen (1991)
γ-Linolenic acid (GLA)	*Spirulina sp.*	Infant formulas, nutritional supplements	Sajilata et al. (2008)
Ascorbic acid	*Prototheca moriformis, Chlorella sp.*	Human nutrition	Olaizola (2003)
Astaxanthin	*H. pluvialis*	Nutraceuticals food industries	Huang et al. (2006); Vega-Estrada et al. (2005)
Biomass, ascorbic acid	*Chlorella vulgaris*	Food supplement	Apt and Behrens (1999)
Emulsions (oil/water)	Green and orange *C. vulgaris* (after carotenogenesis), green and orange *C. vulgaris* and red *H. pluvialis* (after carotenogenesis)	Coloring agent and nutritional properties such as antioxidant activity	Gouveia et al. (2006)
Fermented milk	*Arthospira platensis*	Nutritional properties like antioxidant, omega-3 polyunsaturated fatty acids, etc.	Martelli et al. (2020)
Frozen yogurt	*Arthrospira sp., A. maxima, C. vulgaris*		Torres-Duran et al. (2007); Selmi et al. (2011)
Pasta	*D. salina, D. vlkianum and I. galbana, S. platensis, A. platensis*		Capelli and Cysewski (2010); Parmar and Singh (2018)

(Continued)

TABLE 13.2   (Continued)

Product	Microalgal species	Advantages	References
Probiotic yogurt	*A. platensis*		Cai et al. (2013)
Processed cheese	*Chlorella sp., A. maxima and D. vlkianum*		Farré et al. (2010)
Vegetarian food gels	*A. maxima, C. vulgaris, D. vlkianum H. pluvialis*		Brennan and Owende (2010); Wang et al. (2008)
Yogurt	*Chlorella sp.*		Hanagata et al. (1992)

poultry feeds (Giorgia et al., 2009; Kanagaraju and Omprakash, 2016). *Spirulina* constitutes around 50%−70% proteins (Soni et al., 2017) and other essential amino acids (Devi et al., 1981) and is hence exploited as an efficient alternative of protein sources in poultry feeds (Austic et al., 2013). Additionally, they constitute FAs; carbohydrates; vitamin B complex (thiamine, riboflavin, pyridoxine, and vitamin B12); vitamin C, A, and E, minerals like calcium (Ca), iron (Fe), zinc (Zn), manganese (Mn), magnesium (Mg), phosphorus (P), copper (Cu), chromium (Cr), sodium (Na), and potassium (K); pigments like carotenoids, phycocyanin, and xanthophylls; and antioxidants (Aly, 2015; Beheshtipour et al., 2013; Jafari et al., 2014; Hynstova et al., 2018). Tables 13.3 and 13.4 discuss the significant effect of microalgae-based diets on the growth performance and meat quality of ruminants, pigs, rabbits, and poultry, respectively.

13.3.3 Microalgal-based immunomodulatory properties

Globally, cancer has become the second major reason for several deaths. Nearly one out of six deaths has been accounted for a variety of cancers. Malignant tumors are defined as the proliferation of cells that deal with significant alterations in the pathology (Martínez Andrade et al., 2018). There are more than 200 diverse types of malignant tumors that can metastasize into nearby tissues, which further results in fatal metastatic tumors. According to the WHO reports, by the year 2030, there will be around 21 million new cancer cases and approximately 13 million deaths (Martínez Andrade et al., 2018). The researchers have given considerable attention to completely rule out cancer (Chen et al., 2019). Though, various newly designed anticancerous substances can extend the survival rate of patients with minimal side effects to compromise the quality of life of the patients (Riccio and Lauritano, 2020; Martínez et al., 2019).

The currently available anticancerous drugs have serious implications, and hence there is the utmost requirement of new and safe anticancerous drugs. In the recent past, there has been a significant reduction in the exploitation of natural products by numerous pharmaceutical industries, yet they are recognized as a suitable platform to produce novel and efficient drugs to eliminate or prevent a variety of cancers (Mondal et al., 2020; Newman and Cragg, 2012). The currently available treatments to treat cancers are ionizing radiation,

TABLE 13.3 Effects of dietary inclusion of microalgae on growth performance of ruminants, pigs, poultry, and rabbits.

Microalgal species	% Dry matter of diet and duration of the experiment	Animal model and initial weight or age	Significant observations	References
Ruminants				
Arthrospira platensis	1.18%, 90 days	Dairy cows	Improvement in the body conditions by 8.5%−11%	Kulpys et al. (2009)
	0.01% 5 weeks	Lambs (46.5 kg)	Enhanced ADG, ADFI, final body weight, and reduction of FCR	EL-Sabagh et al. (2014)
Isochrysis sp.	4% 14.7−26.2 kg	Weaned male lambs (14.7 kg)	No impact on ADG, ADFI, FCR, carcass weight, and yield, and thickness of backfat	De la Fuente-Vazquez et al. (2014)
Schizochytrium sp	3.97% 6 weeks	Dairy cows	Reduction in the intake of a particular feed	Franklin et al. (1999)
Schizochytrium spp. (DHA-gold extract)	1%−3% 18 weeks	Lambs (22.7 kg)	No significant impact on ADG, ADFI, FCR, carcass traits, and increased backfat thickness	Meale et al. (2014)
	1.92% 6 weeks	Lambs (34.8 kg and 3 months)	No significant impact ADG, FCR, and carcass weight	Hopkins et al. (2014)
	3.89% 16.3−26.7 kg	Lambs (16.3 kg and 55.1 days)	Reduction of ADG and intake of feed, enhanced slaughter age, and no impact on carcass traits	Urrutia et al. (2016)
	2% 15.3−26.0 kg	Weaned male lambs (15.3 kg and 7−8 weeks)	Reduced ADG and intake of feed with more fattening period, no impact on final live weight	Diaz et al. (2017)
Pigs				
A. platensis	**Experiment 1:**0.2%−2% 28 days	**Experiment 1:** Weaned pigs (3.7 kg and 11−12 days)	**Experiment 1:** No impact on ADG and FCR from 0 to 14 days, increased ADFI and ADG by 0.2%−2% from 14 to 28 days	Grinstead et al. (2000)
	Experiment 2:0.1%, 6 weeks 0.2%, 1, 2, 4, and 6 weeks	**Experiment 2:** Weaned pigs (5.6 kg and 18 days)	**Experiment 2:** No impact on ADG and ratio of F: G from 0 to 14 days and 28 to 42 days, increased DG and ADFI from 14 to 28 days and 0 to 28 days by 0.2% and F: G from 0 to 28 days by 0.1%−0.2%	

Arthrospira maxima	Copper fortified biomass (20–105 kg)	Piglets (20.9 kg)	No significant impact on ADG, ADFI, and FCR	Saeid et al. (2013)
A. platensis	0.2% 30.6–96.4 kg	Fattening pigs (30.6 kg and 85 days)	Increased ADG and FCR with no impact on backfat thickness	Šimkus et al. (2013)
A. platensis Chlorella vulgaris	1% 14 days	Weaned piglets (9.1 kg and 28 days)	No significant impact on ADG, ADFI, and FCR	Furbeyre et al. (2017)
Chlorella spp.	0.0002% 30–55 kg	Female pigs (30 kg and 3 months)	No effect on ADG, body weight, hot carcass weight, the thickness of lean muscles, and backfat thickness	Baňoch et al. (2012)
Chlorella vulgaris	0.1%–0.2% 6 weeks	Growing pigs (26.6 kg)	Increased ADG (0.1% dose) but no impact on ADFI, G:F, and body weight	Yan et al. (2012)
Schizochytrium sp.	1.10%–5.51%, 79–106 days, 0.39%–1.94%, 107–120 days	Weaned castrated male swine (9.07 kg)	No impact on ADFI and weight of body and organs, increased ADG and FCR	Abril et al. (2003)
	0.25%, 8 weeks 0.50%, 4 weeks 0.25%, 4 weeks	Barrows (118 kg)	No effect on ADG, FCR, and backfat thickness	Sardi et al. (2006)

Poultry

Arthrospira sp.	4%–8% 16 days	Male chicks (678 g and 21 days)	No impact on ADG	Toyomizu et al. (2001)
A. platensis	1.5%–2.5% 4 weeks	Hens (1.5 kg and 63 weeks)	No impact on ADFI and FCR	Zahroojian et al. (2013)
	6%–21% 21 days	Broiler chickens (30 weeks)	No impact on ADG, ADFI, and FCR	Evans et al. (2015)
	0.5%–1.5% 36 days	Broiler chicks (1 day)	Increased ADG (1% dose) Reduced FCR (1% dose)	Shanmugapriya et al., 2015
	0.5%–1% 42 days	Chicks (49 g)	No impact on ADG and FCR	Bonos et al. (2016)
Chlorella sp.	0.00003% 42 days	Broiler chicks (1 day)	Increased ADG, no impact on the ratio of F:G	Dlouhá et al. (2008)

(Continued)

TABLE 13.3 (Continued)

Microalgal species	% Dry matter of diet and duration of the experiment	Animal model and initial weight or age	Significant observations	References
Chlorella vulgaris	0.07%–0.21% 42 days	Reduced FCR, no impact on ADG	Rezvani et al. (2012)	
Chlorella sp.	1% 28 days	Male broiler chicks (1 day)	Increased ADG, no impact on ADFI and FCR	Kang et al. (2013)
	1.25% 5–39 weeks	Laying hens (25 weeks)	Reduced FCR, no impact on the intake of feed	Englmaierová et al. (2013)
	0.1%–0.2% 42 days	Male Pekin ducks (1 day)	Increased ADG and the intake of feed and no impact on G:F	Oh et al. (2015)
Porphyridium sp.	5%–10% 10 days	Chickens (30 weeks)	No impact on body weight, decreased ADFI	Ginzberg et al. (2000)
Schizochytrium JB5	0.1%–0.2% 35 days	Broiler chicks (42.6 g)	No impact on ADG, ADFI and FCR	Yan and Kim (2013)
Schizochytrium sp. (DHA-gold extract)	3.7%–7.4% 21–35 days	Broilers (21 days)	No impact on FCR and carcass yield, increased ADG (7.4% dose), ADFI (7.4% dose)	Ribeiro et al. (2013)
	7.4% 21–35 days		Increased ADG, ADFI, and carcass yield and reduced FCR	
Rabbits				
A. platensis	5%–15% 24 days	Weaned rabbits (2 kg and 9 weeks)	No significant impact on ADG and FCR, but ADFI increased (10% dose)	Peiretti and Meineri (2009)

ADG, average daily gain; *ADFI*, average daily feed intake; *FCR*, feed conversion ratio.

TABLE 13.4 Effects of dietary inclusion of microalgae on meat quality traits of ruminants, pigs, poultry, and rabbits.

Microalgal species	% Dry matter and experiment duration	Animal and initial weight/age	Significant observations and findings	References
Ruminants				
Isochrysis sp.	4% 14.7–26.2 kg	Weaned male lambs 14.7 kg	No significant impact on chemical composition such as fats, proteins, moisture and ash, color, and pH. There has been a significant increase in mono- and polyunsaturated fatty acids, conjugated linoleic acids, eicosapentaenoic acid, and docosahexaenoic acid, docosapentaenoic acid within a time duration of 7 days. Additionally, there has been a reduction in vitamin E.	De la Fuente-Vazquez et al. (2014)
Schizochytrium spp. (DHA-gold extract)	1%–3% 18 weeks	Lambs ∼22.7 kg	No impact on effect on saturated- and polyunsaturated fatty acids, increase in fatty acids, eicosapentaenoic acid, docosahexaenoic acid.	Meale et al. (2014)
	1.92% 6 weeks	Lambs 34.8 kg and 3 months	No impact on color and pH, more cooking loss, increase in fatty acids, eicosapentaenoic acid, docosahexaenoic acid.	Hopkins et al. (2014)
	3.89% 16.3–26.7 kg	Lambs 16.3 kg and 55.1 days	No significant impact on pH, thiobarbituric acid reactive substances (TBARS), reduction in odor, flavor.	Urrutia et al. (2016)
	2% 15.3–26.0 kg	Weaned male lambs 15.3 kg and 7–8 weeks	Increase of fatty acids, polyunsaturated fatty acids, docosahexaenoic acid of neutral lipids and saturated fatty acids, and polyunsaturated fatty acids of polar lipids.	Ponnampalam et al. (2016)
Pigs				
Arthrospira platensis	0.2% 30.6–96.4 kg	Fattening pigs 30.6 kg and 85 days	No significant consequences on ash, proteins, color, pH, cooking loss. A significant reduction of intramuscular fat.	Šimkus et al. (2013)
Chlorella spp.	0.0002% 30–55 kg	Female pigs 30 kg and 3 months	No significant consequences on color, pH, thiobarbituric acid reactive substances (TBARS), cooking, and drip loss.	Baňoch et al. (2012)
Schizochytrium sp. (DHA-gold extract)	0.25% and 8 weeks 0.50% and 4 weeks 0.25% and 4 weeks	Barrows 118 kg	No impact on the chemical composition, color, and pH. There has been an increase indocosahexaenoic acid.	Sardi et al. (2006)
Schizochytrium sp.	0.3%–1.2% 45 days	Female pigs 75 kg	No significant impact onproximate composition such as dry matter, fat, proteins, color, pH, and thiobarbituric acid reactive substances (TBARS) and increased eicosapentaenoic acid, docosahexaenoic acid.	Vossen et al. (2016)

(Continued)

TABLE 13.4 (Continued)

Microalgal species	% Dry matter and experiment duration	Animal and initial weight/age	Significant observations and findings	References
Poultry				
Arthrospira sp.	4%−8% 16 days	Male chicks 678 g and 21 days	Reduction of L* (4% dose) from *semitendinosus* and *sartorius* muscles. Elevation of a* (4% dose) from *Psoas superficialis* and *Psoas sartorius* muscles and b* (8% dose) from each muscle	Toyomizu et al. (2001)
A. platensis	0.5%−1% 42 days	Chicks 49 g	No effect on saturated-, monounsaturated-, polyunsaturated fatty acids, eicosapentaenoic acid, docosahexaenoic acid, and thiobarbituric acid reactive substances (TBARS) in breast meat. Increase of polyunsaturated fatty acids, eicosapentaenoic acid, and docosahexaenoic acid in thigh meat. Reduction in monounsaturated fatty acids in thigh meat.	Bonos et al. (2016)
Chlorella sp.	0.00003% 42 days	Broiler chicks 1 day-old	Reduction in thiobarbituric acid reactive substances (TBARS)	Dlouhá et al. (2008)
Chlorella vulgaris	0.1%−0.2% 42 days	Male Pekin ducks 1 day-old	Elevation of b*, pH, and shear forces in breast meat and increase of L* and b* in leg meat.	Oh et al. (2015)
Schizochytrium sp.	2.8%−5.5% 3 weeks	Male broilers 4weeks	Increase of saturated fatty acids (5.5% dose), eicosapentaenoic acid, docosahexaenoic acid, and thiobarbituric acid reactive substances (TBARS). Reduced polyunsaturated fatty acidsand flavor scores.	Mooney et al. (1998)
Schizochytrium JB5	0.1%−0.2% 35 days	Broiler chicks 42.6 g	Increase of oleic acid, unsaturated fatty acids, and docosahexaenoic acid. Reduced stearic acid (0.1% dose), saturated fatty acids.	Yan and Kim (2013)

Schizochytrium sp. (DHA-gold extract)	3.7%–7.4% 21–35 days	Broilers 21 days	No consequences on color, pH, cooking loss, shear force, and monounsaturated fatty acids. Increased saturated fatty acids, long-chain polyunsaturated fatty acids, eicosapentaenoic acid, docosahexaenoic acid, and thiobarbituric acid reactive substances (TBARS). Reduced vitamin E, and flavor.	Ribeiro et al. (2013)
	7.4% 21–35 days		No significant impact on pH and cooking loss, Increased saturated fatty acids, long-chain polyunsaturated fatty acids, eicosapentaenoic acid, docosahexaenoic acid, off-flavor, and thiobarbituric acid reactive substances (TBARS). Decreased monounsaturated fatty acids, vitamin E, and flavor.	Ribeiro et al. (2014)
Rabbits				
A. platensis	1% 3 months	Male rabbits with the weight of 2.8 kg	No significant impact on the chemical composition and gross energy from *longissimus dorsi* muscle	Peiretti and Meineri (2009)
Schizochytrium sp. (DHA-Gold extract)	0.4% 56 days	Growing-fattening rabbits (28 days)	No impact on docosapentaenoic acid and increased production of saturated fatty acids like myristic acid, palmitic acid, and stearic acidsand unsaturated fatty acids like linoleic and linolenic acids, eicosapentaenoic acid, and docosahexaenoic acid.	Mordenti et al. (2010)

CIE system values of lightness (L*), redness (a*), and yellowness (b*).

hyperthermia, the use of different alkylating agents, chemotherapy, DNA topoisomerase inhibitors, etc. These can damage DNA by killing normal/healthy cells and cancerous cells, respectively (Beesoo et al., 2014). On the contrary, microalgae protect cells against genetic alterations and result in antiproliferation, cytotoxicity, and proapoptotic activities in cancerous cells, which suggests a plausible usage for treating cancers (Bule et al., 2018; Baudelet et al., 2013). The diverse microalgae hold a wide-spectrum reservoir, constituting natural pharmaceutical compounds such as phenolics, carotenoids, and chlorophyll, which has a potential role in treating life-threatening cancers (Basmadjian et al., 2014; Sarker et al., 2020). Among all the organisms of marine habitat, bacteria, fungi, seaweeds, and sponges have already been exploited for treating cancers (Khalifa et al., 2019; Newman and Cragg, 2014). Nevertheless, the research based on exploiting natural products for manufacturing anticancerous drugs is quite expensive, exhausting, laborious, and time-consuming (Gerwick and Moore, 2012; Montaser and Luesch, 2011). Some of the anticancerous products derived from microalgae are cytarabine (Cytosar), trabectedin (Yondelis), eribulin mesylate (Halaven), and the conjugated antibody brentuximab vedotin (Acentris) (Townsend et al., 2018). Table 13.5 discusses the characteristic activities of a variety of microalgal species against a particular type of cancer.

13.3.4 Microalgae-derived antioxidant properties

In the recent past, the concept of antioxidants has become a significant topic, and extensive research is going on all over the globe due to excessive demands of food and pharmaceutical industries (Hamidi et al., 2019). The oxidative damage due to reactive oxygen species like hydroxyl radical, hydrogen peroxide, nitric acid, or superoxide anions (Hamidi et al., 2020; Fang et al., 2002; Schieber and Chandel, 2014) on lipids, proteins, and nucleic acids during the metabolism of cells results in numerous incurable disorders like coronary heart diseases, lung injuries, aging, diabetes, inflammatory diseases, atherosclerosis, cancers, etc. (Harrison et al., 2003; Finkel, 2011; Waris and Ahsan, 2006). Hence there is a dire need to substitute certain synthetic antioxidants with natural antioxidants due to toxicity and carcinogenic properties in animal models. Additionally, natural antioxidants such as vitamins (vitamin B, C, D, and E) (Hensley et al., 2000), tocopherols, and phenolics protect cells against oxidative damage and strengthen the immune system. Thus it is important to recognize and label safe and cost-effective novel sources of natural antioxidants (Hemalatha et al., 2015; Simioni et al., 2019).

The exploitation of microalgae because of numerous hypometabolic compounds and antioxidative properties with varied chemical structures and bioactive substances like amino acids, carotenoids, FAs, flavonoids, anthocyanins, and polysaccharides have escalated in the recent past. Other significant benefits are (1) source of essential nourishments, (2) PUFAs, (3) sulfated polysaccharides, (4) heat-induced proteins, (5) phenolic compounds, (6) pigments, such as carotenoids, are certain active ingredients that serve human health by rendering physiological and (7) pharmacological benefits (Hemalatha et al., 2015; Mostafa et al., 2019; Rashad et al., 2019; Shao et al., 2018; El-Chaghaby et al., 2019). The appropriate antioxidative property is majorly due to the high content and diversity of various antioxidative molecules in microalgal species (Matos et al., 2017; Trappmann and

TABLE 13.5 Anticancer properties of various microalgal species.

Microalgal species	Characteristic properties and type of cancer	References
Nostoc sp GSV 224	Antitumorigenic properties against solid tumors and human tumor cell lines	Carmichael (1992); Moore (1996)
Nostoc spongiaeforme var. tenue Nostoc linckia	Anticancerous against KB (ubiquitous keratin-forming tumor cell line HeLa) and LoVo human cellosaurus cell lines against human epidermoid carcinoma and colorectal adenocarcinoma	Davidson (1995); Banker and Carmeli (1998)
Lyngbya majuscula (Cucarin-A)	Restrains numerous tumor cells like renal, colon, and breast cell lines, respectively. The action depends on suppressing the binding capacity of tubulin polymerization into the colchicine binding pocket. It inhibits the polymerization of tubulin against particular cancer-based cell lines	Carté (1996)
Anabaena cylindrica, Anabaena variabilis	Antimalignancy	Suzuki et al. (1999)
Protoceratium reticulatum	Enhances the apoptotic death in HeLa cells and induces pathologic tissue alteration and DNA fragmentation. Additionally, it alters the membrane potential of mitochondria.	Oftedal et al. (2010)
Calothrix sp.	Cytotoxic against human HeLa cell lines	Pang et al. (2010)
Stigonema sp,	Antiproliferation against numerous types of human fibroblast and endothelial cells. It has a crucial role in cell cycle regulation and mitotic division, respectively	Rickards et al. (1999)
Leptolyngbya sp.	Mechanistic target of rapamycin (mTOR)-induced cell death, autophagosome formation in MEF cells, human SF-295, and U87-SG glioblastoma cells.	Stevenson et al. (2002)
Leptolyngbya cyanobacterium (Coibamide A)	Cytotoxic against NCI-H460 lung cancer cell line in nM concentrations. It also upregulates G_1 cells in a dose-dependent manner. Also, it is very much efficient against MDA-MB-231 epithelial, human breast cancer cell line	Hau et al. (2013)
Spirulina platensis (C-phycocyanin)	Induces pathologic alterations, DNA fragmentation, upregulation of Fas and ICAM (Intercellular Adhesion Molecule) expression, downregulation of Bcl-2 expression, activation of caspases 2,3,4,6,8,9,10 in HeLa and MCF7 cell line	Medina et al. (2008)
Aphanizomenon flos-aquae	Inhibits AML cell lines at G_0-G_1 stage	Li et al. (2006)
Haematococcus pluvialis	Anticancerous against MV-4–11 and HL (human leukemia)-60 cell lines	Bechelli et al. (2011)
Phormidium molle	Alters adherent cells like HeLa, Jurkat, U-937, A2058, RD, 3T3, FL and destroys monolayer in dose-dependent manners	Dzhambazov et al. (2006)
Symploca sp.	Antitumorigenic which interferes with microtubule synthesis	Kerbrat et al. (2003)

(Continued)

TABLE 13.5   (Continued)

Microalgal species	Characteristic properties and type of cancer	References
Symploca sp. (Largazole)	Antitumorigenic which downregulates type-1 histone deacetylase enzyme (HDAC)	Hong and Luesch (2012)
Undaria pinnatifida	Downregulation of procaspase-3 and Bcl-2, Upregulation of Bax against human lung cancer (A549).	Boo et al. (2011)
Fucus vesiculosus	Inhibits the protein expression of MMP-2 and MMP-9, migration of cells, and invasive behavior of LLC cells against lung cancer (A549)	Huang et al. (2015)
Bryopsis	Target the lysosomal bodies and is active against lung cancer A549 cell lines	Smit (2004)
Hypnea musciformis	Effective against NCI-H292 (human lung mucoepidermoid carcinoma), NCI-H292	Guedes et al. (2013)
Gracilariopsis lemaneiformis	Effective against NSCLC cell line, A549	Kang et al. (2017)
Sargassum fusiforme	Restrains VEGF-A-based angiogenesis and proliferation against lung adenocarcinoma SPC-A-1 cells	Chen et al. (2017)
Sargassum macrocarpum	Advances in the production of reactive oxygen species mediate the inhibition of STAT3 signaling, reduces the expression of Bcl2, accelerates the cleavage of caspase-3 and poly (ADP-ribose polymerase) (PARP), effective against A549 (lung cancer) and H1299 (human NSCLC cell line)	Choi et al. (2017)
Plocamium cartilagineum	NCI-H460 (lung cancer)	Sabry et al. (2017)
Codium decorticatum	Apoptosis and is effective against A549 lung cancer	Senthilkumar and Jayanthi (2016)
Gracilaria edulis	A549 lung cancer	Sakthivel et al. (2016)
Haematococcus sp.	Downregulation of MKK1/2-ERK1/2 which induces cytotoxicity against NSCLC cell line	Zhang et al. (2016)
Chlorella zofingiensis	Downregulation of mRNA and protein levels of cyclin E checkpoint of the cell cycle. This strain is effective against A549 cancer cell line	Cordero et al. (2012)
Dunaliella salina	Downregulation of proliferation of cells and induces apoptosis and cell cycle arrest at G_0/G_1phase. It is effective against the A549 lung cancer cell line	Sheu et al. (2008)
Tetraselmis; Nannochloropsis	Generation of prostaglandin E3 (PGE3) via COX-2 enzyme downregulation of Akt phosphorylation. It works effectively against A549 (lung cancer) and H1299 (lymph node)	Yam et al. (2001); Yang et al. (2015)

(Continued)

TABLE 13.5 (Continued)

Microalgal species	Characteristic properties and type of cancer	References
Chlorella vulgaris	Restrains cell migration and impedes the metastasis process. It is effective against H1299 (lymph node), A549, and H1437 (lung carcinoma)	Wang et al. (2010)
Lyngbya sp.	Cell cycle arrest at G_0/G_1 phase, downregulation of viability of cells and membrane potential of mitochondria, increases the release of LDH. It is effective against lung carcinoma (A549).	Stengel et al. (2011)
Arthrospira platensis, Oscillatoria tenuis	Induces apoptosis and blebbing necrosis of cells and is effective A549 lung cancer cell lines	Thangam et al. (2013)
Lyngbya majuscula; Lyngbya sordida	Apoptosis of cells and is effective against H-460, NCI-H460	Soria-Mercado et al. (2009)
Leptolyngbya	Impedes proliferation of cells and is effective against NCI-H460	Medina-Gundrum et al. (2003)
Caldora penicillata	Inhibits the formation of tubulin and is effective against H-460	Zhang et al. (2016)
Gymnodinium	Impedes type 1 and 2 topoisomerases. It is effective against a range of cancer cell lines like NCI-H23, NCI-H226, NCI-H522, NCI-H460, DMS273, DMS11	Umemura et al. (2003)
Phaeodactylum tricornutum	Apoptosis of cell via p53 and caspase-3 mediated pathway. It works against lung cancer cell lines (A549)	Samarakoon et al. (2014)
Thalassiosira rotula, Skeletonema costatum	The mode of action is effective against colon adenocarcinoma (Caco-2, COLO 205), lung adenocarcinoma (A549)	Sansone et al. (2014)
Chlorella ellipsoidea	It works against colon carcinoma; HCT-116	Cha et al. (2008)
Synedra acus	It is effective against colorectal adenocarcinoma; HT-29 and DLD-1	Kusaykin et al. (2010)
Dunaliella tertiolecta	The strain is effective against breast adenocarcinoma; MCF-7	Pasquet et al. (2011)
Cocconeis scutellum	It works against breast carcinoma; BT20	Nappo et al. (2011)
Odontella aurita, Chaetoseros sp., Cylindrotheca closterium, P. tricornutum	It works against promyelocytic leukemia (HL-60), Caco-2, colon adenocarcinoma (HT-29), DLD-1, and range of prostate cancer (PC-3, DU145, and LNCaP)	Peng et al. (2011)
Chaetoceros calcitrans	It works against MCF-7 (human breast cancer cell line), breast adenocarcinoma (MDA-MB-231)	Ebrahimi Nigjeh et al. (2013); Goh et al. (2014)
Amphidinium carterae	Human leukemia cell lines (HL-60), skin melanoma (B16F10), lung cancer cell line (A549)	Samarakoon et al. (2014)

(Continued)

TABLE 13.5   (Continued)

Microalgal species	Characteristic properties and type of cancer	References
Ostreopsis ovata, Amphidinium operculatum	Human leukemia cell lines (HL-60)	Shah et al. (2014)
Navicula incerta	Liver hepatocellular carcinoma (HepG2)	Kim et al. (2014)
P. tricornutum	Human leukemia cell lines (HL-60), mouse epithelial cell lines (W2, D3)	Andrianasolo et al. (2008); Samarakoon et al. (2014)
Skeletonema costatum, Skeletonema marinoi	Colon cancer (Caco-2), skin melanoma (A2058)	Miralto et al. (1999); Lauritano et al. (2016)
Chlorella sorokiniana	Lung cancer (A549) and lung adenocarcinoma (CL1−5)	Lin et al. (2017)
S. marinoi, Alexandrium minutum, A. tamutum, A. andersoni	Antiproliferation on human melanoma A2058 cell lines	Lauritano et al. (2016)
Dolichospermum crassum HSSASE20	Colon cancer (Caco-2), prostate cancer (PC3)	Senousy et al. (2020)
Oscillatoria sancta HSSASE19	Breast cancer (MCF-7)	
Amphidinium carterae	Human leukemia cell lines (HL-60), B16F10 (murine melanoma cell line), and lung cancer(A549)	Samarakoon et al. (2014)
P. tricornutum	Induces apoptosis in two genetically-similar immortal mouse epithelial cell lines	
Botryidiopsidaceae sp.	Inhibits the growth of numerous cancer cell lines such as HeLa, HCT116, Hs678T, and A537. It also induces apoptosis of cells by modulating apoptotic genes like p53, Bcl-2, and caspase-3 genes	Suh et al. (2017)
Sargassum oligocystum, Gracilaria corticata	Inhibits the growth over leukemic cells	Somasekharan et al. (2016)
Gracilaria tenuistipitata	Inhibits the proliferation and induction of apoptosis in human oral cancer Ca9−22	
Granulocystopsis sp.	Induces cytotoxicity as opposed to varied cancers like prostate cancer, breast cancer, colorectal cancer, and skin melanoma. It has the potential to restrains the formation of colonies of the tumor under an in vitro environment	Tavares-Carreón et al. (2020)
Tribonema sp.	Exhibits potential immune-modulatory activities via stimulation of macrophage cells like upregulation IL-6, IL-10, and TNF-α. Additionally, it has anticancerous activity on HepG2 cells	Chen et al. (2017)

AML, Acute myeloid leukemia; *HDAC*, histone deacetylase enzyme; *IL-10*, interleukin 10; *IL-6*, interleukin 6; *LDH*, lactate dehydrogenase; *LLC*, Lewis lung carcinoma; *MEF*, mouse embryonic fibroblasts; *nM*, nanomolar; *NSCLC*, non−small cell lung carcinoma; *TNF-α*, tumor necrosis factor α.

TABLE 13.6 Antioxidant properties and bioactive compounds of microalgal species.

Microalgal species	Bioactive compounds/functions	References
Haematococcus pluvialis, Chlorella zofingiensis, Chlorococcum sp	Astaxanthin	Higuera-Ciapara et al. (2006); Yuan et al. (2011); Matos et al. (2017)
Haematococcus pluvialis	β-Carotene	
Chlorella	β-1,3-Glucan	
C. vulgaris	Oleic, palmitic, and linolenic acids	
	Vitamin E (α-tocopherol)	
	β-Carotene	
	Astaxanthin	
Dunaliella salina	β-carotene	
	Vitamin E (α-tocopherol)	
	Vitamin C (ascorbic acid)	
	α-Carotene	
	Palmitic, linolenic, and oleic acids	
Isochrysis galbana	Vitamin C (ascorbic acid)	
Spirulina maxima	β-Carotene and astaxanthin	
Spirulina sp.	γ-Linolenic acids (GLAs)	
	Sterols (clionasterol)	
Bellerochea malleus (diatoms)	The protein hydrolysates constitute inhibitory activities of ACE rendering better antioxidant properties	Barkia et al. (2019)
Chlorella sorokiniana	It consists of carotenoid pigment. It increases resistance against oxidative stress with ROS inductor, hydrogen peroxide(H_2O_2)	Qiu et al. (2020)
Euglena cantabrica	Phenolics	Jerez-Martel et al. (2017)

ACE, Angiotensin-converting enzymes; *ROS*, reactive oxygen species.

Hawk, 2011; Cardoso et al., 2016; Raposo et al., 2015). Table 13.6 represents the antioxidative properties and bioactive compounds produced from diverse microalgal species.

13.3.5 Microalgae-derived cosmetic products

The branch of cosmeceuticals constitutes potent ingredients such as vitamins, plant-based chemicals, antioxidants, and essential oils, which are further combined into diverse cosmetic products such as creams and moisturizers. Presently, there is broad diversification concerning the variety of bioactive compounds present in marine microalgal species, which aids in envisaging their exploitation in manufacturing numerous cosmetics and therapeutic products (Kim et al., 2005). Bioactive compounds like carotenoids, terpenoids, phenolic compounds, polysaccharides, sterols, phytols, FAs, etc. have a crucial role in manufacturing cosmetics and personal care products (Table 13.7). Skin-based cosmetics follow up the epidermal layer and do not require any logical scrutiny as a viable product and are further promoted for long-term usage such as body lotions (Cannell, 1993; Paul and Pohnert, 2011).

TABLE 13.7 Applications of microalgal species in cosmetics and personal care products.

Microalgae	Applications
Ulva lactuca	Antiwrinkle face cream and moisturizer
Postelsia palmaeformis	Softens skins and antiwrinkle face cream
Porphyra umbilicalis	Provides proper condition to skin
Spirulina	Antiaging lotions and creams and synthesis of collagen
Dunaliella salina	Skin smoothing
Chlorogloeopsis spp.	Protects formation of wrinkles and sagging of skin
Nannochloropsis	Act as sunscreen lotions
Nannochloropsis oculata	Reduces pigmentation of the skin and whitens the skin
Phaeodactylum tricornutum	Protection of skin against exposure to hazardous ultraviolet light decreases the formation of wrinkles
Muriellopsis sp., D. salina, Galdieria sulphuraria, Chlorella zofingiensis, Chlorella protothecoides, Chlorococcum citriforme, Scenedesmus almeriensis, Neospongiococcus gelatinosum	Protection to skin harmful radiations of ultraviolet light
Chlorella protothecoides (UTEX 31)	Enhances the appearance of skin
Nostoc spp., Lyngbya sp., Calothrix spp., Anabaena spp, Rivularia spp., Schizothrix spp., Scytonema spp., Tolypothrix spp., Diplocolon spp., Gloeocapsa spp., Phormidium spp., Pleurocapsa spp.,	Act as sunscreen lotions
Lyngbya sp., Nostoc commune, Nostoc sphaericum, Nostoc sp. R76DM	Act as sunscreen lotions and behaves as photoprotector
Nostoc commune, Anabaena variabilis, Aphanothece halophytica	Delays skin aging process by inhibiting the formation of end products of the glycation process
Olisthodiscus luteus, Nannochloropsis salina	Delays skin aging by reducing the expression of matrix MMPs and increasing the production of collagen protein
Pediastrum duplex, Chlamydomonas reinhardtii, Chlorella pyrenoidosa, Cyanidium caldarium, Anabaena variabilis, Anacystis nidulans, Oscillatoria species, Phormidium foveolarum	Skin moisturizer

(Continued)

TABLE 13.7   (Continued)

Microalgae	Applications
D. salina, Scenedesmus sp., Spirulina spp.	Reduces skin damage by photo-oxidation and renders protection against sunburn, skin aging, and formation of wrinkles
Botryococcus braunii, Schizochytrium mangrovei, Aurantiochytrium spp., Thraustochytrium spp.	Skin hydration lotions, reduced age spots, and hyperpigmentation
Chlorella	Skin moisturizer
Haematococcus pluvialis	Act as sunscreen lotions
Spirulina, Porphyridium	Source of pigments for manufacturing eye-liners and lipsticks
Nannochloropsis salina, Nannochloropsis oculate, Nannochloropsis gaditana	Manufacturing of tanning cosmetics
Porphyridium cruentum, Spirulina platensis	Manufacturing of diverse eye shadows
Anabaena vaginicola	Antiaging and sunscreen lotions
Halomonas elongata, Halomonas boliviensis, Brevibacterium epidermis, Chromohalobacter israelensis, Chromohalobacter salexigens	Protector against harmful ultraviolet light, skin moisturizer
Chlorella vulgaris	Delays skin aging process by synthesizing collagen

MMPs, Metalloproteinases.

Although cosmetics are not endorsed by Food and Drug Administration, the aim of the cosmeceuticals is to enhance and improve the level of attractiveness, skin cleansing and can be easily purchased (Pereira, 2018). In the present era, every individual is showing a deep interest in exploiting natural products such as microalgae-derived cosmetics (Ariede et al., 2017). Approximately 221 microalgal species such as *Saccharina japonica, Gracilaria* sp., *Monostroma nitidum, Spirulina, Arthrospira* spp are exploited commercially in food, cosmetics, and pharmaceutical industries, with the majority of them being strongly cultivated in countries like Australia, India, Israel, Japan, Malaysia, and Myanmar. All over the globe, the microalgal industries have a worth of more than USD 6 million per year, out of which 85% of the items are designated as nourishment items for the exploitation by *H. sapiens* (Joshi et al., 2018).

Concerning the past two decades, microalgal species like *Chlorella, Spirulina, Anacystis, Nannochloropsis, Dunaliella,* and *Halymenia* have been significantly exploited to manufacture cosmetics products over the globes particularly in Europe and the United States. Companies of Great Britain, such as Optimum Derma Aciditate, have launched a *Spirulina*-based firming mask in the year 2012. The company claims to enhance the skin moisture balance and bolster the immunity of the skin. A French-based company, Phytomer, has launched a microalgae-based cosmetic product that constitutes the extraction of species *C. vulgaris*. It aims to neutralize skin inflammation and hence improve the natural protection process of the skin in combination with the antioxidative properties of the species. Researchers have exploited the extracts of *Porphyridium cruentum* and further claim to enhance the texture of the skin. The product is sold under the name SILIDINE by Greentech. Two significant microalgal strains *Dunaliella salina* and *Haematococcus pluvialis* have been exploited to manufacture a product named REVEAL by Algenist, which is usually exploited as a skin toner (Mourelle et al., 2017). Table 13.8 illustrates numerous other significant commercialized microalgae-based cosmetic products.

13.4 Environmental services of microalgae

With a better understanding of the declining conditions of the environment in the 21st century, a sophisticated and environment-friendly approach dealing with the exploitation of microalgae has grabbed the attention of global researchers and stimulated the management of environmental sustainability. The extensive escalation of the human population and rapid development of industrialization has resulted in utilizing natural resources and thus exhaustion and scarcity of them. Concerning this, microalgae have resolved numerous catastrophic events that generally damage the environment. This segment discusses applications of diverse microalgal strains in WWT, biodegradation of plastic, sequestration of carbon dioxide, production of biofuels such as biohydrogen, bioethanol, biogas, biomethanol, biodiesel (Chowdhury et al., 2016) (Fig. 13.3).

13.4.1 Microalgae: a tool for wastewater treatment

Globally, the aim of rendering clear water for human consumption has become a global issue. Water is an important natural resource of the earth, but the meager supply of pure water in various developing and industrialized nations has become a worldwide issue and has entered a worrisome phase (Sousa et al., 2018). The different sources of pollution such as municipal wastes, agricultural, and industrial wastewaters constituting contamination from organic and inorganic pollutants such as microplastics (Koelmans et al., 2019), xenobiotics (Sousa et al., 2018), heavy metals (mercury, cadmium, zinc, lead, chromium, copper, etc.) (Chowdhury et al., 2016), nitrates (NO_2) (Menció et al., 2016), phosphates (P), and carbon (C) (Sousa et al., 2018) have jeopardized the food supply chain and risked the life of *H. sapiens* (Fig. 13.4). The WWT has become a global issue that cannot be managed by applying single obsolete technology due to less variability of scales at far-off levels, sources and types of contamination, regional conditions, etc. (Wollmann et al., 2019; Abdel-Raouf et al., 2012).

The conventional WWT plants usually target (1) the eradication of suspended solids through mechanical approaches and (2) the dropping of biological oxygen demand through activated sludge procedure (Wang et al., 2017). The biodegradation process consists of the disintegration of organic and inorganic compounds such as nitrogen (N) and phosphorous (P), which are significant for inhibiting the eutrophication of downstream water such as rivers lakes. These old technologies have very defined degradation potential precisely with heavy metals, excess nutrients, and xenobiotics, resulting in improved accumulation of substances in groundwater (Chowdhury et al., 2016).

TABLE 13.8 Commercialized microalgae-based cosmetic products.

Name of company	Commercialized product	Microalgae	Application of product
Optimum Derma Aciditate	Spirulina firming algae mask	*Arthrospira maxima* *Spirulina platensis*	Advances the skin moisture balance and enhances the immunity of the skin
Ferenes Cosmetics	Spirulina whitening facial mask	*Spirulina platensis*	Decreases the formation of wrinkles and enhances the complexion of wrinkles
Phytomer		*Chlorella vulgaris*	Enhances the natural protection mechanism of the skin by neutralizing the skin inflammation
Skinicier		*Spirulina platensis*	Improves the natural production of skin cells and regeneration process
Pentapharm	Pepha-Tight	*Nannochloropsis oculata*	Renders skin tightening characteristics
Pentapharm	Pepta-Ctive	*Dunaliella salina*	Stimulation of proliferation of cells
Biotherm	Blue Retinol		
Exsymol	Protulines	*Spirulina platensis*	Reduces formation of wrinkles, provides skin tightening, and thus delays skin aging process
Solazyme	Algenist	*Anacystis nidulans*	Bolsters the immunity of skin cells
Frutarom	Alguard	*Porphyridium spp.*	Skin protection against photoaging and exhibits antimicrobial properties
Codif Recherche and Nature	Dermochlorella	*Chlorella vulgaris*	Enhances the formation of collagen
Dainippon Ink and Chemicals Inc.		*Spirulina, Porphyra spp., Porphyridium spp*	Coloring agent manufacturing in eye shadows
Pentapharm Ltd.	Pepha-Ctive	*Dunaliella salina*	Stimulation of proliferation of cells
Codif Recherche and Nature	DermochlorellaD	*Chlorella vulgaris*	Ultraviolet light protector, protection against oxidative damage, bolsters the synthesis of collagen
Daniel Jouvance	OceaRides	*Odontella spp.*	Stimulates the synthesis of elastin protein
Algatech	AstaPure	*Haematococcus pluvialis*	Exhibits antioxidative properties
	FucoVital	*Phaeodactylum tricornutum*	
Soliance	Megassane	*Phaeodactylum tricornutum*	Skin cells protector against UV light damage decreases photoaging and age spots
Dainippon Ink & Chemicals Inc.	Linablue	*Arthrospira spp.*	Manufacturing of variety of eye shadows

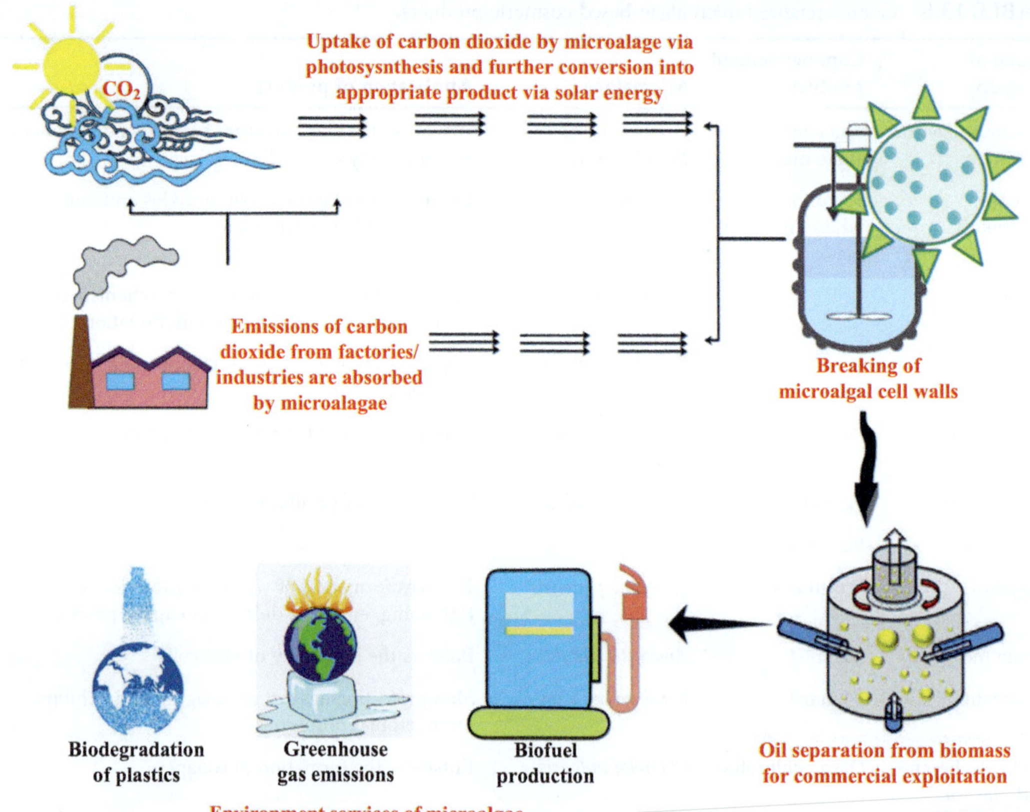

FIGURE 13.3   Various environmental services of microalgal species.

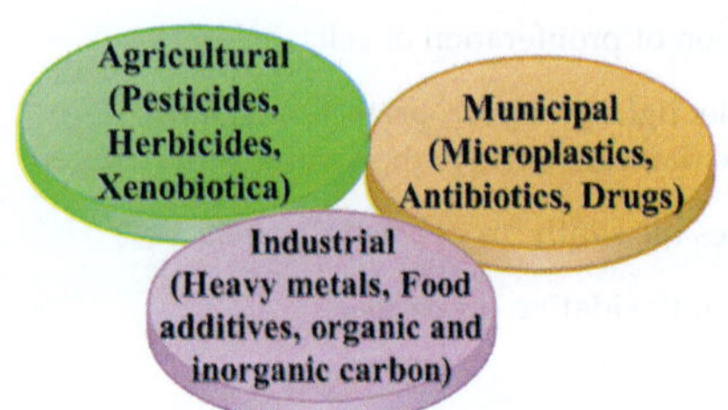

FIGURE 13.4   Diagrammatic representation of sources of wastewaters and their respective impurities. *Source: From Wollmann, F., Dietze, S., Ackermann, J., Bley, T., Walther, T., Steingroewer, J., Krujatz, F., 2019. Microalgae wastewater treatment: biological and technological approaches. Engineering in Life Sciences 19 (12), 860–871. https://doi.org/10.1002/elsc.201900071.*

Due to metabolic flexibilities, diverse microalgal species have significant capabilities for implementing photoautotrophic, mixotrophic, and heterotrophic metabolism (Hu et al., 2018; Subashchandrabose et al., 2013), which further represents those biological systems that treat numerous sources of wastewaters (Wollmann et al., 2019; Molazadeh et al., 2019). Notably, microalgae have developed sustainable biological products such as proteins (Soto-Sierra et al., 2018), FAs (Kumar et al., 2010), pigments (D'Alessandro and Antoniosi Filho, 2016), biofertilizers (Santos and Pires, 2018), and animal feed (Madeira et al., 2017) concerning circular economy, economy, and biorefineries (Venkata Mohan et al., 2016), respectively. Also, they are capable enough to fix inorganic nitrogen and phosphorous efficiently (Wollmann et al., 2019).

Majorly, there are two significant goals of treatment of wastewaters by exploiting microalgae: (1) the direct uptake or conversion of various contaminants in polluted water and (2) improving the performance of bacterial purification systems by providing additional oxygen from photosynthesis that reduces the total energy costs of direct and indirect oxygen supply (Quijano et al., 2017). Presently, research work majorly focuses on conventional microalgal species such as *Chlorella* spp. (Chiu et al., 2015), *Arthrospira* spp. (Zhai et al., 2017), *Scenedesmus* spp. (Ansari et al., 2019), and *Nannochloropsis* spp. (Mitra et al., 2016) due to their capability of accumulating high levels of lipids and starch, respectively. Table 13.9 discusses a few microalgal species being exploited for WWT (Wollmann et al., 2019).

13.4.2 Microalgae for degradation of plastics

There has been a significant rise in the production of plastics all over the globe from 245 million metric tons in the year 2008 to near about 359 million metric tons in the year 2018, which is expected to triple by 2050 (Chia et al., 2020). Since the 1950s, there has been a bulk production of plastics, but there is still a dire need to implement an effective strategy to tackle the menace of disposing of plastic waste. From Fig. 13.5 the recycling rate of plastics is low compared to the generation of plastic, whereas most are disposed of in landfills. Nevertheless, the decomposition of plastics is toughest compared to the decomposition of fruits, papers, leathers, aluminum, etc., due to their persistent nature before the onset of their decay process (Chen and Yan, 2020).

The excessive plastic-generated waste has created havoc in aquatic biodiversity by creating serious implications on marine habitats like coral life, animals, etc., for example, ingestion, entanglement, debilitation, and suffocation, which results in decreased quality of life, drowning, restricted capacities to refrain predator, less reproductive potential, impaired feeding power, and ultimately death (de Stephanis et al., 2013). Furthermore, there are rising concerns regarding the impact of microplastics on humans as researchers have detected the presence of microplastics in numerous food and air samples. Since the use of plastics is inevitable in the human life, keeping in mind the harmful effects, researchers are currently being employed to develop an alternative of traditional plastics with environment-friendly degradation (Chia et al., 2020).

Through the mechanistic action of numerous biological actions, plastic rapidly gets degraded into its basic components (Ahmed et al., 2018). These biological agents exploit organic polymer and render them as a nutritional substrate for their growth and energy which results in the production of microbial biomass as the end product of complete biodegradation (Gopal, 2017). In the recent past, it has been observed that numerous microalgal species boost the degradation of polymers following the reduction of energy required for degradation by weakening their chemical bonds (Bhuyar et al., 2019). Presently, the common fossil-derived polymers are polyethylene terephthalate, polyethylene (PE), low-density PE, and high-density PE, polyvinylchloride, polypropylene, and polystyrene (Chen and Yan, 2020).

Microalgae have the potential to grow onto waste resources with a high accumulation of lipids and thus have become a potential candidate for producing bioplastics by

TABLE 13.9 Microalgal species for wastewater treatment.

Microalgae	Cultivation method	Growth conditions	Removal rate/heavy metal removal	Product	References
Galdieria sulphuraria	Open system (700 L)	The mixotrophic method with diluted raw primary effluent diluted following carbon dioxide fortified headspace	After 3 days: BOD5 = 36–13 mg L^{-1}, N = 23–2.6 mg L^{-1}, P = 4.5–0.6 mg L^{-1}	Biomass	Henkanatte-Gedera et al. (2017)
	Bioreactor (3 L)	Heterotrophic on complex media and glucose	After 100 h: Ammonia (NH3) = 0.31–0.15 g L^{-1}	c-Phycocyanin (250–400 mg L^{-1})	Schmidt et al. (2005)
	Glass tubes	Heterotrophic in media with primary effluent	After 7 days: Ammonia (NH3) = 4.85 mg L^{-1} day^{-1}, phosphates (PO4) = 1.21 mg L^{-1} day^{-1}.	Biomass	Selvaratnam et al. (2014)
	Closed outdoor reactor	Mixotrophic in media with primary effluent and 1%–2% carbon dioxide sparged			
	Shake flasks (500 L)	Hydrolysate from heterotrophic bakery and restaurant wastes with supplemented nitrogen sources			Sloth et al. (2017)
Chlamydomonas acidophila	Batch reactor (1 L)	Mixotrophic method on numerous carbon sources		Lutein (9–10 mg g^{-1}) Zeaxanthin (7–8 mg g^{-1})	Cuaresma et al. (2011)
Chlorella sorokiniana	Batch reactor (1 L)	Phototrophic cultivation, cells immobilized in alginate beads	After 4 days: Ammonia (NH3) from 10 to 0 mg L^{-1}		de-Bashan et al. (2008)
	Shake flasks	The phototrophic method with growth on post-chlorinated wastewater supplemented with diverse nitrogen sources		Urea supplementation	Ramanna et al. (2014)
	Batch reactor (1 L)	Phototrophic growth over anaerobic digester centrate	Carbon monoxide = 20%–30%, carbon dioxide = 30%–45%, nitrogen oxide = 95%–100%	Biomass (250 mg L^{-1} day^{-1})	Lizzul et al. (2014)
	Hanging bags	Phototrophic growth in 10% anaerobic digester effluent supplemented with cattle wastes	Orthophosphorous = 57.70%, total phosphorous = 64.10%, ammoniacal nitrogen = 72.17%	Biomass (13–17 mg L^{-1} day^{-1})	Kobayashi et al. (2013)
	Shake flasks (2 L)	Phototrophic growth onto filtered raw sewages	Chemical oxygen demand = 69.38%, nitrogen = 86.93%, phosphorous = 68.24%, coliforms 99.78%, fecal coliforms 100%	Biomass with 22.36% lipids	Gupta et al. (2016)

Microalgae	System/condition	Condition	Removal/performance	Form	Reference
Asterarcys quadricellulare, Desmodesmus sp., Pseudopediastrum sp.	Open pond raceways	25°C, 8 h light, 160 rpm	Chemical oxygen demand = 12.4 mg L^{-1}, Total nitrogen = 30.9 mg L^{-1}, Ammonia = 29.19 mg L^{-1}, Total phosphorous = 2.29 mg L^{-1}	Biomass with proteins, carbohydrates, and lipids	Do et al. (2019)
Chlamydomonas reinhardtii		pH = 3–7	Removal of cadmium ions (28.8 mg g^{-1})	Biomass (Ca-alginate)	Bayramoğlu et al. (2006)
Spirulina spp.		pH = 7.5	Removal of cobalt ions (0.01 mg g^{-1})	Biomass	Chojnacka et al. (2004)
Chlorella miniata		pH = 4.5	Removal of chromium ions (Cr^{+3}) (41.12 mg g^{-1})		Han et al. (2006)
Chlorella vulgaris		pH = 4.0	Removal of copper ions (Cu^{+2}) (38.89 mg g^{-1})		Romera et al. (2006)
Scenedesmus obliquus		pH = 7.0	Removal of copper ions (Cu^{+2}) (1.8 mg g^{-1})		Yan and Pan (2002)
Chlorella vulgaris		pH = 2.0	Removal of iron ions (Fe^{+3}) (1.8 mg g^{-1})		Romera et al. (2006)
Phaeodactylum tricornutum		pH = 4.0	Removal of mercury ions (Hg^{+2}) (0.51 mg g^{-1})		Schmitt et al. (2001)
Arthrospira platensis		pH = 5.0–5.5	Removal of nickel ions (Ni^{+2}) (20.78 mg g^{-1})		Ferreira et al. (2011)
Phaeodactylum tricornutum		pH = 6.0	Removal of lead ions (Pb^{+2}) (1.49 mg g^{-1})		Schmitt et al. (2001)
Spirulina platensis		pH = 6.0	Removal of zinc ions (Zn^{+2}) (7.36 mg g^{-1})		Sandau et al. (1996)
Nostoc muscorum		pH = 3.0	Removal of chromium ions (Cr^{+6}) (22.92 mg g^{-1})		Gupta and Rastogi (2008)

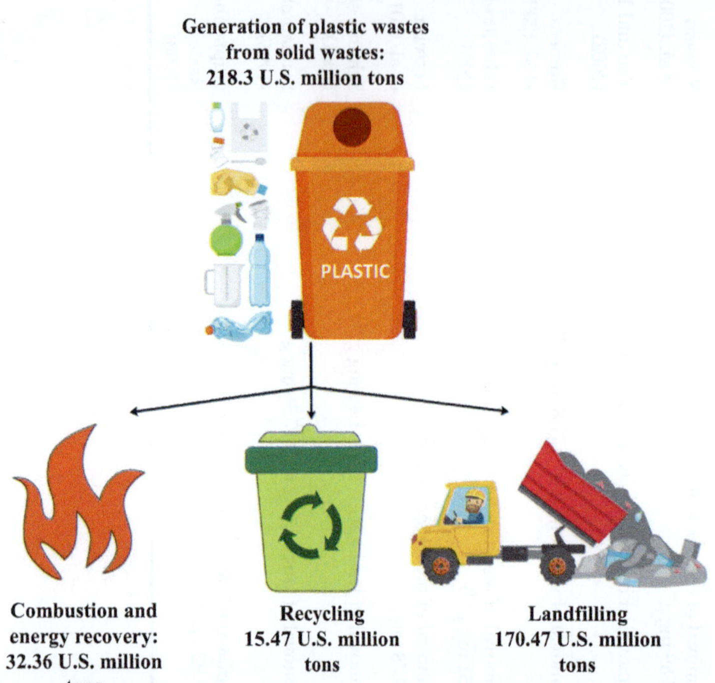

FIGURE 13.5 Generation of plastics from MSW in the United States from the year 1960 to 2017. *MSW*, Municipal solid wastes. Source: *From Chia, W.Y., Ying Tang, D.Y., Khoo, K.S., Kay Lup, A. N., Chew, K.W. 2020. Nature's fight against plastic pollution: algae for plastic biodegradation and bioplastics production. Environmental Science and Ecotechnology 4, 100065. https://doi.org/10.1016/j. ese.2020.100065.*

numerous approaches like biomass blending, additives, intracellular cultivation within the cells of microalgae (Onen Cinar et al., 2020). Additionally, the process of production usually does not compete with diverse food resources (Khoo et al., 2020; Yew et al., 2020) (Fig. 13.6). The most researched microalgal species are *C. reinhardtii* (Barone et al., 2020; Tran and Kaldenhoff, 2020), *Chlorella scenedesmus*, *Raphidocelis subcapitata*, *Anabaena* sp. (PCC 7120), etc. (González-Pleiter et al., 2019; Nolte et al., 2017). Microplastics possess significant serious risks to biota and imbalance the ecological niche. Table 13.10 discusses the variety of microalgal species and the severe impact of toxicity of microplastics like DNA damage, oxidative stress (Anbumani and Kakkar, 2018).

13.4.3 Microalgae for carbon dioxide sequestration

Recently, with escalating awareness of the global energy crisis, along with issues of sustainability and the environment, significant interest has been developed in carbon dioxide sequestration (Uduman et al., 2010). The increase in the concentration of atmospheric carbon dioxide has rendered significant challenges regarding the imbalance of sustainability and the environment, respectively. Currently available technologies for capturing carbon dioxide constitute physicochemical adsorption (costly and difficult management), geological formations (requirement of space in bulk, leakage issues, etc.), and improved biological fixation (Lackner, 2003). The sustainable environment-friendly approach to decrease GHGs is to produce energy from those sources that have less or decreased carbon emissions.

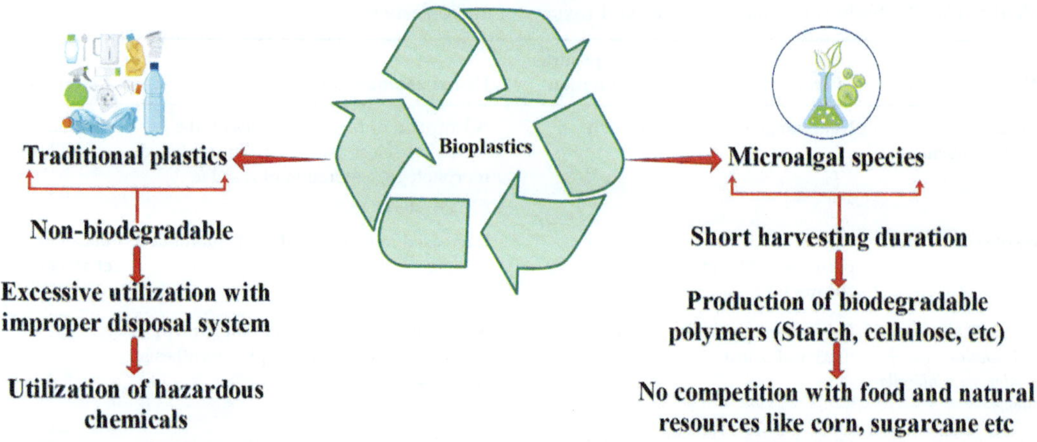

FIGURE 13.6 Diagrammatic representation of the significant potential of microalgal species in the production of bioplastics. Source: *From Chia, W.Y., Ying Tang, D.Y., Khoo, K.S., Kay Lup, A.N., Chew, K.W. 2020. Nature's fight against plastic pollution: algae for plastic biodegradation and bioplastics production. Environmental Science and Ecotechnology 4, 100065. https://doi.org/10.1016/j.ese.2020.100065.*

With the advancement of the R&D sector, biofuel has become a potential candidate as an efficient alternative to conventional fuels which further have a significant role in reducing carbon dioxide emissions from various sources of transportation (Sharma et al., 2020; Zeng et al., 2011).

The concept of biological fixation of carbon dioxide is a long-lasting technology to manage the sustainability of the environment. The substantial need to reduce carbon dioxide emissions, microalgae have proven to be the best fit for carbon fixation and production of diverse biofuels by converting atmospheric carbon dioxide into biomass via autotrophs, though the utilization of nutrients and feedstocks for energy productions have been accomplished sustainably (Kumar et al., 2010). In comparison to other plant-based feedstocks, microalgae have numerous advantages in capturing carbon dioxide and generation of biofuels such as high photosynthetic conversion efficacies, accelerated production rates of biomass, the capacity of producing diverse feedstocks for the production of biofuels, easy flourishment in various ecosystems, distinguished bioremediation like atmospheric carbon dioxide fixation, water purification (Wang et al., 2008; Chisti, 2007), noncompetition for land usage with crops and food market (Fig. 13.7) (Zeng et al., 2011; Comitre et al., 2021).

Such biomass is further converted into methane (CH_4) (Li et al., 2008) and hydrogen (H_2) (Sharma and Arya, 2019; Sharma and Arya, 2017) by exploiting certain processes and bacterial cultures (Haiduc et al., 2009). Due to easy synthesis, microalgae have been majorly exploited for producing oils. Certain approaches such as lipid extraction, reesterification of short-chain alcohols, and certain other by-products of PUFA, β-carotenes (Li et al., 2008) can be hydrolyzed following reesterification to produce biodiesel (Kumar et al., 2010). Table 13.11 discusses various important microalgal strains for sequestering carbon dioxide via numerous bioreactor systems.

TABLE 13.10 Various microalgal species and toxicity of microplastics.

Microalgae system	Microplastics	Exposure duration	Discussion/observations	References
Chlorella spp. Scenedesmus spp	Nano-range plastic beads	72 h	Adsorption of plastic particles on the surface via electrostatic attraction, altered morphology, generation of reactive oxygen species	Bhattacharya et al. (2010)
Scenedesmus obliquus	Nanosized polystyrene particles (0.22 and 103 mg L^{-1})	96 h	Decreased rate of growth and chlorophyll	Besseling et al. (2014)
Dunaliella tertiolecta, Thalassiosira pseudonana, Chlorella vulgaris	Polystyrene particles (0.05 and 6 μm)	72 h	No significant alterations in the growth with a major reduction in photosynthesis (2.5%–45%)	
Chlamydomonas reinhardtii	High-density polyethylene, polypropylene particles	20 days	Transfer via the vertical pathway, major growth inhibition, increased expression of glucose-6-dehydrogenase and galacturonate 4-epimerasegenes which represent exopolysaccharide biosynthesis pathway	
Skeletonema costatum	PVC (1 μm and 1 mm)	96 h	Growth inhibition in particle with size 1 μm but no significant effect on the growth of microalgae with size 1 mm size	
Tetraselmis chuii	Fluorescent red polyethylene microspheres (1–5 μm, 0.046–0.472 mg L^{-1})	96 h	No growth rate inhibition	
Rhodomonas baltica	Fluorescent polystyrene particles (1–5 μm)	60 min	Elevated uptake and formation of biofouling	
Chlamydomas reinhardtii	High density polyethylene, polypropylene(400–1000 μm, 400 mg L^{-1})		No decrease in the growth of HDPE, no change in the expression of chloroplastic genes, no significant effect on apoptotic genes, reduction in the PP growth by 18%	Lagarde et al. (2016)
Dunaliella tertiolecta	Polystyrene (uncharged) (0.05–6 μm, 25 and 250 mg L^{-1})		No significant impact on the photosynthesis, Inhibition of growth at 250 mg L^{-1} (57% for 0.05 μm and 13% for 0.5 μm)	Sjollema et al. (2016)
T. pseudonana	Polystyrene (negatively charged)		No significant impact on the photosynthesis, Inhibition of growth by 13% at 250 mg L^{-1}	
C. vulgaris				
Skeletonema costatum	PVC (1 μm; 5–50 mg L^{-1})		Growth inhibition by 40%, reduced content of chlorophyll, and decreased efficacy of photosynthesis	Zhang et al. (2017)
Phaeodactylum tricornutum	PET	2–6 weeks 210 C	Biological decomposition of PET waste in a saltwater environment through eukaryotic microalga	Moog et al. (2019)

PET, Polyethylene terephthalate; *PVC*, polyvinylchloride.

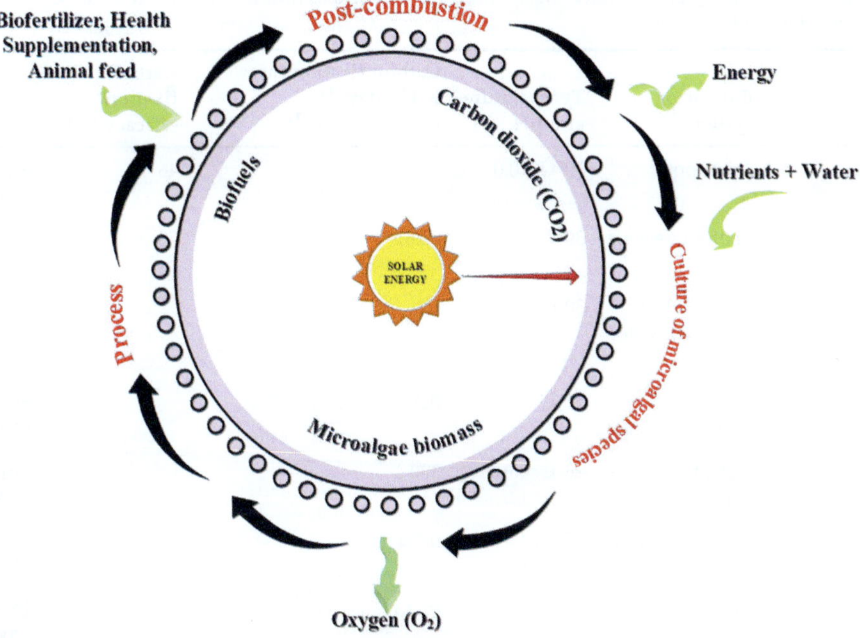

FIGURE 13.7 Diagrammatic representation of unification of microalgae and carbon dioxide utilization (**Sayre, 2010**). Source: *From Sayre, R., 2010. Microalgae: the Potential for carbon capture. BioScience 60 (9), 722–727. https://doi. org/10.1525/bio.2010.60.9.9.*

13.4.4 Microalgae for producing biofuels

Considering the revolutionized era of renewable energy, biofuels are designated as an alternative to traditional fossil fuels (Mobin and Alam, 2018). Biofuels are majorly produced from different biomass, which is further cultivated and harvested from numerous biological sources (Mobin and Alam, 2018; Dutta et al., 2014). Biofuels are majorly categorized into three groups, (1) first-generation biofuels require more agricultural lands. First-generation biofuel feedstocks constitute corn, sugarcane, soybean, potato, beet, soybeans, coconut, sunflower, rapeseed, palm oil, switch grass, jatropha, camellia, cassava, etc., which are accountable for degradation of ecological balance. Hence, to maintain the balance and integrity of ecology and environment, the production of first-generation biomass decreases, (2) second-generation biofuels address the practicality issues of first-generation biomass. The source of second-generation biomass is plenty such as cellulose, nonedible portions of plants, straw, manure, used cooking oil, wood, sawdust, etc. However, second-generation biofuels have not proved much profitable to industries of concern due to expensive production processes (Brennan and Owende, 2010; Dragone et al., 2010) and thus (3) third-generation biofuels have been developed (Chowdhury and Loganathan, 2019).

These types of biofuels are majorly produced from microalgal species which is nowadays designated as a sustainable energy source (Fig. 13.8) by overcoming and addressing

TABLE 13.11 Various significant microalgal strains for their exploitation in carbon dioxide sequestrations through diverse bioreactor system.

Microalgae	Bioreactor system	Temperature and pH	Carbon fixation rate (g L^{-1} day^{-1})/ (mgC L^{-1} day^{-1})	Carbon fixation efficacy (%)	References
Chlorella sp.	Open pond	30°C, 10.0		46	Ramanan et al. (2010)
	Photobioreactor	30°C	0.261	58	Chiu et al. (2008)
Chlorella kessleri	Conical flask photobioreactor	30°C			de Morais and Costa (2007)
Chlorella vulgaris	Tubular photobioreactor	25°C	0.075		Scragg et al. (2002)
Chlorococcum littorale	Flat-plate reactor	25°C, 6.0–7.2	200.4		Hu et al. (1998)
Dunaliella tertiolecta	Bubble column reactor	25°C, 7.2 ± 0.2	0.271		Sydney et al. (2010)
Dunaliella sp.	Photobioreactor	3°C	0.313		Li et al. (2008)
Botryococcus braunii	Bubble column reactor	25°C, 7.2 ± 0.2	0.497		Sydney et al. (2010)
Scendesmus obliquus	Tubular photobioreactor	30°C		28.08	de Morais and Costa (2007)
Nannochlorophsis oculata	Photobioreactor	26°C	0.393	15	Chiu et al. (2008)
Haematococcus pluvialis	Tubular photobioreactor	16°C–18°C	0.143		Huntley and Redalje (2007)
Aphanothece m. Nageli	Airlift bioreactor	25°C, 6.8	14.5		Jacob-Lopes et al. (2008)
	Bubble column reactor	35°C	2.621		Jacob-Lopes et al. (2009)
Thermosynechococcus sp.	Bubble column reactor	55°C			Hsueh et al. (2009)
Spirulina sp.	Tubular photobioreactor	30°C		45.61	Wang et al. (2008)
Spirulina platensis	Open pond			39	Ramanan et al. (2010)
Anabaena sp.	Bubble column reactor	27°C, 8.5	1.45		González López et al. (2009)

(Continued)

TABLE 13.11   (Continued)

Microalgae	Bioreactor system	Temperature and pH	Carbon fixation rate (g L^{-1} day^{-1})/ (mgC L^{-1} day^{-1})	Carbon fixation efficacy (%)	References
Chlorogloeopsis sp.	Photobioreactor	50°C	20.45		Ono and Cuello (2007)
Synechococcus sp.	Airlift bioreactor	30°C, 6.8	0.6		Heasman et al. (2000)
Spirulina sp.	Vertical tubular photobioreactor	30°C	202.3 ± 5	79.3 ± 2.3	Moraes et al. (2016)
Haematococcus pluvialis	Photobioreactor	4°C and 105°C	0.127	36.8	Hong et al. (2019)
Isochrysis sp.	Photobioreactor	30°C, 7.0–8.0	0.35	4.73	Yahya et al. (2020)

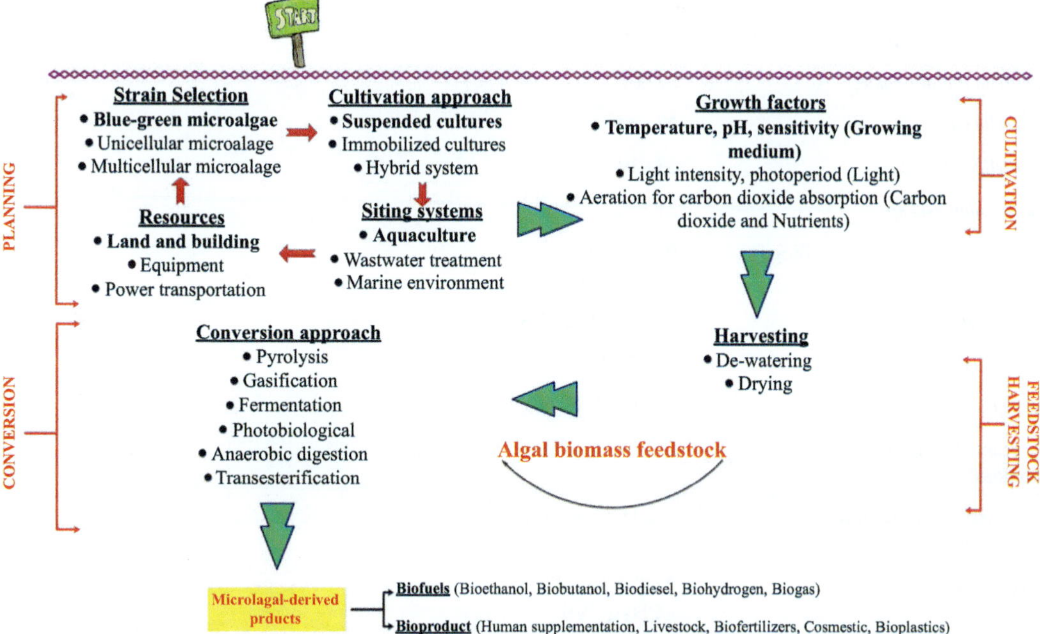

FIGURE 13.8   Schematic flowchart of production of diverse biofuels from microalgal species. Source: *From Chowdhury, H. & Loganathan, B. (2019). Third-generation biofuels from microalgae: a review. Current Opinion in Green and Sustainable Chemistry, 20, 39–44. https://doi.org/10.1016/j.cogsc.2019.09.003.*

the issues of both first- and second-generation biofuels (Chisti, 2007; Nigam and Singh, 2011; Dragone et al., 2010). The sustainable energy source is quickly escalating because of numerous concerns such as increasing global population, the rapid development of industrialization, strong demand for fuels for the transportation sector, extensive environmental damage by using conventional or nonrenewable energy sources (oil, natural gas, coal, etc.) such as a rampant load of GHGs (carbon dioxide, methane, nitrous oxide, etc.) which results in global climate issues (Faried et al., 2017; Abbaszaadeh et al., 2012). Table 13.12 discusses the production of different biofuel from various microalgae species.

TABLE 13.12 Production of biofuels from various microalgal strains.

Microalgae	Biofuel	Yield	References
Chlamydomonas reinhardtii	Biohydrogen	102 mL/1.2 L	Nagappan et al. (2019); Sun et al. (2014); Ho et al. (2010); Bermúdez et al. (2014)
Chlorella vulgaris MSU 01		26 mL/0.5 L	
Scenedesmus obliquus		3.6 mL μg^{-1} Chl a	
Platymonas subcordiformis		11,720 nL h^{-1}	
Dunaliella tertioletca	Bio-oil	34 MJ kg^{-1}	
Chlorella protothecoides		52.00%	
Chlorella sp.		28.60%	
Chlorella vulgaris		35.83%	
Nannochloropsis sp.		31.10%	
Chlorella vulgaris	Biogas	0.63–0.79 LCH$_4$ g^{-1}VS	
Dunaliella salina		0.68 LCH$_4$ g^{-1}VS	
E. gracilis		0.53 LCH$_4$ g^{-1}VS	
Scenedesmus		140 LCH$_4$ g^{-1}VS	
Scenedesmus (lipid-free biomass)		212 LCH$_4$ g^{-1}VS	
Scenedesmus (amino acids free biomass)		272 LCH$_4$ g^{-1}VS	
S. obliquus		0.59–0.69 LCH$_4$ g^{-1}VS	
Botryococcus braunii	Biodiesel	25%–75%	
Chlorella sp.		28%–32%	
Chlorella vulgaris		56%	
Crypthecodinium cohnii		20%	
Nannochloropsis sp.		31%–68%	

(Continued)

TABLE 13.12   (Continued)

Microalgae	Biofuel	Yield	References
Neochloris oleoabundans		35%−54%	
Scenedesmus dimorphus		6%−40%	
S. obliquus		11%−55%	
Schizochytrium sp.		77%	
S. obliquus CNW-N		79 mg L^{-1} day^{-1}	
Neochloris oleoabundans HK-129		69.5 mg L^{-1} day^{-1}	
Chlorella pyrenoidosa	Bioethanol	26%	
Chlorella vulgaris		12%−17%	
Dunaliella salina		32%	
S. obliquus		10%−17%	
Porphyridium cruentum		40%−57%	
E. gracilis		14%−18%	
Spirulina sp.	Biomethanol	38%	Peng et al. (2020)

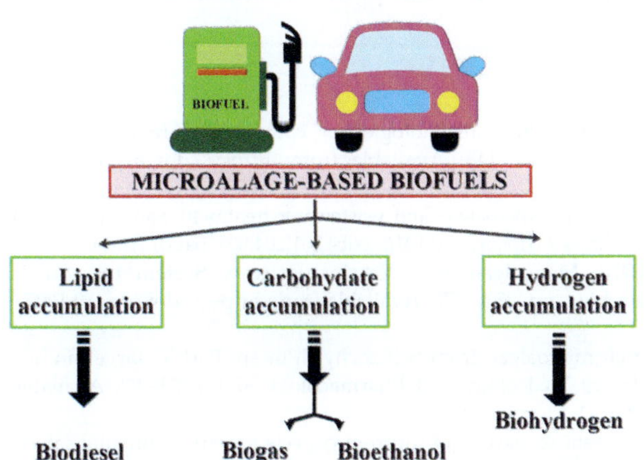

FIGURE 13.9   Schematic representation of the production of various biofuels via different accumulation methods. Source: *From Nagappan, S., Devendran, S., Tsai, P.-C., Dahms, H.-U. & Ponnusamy, V. K. (2019). Potential of two-stage cultivation in microalgae biofuel production. Fuel, 252, 339−349. https://doi.org/10.1016/j.fuel.2019.04.138.*

Biofuels derived from diverse microalgal species have grabbed a lot of attention from all over the globe due to rapid growth rate, high efficacy, no restriction of feedstock supply, high photosynthetic efficiency, less cost of production, high content of lipids, carbohydrates, etc. (Peng et al., 2016). Significantly, microalgal strains are transforming nutrients into appropriate biomass and different cellular components (Bouabidi et al., 2019; Mata et al., 2010).

Fig. 13.9 represents the accumulation of different contents such as lipids, carbohydrates, and hydrogen for producing different biofuels such as biodiesel, biogas, bioethanol, and biohydrogen under appropriate reaction conditions and extraction process (Enamala et al., 2018; Peng et al., 2020)

13.5　Conclusions and future perspectives

The proper functioning of the ecosystem is one of the major concerns of the 21st century. Every *H. sapiens* of planet earth is dependent on the ecosystem, directly or indirectly. There are plenty of services of the ecosystem such as arable agricultural land, clean water for drinking and other purposes, aquatic habitat, terrestrial habitat, and clear air. This chapter attempts to provide significant applications of microalgae that benefit humanity and, thus, the ecosystem. Microalgae have grabbed a significant hierarchy by providing important services such as WWT, GHG emissions, plastic degradation, health supplements, and immunomodulatory properties which bolster in improving the condition of degraded ecology and environment, respectively. Nevertheless, to ensure the sustainability of the ecosystem, numerous bottlenecks require dire attention for futuristic applications such as costly production process of microalgae, less efficiency, the extraction process, contamination, no exploitation of less explored microalgal species, dependency on traditional or obsolete technologies, and less or no technical expertise. Additionally, the microalgal-based technology must be robust for their commercial exploitation. Their demands are dependent on cohesive ideas and a marketing approach picking suitable customers for their suitable products with legitimate pricing rates. The dawn of microalgal-based bioproducts will have a profound effect on increasing population and rampant industrialization era.

References

Abbaszaadeh, A., et al., 2012. Current biodiesel production technologies: a comparative review, Energy Conversion and Management, 63. Elsevier BV, pp. 138–148. Available from: https://doi.org/10.1016/j.enconman.2012.02.027.

Abdel-Raouf, N., Al-Homaidan, A.A., Ibraheem, I.B.M., 2012. Microalgae and wastewater treatment. Saudi Journal of Biological Sciences 19 (3), 257–275. Available from: https://doi.org/10.1016/j.sjbs.2012.04.005. Saudi Arabia.

Abidin, A.A.Z., Suntarajh, M., Yusof, Z.N.B., 2020. Microalgae as a Vaccine Delivery System to Aquatic Organisms. Springer Science and Business Media LLC, pp. 353–372. Available from: https://doi.org/10.1007/978-981-15-0169-2_10.

Abril, R., et al., 2003. Safety assessment of DHA-rich microalgae from Schizochytrium sp. Part V: target animal safety/toxicity study in growing swine. Regulatory Toxicology and Pharmacology 37 (1), 73–82. Available from: https://doi.org/10.1016/s0273-2300(02)00030-2. Elsevier BV.

Ahmed, T., et al., 2018. Biodegradation of plastics: current scenario and future prospects for environmental safety. Environmental Science and Pollution Research 25 (8), 7287–7298. Available from: https://doi.org/10.1007/s11356-018-1234-9. Springer Science and Business Media LLC.

Acién Fernández, F.G., Fernández Sevilla, J.M., Molina Grima, E., 2019. Chapter 21 - Costs analysis of microalgae production. In: Pandey, A., Chang, J.-S., Soccol, C.R., Lee, D.-J., Yusuf, B.T., (Eds.). Biomass, Biofuels, Biochemicals. Biofuels from Algae. 2nd ed. Chisti, 551–566. Elsevier. Available from: https://doi.org/10.1016/B978-0-444-64192-2.00021-4.

Alam, M.A., Wang, Z., 2019. Microalgae Biotechnology for Development of Biofuel and Wastewater Treatment. Springer Singapore, China, pp. 1–655. Available from: https://doi.org/10.1007/978-981-13-2264-8.

Aly, E., 2015. Hepatoprotective, DNA damage prevention and antioxidant potential of *Spirulina platensis* on CCl4-induced hepatotoxicity in mice. American Journal of Biomedical Research 3, 29–34.

Anbumani, S., Kakkar, P., 2018. Ecotoxicological effects of microplastics on biota: a review. Environmental Science and Pollution Research 25 (15), 14373–14396. Available from: https://doi.org/10.1007/s11356-018-1999-x. Springer Nature.

Andremont, A., 2000. Conséquences de l'antibiothérapie sur l'écosystème intestinal. Annales Françaises d'Anesthésie et de Réanimation 19 (5), 395–402. Available from: https://doi.org/10.1016/s0750-7658(00) 90209-0. Elsevier BV.

Andrianasolo, E.H., et al., 2008. Apoptosis-inducing galactolipids from a cultured marine diatom, *Phaeodactylum tricornutum*. Journal of Natural Products 71 (7), 1197–1201. Available from: https://doi.org/10.1021/np800124k. American Chemical Society (ACS).

Ansari, F.A., et al., 2019. Techno-economic estimation of wastewater phycoremediation and environmental benefits using *Scenedesmus obliquus* microalgae. Journal of Environmental Management 240, 293–302. Available from: https://doi.org/10.1016/j.jenvman.2019.03.123. Elsevier BV.

Apt, K.E., Behrens, P.W., 1999. Commercial developments in microalgal biotechnology. Journal of Phycology 35 (2), 215–226. Available from: https://doi.org/10.1046/j.1529-8817.1999.3520215.x. Wiley.

Ariede, M.B., et al., 2017. Cosmetic attributes of algae − a review. Algal Research 25, 483–487. Available from: https://doi.org/10.1016/j.algal.2017.05.019. Elsevier BV.

Austic, R.E., et al., 2013. Potential and limitation of a new defatted diatom microalgal biomass in replacing soybean meal and corn in diets for broiler chickens. Journal of Agricultural and Food Chemistry 61 (30), 7341–7348. Available from: https://doi.org/10.1021/jf401957z. American Chemical Society (ACS).

Bacellar Mendes, L., Vermelho, A., 2013. Allelopathy as a potential strategy to improve microalgae cultivation. Biotechnology for Biofuels 6 (1), 152. Available from: https://doi.org/10.1186/1754-6834-6-152. Springer Nature.

Ball, S.G., 2005. Eukaryotic microalgae genomics. The essence of being a plant. Plant Physiology 137 (2), 397–398. Available from: https://doi.org/10.1104/pp.104.900136. Oxford University Press (OUP).

Banker, R., Carmeli, S., 1998. Tenuecyclamides A − D, cyclic hexapeptides from the Cyanobacterium Nostoc spongiaeforme var. t enue. Journal of Natural Products 61 (10), 1248–1251. Available from: https://doi.org/10.1021/np980138j. American Chemical Society (ACS).

Baňoch, T., et al., 2012. The effect of iodine from iodine-enriched alga *Chlorella* spp. on the pork iodine content and meat quality in finisher pigs. Acta Veterinaria Brno 81 (4), 339–346. Available from: https://doi.org/10.2754/avb201281040339. University of Veterinary and Pharmaceutical Sciences.

Barkia, I., et al., 2019. Indigenous marine diatoms as novel sources of bioactive peptides with antihypertensive and antioxidant properties. International Journal of Food Science & Technology 54 (5), 1514–1522. Available from: https://doi.org/10.1111/ijfs.14006. Wiley.

Barone, G.D., et al., 2020. Hints at the applicability of microalgae and *Cyanobacteria* for the biodegradation of plastics. Sustainability 12 (24), 10449. Available from: https://doi.org/10.3390/su122410449. MDPI AG.

Barrow, C., Shahidi, F. 2007. Marine Nutraceuticals and Functional Foods. CRC Press. https://doi.org/10.1201/9781420015812.

Basmadjian, C., et al., 2014. Cancer wars: natural products strike back. Frontiers in Chemistry 2 (MAY). Available from: https://doi.org/10.3389/fchem.2014.00020Frontiers Media SA.

Baudelet, P.-H., et al., 2013. Antiproliferative activity of *Cyanophora paradoxa* pigments in melanoma, breast and lung cancer cells. Marine Drugs 11 (11), 4390–4406. Available from: https://doi.org/10.3390/md11114390. MDPI AG.

Balia Yusof, Z.N., Azim, N., Subki., A., 2018. Abiotic Stresses Induce Total Phenolic, Total Flavonoid and Antioxidant Properties in Malaysian Indigenous Microalgae and Cyanobacterium. Malaysian Journal of Microbiology 14 (March). Available from: https://doi.org/10.21161/mjm.100317.

Bayramoğlu, G., et al., 2006. Biosorption of mercury(II), cadmium(II) and lead(II) ions from aqueous system by microalgae *Chlamydomonas reinhardtii* immobilized in alginate beads. International Journal of Mineral Processing 81 (1), 35–43. Available from: https://doi.org/10.1016/j.minpro.2006.06.002. Elsevier BV.

Bechelli, J., et al., 2011. Cytotoxicity of algae extracts on normal and malignant cells. Leukemia Research and Treatment 1–7. Available from: https://doi.org/10.4061/2011/373519. Hindawi Limited.

Becker, E.W., 2013. Microalgae for Human and Animal Nutrition. Wiley, pp. 461–503. Available from: https://doi.org/10.1002/9781118567166.ch25.

Beesoo, R., et al., 2014. Apoptosis inducing lead compounds isolated from marine organisms of potential relevance in cancer treatment. Mutation Research/Fundamental and Molecular Mechanisms of Mutagenesis 768 (C), 84–97. Available from: https://doi.org/10.1016/j.mrfmmm.2014.03.005. Elsevier BV.

Beheshtipour, H., et al., 2013. Supplementation of *Spirulina platensis* and chlorella vulgaris algae into probiotic fermented milks. Comprehensive Reviews in Food Science and Food Safety 12 (2), 144–154. Available from: https://doi.org/10.1111/1541-4337.12004. Iran.

Ben-Amotz, A., Avron, M. 1992. Dunaliella: Physiology, Biochemistry, and Biotechnology. CRC Press, Inc., 2000 Corporate Blvd., N.W., Boca Raton, Florida, 33431.

Bermúdez, S., Ogawa, M.R., Rittmannb, B., Parra, R., 2014. Algae Biofuels Production Processes, Carbon Dioxide Fixation and Biorefinery Concept. Journal of Petroleum & Environmental Biotechnology 5 (January), 2157–7463. Available from: https://doi.org/10.4172/2157-7463.1000185.

Besseling, E., et al., 2014. Nanoplastic affects growth of *S. obliquus* and reproduction of *D. magna*. Environmental Science & Technology 48 (20), 12336–12343. Available from: https://doi.org/10.1021/es503001d. American Chemical Society (ACS).

Bhattacharya, P., et al., 2010. Physical adsorption of charged plastic nanoparticles affects algal photosynthesis. The Journal of Physical Chemistry C 114 (39), 16556–16561. Available from: https://doi.org/10.1021/jp1054759. American Chemical Society (ACS).

Bhuyar, P., Muniyasamy, S., Govindan, N., 2019. Biodegradation of Plastic Waste by Using Microalgae and Their Toxins. 21st European Biotechnology Congress. Available from: https://doi.org/10.4172/2155-952X-C5-101.

Bonos, E., et al., 2016. Spirulina as a functional ingredient in broiler chicken diets. South African Journal of Animal Science 94. Available from: https://doi.org/10.4314/sajas.v46i1.12. African Journals Online (AJOL).

Boo, H.-J., et al., 2011. Fucoidan from *Undaria pinnatifida* induces apoptosis in A549 human lung carcinoma cells. Phytotherapy Research 25 (7), 1082–1086. Available from: https://doi.org/10.1002/ptr.3489. Wiley.

Bouabidi, Z.B., El-Naas, M.H., Zhang, Z., 2019. Immobilization of microbial cells for the biotreatment of wastewater: a review. Environmental Chemistry Letters 17 (1), 241–257. Available from: https://doi.org/10.1007/s10311-018-0795-7. Springer Science and Business Media LLC.

Brennan, L., Owende, P., 2010. Biofuels from microalgae—a review of technologies for production, processing, and extractions of biofuels and co-products. Renewable and Sustainable Energy Reviews 14 (2), 557–577. Available from: https://doi.org/10.1016/j.rser.2009.10.009. Elsevier BV.

Bruton, T. et al., 2009. Report-A Review of the Potential of Marine Algae as a Source of Biofuel in Ireland. Sustainable Energy Ireland.

Bule, M.H., et al., 2018. Microalgae as a source of high-value bioactive compounds. Frontiers in Bioscience – Scholar 10 (2), 197–216. Available from: https://doi.org/10.2741/s509. Ethiopia: Frontiers in Bioscience.

Cai, T., Park, S.Y., Li, Y., 2013. Nutrient recovery from wastewater streams by microalgae: status and prospects. Renewable and Sustainable Energy Reviews 19, 360–369. Available from: https://doi.org/10.1016/j.rser.2012.11.030. Elsevier BV.

Cannell, R.J.P., 1993. Algae as a source of biologically active products. Pesticide Science 39 (2), 147–153. Available from: https://doi.org/10.1002/ps.2780390208. Wiley.

Capelli, B., Cysewski, G.R., 2010. Potential health benefits of spirulina microalgae*. Nutrafoods 19–26. Available from: https://doi.org/10.1007/bf03223332. Springer Science and Business Media LLC.

Caporgno, M.P., Mathys, A., 2018. Trends in microalgae incorporation into innovative food products with potential health benefits. Frontiers in Nutrition 5. Available from: https://doi.org/10.3389/fnut.2018.00058. Frontiers Media SA.

Cardoso, C., Afonso, C., Bandarra, N.M., 2016. Dietary DHA and health: cognitive function ageing. Nutrition Research Reviews 29 (2), 281–294. Available from: https://doi.org/10.1017/s0954422416000184. Cambridge University Press (CUP).

Carmichael, W.W., 1992. Cyanobacteria secondary metabolites—the cyanotoxins. Journal of Applied Bacteriology 72 (6), 445–459. Available from: https://doi.org/10.1111/j.1365-2672.1992.tb01858.x. United States.

Carté, B.K., 1996. Biomedical potential of marine natural products. Bioscience 46 (4), 271–286. Available from: https://doi.org/10.2307/1312834. United States: American Institute of Biological Sciences.

Cha, K.H., Koo, S.Y., Lee, D.-U., 2008. Antiproliferative effects of carotenoids extracted from *Chlorella ellipsoidea* and *Chlorella vulgaris* on human colon cancer cells. Journal of Agricultural and Food Chemistry 56 (22), 10521–10526. Available from: https://doi.org/10.1021/jf802111x. American Chemical Society (ACS).

Chen, X., Song, L., Wang, H., Liu, S., Yu, H., Wang, X., Li, R., Liu, T., Li, P., 2019. Partial Characterization, the Immune Modulation and Anticancer Activities of Sulfated Polysaccharides from Filamentous Microalgae Tribonema Sp. Molecules (Basel, Switzerland) 24 (2). Available from: https://doi.org/10.3390/molecules24020322.

Chen, H., et al., 2017. Sargassum fusiforme polysaccharides inhibit VEGF-A-related angiogenesis and proliferation of lung cancer in vitro and in vivo. Biomedicine & Pharmacotherapy 85, 22–27. Available from: https://doi.org/10.1016/j.biopha.2016.11.131. Elsevier BV.

Chen, X., Yan, N., 2020. A brief overview of renewable plastics. Materials Today Sustainability 7–8, 100031. Available from: https://doi.org/10.1016/j.mtsust.2019.100031. Elsevier BV.

Chia, W.Y., et al., 2020. Nature's fight against plastic pollution: algae for plastic biodegradation and bioplastics production. Environmental Science and Ecotechnology 4, 100065. Available from: https://doi.org/10.1016/j.ese.2020.100065. Elsevier BV.

Chisti, Y., 2007. Biodiesel from microalgae. Biotechnology Advances 25 (3), 294–306. Available from: https://doi.org/10.1016/j.biotechadv.2007.02.001. New Zealand.

Chiu, S.-Y., et al., 2008. Reduction of CO_2 by a high-density culture of *Chlorella* sp. in a semicontinuous photobioreactor. Bioresource Technology 99 (9), 3389–3396. Available from: https://doi.org/10.1016/j.biortech.2007.08.013. Elsevier BV.

Chiu, S.-Y., et al., 2015. Cultivation of microalgal *Chlorella* for biomass and lipid production using wastewater as nutrient resource. Bioresource Technology 184, 179–189. Available from: https://doi.org/10.1016/j.biortech.2014.11.080. Elsevier BV.

Choi, Y., et al., 2017. Tuberatolide B suppresses cancer progression by promoting ROS-mediated inhibition of STAT3 signaling. Marine Drugs 15 (3), 55. Available from: https://doi.org/10.3390/md15030055. MDPI AG.

Chojnacka, K., Chojnacki, A., Górecka, H., 2004. Trace element removal by *Spirulina* sp. from copper smelter and refinery effluents. Hydrometallurgy 73 (1–2), 147–153. Available from: https://doi.org/10.1016/j.hydromet.2003.10.003. Poland.

Chowdhury, H., Loganathan, B., 2019. Third-generation biofuels from microalgae: a review. Current Opinion in Green and Sustainable Chemistry 20, 39–44. Available from: https://doi.org/10.1016/j.cogsc.2019.09.003. Elsevier BV.

Chowdhury, S., et al., 2016. Heavy metals in drinking water: occurrences, implications, and future needs in developing countries. Science of the Total Environment 569–570, 476–488. Available from: https://doi.org/10.1016/j.scitotenv.2016.06.166. Elsevier BV.

Christaki, E., Bonos, E., Florou-Paneri, P., 2015. Innovative microalgae pigments as a functional ingredients in nutrition. In: Handbook of marine microalgae. Biotehnology Advances. Elsevier BV, Academic Press, pp. 233–243. https://doi.org/10.1016/b978-0-12-800776-1.00014-5.

Christenson, L., Sims, R., 2011. Production and harvesting of microalgae for wastewater treatment, biofuels, and bioproducts. Biotechnology Advances 29 (6), 686–702. Available from: https://doi.org/10.1016/j.biotechadv.2011.05.015. Elsevier BV.

Chu, W.L., et al., 2002. Influence of irradiance and inoculum density on the pigmentation of *Spairulina platensis*. Asia-Pacific Journal of Molecular Biology and Biotechnology 10 (2), 109–117Malaysia: University of Malaya. Available from: http://www.msmbb.org.my/apjhome.htm.

Cohen, Z., Cohen, S., 1991. Preparation of eicosapentaenoic acid (EPA) concentrate from *Porphyridium cruentum*. Journal of the American Oil Chemists' Society 68 (1), 16–19. Available from: https://doi.org/10.1007/bf02660301. Wiley.

Comitre, A.A., et al., 2021. Renewal of nanofibers in *Chlorella fusca* microalgae cultivation to increase CO_2 fixation. Bioresource Technology 321, 124452. Available from: https://doi.org/10.1016/j.biortech.2020.124452. Elsevier BV.

Cordero, B.F., et al., 2012. Isolation and characterization of a lycopene ε-cyclase gene of *Chlorella* (*Chromochloris*) zofingiensis. Regulation of the carotenogenic pathway by nitrogen and light. Marine Drugs 10 (9), 2069–2088. Available from: https://doi.org/10.3390/md10092069. MDPI AG.

Cuaresma, M., et al., 2011. Productivity and selective accumulation of carotenoids of the novel extremophile microalga *Chlamydomonas acidophila* grown with different carbon sources in batch systems. Journal of Industrial Microbiology & Biotechnology 38 (1), 167–177. Available from: https://doi.org/10.1007/s10295-010-0841-3. Oxford University Press (OUP).

D'Alessandro, E.B., Antoniosi Filho, N.R., 2016. Concepts and studies on lipid and pigments of microalgae: a review. Renewable and Sustainable Energy Reviews 58, 832–841. Available from: https://doi.org/10.1016/j.rser.2015.12.162. Elsevier BV.

Daeseleire, E., et al., 2017. Veterinary drug residues in foods, Chemical Contaminants and Residues in Food, Second Edition Elsevier Inc, Belgium, pp. 117–153. Available from: https://doi.org/10.1016/B978-0-08-100674-0.00006-0.

Davidson, B.S., 1995. New dimensions in natural products research: cultured marine microorganisms. Current Opinion in Biotechnology 6 (3), 284–291. Available from: https://doi.org/10.1016/0958-1669(95)80049-2. United States.

de Carvalho, C.C.C.R., Caramujo, M.J., 2017. Carotenoids in aquatic ecosystems and aquaculture: a colorful business with implications for human health. Frontiers in Marine Science 4. Available from: https://doi.org/10.3389/fmars.2017.00093. Frontiers Media SA.

De la Fuente-Vazquez, J., et al., 2014. Linseed, microalgae or fish oil dietary supplementation affects performance and quality characteristics of light lambs. Spanish Journal of Agricultural Research 12 (2), 436. Available from: https://doi.org/10.5424/sjar/2014122-4639. Instituto Nacional de Investigacion y Tecnologia Agraria y Alimentaria (INIA).

de Morais, M.G., Costa, J.A.V., 2007. Biofixation of carbon dioxide by *Spirulina* sp. and *Scenedesmus obliquus* cultivated in a three-stage serial tubular photobioreactor. Journal of Biotechnology 129 (3), 439–445. Available from: https://doi.org/10.1016/j.jbiotec.2007.01.009. Elsevier BV.

de Stephanis, R., et al., 2013. As main meal for sperm whales: plastics debris. Marine Pollution Bulletin 69 (1–2), 206–214. Available from: https://doi.org/10.1016/j.marpolbul.2013.01.033. Elsevier BV.

de-Bashan, L.E., et al., 2008. *Chlorella sorokiniana* UTEX 2805, a heat and intense, sunlight-tolerant microalga with potential for removing ammonium from wastewater. Bioresource Technology 99 (11), 4980–4989. Available from: https://doi.org/10.1016/j.biortech.2007.09.065. Elsevier BV.

Demirel, Z., et al., 2018. Influence of media and temperature on the growth and the biological activities of Desmodesmus protuberans (F.E. Fritsch & M.F. Rich) E. Hegewald. Turkish Journal of Fisheries and Aquatic Sciences 18 (10), 1195–1203. Available from: https://doi.org/10.4194/1303-2712-v18_10_06. Turkey: Central Fisheries Research Inst.

Devi, M.A., et al., 1981. Studies on the proteins of mass-cultivated, blue-green alga (*Spirulina platensis*). Journal of Agricultural and Food Chemistry 29 (3), 522–525. Available from: https://doi.org/10.1021/jf00105a022. American Chemical Society (ACS).

Diaz, M.T., Pérez, C., Sánchez González, C.I., Lauzurica, S., Cañeque, V., González, C., de la Fuente, J., 2017. Feeding microalgae increases omega 3 fatty acids of fat deposits and muscles in light lambs. Journal of Food Composition and Analysis 56 (March), 115–123. Available from: https://doi.org/10.1016/j.jfca.2016.12.009.

Dlouhá, G., et al., 2008. Effect of dietary selenium sources on growth performance, breast muscle selenium, glutathione peroxidase activity and oxidative stability in broilers. Czech Journal of Animal Science 53 (6), 265–269. Available from: https://doi.org/10.17221/361-cjas. Czech Republic: Institute of Agricultural and Food Information.

Do, J.-M., et al., 2019. A feasibility study of wastewater treatment using domestic microalgae and analysis of biomass for potential applications. Water 11 (11), 2294. Available from: https://doi.org/10.3390/w11112294. MDPI AG.

Dolganyuk, V., et al., 2020. Microalgae: a promising source of valuable bioproducts. Biomolecules 10 (8), 1153. Available from: https://doi.org/10.3390/biom10081153. MDPI AG.

Dragone, G., et al., 2010. Third generation biofuels from microalgaein Current Research 1355–1366.

Dutta, K., Daverey, A., Lin, J.-G., 2014. Evolution retrospective for alternative fuels: first to fourth generation. Renewable Energy 69, 114–122. Available from: https://doi.org/10.1016/j.renene.2014.02.044. Elsevier BV.

Dzhambazov, B., et al., 2006. Vitro cytotoxicity and anticancer properties of two Phormidium molle strains (cyanoprokaryota). Travaux Scientifiques Universite de Plovdiv – Biologie 39 (6), 16.

Ebenezer, V., Medlin, L.K., Ki, J.-S., 2012. Molecular detection, quantification, and diversity evaluation of microalgae. Marine Biotechnology 14 (2), 129–142. Available from: https://doi.org/10.1007/s10126-011-9427-y. Springer Science and Business Media LLC.

Ebrahimi Nigjeh, S., et al., 2013. Cytotoxic effect of ethanol extract of microalga, *Chaetoceros calcitrans*, and its mechanisms in inducing apoptosis in human breast cancer cell line. BioMed Research International 2013, 1–8. Available from: https://doi.org/10.1155/2013/783690. Hindawi Limited.

Egli, T., 2015. Microbial growth and physiology: a call for better craftsmanship. Frontiers in Microbiology 6. Available from: https://doi.org/10.3389/fmicb.2015.00287. Frontiers Media SA.

El-Chaghaby, G.A., et al., 2019. Assessment of phytochemical components, proximate composition and antioxidant properties of scenedesmus obliquus, chlorella vulgaris and *Spirulina platensis* algae extracts. Egyptian Journal of Aquatic Biology and Fisheries 23 (4), 521–526. Available from: https://doi.org/10.21608/ejabf.2019.57884. Egypt: Egyptian Society for the Development of Fisheries and Human Health.

El-Ghany, W.A.A., 2020. Microalgae in poultry field: a comprehensive perspectives. Advances in Animal and Veterinary Sciences 8 (9), 888–897. Available from: https://doi.org/10.17582/JOURNAL.AAVS/2020/8.9.888.897. Egypt: Nexus Academic Publishers.

EL-Sabagh, M.R., et al., 2014. Effects of *Spirulina Platensis* algae on growth performance, antioxidative status and blood metabolites in fattening lambs. Journal of Agricultural Science. Canadian Center of Science and Education . Available from: https://doi.org/10.5539/jas.v6n3p92.

Enamala, M.K., et al., 2018. Production of biofuels from microalgae — a review on cultivation, harvesting, lipid extraction, and numerous applications of microalgae. Renewable and Sustainable Energy Reviews 94, 49—68. Available from: https://doi.org/10.1016/j.rser.2018.05.012. Elsevier BV.

Englmaierová, M., Skřivan, M., Bubancová, I., 2013. A comparison of lutein, spray-dried *Chlorella*, and synthetic carotenoids effects on yolk colour, oxidative stability, and reproductive performance of laying hens. Czech Journal of Animal Science 58 (9), 412—419. Available from: https://doi.org/10.17221/6941-cjas. Czech Republic: Springer-Verlag.

European Commision, 2001. 2nd Opinion on Anti-Microbial Resistance.

Evans, A.M., Smith, D.L., Moritz, J.S., 2015. Effects of algae incorporation into broiler starter diet formulations on nutrient digestibility and 3 to 21 d bird performance. Journal of Applied Poultry Research 24 (2), 206—214. Available from: https://doi.org/10.3382/japr/pfv027. United States: Oxford University Press.

Fagerstone, K.D., et al., 2011. Quantitative measurement of direct nitrous oxide emissions from microalgae cultivation. Environmental Science & Technology 45 (21), 9449—9456. Available from: https://doi.org/10.1021/es202573f. American Chemical Society (ACS).

Fang, Y.-Z., Yang, S., Wu, G., 2002. Free radicals, antioxidants, and nutrition. Nutrition (Burbank, Los Angeles County, Calif.) 18 (10), 872—879. Available from: https://doi.org/10.1016/s0899-9007(02)00916-4. Elsevier BV.

Faried, M., et al., 2017. Biodiesel production from microalgae: processes, technologies and recent advancements. Renewable and Sustainable Energy Reviews 79, 893—913. Available from: https://doi.org/10.1016/j.rser.2017.05.199. Egypt: Elsevier Ltd.

Farré, G., et al., 2010. Travel advice on the road to carotenoids in plants. Plant Science 179 (1—2), 28—48. Available from: https://doi.org/10.1016/j.plantsci.2010.03.009. Elsevier BV.

Ferreira, L.S., et al., 2011. Adsorption of Ni2 + , Zn2 + and Pb2 + onto dry biomass of *Arthrospira* (*Spirulina*) platensis and *Chlorella vulgaris*. I. Single metal systems. Chemical Engineering Journal 173 (2), 326—333. Available from: https://doi.org/10.1016/j.cej.2011.07.039. Elsevier BV.

Ferrón, S., et al., 2012. Air—water fluxes of N₂O and CH4 during microalgae (*Staurosira* sp.) Cultivation in an open raceway pond. Environmental Science & Technology 46 (19), 10842—10848. Available from: https://doi.org/10.1021/es302396j. American Chemical Society (ACS).

Finkel, T., 2011. Signal transduction by reactive oxygen species. Journal of Cell Biology 194 (1), 7—15. Available from: https://doi.org/10.1083/jcb.201102095. United States.

Franklin, S.T., et al., 1999. Dietary marine algae (*Schizochytrium* sp.) increases concentrations of conjugated linoleic, docosahexaenoic and transvaccenic acids in milk of dairy cows. Journal of Nutrition 129 (11), 2048—2052. Available from: https://doi.org/10.1093/jn/129.11.2048. United States: American Institute of Nutrition.

Furbeyre, H., et al., 2017. Effects of dietary supplementation with freshwater microalgae on growth performance, nutrient digestibility and gut health in weaned piglets. Animal 11 (2), 183—192. Available from: https://doi.org/10.1017/S1751731116001543. France: Cambridge University Press.

Gangl, D., et al., 2015. Biotechnological exploitation of microalgae. Journal of Experimental Botany 66 (22), 6975—6990. Available from: https://doi.org/10.1093/jxb/erv426. Oxford University Press (OUP).

García, J.L., de Vicente, M., Galán, B., 2017. Microalgae, old sustainable food and fashion nutraceuticals. Microbial Biotechnology 10 (5), 1017—1024. Available from: https://doi.org/10.1111/1751-7915.12800. Wiley.

Gerwick, W.H., Moore, B.S., 2012. Lessons from the past and charting the future of marine natural products drug discovery and chemical biology. Chemistry & Biology 19 (1), 85—98. Available from: https://doi.org/10.1016/j.chembiol.2011.12.014. Elsevier BV.

Ginzberg, A., et al., 2000. Chickens fed with biomass of the red microalga *Porphyridium* sp. have reduced blood cholesterol level and modified fatty acid composition in egg yolk. Journal of Applied Phycology. Available from: https://doi.org/10.1023/a:1008102622276. Israel: Springer Netherlands.

Giorgia, M., Ingravalle, F., Radice, E., Aragno, M., Peiretti, P.G., 2009. Effects of High Fat Diets and Spirulina Platensis Supplementation in New Zealand White Rabbits. Journal of Animal and Veterinary Advances 8 (January), 2735—2744. Available from: https://doi.org/10.3923/javaa.2009.2735.2744.

Global Chlorella Market Information: By Origin (Organic, Conventional), ByForm (Powder, Liquid, Tablet Capsules and Others), by Application (Food & Beverages, Pharmaceuticals, Animal Feed, Others) and Region Forecast till 2023. https://www.marketresearchfuture.com/report info.pdf?report id=4413.

Goh, S., et al., 2014. Crude ethyl acetate extract of marine microalga, *Chaetoceros calcitrans*, induces Apoptosis in MDA-MB-231 breast cancer cells. Pharmacognosy Magazine. Malaysia 10 (37), 1–8. Available from: https://doi.org/10.4103/0973-1296.126650.

Gong, Y., et al., 2011. Microalgae as platforms for production of recombinant proteins and valuable compounds: progress and prospects. Journal of Industrial Microbiology & Biotechnology 38 (12), 1879–1890. Available from: https://doi.org/10.1007/s10295-011-1032-6. Oxford University Press (OUP).

González López, C.V., et al., 2009. Utilization of the cyanobacteria Anabaena sp. ATCC 33047 in CO_2 removal processes. Bioresource Technology 100 (23), 5904–5910. Available from: https://doi.org/10.1016/j.biortech.2009.04.070. Spain.

González-Pleiter, M., et al., 2019. Secondary nanoplastics released from a biodegradable microplastic severely impact freshwater environments. Environmental Science 6 (5), 1382–1392. Available from: https://doi.org/10.1039/c8en01427b. Nano. Royal Society of Chemistry (RSC).

Gopal, R., 2017. Biodegradation of Polyethylene by Green Photosynthetic Microalgae. Journal of Bioremediation & Biodegradation 8, (January). Available from: https://doi.org/10.4172/2155-6199.1000381.

Gouveia, L., et al., 2006. Chlorella vulgaris and *Haematococcus pluvialis* biomass as colouring and antioxidant in food emulsions. European Food Research and Technology 222 (3–4), 362–367. Available from: https://doi.org/10.1007/s00217-005-0105-z. Springer Science and Business Media LLC.

Grinstead, G.S., et al., 2000. Effects of *Spirulina platensis* on growth performance of weanling pigs. Animal Feed Science and Technology 83 (3–4), 237–247. Available from: https://doi.org/10.1016/S0377-8401(99)00130-3. United States.

Guedes, É.A.C., et al., 2013. Cytotoxic activity of marine algae against cancerous cells. Revista Brasileira de Farmacognosia 23 (4), 668–673. Available from: https://doi.org/10.1590/s0102-695x2013005000060. Springer Science and Business Media LLC.

Gupta, S.K., et al., 2016. Dual role of *Chlorella sorokiniana* and *Scenedesmus obliquus* for comprehensive wastewater treatment and biomass production for bio-fuels. Journal of Cleaner Production 115, 255–264. Available from: https://doi.org/10.1016/j.jclepro.2015.12.040. Elsevier BV.

Gupta, V.K., Rastogi, A., 2008. Equilibrium and kinetic modelling of cadmium(II) biosorption by nonliving algal biomass Oedogonium sp. from aqueous phase. Journal of Hazardous Materials 153 (1–2), 759–766. Available from: https://doi.org/10.1016/j.jhazmat.2007.09.021. India.

Haiduc, A.G., et al., 2009. SunCHem: an integrated process for the hydrothermal production of methane from microalgae and CO_2 mitigation. Journal of Applied Phycology 21 (5), 529–541. Available from: https://doi.org/10.1007/s10811-009-9403-3. Springer Science and Business Media LLC.

Hamidi, M., et al., 2020. Marine bacteria vs microalgae: who is the best for biotechnological production of bioactive compounds with antioxidant properties and other biological applications? Marine Drugs 18 (1), 28. Available from: https://doi.org/10.3390/md18010028. MDPI AG.

Hamidi, M., Kozani, P.S., Pooria Safarzadeh Kozani, P.S., Pierre, G., Michaud, P., Delattre, C., 2019. Marine bacteria versus microalgae: who is the best for biotechnological production of bioactive compounds with antioxidant properties and other biological applications? Marine Drugs 18 (1). Available from: https://doi.org/10.3390/md18010028.

Han, X., Wong, Y.S., Tam, N.F.Y., 2006. Surface complexation mechanism and modeling in Cr(III) biosorption by a microalgal isolate, *Chlorella miniata*. Journal of Colloid and Interface Science 303 (2), 365–371. Available from: https://doi.org/10.1016/j.jcis.2006.08.028. Elsevier BV.

Hanagata, N., et al., 1992. Tolerance of microalgae to high CO_2 and high temperature. Phytochemistry 31 (10), 3345–3348. Available from: https://doi.org/10.1016/0031-9422(92)83682-o. Elsevier BV.

Harrison, D., et al., 2003. Role of oxidative stress in atherosclerosis. The American Journal of Cardiology 91 (3), 7–11. Available from: https://doi.org/10.1016/s0002-9149(02)03144-2. Elsevier BV.

Hau, A.M., et al., 2013. Coibamide A induces mTOR-independent autophagy and cell death in human glioblastoma cells. PLoS One 8 (6), e65250. Available from: https://doi.org/10.1371/journal.pone.0065250. Public Library of Science (PLoS).

Heasman, M., et al., 2000. Development of extended shelf-life microalgae concentrate diets harvested by centrifugation for bivalve molluscs – a summary. Aquaculture Research 31 (8–9), 637–659. Available from: https://doi.org/10.1046/j.1365-2109.2000.318492.x. Wiley.

Hemalatha, A., et al., 2015. Evaluation of antioxidant activities and total phenolic contents of different solvent extracts of selected marine diatoms. Indian Journal of Geo-Marine Sciences 44 (10), 1630–1636India: National Institute of Science Communication and Information Resources (NISCAIR). Available from: http://nopr.niscair.res.in/bitstream/123456789/34993/1/IJMS%2044%2810%29%201630-1636.pdf.

Henkanatte-Gedera, S.M., et al., 2017. Removal of dissolved organic carbon and nutrients from urban wastewaters by *Galdieria sulphuraria*: laboratory to field scale demonstration. Algal Research 24, 450–456. Available from: https://doi.org/10.1016/j.algal.2016.08.001. Elsevier BV.

Hensley, K., et al., 2000. Reactive oxygen species. Cell Signaling, and Cell Injury 28 (10), 252–255. Available from: https://doi.org/10.1016/s0891-5849(00. Free Radical Biology & Medicine.

Higuera-Ciapara, I., Félix-Valenzuela, L., Goycoolea, F.M., 2006. Astaxanthin: a review of its chemistry and applications. Critical Reviews in Food Science and Nutrition 46 (2), 185–196. Available from: https://doi.org/10.1080/10408690590957188. Mexico.

Ho, S.-H., Chen, W.-M., Chang, J.-S., 2010. Scenedesmus obliquus CNW-N as a potential candidate for CO2 mitigation and biodiesel production. Bioresource Technology 101 (22), 8725–8730. Available from: https://doi.org/10.1016/j.biortech.2010.06.112. Elsevier BV.

Hong, J., Luesch, H., 2012. Largazole: from discovery to broad-spectrum therapy. Natural Product Reports 29 (4), 449. Available from: https://doi.org/10.1039/c2np00066k. Royal Society of Chemistry (RSC).

Hong, M.E., et al., 2019. Microalgal-based carbon sequestration by converting LNG-fired waste CO_2 into red gold astaxanthin: the potential applicability. Energies 12 (9), 1718. Available from: https://doi.org/10.3390/en12091718. MDPI AG.

Hopkins, D.L., et al., 2014. The impact of supplementing lambs with algae on growth, meat traits and oxidative status. Meat Science 98 (2), 135–141. Available from: https://doi.org/10.1016/j.meatsci.2014.05.016. Australia: Elsevier Ltd.

Hoskisson, P.A., Hobbs, G., 2005. Continuous culture – making a comeback? Microbiology. Microbiology Society 151 (10), 3153–3159. Available from: https://doi.org/10.1099/mic.0.27924-0.

Hsueh, H.T., et al., 2009. Carbon bio-fixation by photosynthesis of *Thermosynechococcus* sp. CL-1 and Nannochloropsis oculta. Journal of Photochemistry and Photobiology B: Biology 95 (1), 33–39. Available from: https://doi.org/10.1016/j.jphotobiol.2008.11.010. Taiwan.

Hu, Q., et al., 1998. Ultrahigh-cell-density culture of a marine green alga Chlorococcum littorale in a flat-plate photobioreactor. Applied Microbiology and Biotechnology 49 (6), 655–662. Available from: https://doi.org/10.1007/s002530051228. Japan.

Hu, J., et al., 2018. Heterotrophic cultivation of microalgae for pigment production: a review. Biotechnology Advances 36 (1), 54–67. Available from: https://doi.org/10.1016/j.biotechadv.2017.09.009. Elsevier BV.

Huang, J.-C., Chen, F., Sandmann, G., 2006. Stress-related differential expression of multiple β-carotene ketolase genes in the unicellular green alga *Haematococcus pluvialis*. Journal of Biotechnology 122 (2), 176–185. Available from: https://doi.org/10.1016/j.jbiotec.2005.09.002. Elsevier BV.

Huang, T.-H., et al., 2015. Prophylactic administration of fucoidan represses cancer metastasis by inhibiting vascular endothelial growth factor (VEGF) and matrix metalloproteinases (MMPs) in Lewis tumor-bearing mice. Marine Drugs 13 (4), 1882–1900. Available from: https://doi.org/10.3390/md13041882. MDPI AG.

Huntley, M.E., Redalje, D.G., 2007. CO_2 mitigation and renewable oil from photosynthetic microbes: a new appraisal. Mitigation and Adaptation Strategies for Global Change 12 (4), 573–608. Available from: https://doi.org/10.1007/s11027-006-7304-1. Springer Science and Business Media LLC.

Hynstova, V., et al., 2018. Separation, identification and quantification of carotenoids and chlorophylls in dietary supplements containing *Chlorella vulgaris* and *Spirulina platensis* using high performance thin layer chromatography. Journal of Pharmaceutical and Biomedical Analysis 148, 108–118. Available from: https://doi.org/10.1016/j.jpba.2017.09.018. Elsevier BV.

Jacob-Lopes, E., et al., 2009. Development of operational strategies to remove carbon dioxide in photobioreactors. Chemical Engineering Journal 153 (1–3), 120–126. Available from: https://doi.org/10.1016/j.cej.2009.06.025. Elsevier BV.

Jacob-Lopes, E., Cacia Ferreira Lacerda, L.M., Franco, T.T., 2008. Biomass production and carbon dioxide fixation by Aphanothece microscopica Nägeli in a bubble column photobioreactor. Biochemical Engineering Journal 40 (1), 27–34. Available from: https://doi.org/10.1016/j.bej.2007.11.013. Elsevier BV.

Jacob-Lopes, E., et al., 2019. Bioactive food compounds from microalgae: an innovative framework on industrial biorefineries. Current Opinion in Food Science 25, 1–7. Available from: https://doi.org/10.1016/j.cofs.2018.12.003. Elsevier BV.

Jafari, S.M.A., et al., 2014. Effect of dietary *Spirulina platensis* on fatty acid composition of rainbow trout (*Oncorhynchus mykiss*) fillet. Aquaculture International 22 (4), 1307–1315. Available from: https://doi.org/10.1007/s10499-013-9748-0. Springer Science and Business Media LLC.

Janoska, A., Andriopoulos, V., et al., 2018a. Potential of a liquid foam-bed photobioreactor for microalgae cultivation. Algal Research 36, 193–208. Available from: https://doi.org/10.1016/j.algal.2018.09.029. Elsevier BV.

Janoska, A., Barten, R., et al., 2018b. Improved liquid foam-bed photobioreactor design for microalgae cultivation. Algal Research 33, 55–70. Available from: https://doi.org/10.1016/j.algal.2018.04.025. Elsevier BV.

Jerez-Martel, I., et al., 2017. Phenolic profile and antioxidant activity of crude extracts from microalgae and cyanobacteria strains. Journal of Food Quality 2017, 1–8. Available from: https://doi.org/10.1155/2017/2924508. Hindawi Limited.

Joshi, S., Kumari, R., Upasani, V., 2018. Applications of algae in cosmetics: an overview. International Journal of Innovative in Science, Engineering and Technology 7, 1269–1278. Available from: https://doi.org/10.15680/IJIRSET.2018.0702038.

Kanagaraju, P., Omprakash, A.V., 2016. Effect of *Spirulina platensis* algae powder supplementation as a feed additive on the growth performance of Japanese quails. Indian Veterinary Journal 93 (10), 31–33. India: Indian Veterinary Assocaition. Available from: http://ivj.org.in/downloads/341710pg%2031-33.pdf.

Kang, H.K., et al., 2013. Effect of various forms of dietary *Chlorella* supplementation on growth performance, immune characteristics, and intestinal microflora population of broiler chickens. Journal of Applied Poultry Research 22 (1), 100–108. Available from: https://doi.org/10.3382/japr.2012-00622. South Korea: Oxford University Press.

Kang, Y., et al., 2017. Characterization and Potential antitumor activity of polysaccharide from *Gracilariopsis lemaneiformis*. Marine Drugs 15 (4), 100. Available from: https://doi.org/10.3390/md15040100. MDPI AG.

Kerbrat, P., et al., 2003. Phase II study of LU 103793 (dolastatin analogue) in patients with metastatic breast cancer. European Journal of Cancer 39 (3), 317–320. Available from: https://doi.org/10.1016/S0959-8049(02)00531-2. France.

Khalifa, S.A.M., et al., 2019. Marine natural products: a source of novel anticancer drugs. Marine Drugs 17 (9), 491. Available from: https://doi.org/10.3390/md17090491. MDPI AG.

Khanra, S., et al., 2018. Downstream processing of microalgae for pigments, protein and carbohydrate in industrial application: a review. Food and Bioproducts Processing 110, 60–84. Available from: https://doi.org/10.1016/j.fbp.2018.02.002. Elsevier BV.

Khoo, K.S., et al., 2020. Nanomaterials utilization in biomass for biofuel and bioenergy production. Energies 13 (4), 892. Available from: https://doi.org/10.3390/en13040892. MDPI AG.

Kim, H.H., et al., 2005. Eicosapentaenoic acid inhibits UV-induced MMP-1 expression in human dermal fibroblasts. Journal of Lipid Research 46 (8), 1712–1720. Available from: https://doi.org/10.1194/jlr.m500105-jlr200. Elsevier BV.

Kim, Y.-S., et al., 2014. Stigmasterol isolated from marine microalgae *Navicula* incerta induces apoptosis in human hepatoma HepG2 cells. BMB Reports. Korean Society for Biochemistry and Molecular Biology - BMB Reports 47 (8), 433–438. Available from: https://doi.org/10.5483/bmbrep.2014.47.8.153.

Kobayashi, N., et al., 2013. Characterization of three *Chlorella sorokiniana* strains in anaerobic digested effluent from cattle manure. Bioresource Technology 150, 377–386. Available from: https://doi.org/10.1016/j.biortech.2013.10.032. Elsevier BV.

Koelmans, A.A., et al., 2019. Microplastics in freshwaters and drinking water: critical review and assessment of data quality. Water Research 155, 410–422. Available from: https://doi.org/10.1016/j.watres.2019.02.054. Elsevier BV.

Koyande, A.K., et al., 2019. Microalgae: a potential alternative to health supplementation for humans. Food Science and Human Wellness 8 (1), 16–24. Available from: https://doi.org/10.1016/j.fshw.2019.03.001. Elsevier BV.

Kulpys, J., et al., 2009. Influence of *Cyanobacteria Arthrospira* (*Spirulina*) platemis biomass additives towards the body condition of lactation cows and biochemical milk indexes. Agronomy Research 7, 823–835.

Kumar, A., et al., 2010. Enhanced CO_2 fixation and biofuel production via microalgae: recent developments and future directions. Trends in Biotechnology 28 (7), 371–380. Available from: https://doi.org/10.1016/j.tibtech.2010.04.004. Elsevier BV.

Kusaykin, M., Ermakova, S., Shevchenko, N., Isakov, V., Gorshkov, A., Vereshchagin, A., Grachev, M., Zvyagintseva, T., 2010. Structural Characteristics and Antitumor Activity of a New Chrysolaminaran from the Diatom Alga Synedra Acus. Chemistry of Natural Compounds - CHEM NAT COMPD 46 (March), 1–4. Available from: https://doi.org/10.1007/s10600-010-9510-z.

Lackner, K.S., 2003. A guide to CO_2 sequestration. Science (New York, N.Y.) 300 (5626), 1677–1678. Available from: https://doi.org/10.1126/science.1079033. United States.

Lagarde, F., et al., 2016. Microplastic interactions with freshwater microalgae: hetero-aggregation and changes in plastic density appear strongly dependent on polymer type. Environmental Pollution 215, 331–339. Available from: https://doi.org/10.1016/j.envpol.2016.05.006. Elsevier BV.

Lauritano, C., et al., 2016. Bioactivity screening of microalgae for antioxidant, anti-inflammatory, anticancer, anti-diabetes, and antibacterial activities. Frontiers in Marine Science 3 (MAY). Available from: https://doi.org/10.3389/fmars.2016.00068. Frontiers Media SA.

Li, B., et al., 2006. Molecular immune mechanism of C-phycocyanin from *Spirulina platensis* induces apoptosis in HeLa cells in vitro. Biotechnology and Applied Biochemistry 43 (3), 155. Available from: https://doi.org/10.1042/ba20050142. Wiley.

Li, Y., et al., 2008. Biofuels from microalgae. Biotechnology Progress. Available from: https://doi.org/10.1021/bp.070371k. Canada.

Liang, S., et al., 2004. Current microalgal health food R&D activities in China. Hydrobiologia. Available from: https://doi.org/10.1023/B:HYDR.0000020366.65760.98. China.

Lim, H.C., Shin, H., 2013. Fed-Batch Cultures: Principles and Applications of Semi-Batch Bioreactors. In: Cambridge Series in Chemical Engineering, Cambridge University Press. Available from: https://doi.org/10.1017/CBO9781139018777.

Lin, P.-Y., et al., 2017. Chlorella sorokiniana induces mitochondrial-mediated apoptosis in human non-small cell lung cancer cells and inhibits xenograft tumor growth in vivo. BMC Complementary and Alternative Medicine 17 (1). Available from: https://doi.org/10.1186/s12906-017-1611-9. Springer Nature.

Lizzul, A.M., et al., 2014. Combined remediation and lipid production using *Chlorella sorokiniana* grown on wastewater and exhaust gases. Bioresource Technology 151, 12–18. Available from: https://doi.org/10.1016/j.biortech.2013.10.040. United Kingdom: Elsevier Ltd.

Lu, Y., et al., 2014. Antagonistic roles of abscisic acid and cytokinin during response to nitrogen depletion in oleaginous microalga *Nannochloropsis oceanica* expand the evolutionary breadth of phytohormone function. The Plant Journal 80 (1), 52–68. Available from: https://doi.org/10.1111/tpj.12615. Wiley.

Madeira, M.S., et al., 2017. Microalgae as feed ingredients for livestock production and meat quality: a review. Livestock Science 205, 111–121. Available from: https://doi.org/10.1016/j.livsci.2017.09.020. Elsevier BV.

Manyi-Loh, C., et al., 2018. Antibiotic use in agriculture and its consequential resistance in environmental sources: potential public health implications. Molecules (Basel, Switzerland) 23 (4), 795. Available from: https://doi.org/10.3390/molecules23040795. MDPI AG.

Martelli, F., et al., 2020. *Arthrospira platensis* as natural fermentation booster for milk and soy fermented beverages. Foods 9 (3), 350. Available from: https://doi.org/10.3390/foods9030350. MDPI AG.

Martínez Andrade, K., et al., 2018. Marine microalgae with anti-cancer properties. Marine Drugs 16 (5), 165. Available from: https://doi.org/10.3390/md16050165. MDPI AG.

Martínez, K.A., et al., 2019. Amphidinol 22, a new cytotoxic and antifungal amphidinol from the dinoflagellate *Amphidinium carterae*. Marine Drugs 17 (7), 385. Available from: https://doi.org/10.3390/md17070385. MDPI AG.

Mata, T.M., Martins, A.A., Caetano, N.S., 2010. Microalgae for biodiesel production and other applications: a review. Renewable and Sustainable Energy Reviews 14 (1), 217–232. Available from: https://doi.org/10.1016/j.rser.2009.07.020. Elsevier BV.

Matos, J., et al., 2017. Microalgae as healthy ingredients for functional food: a review. Food and Function 8 (8), 2672–2685. Available from: https://doi.org/10.1039/c7fo00409e. Portugal: Royal Society of Chemistry.

Mazokopakis, E.E., et al., 2014. The hypolipidaemic effects of *Spirulina* (*Arthrospira platensis*) supplementation in a Cretan population: a prospective study. Journal of the Science of Food and Agriculture 94 (3), 432–437. Available from: https://doi.org/10.1002/jsfa.6261. Wiley.

Meale, S.J., et al., 2014. Dose-response of supplementing marine algae (*Schizochytrium* spp.) on production performance, fatty acid profiles, and wool parameters of growing lambs. Journal of Animal Science 92 (5), 2202–2213. Available from: https://doi.org/10.2527/jas.2013-7024. Australia: American Society of Animal Science.

Medina, R.A., et al., 2008. Coibamide A, a potent antiproliferative cyclic depsipeptide from the panamanian marine *Cyanobacterium Leptolyngbya* sp. Journal of the American Chemical Society 130 (20), 6324–6325. Available from: https://doi.org/10.1021/ja801383f. American Chemical Society (ACS).

Medina-Gundrum, L., et al., 2003. AMD473 (ZD0473) exhibits marked in vitro anticancer activity in human tumor specimens taken directly from patients. Anti-Cancer Drugs. Ovid Technologies (Wolters Kluwer Health) 14 (4), 275–280. Available from: https://doi.org/10.1097/00001813-200304000-00004.

Menció, A., et al., 2016. Nitrate pollution of groundwater; all right…, but nothing else? Science of The Total Environment 539, 241–251. Available from: https://doi.org/10.1016/j.scitotenv.2015.08.151. Elsevier BV.

Miralto, A., et al., 1999. The insidious effect of diatoms on copepod reproduction. Nature 402 (6758), 173–176. Available from: https://doi.org/10.1038/46023. Italy.

Mitra, M., et al., 2016. Cultivation of *Nannochloropsis oceanica* biomass rich in eicosapentaenoic acid utilizing wastewater as nutrient resource. Bioresource Technology 218, 1178–1186. Available from: https://doi.org/10.1016/j.biortech.2016.07.083. Elsevier BV.

Mobin, S.M.A., Alam, F., 2018. A review of microalgal biofuels, challenges and future directions. Green Energy and Technology. Springer Verlag, Australia, pp. 83–108. Available from: https://doi.org/10.1007/978-981-10-0697-5_4.

Molazadeh, M., et al., 2019. The use of microalgae for coupling wastewater treatment with CO_2 biofixation. Frontiers in Bioengineering and Biotechnology 7. Available from: https://doi.org/10.3389/fbioe.2019.00042. Frontiers Media SA.

Mondal, A., et al., 2020. Marine *Cyanobacteria* and microalgae metabolites—a rich source of potential anticancer drugs. Marine Drugs 18 (9), 476. Available from: https://doi.org/10.3390/md18090476. MDPI AG.

Montaser, R., Luesch, H., 2011. Marine natural products: a new wave of drugs? Future Medicinal Chemistry 3 (12), 1475–1489. Available from: https://doi.org/10.4155/fmc.11.118. Future Science Ltd.

Moog, D., et al., 2019. Using a marine microalga as a chassis for polyethylene terephthalate (PET) degradation. Microbial Cell Factories 18 (1). Available from: https://doi.org/10.1186/s12934-019-1220-z. Springer Science and Business Media LLC.

Mooney, J.W., et al., 1998. Lipid and flavour quality of stored breast meat from broilers fed marine algae. Journal of the Science of Food and Agriculture 78 (1), 134–140. Available from: https://doi.org/10.1002/(SICI)1097-0010(199809)78:1<134::AID-JSFA96>3.0.CO;2-0.

Moore, R.E., 1996. Cyclic peptides and depsipeptides from cyanobacteria: a review. Journal of Industrial Microbiology 16 (2), 134–143. Available from: https://doi.org/10.1007/BF01570074. United States.

Moraes, L., et al., 2016. Microalgal biotechnology for greenhouse gas control: carbon dioxide fixation by *Spirulina* sp. at different diffusers. Ecological Engineering 91, 426–431. Available from: https://doi.org/10.1016/j.ecoleng.2016.02.035. Elsevier BV.

Mordenti, A.L., et al., 2010. Influence of marine algae (*Schizochytrium* spp.) dietary supplementation on doe performance and progeny meat quality. Livestock Science 128 (1–3), 179–184. Available from: https://doi.org/10.1016/j.livsci.2009.12.003. Italy.

Mostafa, S., et al., 2019. Microalgae growth in effluents from olive oil industry for biomass production and decreasing phenolics content of wastewater. Egyptian Journal of Aquatic Biology and Fisheries 23 (1), 359–365. Available from: https://doi.org/10.21608/ejabf.2019.28265. Egypts Presidential Specialized Council for Education and Scientific Research.

Mourelle, M., Gómez, C., Legido, J., 2017. The potential use of marine microalgae and *Cyanobacteria* in cosmetics and thalassotherapy. Cosmetics 4 (4), 46. Available from: https://doi.org/10.3390/cosmetics4040046. MDPI AG.

Nagappan, S., et al., 2019. Potential of two-stage cultivation in microalgae biofuel production. Fuel 252, 339–349. Available from: https://doi.org/10.1016/j.fuel.2019.04.138. Elsevier BV.

Nappo, M., et al., 2011. Apoptotic activity of the marine diatom *Cocconeis scutellum* and eicosapentaenoic acid in BT20 cells. Pharmaceutical Biology 50 (4), 529–535. Available from: https://doi.org/10.3109/13880209.2011.611811. Informa UK Limited.

Newman, D.J., Cragg, G.M., 2012. Natural products as sources of new drugs over the 30 Years from 1981 to 2010. Journal of Natural Products 75 (3), 311–335. Available from: https://doi.org/10.1021/np200906s. American Chemical Society (ACS).

Newman, D.J., Cragg, G.M., 2014. Marine-sourced anti-cancer and cancer pain control agents in clinical and late preclinical development. Marine Drugs 12 (1), 255–278. Available from: https://doi.org/10.3390/md12010255. United States: MDPI AG.

Nigam, P.S., Singh, A., 2011. Production of liquid biofuels from renewable resources. Progress in Energy and Combustion Science 37 (1), 52–68. Available from: https://doi.org/10.1016/j.pecs.2010.01.003. Elsevier BV.

Nolte, T.M., et al., 2017. The toxicity of plastic nanoparticles to green algae as influenced by surface modification, medium hardness and cellular adsorption. Aquatic Toxicology 183, 11–20. Available from: https://doi.org/10.1016/j.aquatox.2016.12.005. Elsevier BV.

Oftedal, L., et al., 2010. Marine benthic *Cyanobacteria* contain apoptosis-inducing activity synergizing with daunorubicin to kill leukemia cells, but not cardiomyocytes. Marine Drugs 8 (10), 2659–2672. Available from: https://doi.org/10.3390/md8102659. MDPI AG.

Oh, S.T., et al., 2015. Effects of dietary fermented *Chlorella vulgaris* (CBT®) on growth performance, relative organ weights, cecal microflora, tibia bone characteristics, and meat qualities in pekin ducks. Asian-Australasian Journal of Animal Sciences 28 (1), 95–101. Available from: https://doi.org/10.5713/ajas.14.0473. South Korea: Asian-Australasian Association of Animal Production Societies.

Olaizola, M., 2003. Commercial development of microalgal biotechnology: from the test tube to the marketplace. Biomolecular Engineering . Available from: https://doi.org/10.1016/S1389-0344(03)00076-5United States: Elsevier.

Olasehinde, T., Olaniran, A., Okoh, A., 2017. Therapeutic potentials of microalgae in the treatment of Alzheimer's disease. Molecules (Basel, Switzerland) 22 (3), 480. Available from: https://doi.org/10.3390/molecules22030480. MDPI AG.

Onen Cinar, S., et al., 2020. Bioplastic production from microalgae: a review. International Journal of Environmental Research and Public Health 17 (11), 3842. Available from: https://doi.org/10.3390/ijerph17113842. MDPI AG.

Ono, E., Cuello, J.L., 2007. Carbon dioxide mitigation using thermophilic *Cyanobacteria*. Biosystems Engineering 96 (1), 129–134. Available from: https://doi.org/10.1016/j.biosystemseng.2006.09.010. Japan.

Pang, M., et al., 2010. Apoptosis induced by yessotoxins in Hela human cervical cancer cells in vitro. Molecular Medicine Reports 3 (4), 629–634. Available from: https://doi.org/10.3892/mmr_00000307. China: Spandidos Publications.

Parikh, P., Mani, U., Iyer, U., 2001. Role of *Spirulina* in the control of glycemia and lipidemia in type 2 diabetes mellitus. Journal of Medicinal Food 4 (4), 193–199. Available from: https://doi.org/10.1089/10966200152744463. Mary Ann Liebert Inc.

Parmar, R.S., Singh, C., 2018. A comprehensive study of eco-friendly natural pigment and its applications. Biochemistry and Biophysics Reports 13, 22–26. Available from: https://doi.org/10.1016/j.bbrep.2017.11.002. Elsevier BV.

Pasquet, V., et al., 2011. Antiproliferative activity of violaxanthin isolated from bioguided fractionation of *Dunaliella tertiolecta* Extracts. Marine Drugs 9 (5), 819–831. Available from: https://doi.org/10.3390/md9050819. MDPI AG.

Paul, C., Pohnert, G., 2011. Production and role of volatile halogenated compounds from marine algae. Natural Product Reports 28 (2), 186–195. Available from: https://doi.org/10.1039/c0np00043d. Germany.

Peiretti, P.G., Meineri, G., 2009. Effects of two antioxidants on the morpho-biometrical parameters, apparent digestibility and meat composition in rabbits fed low and high fat diets. Journal of Animal and Veterinary Advances 8 (11), 2299–2304Italy. Available from: http://medwelljournals.com/fulltext/java/2009/2299-2304.pdf.

Peng, J., et al., 2011. Fucoxanthin, a marine carotenoid present in brown seaweeds and diatoms: metabolism and bioactivities relevant to human health. Marine Drugs 9 (10), 1806–1828. Available from: https://doi.org/10.3390/md9101806. MDPI AG.

Peng, L., et al., 2016. Cultivation of *Neochloris oleoabundans* in bubble column photobioreactor with or without localized deoxygenation. Bioresource Technology 206, 255–263. Available from: https://doi.org/10.1016/j.biortech.2016.01.081. Elsevier BV.

Peng, L., et al., 2020. Biofuel production from microalgae: a review. Environmental Chemistry Letters 18 (2), 285–297. Available from: https://doi.org/10.1007/s10311-019-00939-0. Springer Science and Business Media LLC.

Pereira L., 2018. Therapeutic and Nutritional Use of Algae-Promo. CRC. Science Publishers' (SP), An Imprint of CRC Press/Taylor & Francis Group, ISBN 9781498755382.

Ponnampalam, E.N., et al., 2016. Muscle antioxidant (vitamin E) and major fatty acid groups, lipid oxidation and retail colour of meat from lambs fed a roughage based diet with flaxseed or algae. Meat Science 111, 154–160. Available from: https://doi.org/10.1016/j.meatsci.2015.09.007. Elsevier BV.

Prestinaci, F., Pezzotti, P., Pantosti, A., 2015. Antimicrobial resistance: a global multifaceted phenomenon. Pathogens and Global Health 109 (7), 309–318. Available from: https://doi.org/10.1179/2047773215Y.0000000030. Informa UK Limited.

Pulz, O., Gross, W., 2004. Valuable products from biotechnology of microalgae. Applied Microbiology and Biotechnology 65 (6), 635–648. Available from: https://doi.org/10.1007/s00253-004-1647-x. Springer Science and Business Media LLC.

Qiu, S., et al., 2020. Antioxidant assessment of wastewater-cultivated *Chlorella sorokiniana* in *Drosophila melanogaster*. Algal Research 46, 101795. Available from: https://doi.org/10.1016/j.algal.2020.101795. Elsevier BV.

Quijano, G., Arcila, J.S., Buitrón, G., 2017. Microalgal-bacterial aggregates: applications and perspectives for wastewater treatment. Biotechnology Advances 35 (6), 772–781. Available from: https://doi.org/10.1016/j.biotechadv.2017.07.003. Elsevier BV.

Ramanan, R., et al., 2010. Enhanced algal CO_2 sequestration through calcite deposition by *Chlorella* sp. and *Spirulina platensis* in a mini-raceway pond. Bioresource Technology 101 (8), 2616–2622. Available from: https://doi.org/10.1016/j.biortech.2009.10.061. Elsevier BV.

Ramanna, L., et al., 2014. The optimization of biomass and lipid yields of *Chlorella sorokiniana* when using wastewater supplemented with different nitrogen sources. Bioresource Technology 168, 127–135. Available from: https://doi.org/10.1016/j.biortech.2014.03.064. Elsevier BV.

Rani, K., Sandal, P., Sahoo, 2018. A comprehensive review on chlorella-its composition, health benefits, market and regulatory scenario. The Pharma Innovation Journal 7.

Raposo, M.F., de, J., de Morais, A.M.M.B., 2015. Microalgae for the prevention of cardiovascular disease and stroke. Life Sciences 125, 32–41. Available from: https://doi.org/10.1016/j.lfs.2014.09.018. Elsevier BV.

Rashad, S., et al., 2019. Cyanobacteria cultivation using olive milling wastewater for bio-fertilization of celery plant. Global Journal of Environmental Science and Management 5 (2), 167–174. Available from: https://doi.org/10.22034/gjesm.2019.02.03. Egypt: Iran Solid Waste Association.

Rezvani, Zaghari, Moravej, 2012. A survey on *Chlorella vulgaris* effect's on performance and cellular immunity in broilers. International Journal of Agricultural Science and Research 3, 9–15.

Ribeiro, T., et al., 2013. Direct supplementation of diet is the most efficient way of enriching broiler meat with n-3 long-chain polyunsaturated fatty acids. British Poultry Science 54 (6), 753–765. Available from: https://doi.org/10.1080/00071668.2013.841861. Portugal.

Ribeiro, T., et al., 2014. Effect of reduced dietary protein and supplementation with a docosahexaenoic acid product on broiler performance and meat quality. British Poultry Science 55 (6), 752–765. Available from: https://doi.org/10.1080/00071668.2014.971222. Portugal: Taylor and Francis Ltd.

Ribeiro, C.F.A., et al., 2020. Effects of antibiotic treatment on gut microbiota and how to overcome its negative impacts on human health. ACS Infectious Diseases 6 (10), 2544–2559. Available from: https://doi.org/10.1021/acsinfecdis.0c00036. American Chemical Society (ACS).

Riccio, G., Lauritano, C., 2020. Microalgae with immunomodulatory activities. Marine Drugs 18 (1). Available from: https://doi.org/10.3390/md18010002. Italy: MDPI AG.

Rickards, R.W., et al., 1999. Calothrixins A and B, novel pentacyclic metabolites from calothrix *Cyanobacteria* with potent activity against malaria parasites and human cancer cells. Tetrahedron 55 (47), 833–839. Available from: https://doi.org/10.1016/S0040-4020(99.

Rizwan, M., et al., 2018. Exploring the potential of microalgae for new biotechnology applications and beyond: a review. Renewable and Sustainable Energy Reviews 92, 394–404. Available from: https://doi.org/10.1016/j.rser.2018.04.034. Elsevier BV.

Romera, E., et al., 2006. Biosorption with algae: a statistical review. Critical Reviews in Biotechnology 26 (4), 223–235. Available from: https://doi.org/10.1080/07388550600972153. Spain.

Sabry, O.M.M., et al., 2017. Cytotoxic halogenated monoterpenes from *Plocamium cartilagineum*. Natural Product Research 31 (3), 261–267. Available from: https://doi.org/10.1080/14786419.2016.1230115. Informa UK Limited.

Saeid, A., et al., 2013. Biomass of *Spirulina maxima* enriched by biosorption process as a new feed supplement for swine. Journal of Applied Phycology 25 (2), 667–675. Available from: https://doi.org/10.1007/s10811-012-9901-6. Poland: Kluwer Academic Publishers.

Sajilata, M.G., Singhal, R.S., Kamat, M.Y., 2008. Supercritical CO_2 extraction of γ-linolenic acid (GLA) from *Spirulina platensis* ARM 740 using response surface methodology. Journal of Food Engineering 84 (2), 321–326. Available from: https://doi.org/10.1016/j.jfoodeng.2007.05.028. Elsevier BV.

Sakthivel, R., et al., 2016. *Gracilaria edulis* exhibit antiproliferative activity against human lung adenocarcinoma cell line A549 without causing adverse toxic effect in vitro and in vivo. Food & Function 7 (2), 1155–1165. Available from: https://doi.org/10.1039/c5fo01094b. Royal Society of Chemistry (RSC).

Samarakoon, K.W., et al., 2014. Apoptotic anticancer activity of a novel fatty alcohol ester isolated from cultured marine diatom, *Phaeodactylum tricornutum*. Journal of Functional Foods 6 (1), 231–240. Available from: https://doi.org/10.1016/j.jff.2013.10.011. Elsevier BV.

Sandau, E., Sandau, P., Pulz, O., 1996. Heavy metal sorption by microalgae. Acta Biotechnologica 16 (4), 227–235. Available from: https://doi.org/10.1002/abio.370160402. Germany.

Sansone, C., et al., 2014. Diatom-derived polyunsaturated aldehydes activate cell death in human cancer cell lines but not normal cells. PLoS One 9 (7), e101220. Available from: https://doi.org/10.1371/journal.pone.0101220. Public Library of Science (PLoS).

Santos, F.M., Pires, J.C.M., 2018. Nutrient recovery from wastewaters by microalgae and its potential application as bio-char. Bioresource Technology 267, 725–731. Available from: https://doi.org/10.1016/j.biortech.2018.07.119. Elsevier BV.

Santos-Sánchez, N.F., et al., 2016. Lipids rich in ω-3 polyunsaturated fatty acids from microalgae. Applied Microbiology and Biotechnology 100 (20), 8667–8684. Available from: https://doi.org/10.1007/s00253-016-7818-8. Mexico: Springer Verlag.

Sardi, L., et al., 2006. Effects of a dietary supplement of DHA-rich marine algae on Italian heavy pig production parameters. Livestock Science 103 (1–2), 95–103. Available from: https://doi.org/10.1016/j.livsci.2006.01.009. Italy.

Sarker, S.D., et al., 2020. Anticancer natural products. Annual Reports in Medicinal Chemistry. Academic Press Inc, United Kingdom, pp. 45–75. Available from: https://doi.org/10.1016/bs.armc.2020.02.001.

Sathasivam, R., et al., 2019. Microalgae metabolites: a rich source for food and medicine. Saudi Journal of Biological Sciences 26 (4), 709–722. Available from: https://doi.org/10.1016/j.sjbs.2017.11.003. Elsevier BV.

Sayre, R., 2010. Microalgae: the potential for carbon capture. Bioscience 60 (9), 722–727. Available from: https://doi.org/10.1525/bio.2010.60.9.9. Oxford University Press (OUP).

Schieber, M., Chandel, N.S., 2014. ROS function in redox signaling and oxidative stress. Current Biology 24 (10), R453–R462. Available from: https://doi.org/10.1016/j.cub.2014.03.034. Elsevier BV.

Schmidt, R.A., Wiebe, M.G., Eriksen, N.T., 2005. Heterotrophic high cell-density fed-batch cultures of the phycocyanin-producing red alga *Galdieria sulphuraria*. Biotechnology and Bioengineering 90 (1), 77–84. Available from: https://doi.org/10.1002/bit.20417. Wiley.

Schmitt, D., et al., 2001. The adsorption kinetics of metal ions onto different microalgae and siliceous earth. Water Research 35 (3), 779–785. Available from: https://doi.org/10.1016/S0043-1354(00)00317-1. Germany.

Scragg, A.H., et al., 2002. Growth of microalgae with increased calorific values in a tubular bioreactor. Biomass and Bioenergy 23 (1), 67–73. Available from: https://doi.org/10.1016/S0961-9534(02)00028-4. United Kingdom.

Selmi, C., et al., 2011. The effects of *Spirulina* on anemia and immune function in senior citizens. Cellular & Molecular Immunology 8 (3), 248–254. Available from: https://doi.org/10.1038/cmi.2010.76. Springer Science and Business Media LLC.

Selvaratnam, T., et al., 2014. Evaluation of a thermo-tolerant acidophilic alga, *Galdieria sulphuraria*, for nutrient removal from urban wastewaters. Bioresource Technology 156, 395–399. Available from: https://doi.org/10.1016/j.biortech.2014.01.075. United States: Elsevier Ltd.

Senousy, H.H., Abd Ellatif, S., Ali, S., 2020. Assessment of the antioxidant and anticancer potential of different isolated strains of cyanobacteria and microalgae from soil and agriculture drain water. Environmental Science and Pollution Research 27 (15), 18463–18474. Available from: https://doi.org/10.1007/s11356-020-08332-z. Springer Science and Business Media LLC.

Senthilkumar, D., Jayanthi, S., 2016. Partial characterization and anticancer activities of purified glycoprotein extracted from green seaweed *Codium decorticatum*. Journal of Functional Foods 25, 323–332. Available from: https://doi.org/10.1016/j.jff.2016.06.010. Elsevier BV.

Shanmugapriya, B, Babu, S.S, Hariharan, T, Siwaneswaran, S, Anusha, M.B, College, C.N, 2015. Research article dietary administration 0f spirulina platensis as probiotics on growth performance and histopathology in broiler chicks. International Journal of Recent Scientific Research 6, 2650–2653.

Shao, W., et al., 2018. Enhancement of *Spirulina* biomass production and cadmium biosorption using combined static magnetic field. Bioresource Technology 265, 163–169. Available from: https://doi.org/10.1016/j.biortech.2018.06.009. Elsevier BV.

Shah, M., Samarakoon, K., Ko, J.-Y., Lakmal, H., Lee, J.-H., An, S.-J., Jeon, Y.-J., Lee, J.-B., 2014. Potentiality of Benthic Dinoflagellate Cultures and Screening of Their Bioactivities in Jeju Island, Korea. African Journal of Biotechnology 13 (February), 792–805. Available from: https://doi.org/10.5897/AJB2013.13250.

Sharma, A., Arya, S.K., 2017. Hydrogen from algal biomass: a review of production process. Biotechnology Reports 15, 63–69. Available from: https://doi.org/10.1016/j.btre.2017.06.001. Elsevier BV.

Sharma, A., Arya, S.K., 2019. Photobiological Production of Biohydrogen: Recent Advances and Strategy. In: Rastegari, A., Yadav, A., Gupta, A. (Eds.), Prospects of Renewable Bioprocessing in Future Energy Systems. Biofuel and Biorefinery Technologies, vol 10. Springer, Cham. Available from: https://doi.org/10.1007/978-3-030-14463-0_3.

Sharma, A., Singh, G., Arya, S.K., 2020. Biofuel from rice straw. Journal of Cleaner Production 277, 124101. Available from: https://doi.org/10.1016/j.jclepro.2020.124101. Elsevier BV.

Sheu, M.J., et al., 2008. Ethanol extract of *Dunaliella salina* induces cell cycle arrest and apoptosis in A549 human non-small cell lung cancer cells. In Vivo 22 (3), 369–378. Taiwan: International Institute of Anticancer Research. Available from: http://iv.iiarjournals.org/.

Shokri, H., Khosravi, A.R., Taghavi, M., 2014. Efficacy of *Spirulina platensis* on immune functions in cancer mice with systemic candidiasis. Journal of Mycology Research 1, 7–13.

Silva, S.C., et al., 2020. Microalgae-derived pigments: a 10-year bibliometric review and industry and market trend analysis. Molecules (Basel, Switzerland) 25 (15), 3406. Available from: https://doi.org/10.3390/molecules25153406. MDPI AG.

Simioni, T., Quadri, M.B., Derner, R.B., 2019. Drying of *Scenedesmus obliquus*: experimental and modeling study. Algal Research 39, 101428. Available from: https://doi.org/10.1016/j.algal.2019.101428. Elsevier BV.

Šimkus, A., et al., 2013. The effect of blue algae *Spirulina platensis* on pig growth performance and carcass and meat quality. Veterinarija ir Zootechnika 61 (83), 70–74. Lithuania. Available from: http://vetzoo.lva.lt/data/vols/2013/61/pdf/simkus.pdf.

Sjollema, S.B., et al., 2016. Do plastic particles affect microalgal photosynthesis and growth? Aquatic Toxicology 170, 259–261. Available from: https://doi.org/10.1016/j.aquatox.2015.12.002. Elsevier BV.

Sloth, J.K., et al., 2017. Growth and phycocyanin synthesis in the heterotrophic microalga *Galdieria sulphuraria* on substrates made of food waste from restaurants and bakeries. Bioresource Technology 238, 296–305. Available from: https://doi.org/10.1016/j.biortech.2017.04.043. Elsevier BV.

Smit, A.J., 2004. Medicinal and pharmaceutical uses of seaweed natural products: a review. Journal of Applied Phycology 16 (4), 245–262. Available from: https://doi.org/10.1023/B:JAPH.0000047783.36600.ef. Springer Nature.

Smith, V.H., et al., 2010. The ecology of algal biodiesel production. Trends in Ecology & Evolution 25 (5), 301–309. Available from: https://doi.org/10.1016/j.tree.2009.11.007. Elsevier BV.

Solana, M., Rizza, C.S., Bertucco, A., 2014. Exploiting microalgae as a source of essential fatty acids by supercritical fluid extraction of lipids: comparison between *Scenedesmus obliquus*, *Chlorella protothecoides* and *Nannochloropsis salina*. The Journal of Supercritical Fluids 92, 311–318. Available from: https://doi.org/10.1016/j.supflu.2014.06.013. Elsevier BV.

Soletto, D., et al., 2005. Batch and fed-batch cultivations of *Spirulina platensis* using ammonium sulphate and urea as nitrogen sources. Aquaculture (Amsterdam, Netherlands) 243 (1–4), 217–224. Available from: https://doi.org/10.1016/j.aquaculture.2004.10.005. Italy.

Somasekharan, S.P., et al., 2016. An aqueous extract of marine microalgae exhibits antimetastatic activity through preferential killing of suspended cancer cells and anticolony forming activity. Evidence-Based Complementary and Alternative Medicine 2016, 1–8. Available from: https://doi.org/10.1155/2016/9730654. Hindawi Limited.

Soni, R.A., Sudhakar, K., Rana, R.S., 2017. *Spirulina* – from growth to nutritional product: a review. Trends in Food Science & Technology 69, 157–171. Available from: https://doi.org/10.1016/j.tifs.2017.09.010. Elsevier BV.

Soria-Mercado, I.E., et al., 2009. Alotamide A, a novel neuropharmacological agent from the marine cyanobacterium Lyngbya bouillonii. Organic Letters 11 (20), 4704–4707. Available from: https://doi.org/10.1021/ol901438b. American Chemical Society (ACS).

SØrum, H., Sunde, M., 2001. Resistance to antibiotics in the normal flora of animals. Veterinary Research. EDP Sciences, Norway. Available from: https://doi.org/10.1051/vetres:2001121.

Soto-Sierra, L., Stoykova, P., Nikolov, Z.L., 2018. Extraction and fractionation of microalgae-based protein products. Algal Research 36, 175–192. Available from: https://doi.org/10.1016/j.algal.2018.10.023. Elsevier BV.

Sousa, J.C.G., et al., 2018. A review on environmental monitoring of water organic pollutants identified by EU guidelines. Journal of Hazardous Materials 344, 146–162. Available from: https://doi.org/10.1016/j.jhazmat.2017.09.058. Elsevier BV.

Stengel, D.B., Connan, S., Popper, Z.A., 2011. Algal chemodiversity and bioactivity: sources of natural variability and implications for commercial application. Biotechnology Advances 29 (5), 483–501. Available from: https://doi.org/10.1016/j.biotechadv.2011.05.016. Elsevier BV.

Stevenson, C.S., et al., 2002. The Identification and characterization of the marine natural product scytonemin as a novel antiproliferative pharmacophore. Journal of Pharmacology and Experimental Therapeutics 303 (2), 858–866. Available from: https://doi.org/10.1124/jpet.102.036350. American Society for Pharmacology & Experimental Therapeutics (ASPET).

Subashchandrabose, S.R., et al., 2013. Mixotrophic cyanobacteria and microalgae as distinctive biological agents for organic pollutant degradation. Environment International 51, 59–72. Available from: https://doi.org/10.1016/j.envint.2012.10.007. Elsevier BV.

Suganya, T., et al., 2016. Macroalgae and microalgae as a potential source for commercial applications along with biofuels production: a biorefinery approach. Renewable and Sustainable Energy Reviews 55, 909–941. Available from: https://doi.org/10.1016/j.rser.2015.11.026. Malaysia: Elsevier Ltd.

Suh, S.-S., et al., 2017. Anticancer activities of ethanol extract from the Antarctic freshwater microalga, *Botryidiopsidaceae* sp. BMC Complementary and Alternative Medicine 17 (1). Available from: https://doi.org/10.1186/s12906-017-1991-x. Springer Science and Business Media LLC.

Sun, X., et al., 2014. Effect of nitrogen-starvation, light intensity and iron on triacylglyceride/carbohydrate production and fatty acid profile of *Neochloris oleoabundans* HK-129 by a two-stage process. Bioresource Technology 155, 204–212. Available from: https://doi.org/10.1016/j.biortech.2013.12.109. Elsevier BV.

Suzuki, T., Ezure, T., Ishida, M., 1999. In-vitro antitumour activity of extracts from cyanobacteria. Pharmacy and Pharmacology Communications 5 (10), 619–622. Available from: https://doi.org/10.1211/1460080899128734244. Japan: Pharmaceutical Press.

Swiatkiewicz, S., Arczewska-Włosek A., Józefiak, D., 2016. Application of Microalgae Biomass in Poultry Nutrition. World's Poultry Science Journal 71 (January), 663–72. Available from: https://doi.org/10.1017/S0043933915002457.

Sydney, E.B., et al., 2010. Potential carbon dioxide fixation by industrially important microalgae. Bioresource Technology 101 (15), 5892–5896. Available from: https://doi.org/10.1016/j.biortech.2010.02.088. Elsevier BV.

Tan, X.B., et al., 2018. Cultivation of microalgae for biodiesel production: a review on upstream and downstream processing. Chinese Journal of Chemical Engineering 26 (1), 17–30. Available from: https://doi.org/10.1016/j.cjche.2017.08.010. Elsevier BV.

Tang, D.Y.Y., et al., 2020. Potential utilization of bioproducts from microalgae for the quality enhancement of natural products. Bioresource Technology 304, 122997. Available from: https://doi.org/10.1016/j.biortech.2020.122997. Elsevier BV.

Tavares-Carreón, F., et al., 2020. In vitro anticancer activity of methanolic extract of *Granulocystopsis* sp., a microalgae from an oligotrophic oasis in the Chihuahuan desert. PeerJ 2020 (3), e8686. Available from: https://doi.org/10.7717/peerj.8686.

Thangam, R., et al., 2013. C-phycocyanin from *Oscillatoria tenuis* exhibited an antioxidant and in vitro antiproliferative activity through induction of apoptosis and G 0/G1 cell cycle arrest. Food Chemistry 140 (1–2), 262–272. Available from: https://doi.org/10.1016/j.foodchem.2013.02.060. India.

Torres-Duran, P.V., Ferreira-Hermosillo, A., Juarez-Oropeza, M.A., 2007. Antihyperlipemic and antihypertensive effects of *Spirulina maxima* in an open sample of Mexican population: a preliminary report. Lipids in Health and Disease 6, 33. Available from: https://doi.org/10.1186/1476-511X-6-33. Springer Nature.

Torres-Tiji, Y., Fields, F.J., Mayfield, S.P., 2020. Microalgae as a future food source. Biotechnology Advances 41, 107536. Available from: https://doi.org/10.1016/j.biotechadv.2020.107536. Elsevier BV.

Townsend, M., et al., 2018. The challenge of implementing the marine ecosystem service concept. Frontiers in Marine Science 5. Available from: https://doi.org/10.3389/fmars.2018.00359. Frontiers Media SA.

Toyomizu, M., et al., 2001. Effects of dietary *Spirulina* on meat colour in muscle of broiler chickens. British Poultry Science 42 (2), 197–202. Available from: https://doi.org/10.1080/00071660120048447. Japan.

Tran, N.T., Kaldenhoff, R., 2020. Achievements and challenges of genetic engineering of the model green alga *Chlamydomonas reinhardtii*. Algal Research 50, 101986. Available from: https://doi.org/10.1016/j.algal.2020.101986. Elsevier BV.

Trappmann, J., Hawk, S.N., 2011. The effects of n-3 fatty acids and bexarotene on breast cancer cell progression. Journal of Cancer Therapy 710–714. Available from: https://doi.org/10.4236/jct.2011.25096. Scientific Research Publishing, Inc.

Tufa, T., 2015. Veterinary Drug Residues in Food-Animal Products: Its Risk Factors and Potential Effects on Public Health. Journal of Veterinary Science & Technology 07 (January). Available from: https://doi.org/10.4172/2157-7579.1000285.

Uduman, N., et al., 2010. Marine microalgae flocculation and focused beam reflectance measurement. Chemical Engineering Journal 162 (3), 935–940. Available from: https://doi.org/10.1016/j.cej.2010.06.046. Elsevier BV.

Umemura, K., et al., 2003. Inhibition of DNA topoisomerases I and II, and growth inhibition of human cancer cell lines by a marine microalgal polysaccharide. Biochemical Pharmacology 66 (3), 481–487. Available from: https://doi.org/10.1016/s0006-2952(03)00281-8. Elsevier BV.

Urrutia, O., et al., 2016. Effects of addition of linseed and marine algae to the diet on adipose tissue development, fatty acid profile, lipogenic gene expression, and meat quality in lambs. PLoS One 11 (6), e0156765. Available from: https://doi.org/10.1371/journal.pone.0156765. Public Library of Science (PLoS).

Usher, P.K., et al., 2014. An overview of the potential environmental impacts of large-scale microalgae cultivation. Biofuels 5 (3), 331–349. Available from: https://doi.org/10.1080/17597269.2014.913925. Informa UK Limited.

Vega-Estrada, J., et al., 2005. *Haematococcus pluvialis* cultivation in split-cylinder internal-loop airlift photobioreactor under aeration conditions avoiding cell damage. Applied Microbiology and Biotechnology 68 (1), 31–35. Available from: https://doi.org/10.1007/s00253-004-1863-4. Mexico.

Venkata Mohan, S., et al., 2016. Waste biorefinery models towards sustainable circular bioeconomy: critical review and future perspectives. Bioresource Technology 215, 2–12. Available from: https://doi.org/10.1016/j.biortech.2016.03.130. Elsevier BV.

Vossen, E., et al., 2016. Production of docosahexaenoic acid (DHA) enriched loin and dry cured ham from pigs fed algae: nutritional and sensory quality. European Journal of Lipid Science and Technology 119 (5), 1600144. Available from: https://doi.org/10.1002/ejlt.201600144. Wiley.

Wang, B., et al., 2008. CO_2 bio-mitigation using microalgae. Applied Microbiology and Biotechnology 79 (5), 707–718. Available from: https://doi.org/10.1007/s00253-008-1518-y. Springer Science and Business Media LLC.

Wang, H.-M., et al., 2010. Identification of anti-lung cancer extract from *Chlorella vulgaris* C-C by antioxidant property using supercritical carbon dioxide extraction. Process Biochemistry 45 (12), 1865–1872. Available from: https://doi.org/10.1016/j.procbio.2010.05.023. Elsevier BV.

Wang, Q., et al., 2017. Technologies for reducing sludge production in wastewater treatment plants: state of the art. Science of The Total Environment 587–588, 510–521. Available from: https://doi.org/10.1016/j.scitotenv.2017.02.203. Elsevier BV.

Waris, G., Ahsan, H., 2006. Reactive oxygen species: role in the development of cancer and various chronic conditions. Journal of Carcinogenesis 5. Available from: https://doi.org/10.1186/1477-3163-5-14. United States.

Wollmann, F., et al., 2019. Microalgae wastewater treatment: biological and technological approaches. Engineering in Life Sciences 19 (12), 860–871. Available from: https://doi.org/10.1002/elsc.201900071. Wiley.

Yahya, L., Harun, R., Abdullah, L.C., 2020. Screening of native microalgae species for carbon fixation at the vicinity of Malaysian coal-fired power plant. Scientific Reports 10 (1). Available from: https://doi.org/10.1038/s41598-020-79316-9. Springer Science and Business Media LLC.

Yam, D., Peled, A., Shinitzky, M., 2001. Suppression of tumor growth and metastasis by dietary fish oil combined with vitamins E and C and cisplatin. Cancer Chemotherapy and Pharmacology 47 (1), 34–40. Available from: https://doi.org/10.1007/s002800000205. Israel.

Yamaguchi, K., 1996. Recent advances in microalgal bioscience in Japan, with special reference to utilization of biomass and metabolites: a review. Journal of Applied Phycology 8 (6), 487–502. Available from: https://doi.org/10.1007/bf02186327. Springer Science and Business Media LLC.

Yan, L., Kim, I.H., 2013. Effects of dietary ω-3 fatty acid-enriched microalgae supplementation on growth performance, blood profiles, meat quality, and fatty acid composition of meat in broilers. Journal of Applied Animal Research 41 (4), 392–397. Available from: https://doi.org/10.1080/09712119.2013.787361. South Korea.

Yan, H., Pan, G., 2002. Toxicity and bioaccumulation of copper in three green microalgal species. Chemosphere 49 (5), 471–476. Available from: https://doi.org/10.1016/s0045-6535(02)00285-0. Elsevier BV.

Yan, L., Lim, S.U., Kim, I.H., 2012. Effect of fermented chlorella supplementation on growth performance, nutrient digestibility, blood characteristics, fecal microbial and fecal noxious gas content in growing pigs. Asian-Australasian Journal of Animal Sciences 25 (12), 1742–1747. Available from: https://doi.org/10.5713/ajas.2012.12352. South Korea.

Yang, X., et al., 2015. Microcystis aeruginosa/pseudomonas pseudoalcaligenes interaction effects on off-flavors in algae/bacteria co-culture system under different temperatures. Journal of Environmental Sciences 31, 38–43. Available from: https://doi.org/10.1016/j.jes.2014.07.034. Elsevier BV.

Yew, G.Y., et al., 2020. A novel lipids recovery strategy for biofuels generation on microalgae *Chlorella* cultivation with waste molasses. Journal of Water Process Engineering 38, 101665. Available from: https://doi.org/10.1016/j.jwpe.2020.101665. Elsevier BV.

Yuan, J.-P., et al., 2011. Potential health-promoting effects of astaxanthin: a high-value carotenoid mostly from microalgae. Molecular Nutrition & Food Research 55 (1), 150–165. Available from: https://doi.org/10.1002/mnfr.201000414. Wiley.

Zahroojian, N., Moravej, H., Shivazad, M., 2013. Effects of dietary marine algae (*Spirulina platensis*) on egg quality and production performance of laying hens. Journal of Agricultural Science and Technology 15, 1353–1360. Iran. Available from: http://jast.modares.ac.ir/?_action = showPDF&article = 10216&_ob = a0e4a75fa200be2fa4cdb336f9040ce7&fileName = full_text.pdf.

Zeng, X., et al., 2011. Microalgae bioengineering: from CO_2 fixation to biofuel production. Renewable and Sustainable Energy Reviews 15 (6), 3252–3260. Available from: https://doi.org/10.1016/j.rser.2011.04.014. Elsevier BV.

Zhai, J., et al., 2017. Optimization of biomass production and nutrients removal by *Spirulina platensis* from municipal wastewater. Ecological Engineering 108, 83–92. Available from: https://doi.org/10.1016/j.ecoleng.2017.07.023. Elsevier BV.

Zhang, D., et al., 2016. Dynamic modelling of *Haematococcus pluvialis* photoinduction for astaxanthin production in both attached and suspended photobioreactors. Algal Research 13, 69–78. Available from: https://doi.org/10.1016/j.algal.2015.11.019. Elsevier BV.

Zhang, C., et al., 2017. Toxic effects of microplastic on marine microalgae *Skeletonema costatum*: interactions between microplastic and algae. Environmental Pollution 220, 1282–1288. Available from: https://doi.org/10.1016/j.envpol.2016.11.005. Elsevier BV.

Zhu, L., 2015. Microalgal culture strategies for biofuel production: a review. Biofuels, Bioproducts and Biorefining 9 (6), 801–814. Available from: https://doi.org/10.1002/bbb.1576. Wiley.

Valorization of microalgae for biogas methane enhancement

Fayaz A. Malla[1], Nazir Ahmad Sofi[2], Navindu Gupta[3] and Suhaib A. Bandh[4]

[1]Assistant Professor, Department of Environmental Science, Govt. Degree College Tral, Jammu and Kashmir, India [2]Department of Agriculture Research Information System, Sher-e-Kashmir University of Agricultural Sciences and Technology, Shalimar Campus, Srinagar, Jammu and Kashmir, India [3]Center for Environment science and climate-resilient agriculture (CESRA), Indian Agricultural Research Institute, Delhi, New Delhi, India [4]Assistant Professor, Environmental Science, Higher Education Department, Government of Jammu and Kashmir, Srinagar, India

14.1 Introduction

Nonrenewable energy sources such as crude oil, coal, and natural gas are being extensively used for a long time. Their known reserves have been depleted with a 1% increase in worldwide primary energy consumption (BP Statistical Review of World, 2017). On the other hand, renewable sources of energy such as biomass, biogas, wind, hydropower, geothermal, and solar, which exist infinitely, are naturally replenished (Bhattacharya, 1998). These nonconventional sources contribute 14% of total global energy demand (Demirbas, 2009). Renewable energies in 2016 were the fastest-growing energy sources at a rate of 12%, which accounts for one-third increase in primary power, although having a stake of only 4%. Out of these, biogas could be tapped as one of the cheapest renewable sources of energy generated from anaerobic digestion of waste materials like kitchen wastes, farm wastes, and cow dung. Biogas is utilized for power generation and in proper solid waste management and generation of slurry, which could be used as a fertilizer. The world and India emphasize the establishment of biogas technology for maximum energy production. More than 5 million small biogas digesters have been constructed in China, and around 20 million households are using biogas as fuel. About 47.5 lakh biogas plants were already installed in India up to March 31, 2014 under National

Valorisation of Microalgal Biomass and Wastewater Treatment
DOI: https://doi.org/10.1016/B978-0-323-91869-5.00015-6

Biogas, and Manure Management Programme, and additionally, 65,180 family type biogases were targeted for 2017–18 (Ministry of New and Renewable Energy, 2017).

The biogas mainly comprises methane (40%–75%) and carbon dioxide (25%–55%). Traces of other gases such as water (<10%), hydrogen sulfide (<5000 ppm), siloxanes (0%–0.02%), halogenated hydrocarbons (<0.6%), ammonia (<1%), oxygen (0%–1%), carbon monoxide (<0.6%), and nitrogen (N_2, 0%–5%) could also be present (Table 14.1). The chemical composition of biogas depends on the nature of the feedstock and the operational parameters used. Anaerobic digestion involves a series of metabolic interactions among various microorganisms, which is carried out at a temperature ranging between 30°C and 65°C. The incombustible gases in biogas like CO_2 lower its calorific value, thus degrading its quality. On average, the calorific value of biogas is 21.5 MJ m^{-3}, which is much lower when compared with natural gas (35.8 MJ m^{-3}). The most common contaminant in biogas is H_2S and other sulfur-containing compounds that come from sulfur-bearing organic matter. This contaminant is highly undesirable in combustion systems due to its conversion to highly corrosive and environmentally hazardous compounds. Its removal is essential before any eventual utilization of biogas. Ammonia (NH_3) is another common impurity resulting from the anaerobic digestion of nitrogen-containing organic molecules. Its properties are the same as H_2S and represent a health risk. Its combustion slightly increases nitrogen oxide (NOx) emissions. Siloxanes are a group of silicon (Si)-containing molecules found in landfill-originated biogases and are considered the third most crucial contaminant. Their presence during combustion is detrimental because they form glassy microcrystalline silica. Their removal is vital to ensure a satisfactory and extended lifecycle of process equipment (Abatzoglou and Boivin, 2009).

Stripping CO_2 and H_2S from biogas, known as biogas upgradation or methane enrichment, increases its calorific value. Biogas can be used for electricity generation in cogeneration or combined heat and power. Under cryogenic conditions, biogas is converted into compressed biogas or liquid biogas that can be injected into the household pipeline systems for distribution. It can also be compressed in gas cylinders only after upgradation. There is more significant potential for biogas if it can be made viable as a transport vehicle fuel. In all cases, the quality of Biogas is crucial in both its CH_4 content and purity. The most common contaminant in Biogas is H_2S and other Sulfur-containing compounds that come from Sulfur bearing organic matter. This contaminant is highly undesirable in combustion systems due to its conversion to highly corrosive and environmentally hazardous compounds. Its removal is essential before any eventual utilization of Biogas. Ammonia (NH_3) is another common impurity resulting from the anaerobic digestion of nitrogen-containing organic molecules. Its properties are the same as that of H_2S, corrosive, and represent a health risk, besides its combustion slightly increases nitrogen oxides (NOx) emissions, and is less harmful than H_2S. Siloxanes are a group of silicon (Si)-containing molecules found in landfill-originated biogases and considered as the third most crucial contaminant. Their presence during combustion is detrimental because they form glassy microcrystalline silica. Their removal is vital to ensure a satisfactory and extended lifecycle of process equipment (Abatzoglou and Boivin, 2009). Reducing CO_2 and H_2S content will significantly improve the quality of Biogas.

Various technologies have been developed to separate CO_2 from gas streams in the past. These include physical absorption, chemical absorption, cryogenic separation, membrane

TABLE 14.1 Cultivation of microalgae for biogas upgradation.

Microalgae	Composition of biogas	Methane content in upgraded biogas (% v/v)	References
Mutant Chlorella sp.	70% CH_4, 20% CO_2, and 8% N_2	84–87	Kao et al. (2012a, b)
	69% CH_4, 20% CO_2, and <50 ppm H_2S	85–90	Kao et al. (2012a, b)
Chlorella sp.	70.65% CH_4, 26.14% CO_2, 0.23% O_2, 3.11% H_2O, and <0.005% H_2S	92.16	Yan and Zheng (2013)
	67.35% CH_4, 28.41% CO_2, 0.73% O_2, 3.48% H_2O, and <0.005% H_2S	92.74	Zhao et al. (2013)
	Synthetic biogas containing 50% CH_4 and 50% CO_2	94.7	Tongprawhan et al. (2014)
	61.38% CH_4, 32.57% CO_2, 0.54% O_2, 5.52% H_2O, and <0.005% H_2S	84.21	Wang et al. (2016)
	34.2% CO_2, 65.79% CH_4	86.4	Khan et al. (2018)
Leptolyngbya sp.	Synthetic biogas containing 25% CO_2 and 75% CH_4	99.56	Choix et al. (2017b)
Scenedesmus obliquus	61.38% CH_4, 32.57% CO_2, 0.54% O_2, 5.52% H_2O, and <0.005% H_2S	84.28	Wang et al. (2016)
	61.75% CH_4, 32.28% CO_2, 0.31% O_2, 2.68% H_2O, and <0.005% H_2S	94.41	Ouyang et al. (2015)
	Synthetic biogas containing 25% CO_2 and 75% CH_4	96.5	Choix et al. (2017a)
Scenedesmus sp.	Synthetic biogas containing 40% CO_2 and 60% CH_4	>90	Srinuanpan et al. (2017)
		>98	Srinuanpan et al. (2018)
Anabaena spiroides	61.38% CH_4, 32.57% CO_2, 0.54% O_2, 5.52% H_2O, and <0.005% H_2S	81	Wang et al. (2016)
Selenastrum capricornutum	61.38% CH_4, 32.57% CO_2, 0.54% O_2, 5.52% H_2O, and <0.005% H_2S	78	Wang et al. (2016)
Selenastrum bibraianum	61.75% CH_4, 32.28% CO_2, 0.31% O_2, 2.68% H_2O, and <0.005% H_2S	79	Ouyang et al. (2015)
Nannochloropsis gaditana	Synthetic biogas containing 28% CO_2 and 72% CH_4	98.6	Meier et al. (2015)

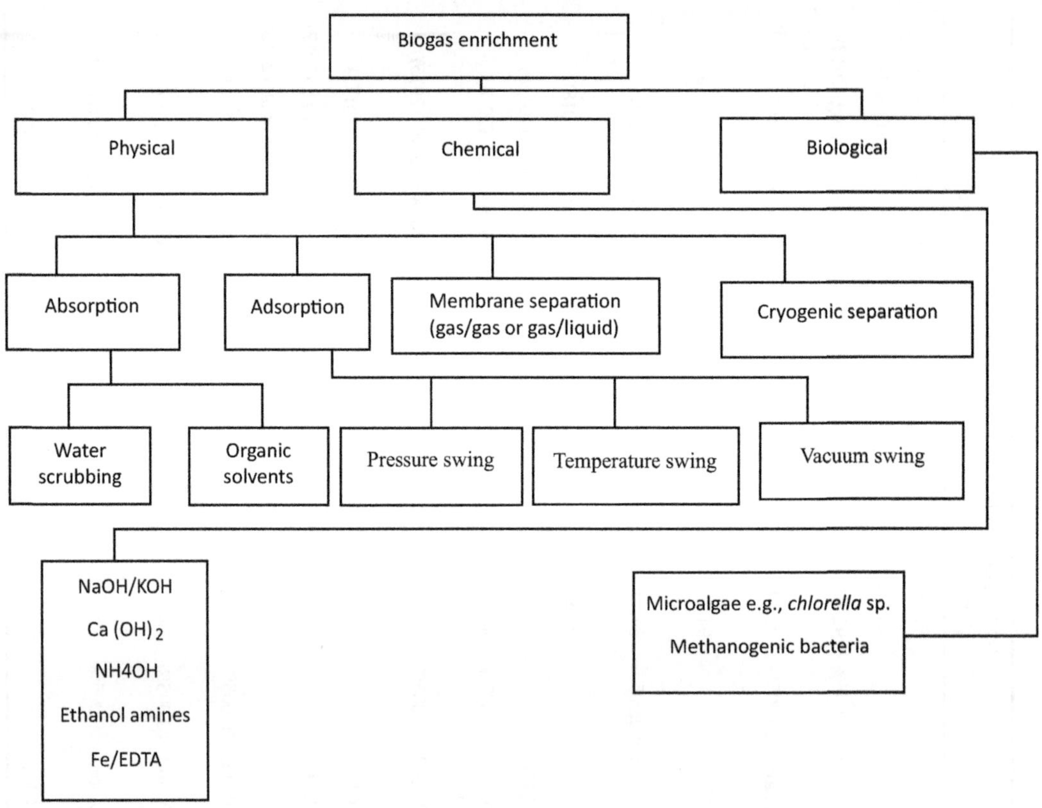

FIGURE 14.1   Various biogas upgrading methods.

separation, and CO_2 fixation by biological or chemical methods (Fig. 14.1). Upgraded biogas or Bio-CNG or biomethane, being eco-friendlier, could be used as a fuel that pollutes the atmosphere least compared to fossil fuels, proving the key source to promoting regional development.

14.2 Microalgae-mediated biogas methane enrichment

Various physical, chemical, and biological techniques have been developed to deracinate CO_2 from biogas. Utilizing the potential of microalgae for fixing the CO_2 content in biogas and producing biomass for later use is an evolving technology that has tremendous scope in the field of renewable energy (Fig. 14.2). One kg of algae (dry weight) can absorb up to 1.83 kg of carbon dioxide per year. The *Chlorella* sp., one of the most common microalgae, has been widely used to capture carbon dioxide from the atmosphere, owing to its relatively higher tolerance to carbon dioxide (Kumar et al., 2011). The microalgal biomass produced from biogas enrichment can be converted into various biofuels by various methods such as direct combustion, torrefaction or thermal chemical liquefaction or pyrolysis. The transesterification method converts the lipids in microalgal biomass into biodiesel. Physical and chemical absorption is the most commonly used techniques for biogas

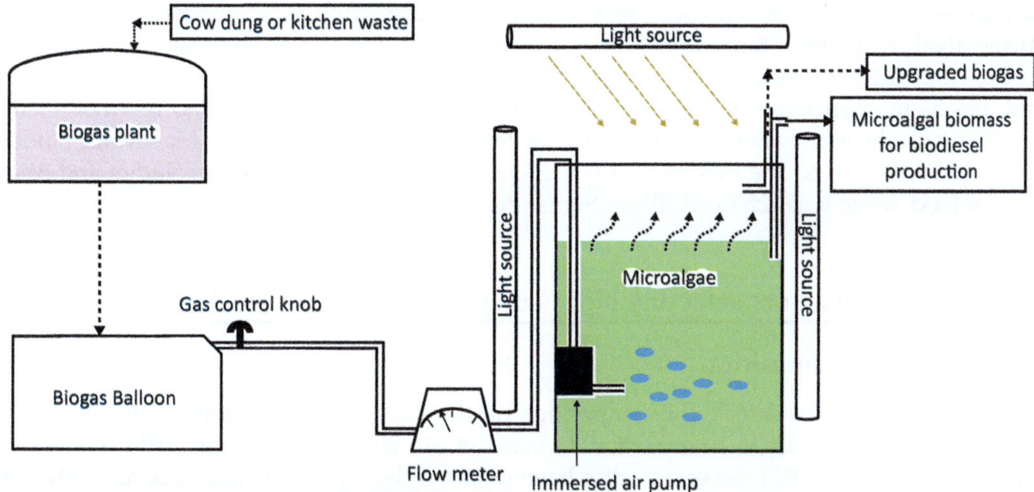

FIGURE 14.2 Scheme for microalgal cultivation coupled with biogas upgrading, biomass production of biodiesel, and phytoremediation of wastewater.

enrichment. These techniques need posttreatment of the waste materials for regeneration of cycling utilization.

Many microalgal species such as *Chlorella* spp.(Kao et al., 2012a, b; Zhao et al., 2013; Tongprawhan et al., 2014), *Leptolyngbya* spp. (Choix et al., 2017b), *Scenedesmus* sp. (Srinuanpan et al., 2017; Kao et al., 2012a, b; Yan and Zheng, 2013; Srinuanpan et al., 2018), *Nannochloropsis* spp. (Meier et al., 2015), *Selenastrum* spp. (Ouyang et al., 2015), and *Anabaena* spp. (Zhao et al., 2013) have been experimented for biogas methane enrichment by reducing the CO_2 content in the biogas. As per various studies mentioned in Table 14.1, microalgae can remove >90% of the CO_2 content in raw biogas and simultaneously enrich the biogas with methane (>90%). Among the experimented microalgae for biogas methane enrichment, few microalgal species such as *Chlorella* spp. (Tongprawhan et al., 2014) and *Scenedesmus* spp. (Srinuanpan et al., 2017; Choix et al., 2017a; Srinuanpan et al., 2018) have the potential to accumulate more lipids in the biomass, which could be converted into biodiesel (Kao et al., 2012a, b).

Sirikulrat et al. (2013) used photobioreactor for microalgal cultivation for upgrading biogas and to deracinate CO_2 (Sirikulrat and Koonaphapdeelert, 2013) and reported 70% reduction in CO_2 content in biogas. Subsequently, the CH_4 concentration was reported to increase by more than 80%. The development of gas cycle in switching mode allowed continuous biogas enrichment, which has a great potential for biogas upgrading applications.

The microalgae-based bag photobioreactor utilized by Zhao et al. effectively enriched biogas and simultaneously reduced the nutrient content in digestate (Zhao et al., 2015). Some authors effectively upgraded biogas that was carried out under the optimum conditions for microalgal growth and biogas effluent nutrient reduction (Yan et al., 2016). The low light intensity with long photoperiod and moderate light intensity (350 μmol m^{-2} s^{-1}) with middle photoperiod (14-hour light: 10-hour dark) obtained the best biogas CO_2 removal and biogas upgrading effects. Mann et al. studied the use of *Chlorella* sp. for conditioning biogas (Mann et al., 2009) and reduced the CO_2 and H_2S content of biogas by up to 97.07% and 100%, respectively. Also, an increase in microalgae cell count was documented, providing exciting alternatives to producing algal ingredients. Sumardiono et al. (2014) studied Biogas utilization as a carbon dioxide provider for *Spirulina platensis* Culture. Fernández et al. studied capturing carbon dioxide from biogas in anaerobic digesters treating sewage sludge and food waste and revealed methane (CH_4) production increase of 11%−16% for food waste and 96%−138% for sewage sludge over the first 24 hours. The Potential CO_2 reductions of 8%−34% for sewage sludge and 3%−11% for food waste was achieved (Bajón Fernández et al., 2014).

14.3 Factors affecting biogas upgrading and lipid production

14.3.1 Biogas composition

Depending on the feedstock, the composition of CO_2 varies in biogas, limiting the growth of microalgae used for biogas upgradation (Amaro et al., 2011; Wang et al., 2014; González-Sánchez and Posten, 2017). Several studies have reported the microalgal tolerance for high CO_2 and CH_4 content in biogas (Tang et al., 2011; Zhao et al., 2015; Thawechai et al., 2016) during which some microalgal species like *Chlorella* and *Scenedesmus* have shown an excellent tolerance which could be effectively used for biogas upgradation. Both high and low concentrations of CO_2 in raw biogas may limit microalgae growth. The high concentration of CO_2 in biogas could lead to acidic pH in the medium, which could reduce the CO_2 capacity of microalgae.

The concentration of CH_4 in the biogas is an important factor for the biogas upgradation by microalgae. Various studies have been conducted on the effect of CH_4 concentration on microalgae growth. Meier et al. reported that *Nannochloropsis gaditana* was not affected even by a 100% increase in CH_4 content (Meier et al., 2015), while Kao et al. (2012a, b) asserted that with the 20%−80% increase in CH_4, the growth of *Chlorella* sp. MM-2 was limited. Besides CO_2 and CH_4, the H_2S content highly affects microalgae's growth, CO_2 fixation, and biomass production ability. Depending on the feedstock for anaerobic digestion, the H_2S content (0.005%−2%) varies greatly in the biogas (González-Sánchez and Posten, 2017). Although only a few studies have been conducted on the effects of H_2S on the biogas enrichment by microalgae, *Chlorella* sp have been reported to perform better (concerning biomass production) when H_2S content is <150 ppm in biogas (Kao et al., 2012a, b). Mera et al. reported that some microalgae convert H_2S into sulfate, microalgae later assimilate that for growth (Mera et al., 2016). However, the H_2S assimilation is strongly pH-dependent, and acidic pH favors the H_2S uptake. González-Sánchez and Posten (2017) reported that the H_2S inhibits the growth of microalgae at a concentration of >200 ppm.

14.3.2 Gas flow rate

The rate at which biogas is injected into the medium for biogas enrichment using microalgae is an important parameter for the CO_2 removal efficiency of microalgae. A suitable gas flow rate is important for the proper mixing of gas in the medium (Yan and Zheng, 2013). Increasing the gas flow rate could increase microalgal biomass productivity and CO_2 removal efficiency. However, if the gas flow rate is much higher than the ability of the microalgae to fix CO_2, the pH of the culture medium would become acidic and result in less availability of nutrients, thus affecting the activity of enzymes and reducing the performance of microalgae (Tang et al., 2011). The higher gas flow rate increases the size of a gas bubble in the medium, which results in less surface area and less absorption of CO_2 by microalgal cells (Kao et al., 2012a, b). In addition to this, the evaporation rate of the culture medium also increases when the gas flow rate increases (Su et al., 2017). Srinuanpan et al. (2017) reported that at biogas' minimal gas flow rate, the microalgae might completely fix CO_2 from biogas. Still, the biogas upgradation will be performed at a lower rate, resulting in less CO_2 removal rate and less productivity of purified biogas. However, Khan et al. (2018) achieved higher biogas CO_2 removal at a lower gas flow rate under optimal conditions. Many authors have reported that the gas flow rate affects the lipid and fatty acid composition in the microalgal biomass (Widjaja et al., 2009; Binnal and Babu, 2017). The enhanced gas flow rate could lead to a high C: N ratio, which plays an important role in improving the lipid content in the cells (Su et al., 2017).

14.3.3 Light Energy

14.3.3.1 *Light intensity and photoperiods*

Microalgal growth is highly dependent on the light intensity and photoperiods. During photorespiration, the microalgae consume their stored carbohydrates and lipids, which results in low biomass productivity (Zhao et al., 2013). An increased photoperiod with low light intensity supports the higher performance of microalgae concerning growth and biomass production (Xue et al., 2011; Yan and Zheng, 2013).

The microalgae could not grow at a higher light intensity due to the light absorption limitation because high light intensity above the saturation point may destroy the photosystems I and II (light adapter system) (Jeong et al., 2013; Yan and Zheng, 2013). Several researchers have reported avoiding photoinhibition by decreasing the photoperiod during high light intensity (Yan and Zheng, 2013; Xue et al., 2011).

The effect of various light intensities on the efficiency of microalgae for the removal of CO_2 from biogas has been widely reported. The response of different microalgal species is different to various light intensities. Khan et al. (2018) reported that the CO_2 fixation by *Chlorella minutissima* was in the range of 75%−85.4% at 1296 m mol m^{-2} s^{-1} light intensity, while some other microalgae require 150−350 mmol m^{-2} s of light intensity (Yan and Zheng, 2013; Ouyang et al., 2015). Pertinently the photoperiod and optimal light intensity may vary with different microalgal species, for which microalgal cell density in the medium could be the main reason. The light intensity in the initial cultivation phase when the microalgal cell density is low may cause photoinhibition and damage the microalgal cells. Using low-intensity light during the initial phase of microalgal cultivation could save electricity. In the later stages of microalgal cultivation, having high cell density may cause the cell shading effect, which could result in poor light penetration (Das et al., 2011; Yan et al., 2013; Pilon et al., 2011). Few authors like Srinuanpan et al. (2018) reported that the gradual increase in light intensity during the biogas upgrading might depend on the microalgal growth curve.

This approach may promote microalgal growth and increase the efficiency to fix CO_2 from the biogas even though many authors reported higher microalgal growth with reduced lipid content during higher light intensities, which might be because of higher cell division rather than lipid accumulation (Yeesang and Cheirsilp, 2011; Cheirsilp and Torpee, 2012; George et al., 2014). However, different studies showed different trends, explaining that during higher light intensities, a saturation of light may occur, resulting in low photosynthetic activity (Liu et al., 2012; Difusa et al., 2015; He et al., 2015).

14.3.3.2 *Light wavelength*

The wavelength of light is a crucial parameter for photosynthesis in the case of microalgae. The light source must comprise the absorption bands for chlorophyll pigments (Yan et al., 2016b). Plants in general and microalgae, in particular, absorb light wavelengths in the range of 400–700 nm for photosynthesis through chlorophyll a (450–475 nm), chlorophyll b (630–675 nm), and carotenoids (400–500 nm). Many authors have reported that the optimum wavelength that supports microalgae biomass production and CO_2 deracination from biogas falls in the range of 600–700 nm (Ho et al., 2011; Kim et al., 2013; Yan et al., 2016b). It has also been reported that the operational cost for indoor cultivation of microalgae could be reduced by using efficient light sources such as narrow-band light-emitting diode. It is reported that microalgae use higher wavelengths (red light) more effectively than any other light wavelength (Zhao et al., 2013), while the shorter wavelengths (blue light) has an inhibiting effect on photosynthesis (Yan et al., 2016a). However, some authors have claimed that using mixed light wavelengths may avoid light saturation and the photoinhibition effect on microalgae (Yan et al., 2016b; Zhao et al., 2013). Zhao et al. (2013) and Yan et al. (2016a, 2016b) reported that the combination of red and blue wavelengths yielded higher microalgal biomass and also affected the lipid composition than in a single wavelength (Teo et al., 2014; Wahidin et al., 2013). In *Nannochloropsis* sp., it is reported that microalgae showed higher fatty acid content when it was exposed to green wavelength (Kurano and Miyachi, 2005). Green light wavelength gives a stimulus for enhancing the fatty acid content on the microalgae (Ra et al., 2016; Anderson et al., 2012; Yamamoto, 2016). However, some controversial reports state that microalgae like *Tetraselmis* sp. and *Nannochloropsis* sp. can produce higher amount of lipid content in biomass when exposed to blue light than any other wave lengths (Teo et al., 2014). The effect supports that the enzymes such as ribulose bisphosphate carboxylase/oxygenase (Rubisco) and carbonic anhydrase are active under blue light and could affect the utilization of CO_2 in microalgae cells (Anderson et al., 2012; Kim et al., 2014; Ra et al., 2016; Yamamoto, 2016).

14.3.4 Concentration and source of nutrients

Nitrogen is the nutrient source for the biological synthesis of proteins, lipids, and nucleic acids in microalgal cells. Nitrogen content within a limit in the growth medium may lead to the healthy growth of microalgae (Arumugam et al., 2013) instead of excessive nitrogen content, which may lead to inhibition of microalgal growth (Li et al., 2008). The commonly used chemical as a nitrogen source for microalga is KNO_3, $NaNO_3$, Urea, $CaNO_3$, NH_4NO_3, and NH_4Cl (Arumugam et al., 2013). Among them, KNO_3 is more commonly used as it sources both nitrogen and potassium. The nitrogen concentration and its sources greatly affect the CO_2 fixing efficiency and lipid production in microalgae. It has been reported that *Chlorella* sp. and *Scenedesmus* sp showed improvement in growth when grown in 0.62–0.8 g L^{-1} of

TABLE 14.2 Microalgal cultivation for biogas upgrading and wastewater treatment.

Microalgae	Composition of biogas	Wastewater characteristics	Methane content after upgrading (%v/v)	Removal efficiency			References
				Removal of COD (%)	Removal of TN (%)	Removal of TP (%)	
Chlorella minutissima	Synthetic biogas containing 70% CH_4, 29.5% CO_2, and 0.5% H_2S	COD (1745 ppm), TN (1815 ppm), TP (48 ppm)	96	85	85	100	Toledo-Cervantes et al. (2017)
Chlorella pyrenoidosa	Crude biogas containing 63.84% CH4, 31.02% CO_2, 0.56% O_2, 3.59% H_2O, and <0.005% H_2S	pH (6.79), COD (962.41 ppm), TN (360.28 ppm), TP (31.46 ppm)	92.87	92.67	80.87	79.33	He et al. (2015)
Chlorella vulgaris	Crude biogas containing 67.32% CH4, 34.45% CO_2, 0.62% O_2, 3.66% H_2O, and <0.005% H2S	pH (6.84), COD (1013.87 ppm), TN (308.75 ppm), TP (9.93 ppm)	80.04	89.27	51.32	63.22	Zhao et al. (2015)
Nannochloropsis oleoabundans	Crude biogas containing 67.32% CH_4, 34.45% CO_2, 0.62% O_2, 3.66% H_2O, and <0.005% H_2S	pH (6.84) COD, (1013.87 ppm), TN (308.75 ppm), TP (9.93 ppm)	80.06	61.03	51.74	54.33	Zhao et al. (2015)

KNO$_3$ (Srinuanpan et al., 2017; Zhao et al., 2013; Tongprawhan et al., 2014). Adding nitrogen gradually to the growth medium improves the CO$_2$ fixing capacity of microalga rather than adding it to one, which may inhibit the growth of microalgae (Srinuanpan et al., 2018). Although growing microalgae under nitrogen-deficient conditions could be effective for lipid productivity (Siaut et al., 2011), the nitrogen-deficient conditions may also limit the microalgal growth and will result in low CO$_2$ removal efficiency.

14.4 Biogas methane enrichment with wastewater treatment

Recently, many studies reported combined biogas methane enrichment with wastewater treatment (Table 14.2). Cultivating microalgae like *Chlorella* spp. on domestic wastewater for biogas methane enrichment achieved >90% methane content in effluent gas and >80% reduction in chemical oxygen demand, >70% reduction in total nitrogen, and >70% reduction in total phosphorus in wastewater. Similarly, using *Scenedesmus* sp. for upgrading the biogas with >70% of methane content on swine wastewater the removal of >60% chemical oxygen demand, >60% total nitrogen, and >70% total phosphorus was achieved (Xu et al., 2015; Prandini et al., 2016; Yan et al., 2016b; Cheirsilp et al., 2017). *Nannochloropsis oleoabundans* also showed around 80% methane content in upgraded biogas grown on biogas slurry (Zhao et al., 2015). The effect of high turbidity and the presence of xenobiotic compounds in wastewater may inhibit the growth of microalgae by reducing the light penetration and photosynthesis, resulting in reduced efficiency of CO$_2$ from biogas. Therefore the wastewater might be pretreated or diluted before using as a growth medium for microalgae for biogas upgradation. Like Xu et al. (2015), some authors reported that chemical oxygen demand below 1600 ppm in wastewater is suitable for growing microalgae for upgrading biogas.

Most of the studies, as discussed earlier, combined two processes, that is, biogas methane enrichment and wastewater treatment. However, Khan et al. (2018) has integrated biogas methane enrichment with biodiesel production and wastewater treatment. The authors have reported an 81.6% increase in methane after biogas upgradation and 26% lipid accumulation in dry biomass. This is a win-win approach in which the high value of upgraded biogas is obtained along with the production of feedstock for biodiesel production.

14.5 Economics in biogas enrichment

The economic analysis of the working of any biogas enrichment process varies depending on the assessment brought into use. The chief financial cost heads include total investment cost, operation and maintenance cost, and cost of CH$_4$ loss during the biogas enrichment process, which are used to determine the biogas/gas-processing cost (GPC) (Hao et al., 2008). GPC indicates the total expenditure of producing 1 m^3 of methane-enriched biogas or biomethane. The GPC depends on the plant capacity, and technology adopted, location, and operating process conditions (Dirkse, 2008; Petersson and Wellinger, 2009; Warren, 2012). The average investment cost for a biogas plant is not

TABLE 14.3 Comparative analysis of the different biogas methane enrichment technologies.

Biogas enrichment method	Mode of operation	Absorbent/ adsorbent	Energy consumption (kWh Nm^{-3})	Gas-processing cost ($ Nm^{-3})	Value addition after upgradation ($ Nm^{-3})	Capital	Maintenance	References
Water scrubbing	Physical absorption	Water	0.2−0.3; 0.4−0.5	0.146	0.2099	1,132,400	16,986	Scholwin and Nelles (2013); Jonsson and Westman (2011); Niesner et al. (2013); Scholwin, Nelles (2013); Jonsson and Westman (2011); Niesner et al. (2013)
Chemical absorption	Chemical absorption	Amines (MEA, DMEA), alkali solutions	0.06−0.17; 0.05−0.18	0.101	0.2549	2,264,800	66,811	Johansson (2008); Jonsson and Westman (2011); Bauer et al. (2013); Kadam and Panwar (2017)
Pressure swing adsorption	Adsorption	Molecular sieves	0.16−0.35; 0.29−0.60	0.282	0.0739	1,981,700	63,414	Scholwin and Nelles (2013); Jonsson and Westman (2011); Singhal et al. (2017)
Membrane	Permeation	Polymer of silicone rubbers, cellulose etc.	0.18−0.35; 0.25 0.20−0.3	0.192	0.1639	2,264,800	28,310	Scholwin and Nelles (2013); Bauer et al. (2013); Vrbová and Ciahotný (2017)
Cryogenic	Multistage compression and condensation	Not required	0.18−0.25; 0.42−0.63	0.495	−0.1401	1,141,459	499,388	Scholwin and Nelles (2013)
Physical absorption	Physical absorption	Organic solvents, polyethylene glycol	0.10−0.5; 0.23−0.33	0.15	0.2	1,132,400	16,986	Bauer et al. (2013); Scholwin and Nelles (2013); Masebinu et al. (2014)
Biological	Absorption	Microalgae	0.00507	0.011	0.3449	3938.6	6.31	Khan et al. (2018)

particular and varies with the equipment installed, the inflow of feedstock, composition, and purity of feedstock [Severn Wye Energy Agency (SWEA)].

Various experiments and estimations have been done for the cost calculation for biogas methane enrichment. As per the SWEA report, the capital cost of membrane-based biogas methane enrichment ranges from $8150 to $8500/(m^3 biomethane/h) with a plant capacity of 100m^3 h^{-1}. With similar specifications the water scrubbing biogas methane enrichment is $11,250/(m^3 biomethane/h) and $11,600/(m^3 biomethane/h) in case of pressure swing absorption. However, the GPC for <2000 m^3 h^{-1} upgrading plant with pressurized water scrubbing technique was < $0.12 m^{-3} of biomethane (Dirkse, 2008). Some authors reported a decrease in gas procession cost with an increase in feed volume ($0.1–$0.7 m^{-3} of biomethane) (Ahmad et al., 2012; Petersson and Wellinger, 2009; Warren, 2012). As per a study done by Khan et al. (2018), the biogas enrichment was $ 0.011 m^{-3} using microalgae. The total cost of the experiment was estimated to be nearly $393.86 on a laboratory scale. The difference in the price of enrichment between traditional methods and employing microalgae is enormous, which indicates that microalgae could be economically suitable. However, there is still a need for research for microalgal feasibility on a large scale. A detailed economic report with different methods of biogas methane enrichment is presented in Table 14.3.

14.6 Conclusions

Coupling the biogas methane enrichment with the microalgal biomass for biodiesel production is a new, eco-friendly, and sustainable approach. Integrating the process of CO$_2$ fixation with biodiesel production is a win-win approach. Using wastewater as a medium for microalgae for biogas upgrading can solve the energy problem and wastewater treatment. Among all the biogas methane enrichment technologies, the microalgal approach is highly economical and user-friendly. The combined techno-economic and environmental impact assessment is a useful tool for policymakers and producers in identifying potential bottlenecks and choosing suitable implementations.

References

Abatzoglou, Boivin, 2009. A review of biogas purification processes. Biofuels, Bioproducts and Biorefining 42–71.

Ahmad, F., Lau, K.K., Shariff, A.M., Murshid, G., 2012. Process simulation and optimal design of membrane separation system for CO$_2$ capture from natural gas. Computers & Chemical Engineering 36, 119–128.

Amaro, H.M., Guedes, A.C., Malcata, F.X., 2011. Advances and perspectives in using microalgae to produce biodiesel. Applied Energy 88 (10), 3402–3410. Available from: https://doi.org/10.1016/j.apenergy.2010.12.014. Portugal: Elsevier Ltd.

Anderson, J.M., et al., 2012. Towards elucidation of dynamic structural changes of plant thylakoid architecture. Philosophical Transactions of the Royal Society B: Biological Sciences 367 (1608), 3515–3524. Available from: https://doi.org/10.1098/rstb.2012.0373. Australia: Royal Society.

Arumugam, M., et al., 2013. Influence of nitrogen sources on biomass productivity of microalgae *Scenedesmus bijugatus*. Bioresource Technology 131, 246–249. Available from: https://doi.org/10.1016/j.biortech.2012.12.159. India: Elsevier Ltd.

Bajón Fernández, Y., et al., 2014. Carbon capture and biogas enhancement by carbon dioxide enrichment of anaerobic digesters treating sewage sludge or food waste. Bioresource Technology 159, 1–7. Available from: https://doi.org/10.1016/j.biortech.2014.02.010. United Kingdom: Elsevier Ltd.

Bauer, F., et al., 2013. Biogas upgrading – technology overview, comparison and perspectives for the future. Biofuels, Bioproducts and Biorefining 7 (5), 499–511. Available from: https://doi.org/10.1002/bbb.1423. Wiley.

Bhattacharya, SC, 1998. State of the art biomass combustion. Energy Sources 20, 113–130.

Binnal, P., Babu, P.N., 2017. Statistical optimization of parameters affecting lipid productivity of microalga *Chlorella protothecoides* cultivated in photobioreactor under nitrogen starvation. South African Journal of Chemical Engineering 23, 26–37. Available from: https://doi.org/10.1016/j.sajce.2017.01.001. India: Elsevier B.V.

BP Statistical Review of World Energy 2017. https://www.bp.com/en/global/corporate/energy-economics/statistical-review-of-world-energy.html.

Cheirsilp, B., Torpee, S., 2012. Enhanced growth and lipid production of microalgae under mixotrophic culture condition: Effect of light intensity, glucose concentration and fed-batch cultivation. Bioresource Technology 110, 510–516. Available from: https://doi.org/10.1016/j.biortech.2012.01.125. Thailand.

Cheirsilp, B., Thawechai, T., Prasertsan, P., 2017. Immobilized oleaginous microalgae for production of lipid and phytoremediation of secondary effluent from palm oil mill in fluidized bed photobioreactor. Bioresource Technology 241, 787–794. Available from: https://doi.org/10.1016/j.biortech.2017.06.016. Thailand: Elsevier Ltd.

Choix, F.J., et al., 2017a. Nutrient composition of culture media induces different patterns of CO_2 fixation from biogas and biomass production by the microalga *Scenedesmus obliquus* U169. Bioprocess and Biosystems Engineering 40 (12), 1733–1742. Available from: https://doi.org/10.1007/s00449-017-1828-5. Mexico: Springer Verlag.

Choix, F.J., et al., 2017b. CO_2 Removal from biogas by *Cyanobacterium* Leptolyngbya sp. CChF1 isolated from the lake Chapala, Mexico: optimization of the temperature and light intensity. Applied Biochemistry and Biotechnology 183 (4), 1304–1322. Available from: https://doi.org/10.1007/s12010-017-2499-z. Mexico: Humana Press Inc.

Das, P., et al., 2011. Enhanced algae growth in both phototrophic and mixotrophic culture under blue light. Bioresource Technology 102 (4), 3883–3887. Available from: https://doi.org/10.1016/j.biortech.2010.11.102. Singapore.

Demirbas, A., 2009. Biofuels securing the planet's future energy needs. Energy Conversion and Management 50 (9), 2239–2249. Available from: https://doi.org/10.1016/j.enconman.2009.05.010. Turkey.

Difusa, A., et al., 2015. Effect of light intensity and pH condition on the growth, biomass and lipid content of microalgae *Scenedesmus* species. Biofuels 6 (1–2), 37–44. Available from: https://doi.org/10.1080/17597269.2015.1045274. India: Taylor and Francis Ltd.

Dirkse, E.H.M., 2008. Biogas upgrading using the DMT TS-PWS technology. DMT Environmental Technology.

George, B., et al., 2014. Effects of different media composition, light intensity and photoperiod on morphology and physiology of freshwater microalgae *Ankistrodesmus falcatus* – a potential strain for bio-fuel production. Bioresource Technology 171, 367–374. Available from: https://doi.org/10.1016/j.biortech.2014.08.086. India: Elsevier Ltd.

González-Sánchez, A., Posten, C., 2017. Fate of H_2S during the cultivation of *Chlorella* sp. deployed for biogas upgrading. Journal of Environmental Management 191, 252–257. Available from: https://doi.org/10.1016/j.jenvman.2017.01.023. Germany: Academic Press.

Hao, J., Rice, P.A., Stern, S.A., 2008. Upgrading low-quality natural gas with H_2S- and CO_2-selective polymer membranes: Part I. Process design and economics of membrane stages without recycle streams. Journal of Membrane Science. 320, 1–2. https://www.sciencedirect.com/science/article/abs/pii/S0376738808002597.

He, Q., et al., 2015. Effect of light intensity on physiological changes, carbon allocation and neutral lipid accumulation in oleaginous microalgae. Bioresource Technology 191, 219–228. Available from: https://doi.org/10.1016/j.biortech.2015.05.021. China: Elsevier Ltd.

Ho, S.H., et al., 2011. Perspectives on microalgal CO_2-emission mitigation systems – a review. Biotechnology Advances 29 (2), 189–198. Available from: https://doi.org/10.1016/j.biotechadv.2010.11.001. Taiwan.

Jeong, H., Lee, J., Cha, M., 2013. Energy efficient growth control of microalgae using photobiological methods. Renewable Energy 54, 161–165. Available from: https://doi.org/10.1016/j.renene.2012.08.030. South Korea.

Johansson, Nina, 2008. Production of liquid biogas, LBG, with cryogenic and conventional upgrading technology - Description of systems and evaluations of energy balances. Master thesis. Department of Technology and scocirty. Lunds University. Available from: https://lup.lub.lu.se/luur/download?func = downloadFile&recordOId = 4468178&fileOId = 4469242.

Jonsson, S., Westman, J., 2011. Cryogenic biogas upgrading using plate heat exchangers. Master's Thesis within the Sustainable Energy Systems Master's programme. Available from: https://publications.lib.chalmers.se/records/fulltext/145544.pdf.

Kadam, R., Panwar, N.L., 2017. Recent advancement in biogas enrichment and its applications. Renewable and Sustainable Energy Reviews 73, 892–903. Available from: https://doi.org/10.1016/j.rser.2017.01.167. India: Elsevier Ltd.

Kao, C.Y., Chiu, S.Y., Huang, T.T., Dai, L., Hsu, L.K., et al., 2012a. Ability of a mutant strain of the microalga *Chlorella* sp. to capture carbon dioxide for biogas upgrading. Applied Energy 93, 176–183. Available from: https://doi.org/10.1016/j.apenergy.2011.12.082. Taiwan: Elsevier Ltd.

Kao, C.Y., Chiu, S.Y., Huang, T.T., Dai, L., Wang, G.H., et al., 2012b. A mutant strain of microalga Chlorella sp. for the carbon dioxide capture from biogas. Biomass and Bioenergy 36, 132–140. Available from: https://doi.org/10.1016/j.biombioe.2011.10.046. Taiwan.

Khan, S.A., et al., 2018. Potential of wastewater treating *Chlorella minutissima* for methane enrichment and CO_2 sequestration of biogas and producing lipids. Energy 150, 153–163. Available from: https://doi.org/10.1016/j.energy.2018.02.126. India: Elsevier Ltd.

Kim, T.H., et al., 2013. The effects of wavelength and wavelength mixing ratios on microalgae growth and nitrogen, phosphorus removal using *Scenedesmus* sp. for wastewater treatment. Bioresource Technology 130, 75–80. Available from: https://doi.org/10.1016/j.biortech.2012.11.134. South Korea: Elsevier Ltd.

Kim, D.G., et al., 2014. Manipulation of light wavelength at appropriate growth stage to enhance biomass productivity and fatty acid methyl ester yield using *Chlorella vulgaris*. Bioresource Technology 159, 240–248. Available from: https://doi.org/10.1016/j.biortech.2014.02.078. South Korea: Elsevier Ltd.

Kumar, K., Dasgupta, C.N., Nayak, B., Lindblad, P., Das, D., 2011. Development of suitable photobioreactors for CO_2 sequestration addressing global warming using green algae and cyanobacteria. Bioresource Technology 102, 4945–4953. Available from: https://www.sciencedirect.com/science/article/abs/pii/S0960852411001234.

Kurano, N., Miyachi, S., 2005. Selection of microalgal growth model for describing specific growth rate-light response using extended information criterion. Journal of Bioscience and Bioengineering 100 (4), 403–408. Available from: https://doi.org/10.1263/jbb.100.403. Japan.

Li, Y., et al., 2008. Effects of nitrogen sources on cell growth and lipid accumulation of green alga Neochloris oleoabundans. Applied Microbiology and Biotechnology 81 (4), 629–636. Available from: https://doi.org/10.1007/s00253-008-1681-1. Canada.

Liu, J., et al., 2012. Effects of light intensity on the growth and lipid accumulation of microalga Scenedesmus sp. 11-1 under nitrogen limitation. Applied Biochemistry and Biotechnology 166 (8), 2127–2137. Available from: https://doi.org/10.1007/s12010-012-9639-2. China.

Mann, G., Schlegel, M., Schumann, R., Sakalauskas, A., 2009. Biogas-conditioning with microalgae. Agronomy Research 7. Available from: https://www.cabdirect.org/cabdirect/abstract/20093231945.

Masebinu, S.O., Aboyade, A., Muzenda, E., 2014. Enrichment of biogas for use as vehicular fuel: a review of the upgrading techniques. International Journal of Research in Chemical, Metallurgical and Civil Engineering 1, 2349.

Meier, L., et al., 2015. Photosynthetic CO_2 uptake by microalgae: an attractive tool for biogas upgrading. Biomass and Bioenergy 73, 102–109. Available from: https://doi.org/10.1016/j.biombioe.2014.10.032. Chile: Elsevier Ltd.

Mera, R., Torres, E., Abalde, J., 2016. Influence of sulphate on the reduction of cadmium toxicity in the microalga *Chlamydomonas moewusii*. Ecotoxicology and Environmental Safety 128, 236–245. Available from: https://doi.org/10.1016/j.ecoenv.2016.02.030. Spain: Academic Press.

Niesner, J., Jecha, D., Stehlík, P., 2013. Biogas upgrading technologies: State of art review in European region. Chemical Engineering Transactions. Czech Republic: Italian Association of Chemical Engineering – AIDIC. Available from: https://doi.org/10.3303/CET1335086.

Ministry of New and Renewable Energy, 2017. National Biogas and Manure Management Programme . Available from: http://mnre.gov.in/file-manager/UserFiles/physical_targets_for-NBMMP-2017-18.pdf.

Ouyang, Y., et al., 2015. Effect of light intensity on the capability of different microalgae species for simultaneous biogas upgrading and biogas slurry nutrient reduction. International Biodeterioration and Biodegradation 104, 157–163. Available from: https://doi.org/10.1016/j.ibiod.2015.05.027. China: Elsevier Ltd.

Petersson, A., Wellinger, A., 2009. Biogas upgrading technologies: Development & innovation. Swedish gas center & Nova Energie.

Pilon, L., Berberoğlu, H., Kandilian, R., 2011. Radiation transfer in photobiological carbon dioxide fixation and fuel production by microalgae. Journal of Quantitative Spectroscopy and Radiative Transfer 112 (17), 2639–2660. Available from: https://doi.org/10.1016/j.jqsrt.2011.07.004. United States.

Prandini, J.M., et al., 2016. Enhancement of nutrient removal from swine wastewater digestate coupled to biogas purification by microalgae *Scenedesmus* spp. Bioresource Technology 202, 67–75. Available from: https://doi.org/10.1016/j.biortech.2015.11.082. Brazil: Elsevier Ltd.

Ra, C.H., et al., 2016. Effects of light-emitting diodes (LEDs) on the accumulation of lipid content using a two-phase culture process with three microalgae. Bioresource Technology 212, 254–261. Available from: https://doi.org/10.1016/j.biortech.2016.04.059. South Korea: Elsevier Ltd.

Scholwin, F., Nelles, M., 2013. Energy flows in biogas plants: analysis and implications for plant design, in The Biogas Handbook: Science. Production and Applications 212–227. Available from: https://doi.org/10.1533/9780857097415.2.212. Germany: Elsevier Inc.

Siaut, M., et al., 2011. Oil accumulation in the model green alga *Chlamydomonas reinhardtii*: characterization, variability between common laboratory strains and relationship with starch reserves. BMC Biotechnology 11. Available from: https://doi.org/10.1186/1472-6750-11-7. France.

Singhal, S., et al., 2017. Upgrading techniques for transformation of biogas to Bio-CNG: a review. International Journal of Energy Research 41 (12), 1657–1669. Available from: https://doi.org/10.1002/er.3719. India: John Wiley and Sons Ltd.

Sirikulrat, K., Koonaphapdeelert, S., 2013. Capture of carbon dioxide in biogas by using *Chlorella* sp. in photobioreactor. In: Proc. International Graduate Research Conference.

Srinuanpan, S., et al., 2017. Strategies to improve methane content in biogas by cultivation of oleaginous microalgae and the evaluation of fuel properties of the microalgal lipids. Renewable Energy 113, 1229–1241. Available from: https://doi.org/10.1016/j.renene.2017.06.108. Thailand: Elsevier Ltd.

Srinuanpan, S., Cheirsilp, B., Prasertsan, P., 2018. Effective biogas upgrading and production of biodiesel feedstocks by strategic cultivation of oleaginous microalgae. Energy 148, 766–774. Available from: https://doi.org/10.1016/j.energy.2018.02.010. Thailand: Elsevier Ltd.

Su, Y., et al., 2017. Progress of microalgae biofuel's commercialization. Renewable and Sustainable Energy Reviews 74, 402–411. Available from: https://doi.org/10.1016/j.rser.2016.12.078. China: Elsevier Ltd.

Sumardiono, S., et al., 2014. Utilization of biogas as carbon dioxide provider for *Spirulina platensis* culture. Current Research Journal of Biological Sciences 6 (1), 53–59. Available from: https://doi.org/10.19026/crjbs.6.5498. Maxwell Scientific Publication Corp.

Tang, D., et al., 2011. CO_2 biofixation and fatty acid composition of *Scenedesmus obliquus* and *Chlorella pyrenoidosa* in response to different CO_2 levels. Bioresource Technology 102 (3), 3071–3076. Available from: https://doi.org/10.1016/j.biortech.2010.10.047. China.

Teo, C.L., et al., 2014. Enhancing growth and lipid production of marine microalgae for biodiesel production via the use of different LED wavelengths. Bioresource Technology 162, 38–44. Available from: https://doi.org/10.1016/j.biortech.2014.03.113. Malaysia: Elsevier Ltd.

Thawechai, T., et al., 2016. Mitigation of carbon dioxide by oleaginous microalgae for lipids and pigments production: Effect of light illumination and carbon dioxide feeding strategies. Bioresource Technology 219, 139–149. Available from: https://doi.org/10.1016/j.biortech.2016.07.109. Thailand: Elsevier Ltd.

Toledo-Cervantes, A., et al., 2017. Influence of the gas-liquid flow configuration in the absorption column on photosynthetic biogas upgrading in algal-bacterial photobioreactors. Bioresource Technology 225, 336–342. Available from: https://doi.org/10.1016/j.biortech.2016.11.087. Spain: Elsevier Ltd.

Tongprawhan, W., Srinuanpan, S., Cheirsilp, B., 2014. Biocapture of CO2 from biogas by oleaginous microalgae for improving methane content and simultaneously producing lipid. Bioresource Technology 170, 90–99. Available from: https://doi.org/10.1016/j.biortech.2014.07.094. Thailand: Elsevier Ltd.

Vrbová, V., Ciahotný, K., 2017. Upgrading biogas to biomethane using membrane separation. Energy and Fuels 31 (9), 9393–9401. Available from: https://doi.org/10.1021/acs.energyfuels.7b00120. Czech Republic: American Chemical Society.

Wahidin, S., Idris, A., Shaleh, S.R.M., 2013. The influence of light intensity and photoperiod on the growth and lipid content of microalgae *Nannochloropsis* sp. Bioresource Technology 129, 7–11. Available from: https://doi.org/10.1016/j.biortech.2012.11.032. Malaysia: Elsevier Ltd.

Wang, X.W., et al., 2014. Biomass, total lipid production, and fatty acid composition of the marine diatom *Chaetoceros muelleri* in response to different CO_2 levels. Bioresource Technology 161, 124–130. Available from: https://doi.org/10.1016/j.biortech.2014.03.012. China: Elsevier Ltd.

Wang, Z., et al., 2016. Selection of microalgae for simultaneous biogas upgrading and biogas slurry nutrient reduction under various photoperiods. Journal of Chemical Technology and Biotechnology 91 (7), 1982–1989. Available from: https://doi.org/10.1002/jctb.4788. China: John Wiley and Sons Ltd.

Warren, K.E.H., 2012. A techno-economic comparison of biogas upgrading technologies in Europe. Master's, Department of Biological and Environmental Science, University of Jyvaskyla, Renewable Energy, Sustainable Energy Technologies.

Widjaja, A., Chien, C.C., Ju, Y.H., 2009. Study of increasing lipid production from fresh water microalgae *Chlorella vulgaris*. Journal of the Taiwan Institute of Chemical Engineers 40 (1), 13–20. Available from: https://doi.org/10.1016/j.jtice.2008.07.007. Indonesia.

Xu, J., et al., 2015. Nutrient removal and biogas upgrading by integrating freshwater algae cultivation with piggery anaerobic digestate liquid treatment. Applied Microbiology and Biotechnology 99 (15), 6493–6501. Available from: https://doi.org/10.1007/s00253-015-6537-x. China: Springer Verlag.

Xue, S., Su, Z., Cong, W., 2011. Growth of *Spirulina platensis* enhanced under intermittent illumination. Journal of Biotechnology 151 (3), 271–277. Available from: https://doi.org/10.1016/j.jbiotec.2010.12.012. China.

Yamamoto, Y., 2016. Quality control of photosystem II: The mechanisms for avoidance and tolerance of light and heat stresses are closely linked to membrane fluidity of the thylakoids. Frontiers in Plant Science 7 (2016). Available from: https://doi.org/10.3389/fpls.2016.01136. Japan: Frontiers Media S.A.

Yan, C., Zheng, Z., 2013. Performance of photoperiod and light intensity on biogas upgrade and biogas effluent nutrient reduction by the microalgae Chlorella sp. Bioresource Technology 139, 292–299. Available from: https://doi.org/10.1016/j.biortech.2013.04.054. China: Elsevier Ltd.

Yan, C., et al., 2013. Effects of various LED light wavelengths and intensities on the performance of purifying synthetic domestic sewage by microalgae at different influent C/N ratios. Ecological Engineering. China 51, 24–32. Available from: https://doi.org/10.1016/j.ecoleng.2012.12.051.

Yan, C., et al., 2016a. The effects of various LED (light emitting diode) lighting strategies on simultaneous biogas upgrading and biogas slurry nutrient reduction by using of microalgae *Chlorella* sp. Energy 106, 554–561. Available from: https://doi.org/10.1016/j.energy.2016.03.033. China: Elsevier Ltd.

Yan, C., Zhu, L., Wang, Y., 2016b. Photosynthetic CO_2 uptake by microalgae for biogas upgrading and simultaneously biogas slurry decontamination by using of microalgae photobioreactor under various light wavelengths, light intensities, and photoperiods. Applied Energy 178, 9–18. Available from: https://doi.org/10.1016/j.apenergy.2016.06.012. China: Elsevier Ltd.

Yeesang, C., Cheirsilp, B., 2011. Effect of nitrogen, salt, and iron content in the growth medium and light intensity on lipid production by microalgae isolated from freshwater sources in Thailand. Bioresource Technology 102 (3), 3034–3040. Available from: https://doi.org/10.1016/j.biortech.2010.10.013. Thailand.

Zhao, Y., et al., 2013. Effects of various LED light wavelengths and intensities on microalgae-based simultaneous biogas upgrading and digestate nutrient reduction process. Bioresource Technology 136, 461–468. Available from: https://doi.org/10.1016/j.biortech.2013.03.051. China: Elsevier Ltd.

Zhao, Y., et al., 2015. Performance of three microalgal strains in biogas slurry purification and biogas upgrade in response to various mixed light-emitting diode light wavelengths. Bioresource Technology 187, 338–345. Available from: https://doi.org/10.1016/j.biortech.2015.03.130. China: Elsevier Ltd.

Aquatic microalgal biofuel production

Fayaz A. Malla[1] and Suhaib A. Bandh[2]

[1]Assistant Professor, Department of Environmental Science, Govt. Degree College Tral, Jammu and Kashmir, India [2]Assistant Professor, Environmental Science, Higher Education Department, Government of Jammu and Kashmir, Srinagar, India

15.1 Introduction

The global energy crisis and rising greenhouse gas emissions have sparked a renewed interest in renewable energy sources that are more ecologically benevolent. With its potential to replace fossil-based fuels in the long term, microalgal biofuels have been recognized as the key renewable energy resources for sustainable development. Technically sound, commercially viable and cost-effective, microalgal biofuels require no extra land, need low water consumption, and reduce greenhouse gas emissions into the atmosphere. However, due to the low biomass concentration and high downstream costs, commercial production of microalgae biodiesel is still in its primary stages. Microalgal biofuel production can be done by building sophisticated photobioreactors, inventing low-cost methods for biomass collection, drying, and oil extraction at a reduced cost. Improvements in the genetic engineering tactics for controlling environmental stress conditions and the development of metabolic pathways that produce high lipids can also lead to their large-scale commercial production.

15.2 Factors affecting the microalgae growth rate

The growth, development, cultivation, and harvesting of microalgae depend on many parameters. To increase algal biofuels' production, the necessary growth elements must be available in an optimum range while being economical and environment-friendly. The ability of microalgae to produce the metabolites like fatty acids, carbohydrates, and proteins also depends on some external factors like nutrient content in the growth medium, duration and intensity of sunlight, temperature, CO_2 concentration, and pH (Walker, 1954). These factors, in combination, decide the photosynthetic efficiency, CO_2 fixation, and fatty acid composition in microalgae (Juneja et al., 2013). Different parameters affecting algal growth are as follows:

15.2.1 Environmental factors

15.2.1.1 Potential of hydrogen (pH)

The potential of hydrogen (pH) measures acidity or alkalinity in the medium. pH has an essential role in cultivating microalgae by impacting metabolism and regulating the solubility and availability of CO_2 and nutrients (micro and macro) (Chen and Durbin, 1994). For most algae the optimum pH is in the range of 6—8. However, during photosynthesis, the pH of the medium tends to increase (10—11) due to CO_2 fixation during light exposure (Hansen, 2002; Pragya et al., 2013a) and decrease in pH during the dark phase due to the respiratory processes. The CO_2 and pH are related by the chemical equilibrium of various chemical species (CO_2, H_2CO_3, HCO_3^-, and CO_3^{-2}) in the medium. Increasing CO_2 concentration promotes higher biomass accumulation but leads to lower pH value, which later harms microalgal physiology. However, if pH increases beyond the limit, photosynthesis can be affected due to CO_2 deficiency.

15.2.1.2 Nutrients

Nutrients, especially nitrogen and phosphorus, are essential for microalgal growth. Silicon is also a vital element, especially for saltwater algae. Besides the macronutrients, trace metals, such as Fe, Mg, Mn, B, Mo, K, Ca, and Zn, are essential for microalgal growth. For lipid and fatty acid production in microalgae, nitrogen is essential. Several authors have reported higher lipid production in nitrogen scare conditions.

Xin et al. (2010) reported a 53% increase in lipid production in limited phosphorus conditions in *Scenedesmus* sp. and a 30% increase in lipid production in nitrogen-limited conditions. Physiological and morphological changes can be triggered in microalgae due to excess or deficiency of nutrients by inhibiting the metabolic pathways. Researchers are now focusing on the two-phase growth system. In the first phase, algae are grown in a nutrient-abundant medium and then transferred to a nutrient-deficient medium where lipid accumulation is boosted (Mercer and Armenta, 2011).

15.2.1.3 Temperature

Temperature is one of the vital parameters that highly influences microalgal growth, nutrient requirements, and biomass composition. In temperate and subtropical climates, microalgae grow at a time of year, while they grow around the year in tropical regions. Microalgae grown for lipids have an optimum temperature range of 15°C—40°C. Various authors have reported (Pragya et al., 2013a) that several oil-producing microalgal species grow best between 25°C and 30°C. However, the optimal temperature varies from species to species and the desired algal response.

Species like *Nannochloropsis oculata* and *Chlorella vulgaris* have an optimum temperature of 25°C. While increasing the temperature from 20°C to 25°C, the lipid content is doubled from 7.90% to 14.92% in *N. oculata*. In the case of *C. vulgaris*, increasing temperature from

25°C to 30°C decreased the lipid content from 14.71% to 5.90% (Converti et al., 2009). Temperature also plays a role in determining the content of proteins, starch, and pigments (Liu and Lee, 2000; Tripathi et al., 2002). Some authors observed the higher temperature responsible for saturated lipid accumulation and lower temperature for unsaturated lipid accumulation (Chaisutyakorn et al., 2018).

15.2.1.4 Carbon dioxide

Carbon dioxide (CO_2) is a vital carbon source for photosynthetic algae for biomass production. In the absence of CO_2, the growth and productivity of microalgae are limited. In microalgae, carbon can be utilized in CO_2, carbonates, or bicarbonates. Based on estimates of the average chemical composition of microalgal biomass, roughly around 1.8 tons of CO_2 are needed to grow 1 ton of biomass. Natural diffusion of CO_2 from the air into the medium may not be enough to overcome by bubbling air into the medium. Other options include using pure CO_2, which is relatively expensive, or a waste source of CO_2 like flue gas from a boiler. Flue gas can grow microalgae without harmful effects (Doucha et al., 2005; Muradyan et al. (2004) reported that lower CO_2 concentrations increase in 22:6 (n53) PUFA, whereas 14:0 fatty acids were found to be predominant at higher CO_2 levels. Along with the change in composition, increased CO_2 was also observed to increase the amount of fatty acid accumulation in *Dunaliella salina*.

15.2.2 Processing parameters

15.2.2.1 Mixing

Regular mixing of growth medium with microalgal cells is an essential part of microalgal cultivation for uniform distribution of light and temperature throughout the medium (Rocha et al., 2003). Without any forced mixing, microalgae at the surface of the medium absorb all the available light and become photo inhibited, while algae deeper in the media becomes devoid of light. Mixing of the medium can be done by many methods like using a mechanical stirrer or a paddlewheel in raceways and bubbling in gas for the homogenous mixture. However, it is also to note that many microalgal species may not tolerate intensive mixing. A study on the effect of mixing on *Spirulina platensis* in three ways (mixing with a magnetic agitator inside the column, bubbling air into the column, and recirculating through a pump) showed that the growth of the microalga was highest using a bubble column. It is reported that the mixing of *Chlorella* sp. showed increased growth of microalgae significantly (up to 30%) (Persoone et al., 1980). A study on *Phaeodactylum tricornutum* revealed that mixing, mass transfer, and carbon dioxide consumption could be the essential factors limiting microalga's growth (Contreras, 1997).

15.2.2.2 Light intensity

For cultivating microalgae, photosynthesis plays an essential role in its growth rate, lipid productivity, and biomass production. Depth of culture medium, algal cell density and type of algae depends on light intensity. The culture's higher depth and cell concentration need higher light intensity to penetrate through the medium. However, direct sunlight or high-density artificial light may prove a photo-inhibitor for microalgae. Besides, photoinhibition overheating caused by natural and artificial illumination is unexpected and may be avoided. For cultivating microalgae in a flask, the suitable light intensity is around 1000 lux which could be increased to 5000–10,000 lux for higher volumes. Light/dark cycles are required for effective photosynthesis in microalgae. Light phase is needed for the photochemical phase to produce adenosine triphosphate and nicotinamide adenine dinucleotide phosphate and dark phase for the biochemical phase to synthesize essential biomolecules for microalgal growth. Microalgae can utilize a part of the visible spectrum for CO_2 fixation during photosynthesis. Depending on the temperature and other growth conditions, around 25% of the biomass produced during the light phase may be utilized during the dark period.

Light intensity is a determining factor for the biochemical composition of microalgae; for example, in *P. tricornutum*, an increase in the synthesis of protein has been observed at a low light intensity. Similarly, using *Desmarestia viridis*, the lipid content increased in the absence of light while triglycerides, free fatty acids, free alcohols, and sterols decreased (Smith et al., 1993). In contrast, Renaud et al. (1991) reported a decrease in lipid and polyunsaturated fatty acids in the presence of high light intensity. Besides the visible light, Ultraviolet (UV) light can impact microalgal growth. UV-B has been reported to cause more damage to microalgal cells than UV-A radiation. UV-B radiations cause DNA damage, while UV-A damage is limited to indirect loss through enhanced production of reactive oxygen and hydroxyl radicals. Algae overcome the damage caused by UV radiation by the development of protective cell walls, increased synthesis of carotenoids, and other pigments (Pessoa, 2012; Rastogi and Incharoensakdi, 2013).

15.3 Feedstock harvesting

Harvesting of microalgal biomass is the biggest challenge in its commercialization. Low-cost harvesting of microalgae from the medium is essential for industrial use. Gravity sedimentation, flocculation, centrifugation, filtration, flotation, and electrocoagulation are the commonly used technologies for harvesting microalgae (Vandamme et al., 2013). Under typical conditions (open ponds or bioreactors), the maximum biomass reached is $1g\,L^{-1}$ because of light limitations in dense cell mass. However, in most industrial processes, this could reach up to $100g\,L^{-1}$. Harvesting accounts for approximately 30% of the total production cost of microalgal products. Since the cell density of microalgae is very low in the medium, it takes a high cost and energy for harvesting the microalgae. To reduce biomass production costs, employing an efficient harvesting technology is feasible. The selection of the harvesting technique depends on the properties of microalgae and the value of the biomass products. Centrifugation for harvesting valuable algal products is essential for culture with 90% microalgal biomass. However, the capital and processing cost is very high, making this method unsuitable for bulk products like biofuels. For processing large volumes like municipal wastewater treatment, gravity sedimentation is used, which is cost-effective and energy-saving. However, this method could not make it viable for harvesting microalgae, as the settling rate is too slow for microalgal cells. Filtration is the other classical separation method, but it is only suitable for large filamentous species such as *Spirulina*.

According to reports, the most economical way for harvesting microalgal biomass is flocculation and flotation (Molina Grima et al., 2003). Flocculants are multivalent cations or cationic polymers that neutralize the negative charges on the surface of microalgal cells, thus allowing them to clump more easily. The choice of flocculants is product dependent; for example, flocculants based on heavy metals may not be suitable for food and feed products. Some flocculants are specific to the condition, such as chitosan, an edible flocculant, but is less effective in seawater or medium having salinities move than seawater (Bilanovic et al., 1988). Another method is the flotation method used in combination with flocculants by employing dissolved or dispersed air to produce bubbles that lift the algal cells to the surface for harvesting. Traditional air flotation used in sewage works is very energy-intensive, thus adding to the harvesting costs if used with microalgae. Generally, the harvesting process could be divided into two steps, bulk harvesting and thickening. The purpose of bulk harvesting technologies, including flocculation, flotation, and gravity sedimentation, is to separate biomass from the culture medium. Thickening, with much higher energy consumption than bulk harvesting, includes filtration and centrifugation.

15.3.1 Sedimentation/centrifugation/flocculation

Gravity sedimentation is the simplest and energy-saving method for microalgal harvesting. However, under certain circumstances, these methods are inefficient and highly time-consuming. Few authors have reported that for the sedimentation method, *S. platensis* and *Cylindrotheca fusiformis* showed maximum sedimentation rate $[11\,g\,(L.h)^{-1}]$ (Griffiths et al., 2012). Sedimentation is feasible for harvesting the microalgae with large cells size and high biomass density. Besides, in phycoremediated wastewater, harvesting microalgal cells is impossible through gravity sedimentation. The centrifugation method could be employed in such cases, which is not economically feasible. The cost of centrifugation harvesting could be reduced by 82% if harvesting is done at multiple stages and increases the algal culture flow rate (Dassey and Theegala, 2013). However, even under improved harvesting methods, the approximate energy required to produce algal oil and the oil price were higher than $9.6\,kW\,h\,L^{-1}$ and $0.86/L, respectively (Dassey and Theegala, 2013). Therefore it is recommended that centrifugation may be employed in an alga with valuable composition (Barros et al., 2015). For harvesting microalgae on a large scale, chemical flocculation is the most efficient method (Chatsungnoen and Chisti, 2016). Chemical flocculants include aluminum salt, ferric salt, and synthetic polyacrylamide polymers. Some authors have reported 95% efficiency in harvesting microalgae using aluminum sulfate and ferric chloride as flocculants (Chatsungnoen and Chisti, 2016). The use of flocculants may cause contamination of biomass if used in excessive quantities, limiting the use of biomass in food, feed, and biofertilizer production. Filtration, which is carried out commonly on membranes with the assistance of a suction pump, is particularly suitable for collecting algae of low densities. However, the solid particles in wastewaters would lead to the eventual clogging of the filter and reduce the harvesting efficiency. Therefore filtration is not feasible in harvesting algae grown on most wastewaters with solids.

15.3.2 Fungi-assisted sedimentation

Some filamentous fungal species can also be used to harvest microalgae by forming large pellets. Denser spherical aggregates (pelleted fungi) could be effectively utilized for algal harvesting (Gultom and Hu, 2013). The fungal pellets, formed by electrostatic forces, hydrophobicity, and specific fungal cell wall composition interactions, are essential in microalgal harvesting. The harvesting by fungal-assisted sedimentation could be done through a sieve which is much more comfortable than gravitational sedimentation. Compared to other harvesting techniques, fungal-assisted sedimentation is cost-effective and energy-efficient. The nontoxicity of these fungal species may not limit their use in industries. The algal biomass entrapped in fungal pellets (2–10 mm) could be separated from solids (<0.2 mm) in wastewaters by a simple mechanism (Chan et al., 2018). Various filamentous fungal species like *Aspergillus* sp., *Mucor* sp., *Rhizopus* sp., and *Penicillium* sp. have been successfully used for harvesting microalgae (Liu et al., 2008; Zhou et al., 2013) with 90%–100% efficiency (Gultom and Hu, 2013). The formation and structure of fungal pellets depend on factors like fungal species used, carbon–nitrogen ratio (C/N ratio), cell wall composition, inoculation ratio, temperature, agitation, pH, and suspended solids in the medium (Chan et al., 2018). Zhou et al. (2013) reported that acidic conditions were favorable to the formation of *Aspergillus oryzae* pellets, while Liu et al. (2008) did not observe a significant difference in the formation of *Rhizopus oryzae* pellets with a pH range of 2.5–7.0. Agitation speed which is a key parameter determining the size of fungal pellets demonstrated the fact that diameters formed in flasks agitated at 50 and 150 rpm were about 3 and 8 mm, respectively (Zhou et al., 2013).

15.3.3 Flotation

Flotation is in use since the 1960s for microalgae production (Wiley, Brenneman, and Jacobson, 2009). In this method the algal cell adheres to the microbubbles formed physically or chemically. It makes them float onto the surface of the culture medium for harvesting. The high energy consumption for compressor and saturator is a severe drawback to this method. Besides this, the algal cells block the nozzle for bubble generation, increasing the maintenance cost. However, some advancement has been made in the floatation method in recent times. Wiley et al. (2009) established a suspended flotation system that used surfactant to generate microbubbles for algal harvesting and reported that the suspended floatation method is energy-saving (Wiley et al., 2009). However still, research needs to be done on the suspended floatation method. In electofloatation, microbubbles are generated using hydrogen and oxygen gas by electrolysis. The favorable part of using electrofloatation is that it neutralizes the negative charge on the surface of algal cells, causing coagulation that is favorable for algal harvesting (Shin et al., 2017). Foreseeing the potential risk of hydrogen gas explosion, this method has many modifications and improvements before its commercial applicability.

15.4 Algae biofuels and conversion process

Algal biomass after the transformation has a vast application in pharmaceutical, food nutrients, chemicals, and various other nonconventional energy products. Fig. 15.1 depicts the multiple products from various conversion processes. Microalgal biomass's main

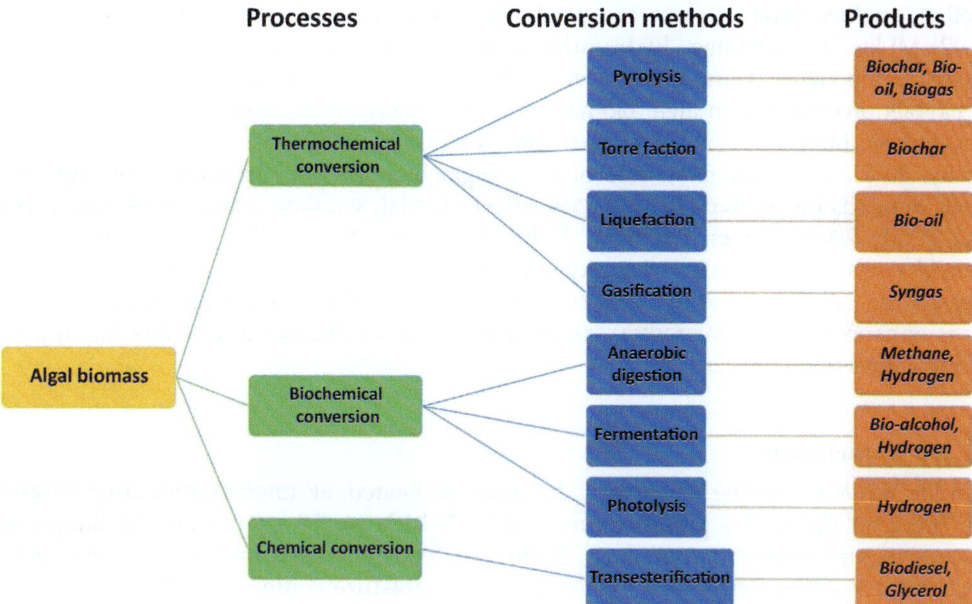

FIGURE 15.1 Algal biomass conversion processes and the products.

components (carbohydrates, lipids, and proteins) could be converted into bioethanol, biodiesel, biogas, biohydrogen, syngas through the biochemical, chemical, thermochemical, and direct combustion routes. The biochemical processes being more eco-friendly and energy-saving may not be suitable for large-scale or commercial production because of their low transformation efficiency, sophisticated process, and noneconomical nature (Nyberg et al., 2015). The different types of the conversion process are discussed next.

15.4.1 Thermochemical conversion

Thermochemical conversion is a decomposition process. The algal biomass is converted into solid, liquid, and gaseous biofuel and classified into pyrolysis, torrefaction, gasification, and liquefaction based on temperature and pressure conditions (Demirbas, 2010). The thermochemical process is the most straightforward process among other processes for converting the algal biomass into biofuels.

15.4.1.1 Pyrolysis

In pyrolysis the microalgal biomass heated at a temperature range of 400°C–600°C at a pressure of 0.1 MPa for 0.5–1.05 hours. under anaerobic conditions yields bio-oil, gas, and biochar. The biochar produced by pyrolysis has a high surface area, which is more suitable for soil fertility improvement and waste treatment. Various authors have explored the behavior of microalgal biomass under fast and slow pyrolysis (in the presence or absence of catalyst) (Miao and Wu, 2004; Grierson et al., 2009; Babich et al., 2011; Harman-Ware et al., 2013).

Bio-oil is another product from the algal biomass pyrolysis having a high calorific value ($31-42$ MJ kg^{-1}) containing hydrocarbons from lipids and nitrogenous compounds as principal components (Harman-Ware et al., 2013; Miao and Wu, 2004). It has been reported that bio-oils produced from algal biomass are much more stable than bio-oils produced from lignocellulosic biomass, which increases the scope of algal bio-oils in the future (Suali and Sarbatly, 2012). The conversion rate of microalgal biomass and production of high-quality bio-oils depends on the type of pyrolysis, effect of catalysts, time, temperature and pressure (Peng et al., 2000; Pan et al., 2010). It has been investigated the catalytic pyrolysis for *Nannochloropsis* sp. decreased the oxygen content in bio-oils (from 30 to 19 wt.%) and increased the calorific value (from 24.6 to 32.5 MJ kg^{-1}) (Pan et al., 2010). The effect of catalysts is enhanced by the microwave-assisted pyrolysis of microalgal biomass for improved, calorific values (Du et al., 2011).

15.4.1.2 Liquefaction

Under liquefaction the microalgal biomass is heated at lower temperatures ranging from 300°C to 350°C at a pressure range of $5-20$ MPa for $50-60$ minutes in the presence of a catalyst and solvent (Goyal et al., 2008). The feedstock used for liquefaction should be in slurry form with microalgal biomass having a moisture content of around $80\%-90\%$, a suitable property for the liquefaction process (Patil t al., 2008). One of the most significant limitations of liquefaction is its complicated and costly process (McKendry, 2002). Liquefaction of *Botryococcus braunii* at 300°C has reported 64% bio-oil with a calorific value of 45.8% MJ kg^{-1} (McKendry, 2002). However, Matsui et al. investigated the effect of Fe (CO)5eS as the catalyst for liquefaction of *Spirulina* sp. at 350°C for 60 minutes under 5 MPa and reported 14.6% increase in bio-oil production and $32-33$ MJ kg^{-1} calorific value (McKendry, 2002).

15.4.1.3 Gasification

Gasification is a process that converts the microalgal biomass into fuel gases like H_2, CO, CH_4 at a high temperature of $800°C-1000°C$ under insufficient oxygen conditions (McKendry, 2002). Typically, gasification of microalgal biomass is done in four stages (drying, pyrolysis, combustion, and reduction). Some authors have explored the impact of temperature on *Spirulina* sp. biomass and reported enhancement in H_2 production and decreased CO_2, CO, and CH_4 production during an increase in temperature (Hirano et al., 1998). Moreover, few authors have investigated cogasification of microalgal biomass with lignocellulosic biomass in bubbling fluidized-bed reactors and fluidized-bed reactors (Yang et al., 2013; Alghurabie et al., 2013). Due to the high moisture content in microalgal biomass, the drying process in a typical gasification process takes lots of energy. Some researchers have come up with supercritical water gasification to reduce energy consumption, which avoids the drying stage and formation of tar or char (Amin, 2009; Haiduc et al., 2009). The supercritical gasification is done beyond the critical point of water (374°C and 22.1 MPa). However, an increase in H_2 and CH_4 production in supercritical gasification of *Nannochloropsis* sp. in the presence of NaOH and KOH (Guan, Wei, Ning, tian, and Gu, 2013) is reported.

15.4.1.4 Direction combustion

Direction combustion is a thermochemical technique in which the algal biomass is burned in the open air or excess air at around 800°C−1000°C in a furnace or a steam turbine mainly for energy production (Suganya et al., 2016). Microalgal biomass with low moisture content (<50%) is suitable for this process (McKendry, 2002). Drying and grinding of microalgal biomass are necessary for effective combustion, which incur additional costs to energy production (Goyal et al., 2008). The cost of energy production from direct combustion is slightly higher than pyrolysis and gasification due to the need for biomass pretreatment, such as dehydrating, cutting, and crushing, before introducing into the combustion chamber. However, efficient energy consumption can reduce the additional cost for drying and grinding (Rafael Kandiyoti Alan Herod Keith Bartle Trevor Morgan, 2017). Kadam suggested that the idea of coal−algae cofiring could reduce the emission of CO_2 by recycling the CO_2 from the combustion process to microalgal cultivation (Kadam, 2001), which could lead to lower emission of GHGs into the atmosphere as a golden opportunity for carbon credit program.

15.4.1.5 Torrefaction

Torrefaction is the thermochemical process in which microalgal biomass is converted into biochar with liquids and gases as by-products (Li et al., 2018). The biochar produced in torrefaction has a higher calorific value than biochar produced from pyrolysis (Sukiran et al., 2017). Torrefaction has been classified into (1) dry torrefaction and (2) wet torrefaction. Dry torrefaction is also known as mild pyrolysis or low-temperature pyrolysis and is done at a temperature range of 200°C−300°C, under atmospheric pressure in anaerobic conditions (Sukiran et al., 2017). Atmospheric nitrogen is commonly used to prevent the oxidation of biochar during processing. Some authors have used noninert gases to make the process cost-effective (Uemura et al., 2017; Chen et al., 2015). In torrefaction, temperature is a critical parameter and could impact the production like in *Chlamydomonas* sp. JSC4, the solid yield had decreased from 93.9% to 51.3% when the torrefaction temperature increased from 200°C to 300°C (Chen et al., 2016). Similarly, in the torrefaction of *S. obliquus*, CNW-N yield decreased from 86.37% to 63.23% when temperature increased from 200°C to 300°C (Chen et al., 2014). The main limitation of dry torrefaction is its requirement of the predrying step for reducing the moisture content <10% (Bach and Skreiberg, 2016). Wet torrefaction, done under inert conditions within the temperature range of 180°C−260°C, is a viable option to reduce energy costs (Bach and Skreiberg, 2016). Subcritical water is used in wet torrefaction. Wet torrefaction is more efficient than dry torrefaction as it requires relatively low temperature and holding due to the highly reactive reaction media (hydrothermal treatment) (Bach and Skreiberg, 2016).

15.4.2 Biochemical conversion

Biochemical conversion of microalgal biomass can be done by various processes like anaerobic digestion, fermentation, and photobiological technique. The microalgal biomass is converted into biofuels by time-consuming enzymatic processes (Naik et al., 2010). The specific properties of microalgal biomass, such as moisture content (80%−90%) and high polysaccharides with no lignin, make it a significant feedstock for biogas production (McKendry, 2002). Yet many other factors like resistance to the cell wall, C/N, and high protein content negatively affect biogas production. Therefore pretreatment of microalgal biomass, which later enhances biogas production, is required (Passos et al., 2014). The microalgal biomass with a low C/N ratio co digested with biomass with high carbon content like white paper can significantly increase biogas production (Suganya et al., 2016; Yen and Brune, 2007). In biomass with high protein content, the ammonium production increases, slowing the anaerobic digestion; however, using salt-resistant microorganisms, such an issue could be overcome (Rafael Kandiyoti Alan Herod Keith Bartle Trevor Morgan, 2017).

The bioethanol production process includes pretreatment followed by enzymatic hydrolysis and fermentation. Microalgal biomass is considered a favorable feedstock for enzymatic hydrolysis because lignin is absent in their cell wall, reducing digestion resistance compared to lignocellulosic biomass. Various pretreatment techniques (mechanical, physical, thermal, chemical, and combined) are used to break the cell wall and release the lipids for enzymatic hydrolysis and bioethanol production (Onumaegbu et al., 2018). Comparison of the ethanol production in pretreated and untreated biomass of *Chlorococcum* sp showed 60% more ethanol production when biomass was pretreated for lipid extraction (Harun et al., 2010). Such investigations indicate that lipid extraction for biodiesel with the fermentation of leftover carbohydrates for bioethanol production can be a favorable route for algal biofuels. *C. vulgaris* showed 65% conversion to bioethanol from biomass (Hirano et al., 1997).

Hydrogen could be produced from microalgal biomass by oxygenic photosynthesis, which involves hydrogenase or nitrogenase enzyme (Hirano et al., 1997). During the photosynthetic process, the water molecule in microalgae is converted into protons, electrons, and oxygen; a proton is then subsequently turned into H_2 by hydrogenase enzymes (Cantrell et al., 2008). However, oxygen generation inhibits the hydrogenase enzyme, mainly H_2 production. The issue can be resolved by properly separating oxygen produced during the photosynthetic processes.

15.4.3 Chemical reaction

The chemical conversion process includes the extraction of lipids from microalgal biomass and converting them into biodiesel by the transesterification process. The transesterification process takes place between lipids (triglyceride) and alcohol in the presence of a suitable catalyst (potassium hydroxide or sodium hydroxide, hydrochloric, or sulfuric acid) which may be acidic, alkaline, or enzyme based to produce fatty acid methyl ester (FAME) and glycerol (Suganya et al., 2016). The catalytic activity of the acid catalyst is low and requires a high temperature during the initial step of the transesterification reaction. The rate of reaction of alkaline catalyst is around 4000 times higher than the acid catalyst, making it suitable for large-scale operations (Kiran et al., 2014; Chen and Durbin, 1994). The glycerol by-products could produce different pharmaceutical and cosmetic products (Suganya et al., 2016). Milano et al. obtained up to 90% FAME conversion from two different microalgal species such as *Spirogyra* and *Oedigonium* in the presence of an alkaline catalyst (Milano et al., 2016). For ex situ transesterification the retreatment is a highly energy-intensive, time taking process and consumes almost 80% of material preparation cost (Jazzar et al., 2015).

In comparison to ex situ, in situ transesterification sidelines the pretreatment requirement, resulting in overall process cost economical (Jazzar et al., 2015). However, some authors have investigated both in situ and ex situ transesterification on *Schizochytrium limacinum* and reported less biodiesel production in in situ than ex situ transesterification (Johnson and Wen, 2009). Salam et al. (2016) suggested that the biogas production from the residual microalgal biomass from the biodiesel production process could provide the energy required to cultivate or separate microalgae from dilute solution before the in situ process. The in situ transesterification is rapid, which combine two steps, including lipid extraction and transesterification which results in higher FAME production than ex situ transesterification. *Chlorella pyrenoidosa* having lipid content of 56.2% produced 95% biodiesel by in situ transesterification in presence of hexane as cosolvent and 0.5 M sulfuric acid at 90°C for 2 hours (Lam and Lee, 2012). However, that moisture content of 37.1% could completely impede the in situ transesterification process by producing fatty acid and diglyceride rather than fatty acid esters (Lam and Lee, 2012).

15.5 Approaches to increase algal biofuel production

The hindrances in microalgal biofuel production are the development of new processing techniques for large-scale commercial biomass production. There is much scope in algal biomass processing, which may integrate biofuel production and add-in products with high economic value. Many approaches were developed to overcome the bottlenecks in microalgal biofuel production, such as the concept of biorefinery, nutrient composition, and photobioreactors. Recent advancements in the technological field have cleared the way forward for the large-scale commercialization of biofuels based on microalgae. Biorefinery with algal biofuel production as combined processes could make the overall process cost-effective and generate high valued products. To increase the algal biofuel productivity and overall development of algal biomass, various approaches like optimum growth factors could be explored; for example, in *C. vulgaris* iron has been reported to be highly effective on growth and lipid productivity (Liu et al., 2008). The contemporary world's biochemical tools and techniques for metabolic and synthetic biology could enhance biofuel production. Some of these approaches for improving biofuel production from microalgae are discussed next.

15.5.1 Selection of species or strain

Microalgae originate from freshwater or marine water and grow in water-based medium or sediment. Even though microalgae are single-celled organisms, they exist in chains or groups. Their size can vary from micrometers (μm) to a few hundred micrometers. Like plants, they also use solar energy, CO_2, and nutrients for growth. Currently, more than 3 million microalgal species have been discovered, and numerous important chemicals have been identified in their biomass (Chisti, 2007). Proteins, carbohydrates, and fatty acids (lipid), antioxidants, enzymes, carotenoids, peptides, toxins, sterols, and polymers are produced by different kinds of microalgal species (Borowitzka and Borowitzka, 1988) and have been quantified (Wu et al., 2005; Zhou et al., 2014, 2015). Proteins in microalgal biomass could be converted into products used as animal feed, nutritional supplements, and biofertilizers, whereas carbohydrates and lipids can be used for biofuel production. Therefore selecting microalgal strain is important concerning end products and economic opportunities. Microalgae have been classified as Rhodophytes, Chlorophytes, Cyanobacteria, and Chromophytes, and each group contains hundreds of species and thousands of strains (Hochman and Zilberman, 2014). Out of these thousands of strains, a few have been explored for valuable uses. Chlorophytes (green algae) and Cyanobacteria (blue and green algae) have been the most widely used microalgae for developing commercial products like biomass, phycocyanin, proteins, fatty acids, carbohydrate, carotenoids, astaxanthin, eicosatetraenoic acid (EPA), docosahexaenoic acid (DHA), polysaccharides, etc. (Varfolomeev and Wasserman, 2011). Some popular products, including the applications of some microalgal species, are listed in Table 15.1.

15.5.2 Exploration of growth conditions and nutrients

Biofuel production from microalgal biomass depends on algal growth. Identifying the parameters to enhance biofuel production could resolve the financial bottlenecks related to algal biofuels. The economics of the nutrients such as nitrogen and phosphorus for

TABLE 15.1 Application of microbial species.

Products	Industrial use	Algal sources
Biomass feedstock	Biofuel production	*Arthrospira (Spirulina) platensis, Chlorella vulgaris, Isochrysis galbana, Phaeodactylum tricornutum, Scenedesmus spp., Nannochloropsis oculata*
Phycocyanin	Dietary supplement with medicinal and nutritional value, cosmetics	*Arthrospira (Spirulina) platensis*
Protein	Dietary supplement	*Arthrospira (Spirulina), Scenedesmus spp*
Vitamin B12	Pharmaceutical industry	*Arthrospira (Spirulina)*
Fatty acids	Biofuel production, pharmaceutical industry	*Aphanizomenon flos-aquae, Odontella aurita, I. galbana, P. tricornutum*
β-Carotene	Pharmaceutical industry	*A. flos-aquae, Dunaliella salina*
Carbohydrate	Biofuel production, dietary supplement, animal feed	*Chlorella spp., C. vulgaris*
Carotenoids	Pharmaceutical industry	*D. salina, Haematococcus pluvialis*
Astaxanthin	Pharmaceutical industry	*H. pluvialis*
EPA	Pharmaceutical industry	*O. aurita, Schizochytrium sp., Nannochloropsis sp.*
DHA	Pharmaceutical industry	*Schizochytrium sp., Crypthecodinium cohnii*
Polysaccharides	Pharmaceutical industry	*Porphyridium cruentum*

microalgae growth needs proper understanding. Using wastewater for microalgal cultivation can prove a sustainable approach with the energy-saving mechanism. The phytoremediation potential and microalgal biomass growth have been widely investigated by various authors (Cabanelas et al., 2013; Cho et al., 2013; Sharma et al., 2020; Saranya and Shanthakumar, 2020). Integrating phytoremediation with biofuel production could prove eco-friendly and economical (Beal et al., 2012). An efficient microalgal cultivation strategy is needed to be identified for higher biomass production and subsequent biofuel production. The most widely used method for large-scale production of algal biomass is raceway ponds. Raceway ponds are shallow, open environments and have paddles wheel for proper mixing and circulation of nutrients and CO_2. Raceway ponds are economical to operate but have limitations, such as the high probability of contamination from water and air (Mata, Martins, and Caetano, 2010). The use of photobioreactors for algal cultivation eliminates contaminants from air or rainwater. Moreover, in a photobioreactor, there is better control on pH, temperature, and light intensity which can give better results in high cell density (Mata et al., 2010). Despite the high efficiency, the high production cost is its biggest limitation.

Microalgae has a great potential to grow in different aquatic media like freshwater, seawater, municipal or industrial wastewaters if proper nutrients are available for growth (Zhou et al., 2014). Marine water enriched with nitrogen, phosphorus, and other micronutrients is commonly used for cultivating marine microalgae (Molina Grima et al., 2003). In contrast to marine or freshwater, wastewater contains many nutrients suitable for microalgal biomass production. Many authors have reported the potential of microalgae cultivation on wastewater for biofuel production (Hoffmann, 1998; Mobin and Alam, 2014; Zhou et al., 2014). However, algal production from wastewater has many bottlenecks like properties of wastewater from various sources, climatic conditions, pretreatment approaches, nutrient content (C:N and C:P ratio), turbidity and contaminants (Zhou et al., 2014). Another important factor is the availability of land for microalgal cultivation, which adds to the cost of cultivation. For low production cost the microalgal cultivation site should be accessible to water, sunlight, CO_2, and nutrients (Slade and Bauen, 2013).

15.5.3 Advancement in harvesting technology

For large-scale commercialization of biofuel production, the major bottleneck is a cost-effective process. Improving the postcultivation techniques could be an important stage for increased productivity and economic production. In this regard, the drying of algal biomass and thermal processes need to be further investigated. The sun drying of microalgal biomass is an inexpensive method that could be adopted. However, sun drying has many limitations, and for this process, various other economical drying methods have been reported (Prakash et al., 1997).

To cultivate microalgae, various methods, including chemical, biological, mechanical, and electrical, have been employed. For better results, many authors have reported a combination of these processes. Chemical flocculation methods have been widely used as a pretreatment technique for improving the buoyancy of algal biomass. Under the mechanical techniques, centrifugation is performed for collecting biomass quickly with high consistency (Kumar et al., 1981).

15.5.4 Metabolic engineering

The metabolic engineering technique could be used for a substantial improvement in microalgal biofuel production. The first report on a full-length gene encoding the acetyl-coenzyme A carboxylase in *Cyclotella cryptica* was reported by Roessler (1990). The limitations related to light harvesting were aptly dealt with by engineering the photosynthetic apparatus targeting the antenna (de Mooij et al., 2015).

15.5.5 Biorefinery approach

In a biorefinery approach, various bioproducts are produced in a cost-effective and eco-friendly way (Yanqun Li et al., 2008; Malla and Khan, 2018). It encompasses the technological and economic parameters for cultivation, processing, and harvesting to develop maximum outputs (Borowitzka, 2013; Khan et al., 2019). The biorefinery is a synergistic and sustainable approach to address phycological constraints. It emphasizes the sustainable approaches for optimum production and generation of various products with complete utilization of algal biomass.

15.6 Economic prospects of microalgal biofuels

In the supply value chain for microalgal feedstock production, several steps (Fig. 15.2) incur biofuel production costs. The various parameters affecting the cost of biofuel production are as follows:

Feedstock cultivation: the various cost included in the feedstock cultivation are (1) capital cost (cost of land, strain development, research, and cultivation system), (2) running cost (cost of human resources, nutrients, electricity, etc.)

Feedstock harvesting and extraction: the various costs that fall under feedstock harvesting and extraction are dewatering and drying, machinery, human resources, electricity, etc.

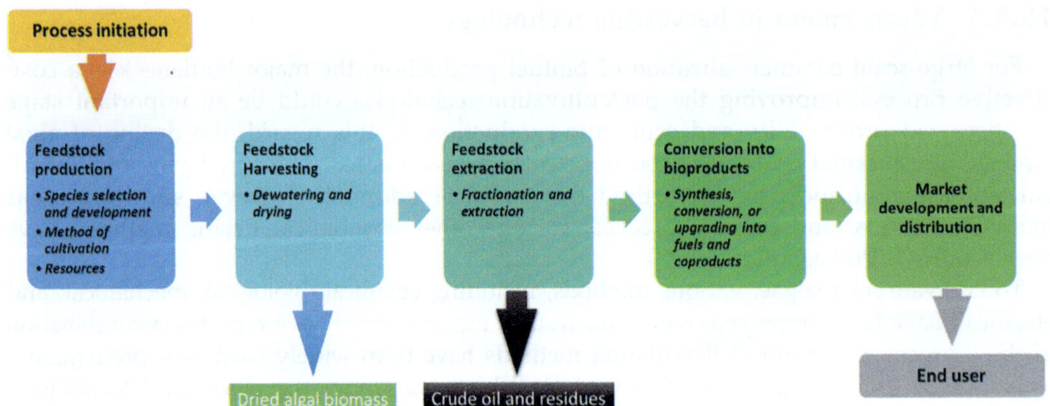

FIGURE 15.2   Supply value chain of microalgal feedstock.

Conversion into products: cost of producing biofuel from feedstock (biodiesel or bioethanol).

The final cost of biofuel will be all the components, including the marketing and distribution costs.

15.6.1 The production cost of algal biomass feedstock

Various cost components could be identified, starting from cultivating microalgal biomass feedstock. The US Department of Energy reported the projected cost for production of dry algal biomass to be around $1227 with approximately 87% of the cost incurred by the cultivation, which includes the cost of species selection/development, cultivation methods, and other resources. Harvesting, which includes dewatering and drying algal biomass, incurs 10% of the cost.

15.6.2 The production cost of crude algal oil

Various institutions such as National Renewable Energy Laboratory, Sandia National Laboratories, New Mexico State University, and Seambiotic worked on estimating the price of algal crude based on a common framework (Brennan and Owende, 2010; Zhou et al., 2015). It was estimated that the cost of production of crude algal oil varies from $2.87/L to $3.51/L. However, another study by DOE, Argonne National Laboratory, Pacific Northwest National Laboratory, and NREL reported $4.40/L to $4.62/L as the cost of crude algal oil. The cost of biofuel production varies with the biomass feedstock. As per the reported data, it is clear that the production cost of crude oil from algal biomass is still much higher compared to other feedstocks.

15.6.3 The production cost of biofuels from algal biomass feedstock

Techno-economic analysis (TEA) is a primary investigation tool used to estimate the cost and determine algal biofuels' economic feasibility. TEA analyzes various processes like engineering and thermodynamic modeling, capital and operating cost, economic flow, sensitivity analysis, and risk assessment (Langholtz et al., 2016). There have been many TEAs on the feasibility of various pathways to produce algal biofuels. The various cost components for biofuel production from algal biomass have been discussed in the preceding sections. The biodiesel from algal biomass has a production cost of around $4.67–4.91/L (Langholtz et al., 2016), which is costlier than other feedstocks. The cost of bioethanol and biodiesel from corn or soybean is around $0.20–0.32/L. The cost of cellulosic ethanol is around $0.62–1.00/L. The most reported method of algal biomass cultivation is open raceway ponds. Photobioreactors have been reported to be around 2–2.5 times more costly compared to open ponds, with biofuel costs ranging between $0.44/L (Benemann, Tillett, and Weissman, 1987) to $8.76/L (Richardson, Johnson, and Outlaw, 2012). The cost of biofuel varies and depends mainly on the scale of experiment, assumptions regarding growth rate, nutrient requirement, lipid productivity and production, and energy requirement.

15.7 Engine performance and emission characteristics using algal biofuel

Microalgal biofuels could be used as an alternative for petroleum fuels for transportation. In 2012 India performed its first test-run of commercial vehicles on 100% marine microalgal biodiesel. However, the exhaust emissions (particulate matter emissions, CO, CO_2, NO, NOx) from algal-based biofuels and the engine characteristics should be considered before opting for algal biofuels as a replacement for petroleum fuels (Wahlen et al., 2013). Few studies have reported reduced cylindrical pressure in the combustion stage and subsequently high pressure in the later stage when using 100% microalgal oil derived from *Crypthecodinium cohnii* (Aminul et al., 2015). When blended with butanol and 100% diesel, the microalgae biodiesel showed decreasing brake power and torque compared with a blend of 100% diesel and butanol due to higher oxygenate content in microalgae biodiesel (Kumar et al., 2018). Another study reported higher brake fuel consumption with lower brake thermal efficiency for microalgal biodiesel blend when compared to diesel oil (Makarevičiene et al., 2014). Another contracting study reported an increase in thermal brake efficiency of microalgal biodiesel blend compared to 100% diesel (Hariram and Mohan Kumar, 2013; Jayaprabakar and Karthikeyan, 2014). Higher NOx emissions of microalgal biodiesel blends were also reported, probably because of the high oxygen content in the microalgal biodiesel (Jayaprabakar and Karthikeyan, 2014). Some studies reported reduced emissions of NO and NOx after blending microalgal biodiesel with butanol or diesel (Fisher, 2018; Kumar, 2018). Increased particulate matter from microalgal biodiesel has also been reported (Rahman et al., 2015) due to low volatility, high boiling point, high density, and viscosity of microalgae biodiesel. Furthermore, reduced CO emissions due to higher oxygen content in the microalgal biodiesel compared to petroleum diesel (Gökhan Tüccar, Tayfun Özgür, 2014; Jayaprabakar and Karthikeyan, 2014; Kumar et al., 2018; Milano et al., 2016).

15.8 Global algal biofuel activities

Since the inception of algal biofuel production in the 19th century, many efforts have been made to develop efficient and cost-effective algal biofuel technologies. During the oil and gas scarcity of the 1970s, the role and production of biodiesel from microalgae gained popularity (Bitog et al., 2011). During 1976−96 the US launched the "Aquatic species Programme" to produce biodiesel from algal lipids. In the United States, under the entitled "Marine Biomass Program" project during 1968−1990, the Department of Energy investigated the technological and economic practicalities of microalgae cultivation for biodiesel and biogas production (Bird and P, 1987). ExxonMobil Corporation, in 2009, financed around US$600 million for algal biofuel-based transportation (Li, Moheimani, and Schenk, 2012). In India the Council of Scientific and Industrial Research (CSIR) launched a project to extract biodiesel from microalgae as part of the council's New Millennium Indian Technology Leadership Initiative in 2010. Currently, 9 CSIR laboratories across India are at work on the project. In the field of algal biofuels, the Japanese government sponsored a research program, "Biological CO_2 Fixation and Utilization": during 1990−99. However, in Japan, most developments in algal biofuels happened only after the tragedy of the Fukushima nuclear power plant in 2011(Murakami and

Ikenouchi, 1997). During 2010–13, the United States sponsored US$48.6 million for Advancement in Biofuels and Bioproducts consortium and during 2011–15 financed US$11 million for Algal Biofuels Commercialization. Many other countries like China, South Korea, the United Kingdom, Italy, the Philippines, and Nigeria are actively researching the field of algal biofuels.

15.9 Challenges and future perspective of biofuel production from algal biomass

Compared to first- and second-generation biofuel feedstocks, the microalgal feedstocks have many advantages. As a result of extensive research in algal biofuels, lots of advancements have been made in algal cultivation, biomass harvesting and subsequent biofuel production. However, the algal biofuel still faces a bottleneck for large-scale commercialization because of techno-economical infeasibilities, and there is a need to look at some technical issues.

15.9.1 Cultivation and harvesting

Compared with other feedstocks, microalgae have a higher cost of cultivation which is one of the biggest challenges in algal biofuel production. Besides this, the harvesting and dewatering of microalgae are high energy-consuming processes that increase the overall capital expenditure. It has been reported that harvesting incurs 20%–30% of the total cost in biomass production (Raheem et al., 2015). To deal with this problem, there is a need for integrating multiple operations for increasing the efficiency of harvesting and dewatering. One of these methods is the preconcentrating the algal biomass (Fasaei et al., 2018). However, low cost and highly efficient preconcentration technologies are yet to be developed, and the growth kinetics of cocultivation flocculants and microalgae are yet to be explored extensively.

15.9.2 Microalgae compositional characteristics

The biofuels from the microalgal biomass are extremely dependent on the composition of carbohydrates, lipids, and proteins accumulated in the biomass; for example, biodiesel production is highly reliant on lipid content. The accumulation of these compounds happens mainly by photosynthesis. It has been reported that the microalgae grown under nitrogen-starved conditions have high carbohydrate and lipid content in biomass compared to biomass produced in nitrogen-rich conditions. Various genetic engineering processes are going on for manipulating carbohydrate metabolism (Radakovits et al., 2010). There is a need for detailed investigation for metabolism mechanism for identifying the nexus between carbohydrate and lipid accumulation.

15.9.3 Pretreatment for fractionation

The inflexibility of the microalgal cell wall is one of the hindrances for the commercialization processing efficiency in algal biofuel production. The pretreatment process of algal biomass is an important step for increasing the conversion rate of algal bioethanol from carbohydrate compounds (Velazquez-Lucio et al., 2018). For biodiesel production, lipids compounds must be removed from algal biomass. Energy-intensiveness, costly, and time-consumption are some limitations in the pretreatment techniques for cell wall disruption. There is a need for comprehensive research on optimizing the pretreatment methods to increase biodiesel production.

15.9.4 Conversion technology and integrated biorefinery approach

Each biomass to biofuel conversion process is different based on chemistry and the type of biofuel produced. Therefore opting for the right method is an important step for developing an economical and eco-friendly biofuel production method from algal biomass. It has been reported that thermochemical conversion is more feasible than biochemical conversion because no pretreatment is required and because of the low efficiency of biochemical processes (Sims et al., 2010). Production of biogas from algal biomass is more economical and easier than biodiesel production, as anaerobic digestion does not require drying of biomass and is less energy-intensive (Wiley et al., 2011). In addition, lipid extraction for biodiesel production could be integrated with biogas production from the residual biomass through anaerobic digestion (Sialve et al., 2009). There is no comprehensive research analysis done from the techno-economic point of view; there is a lot of scope for techno-economic investigations and the development of cultivation methods.

References

Alghurabie, I.K., et al., 2013. Fluidized bed gasification of Kingston coal and marine microalgae in a spouted bed reactor. Chemical Engineering Research and Design 91 (9), 1614–1624. Available from: https://doi.org/10.1016/j.cherd.2013.04.024. Australia.

Amin, S., 2009. Review on biofuel oil and gas production processes from microalgae. Energy Conversion and Management 50 (7), 1834–1840. Available from: https://doi.org/10.1016/j.enconman.2009.03.001. Indonesia.

Babich, I.V., et al., 2011. Catalytic pyrolysis of microalgae to high-quality liquid bio-fuels. Biomass and Bioenergy 35 (7), 3199–3207. Available from: https://doi.org/10.1016/j.biombioe.2011.04.043. the Netherlands.

Bach, Q.V., Skreiberg, O., 2016. Upgrading biomass fuels via wet torrefaction: a review and comparison with dry torrefaction. Renewable and Sustainable Energy Reviews 54, 665–677. Available from: https://doi.org/10.1016/j.rser.2015.10.014. Norway: Elsevier Ltd.

Barros, A.I., et al., 2015. Harvesting techniques applied to microalgae: a review. Renewable and Sustainable Energy Reviews 41, 1489–1500. Available from: https://doi.org/10.1016/j.rser.2014.09.037. Portugal: Elsevier Ltd.

Beal, C.M., et al., 2012. Energy return on investment for algal biofuel production coupled with wastewater treatment. Water Environment Research 84 (9), 692–710. Available from: https://doi.org/10.2175/106143012X13378023685718. United States: Water Environment Federation.

Benemann, J.R., Tillett, D.M., Weissman, J.C., 1987. Microalgae biotechnology. Trends in Biotechnology 5 (2), 47–53. Available from: https://doi.org/10.1016/0167-7799(87)90037-0. United States.

Bilanovic, D., Shelef, G., Sukenik, A., 1988. Flocculation of microalgae with cationic polymers – effects of medium salinity. Biomass. Israel 17 (1), 65–76. Available from: https://doi.org/10.1016/0144-4565(88)90071-6.

Bird, K., Benson, P.H., 1987. Seaweed Cultivation for Renewable Resources. Elsevier.

Bitog, J.P., et al., 2011. Application of computational fluid dynamics for modeling and designing photobioreactors for microalgae production: A review. Computers and Electronics in Agriculture 76 (2), 131–147. Available from: https://doi.org/10.1016/j.compag.2011.01.015. South Korea.

Borowitzka, M.A., 2013. High-value products from microalgae-their development and commercialisation. Journal of Applied Phycology 25 (3), 743–756. Available from: https://doi.org/10.1007/s10811-013-9983-9. Australia.

Borowitzka, M.A., Borowitzka, L.J., 1988. Micro-Algal Biotechnology. Available from: https://doi.org/10.1002/jctb.280470214.

Brennan, L., Owende, P., 2010. Biofuels from microalgae—a review of technologies for production, processing, and extractions of biofuels and co-products. Renewable and Sustainable Energy Reviews 14 (2), 557–577. Available from: https://doi.org/10.1016/j.rser.2009.10.009. Ireland.

Cabanelas, I.T.D., et al., 2013. Comparing the use of different domestic wastewaters for coupling microalgal production and nutrient removal. Bioresource Technology 131, 429–436. Available from: https://doi.org/10.1016/j.biortech.2012.12.152. Spain: Elsevier Ltd.

Cantrell, K.B., et al., 2008. Livestock waste-to-bioenergy generation opportunities. Bioresource Technology 99 (17), 7941–7953. Available from: https://doi.org/10.1016/j.biortech.2008.02.061. United States.

Chaisutyakorn, P., Praiboon, J., Kaewsuralikhit, C., 2018. The effect of temperature on growth and lipid and fatty acid composition on marine microalgae used for biodiesel production. Journal of Applied Phycology. Available from: https://doi.org/10.1007/s10811-017-1186-3. Thailand: Springer Netherlands.

Chan, L.G., Cohen, J.L., De Moura Bell, J.M.L.N., 2018. Conversion of Agricultural Streams and Food-Processing By-Products to Value-Added Compounds Using Filamentous Fungi. Annual Review of Food Science and Technology 9, 503–523. Available from: https://doi.org/10.1146/annurev-food-030117-012626. United States: Annual Reviews Inc.

Chatsungnoen, T., Chisti, Y., 2016. Harvesting microalgae by flocculation-sedimentation. Algal Research 13, 271–283. Available from: https://doi.org/10.1016/j.algal.2015.12.009. New Zealand: Elsevier B.V.

Chen, C.Y., Durbin, E.G., 1994. Effects of pH on the growth and carbon uptake of marine phytoplankton. Marine Ecology Progress Series 109 (1), 83–94. Available from: https://doi.org/10.3354/meps109083. United States: Inter-Research.

Chen, W.H., et al., 2015. Torrefaction operation and optimization of microalga residue for energy densification and utilization. Applied Energy 154, 622–630. Available from: https://doi.org/10.1016/j.apenergy.2015.05.068. Taiwan: Elsevier Ltd.

Chen, W.H., Wu, Z.Y., Chang, J.S., 2014. Isothermal and non-isothermal torrefaction characteristics and kinetics of microalga *Scenedesmus obliquus* CNW-N. Bioresource Technology 155, 245–251. Available from: https://doi.org/10.1016/j.biortech.2013.12.116. Taiwan: Elsevier Ltd.

Chen, Y.C., et al., 2016. Impact of torrefaction on the composition, structure and reactivity of a microalga residue. Applied Energy 181, 110–119. Available from: https://doi.org/10.1016/j.apenergy.2016.07.130. Taiwan: Elsevier Ltd.

Chisti, Y., 2007. Biodiesel from microalgae. Biotechnology Advances 25 (3), 294–306. Available from: https://doi.org/10.1016/j.biotechadv.2007.02.001. New Zealand.

Cho, S., et al., 2013. Microalgae cultivation for bioenergy production using wastewaters from a municipal WWTP as nutritional sources. Bioresource Technology 131, 515–520. Available from: https://doi.org/10.1016/j.biortech.2012.12.176. South Korea: Elsevier Ltd.

Contreras, A., et al., 1997. Interaction Between CO_2–mass transfer, light availability, and hydrodynamic stress in the growth of phaeodactylum tricornutum in a concentric tube airlift photobioreactor. Biotechnology Bioengineering 60 (3), 317–325.

Converti, A., et al., 2009. Effect of temperature and nitrogen concentration on the growth and lipid content of *Nannochloropsis oculata* and Chlorella vulgaris for biodiesel production. Chemical Engineering and Processing: Process Intensification 48 (6), 1146–1151. Available from: https://doi.org/10.1016/j.cep.2009.03.006. Italy.

Dassey, A.J., Theegala, C.S., 2013. Harvesting economics and strategies using centrifugation for cost effective separation of microalgae cells for biodiesel applications. Bioresource Technology 128, 241–245. Available from: https://doi.org/10.1016/j.biortech.2012.10.061. United States: Elsevier Ltd.

Demirbas, A., 2010. Use of algae as biofuel sources. Energy Conversion and Management 51 (12), 2738–2749. Available from: https://doi.org/10.1016/j.enconman.2010.06.010. Turkey.

de Mooij, T., et al., 2015. Antenna size reduction as a strategy to increase biomass productivity: a great potential not yet realized. Journal of Applied Phycology 27 (3), 1063–1077. Available from: https://doi.org/10.1007/s10811-014-0427-y. Netherlands: Kluwer Academic Publishers.

Doucha, J., Straka, F., Lívanský, K., 2005. Utilization of flue gas for cultivation of microalgae (*Chlorella* sp.) in an outdoor open thin-layer photobioreactor. Journal of Applied Phycology 17 (5), 403–412. Available from: https://doi.org/10.1007/s10811-005-8701-7. Czech Republic.

Du, Z., et al., 2011. Microwave-assisted pyrolysis of microalgae for biofuel production. Bioresource Technology 102 (7), 4890−4896. Available from: https://doi.org/10.1016/j.biortech.2011.01.055. United States.

Fasaei, F., et al., 2018. Techno-economic evaluation of microalgae harvesting and dewatering systems. Algal Research 31, 347−362. Available from: https://doi.org/10.1016/j.algal.2017.11.038. Netherlands: Elsevier B.V.

Fisher, B.C., et al., 2018. Measurement of gaseous and particulate emissions from algae-based fatty acid methyl esters. SAE International Journal of Fuels and Lubricants-V119-4.

Goyal, H.B., Seal, D., Saxena, R.C., 2008. Bio-fuels from thermochemical conversion of renewable resources: a review. Renewable and Sustainable Energy Reviews 12 (2), 504−517. Available from: https://doi.org/10.1016/j.rser.2006.07.014. India.

Grierson, S., et al., 2009. Thermal characterisation of microalgae under slow pyrolysis conditions. Journal of Analytical and Applied Pyrolysis 85 (1−2), 118−123. Available from: https://doi.org/10.1016/j.jaap.2008.10.003. Australia: Elsevier.

Griffiths, M.J., van Hille, R.P., Harrison, S.T.L., 2012. Lipid productivity, settling potential and fatty acid profile of 11 microalgal species grown under nitrogen replete and limited conditions. Journal of Applied Phycology 24 (5), 989−1001. Available from: https://doi.org/10.1007/s10811-011-9723-y. South Africa.

Gultom, S.O., Hu, B., 2013. Review of microalgae harvesting via co-pelletization with filamentous fungus. Energies 6 (11), 5921−5939. Available from: https://doi.org/10.3390/en6115921. United States.

Haiduc, A.G., et al., 2009. SunCHem: an integrated process for the hydrothermal production of methane from microalgae and CO2 mitigation. Journal of Applied Phycology 21 (5), 529−541. Available from: https://doi.org/10.1007/s10811-009-9403-3. Switzerland.

Hansen, P.J., 2002. Effect of high pH on the growth and survival of marine phytoplankton: Implications for species succession. Aquatic Microbial Ecology 28 (3), 279−288. Available from: https://doi.org/10.3354/ame028279. Denmark: Inter-Research.

Hariram, V., Mohan Kumar, G., 2013. Combustion analysis of algal oil methyl ester in a direct injection compression ignition engine. Journal of Engineering Science and Technology. India 8 (1), 77−92. Available from: http://jestec.taylors.edu.my/Vol%208%20Issue%201%20February%2013/Vol_8_1_077-092_HARIRAM.pdf.

Harman-Ware, A.E., et al., 2013. Microalgae as a renewable fuel source: fast pyrolysis of *Scenedesmus* sp. Renewable Energy 60, 625−632. Available from: https://doi.org/10.1016/j.renene.2013.06.016. United States: Elsevier Ltd.

Harun, R., Danquah, M.K., Forde, G.M., 2010. Microalgal biomass as a fermentation feedstock for bioethanol production. Journal of Chemical Technology and Biotechnology 85 (2), 199−203. Available from: https://doi.org/10.1002/jctb.2287. Australia.

Hirano, A., et al., 1997. CO_2 fixation and ethanol production with microalgal photosynthesis and intracellular anaerobic fermentation. Energy 22 (2−3), 137−142. Available from: https://doi.org/10.1016/s0360-5442(96)00123-5. Elsevier BV.

Hirano, A., et al., 1998. Temperature effect on continuous gasification of microalgal biomass: theoretical yield of methanol production and its energy balance. Catalysis Today 45 (1−4), 399−404. Available from: https://doi.org/10.1016/S0920-5861(98)00275-2. Japan: Elsevier.

Hochman, G., Zilberman, D., 2014. Algae farming and its bio-products. Plants and BioEnergy 49−64. Available from: https://doi.org/10.1007/978-1-4614-9329-7_4. United States: Springer New York.

Hoffmann, J.P., 1998. Wastewater treatment with suspended and nonsuspended algae. Journal of Phycology 34 (5), 757−763. Available from: https://doi.org/10.1046/j.1529-8817.1998.340757.x. United States: Phycological Society of America.

Jayaprabakar, J., Karthikeyan, A., 2014. Analysis on the performance, combustion and emission characteristics of a CI engine fuelled with algae biodiesel. Applied Mechanics and Materials. Available from: https://doi.org/10.4028/http://www.scientific.net/AMM.591.33. India: Trans Tech Publications Ltd.

Jazzar, S., et al., 2015. A whole biodiesel conversion process combining isolation, cultivation and in situ supercritical methanol transesterification of native microalgae. Bioresource Technology 190, 281−288. Available from: https://doi.org/10.1016/j.biortech.2015.04.097. Tunisia: Elsevier Ltd.

Johnson, M.B., Wen, Z., 2009. Production of biodiesel fuel from the microalga *schizochytrium limacinum* by direct transesterification of algal biomass. Energy and Fuels 23 (10), 5179−5183. Available from: https://doi.org/10.1021/ef900704h. United States.

Juneja, A., Ceballos, R.M., Murthy, G.S., 2013. Effects of environmental factors and nutrient availability on the biochemical composition of algae for biofuels production: A review. Energies 6 (9), 4607–4638. Available from: https://doi.org/10.3390/en6094607. United States: MDPI AG.

Kadam, K.L., 2001. Environmental implications of power generation via coal-microalgae cofiring. Proceedings of the International Joint Power Generation Conference United States.

Khan, S.A., et al., 2019. Microalgae based biofertilizers: a biorefinery approach to phycoremediate wastewater and harvest biodiesel and manure. Journal of Cleaner Production 211, 1412–1419. Available from: https://doi.org/10.1016/j.jclepro.2018.11.281. India: Elsevier Ltd.

Kiran, B., Kumar, R., Deshmukh, D., 2014. Perspectives of microalgal biofuels as a renewable source of energy. Energy Conversion and Management 88, 1228–1244. Available from: https://doi.org/10.1016/j.enconman.2014.06.022. India: Elsevier Ltd.

Kumar, H.D., Yadava, P.K., Gaur, J.P., 1981. Electrical flocculation of the unicellular green alga *Chlorella vulgaris* Beijerinck. Aquatic Botany 11 (C), 187–195. Available from: https://doi.org/10.1016/0304-3770(81)90059-0. India.

Kumar, V., et al., 2018. Production of biodiesel and bioethanol using algal biomass harvested from fresh water river. Renewable Energy 116, 606–612. Available from: https://doi.org/10.1016/j.renene.2017.10.016. India: Elsevier Ltd.

Lam, M.K., Lee, K.T., 2012. Microalgae biofuels: A critical review of issues, problems and the way forward. Biotechnology Advances 30 (3), 673–690. Available from: https://doi.org/10.1016/j.biotechadv.2011.11.008. Malaysia.

Langholtz, M.H., et al., 2016. Potential land competition between open-pond microalgae production and terrestrial dedicated feedstock supply systems in the U.S. Renewable Energy 93, 201–214. Available from: https://doi.org/10.1016/j.renene.2016.02.052. United States: Elsevier Ltd.

Li, S.X., et al., 2018. Torrefaction of corncob to produce charcoal under nitrogen and carbon dioxide atmospheres. Bioresource Technology 249, 348–353. Available from: https://doi.org/10.1016/j.biortech.2017.10.026. China: Elsevier Ltd.

Li, Y., et al., 2008. Effects of nitrogen sources on cell growth and lipid accumulation of green alga *Neochloris oleoabundans*. Applied Microbiology and Biotechnology 81 (4), 629–636. Available from: https://doi.org/10.1007/s00253-008-1681-1. Canada.

Liu, B.H., Lee, Y.K., 2000. Secondary carotenoids formation by the green alga *Chlorococcum* sp. Journal of Applied Phycology. Available from: https://doi.org/10.1023/a:1008185212724. Singapore: Springer Netherlands.

Liu, Y., Liao, W., Chen, S., 2008. Study of pellet formation of filamentous fungi *Rhizopus oryzae* using a multiple logistic regression model. Biotechnology and Bioengineering 99 (1), 117–128. Available from: https://doi.org/10.1002/bit.21531. United States.

Liu, Z.Y., Wang, G.C., Zhou, B.C., 2008. Effect of iron on growth and lipid accumulation in *Chlorella vulgaris*. Bioresource Technology 99 (11), 4717–4722. Available from: https://doi.org/10.1016/j.biortech.2007.09.073. China.

Li, Y., Moheimani, N.R., Schenk, P.M., 2012. Current research and perspectives of microalgal biofuels in Australia. Biofuels 3 (4), 427–439. Available from: https://doi.org/10.4155/bfs.12.32. Australia.

Makarevičiene, V., et al., 2014. Performance and emission characteristics of diesel fuel containing microalgae oil methyl esters. Fuel 120, 233–239. Available from: https://doi.org/10.1016/j.fuel.2013.11.049. Lithuania.

Mata, T.M., Martins, A.A., Caetano, N.S., 2010. Microalgae for biodiesel production and other applications: a review. Renewable and Sustainable Energy Reviews 14 (1), 217–232. Available from: https://doi.org/10.1016/j.rser.2009.07.020. Portugal.

McKendry, P., 2002. Energy production from biomass (part 2): conversion technologies. Bioresource Technology 83 (1), 47–54. Available from: https://doi.org/10.1016/s0960-8524(01)00119-5. Elsevier BV.

Mercer, P., Armenta, R.E., 2011. Developments in oil extraction from microalgae. European Journal of Lipid Science and Technology 113 (5), 539–547. Available from: https://doi.org/10.1002/ejlt.201000455. Canada.

Miao, X., Wu, Q., 2004. High yield bio-oil production from fast pyrolysis by metabolic controlling of *Chlorella prototothecoides*. Journal of Biotechnology 110 (1), 85–93. Available from: https://doi.org/10.1016/j.jbiotec.2004.01.013. China.

Milano, J., et al., 2016. Microalgae biofuels as an alternative to fossil fuel for power generation. Renewable and Sustainable Energy Reviews 58, 180–197. Available from: https://doi.org/10.1016/j.rser.2015.12.150. Malaysia: Elsevier Ltd.

Mobin, S., Alam, F., 2014. "Biofuel production from algae utilizing wastewater." In: Proc. of the 19th Australasian Fluid Mechanics Conference, AFMC 2014. Australia: Australasian Fluid Mechanics Society.

Molina Grima, E., et al., 2003. Recovery of microalgal biomass and metabolites: process options and economics. Biotechnology Advances 20 (7−8), 491−515. Available from: https://doi.org/10.1016/S0734-9750(02)00050-2. Spain: Elsevier Inc.

Muradyan, E.A., et al., 2004. Changes in lipid metabolism during adaptation of the *Dunaliella salina* photosynthetic apparatus to high CO_2 concentration. Russian Journal of Plant Physiology 51 (1), 53−62. Available from: https://doi.org/10.1023/B:RUPP.0000011303.11957.48. Russian Federation.

Murakami, M., Ikenouchi, M., 1997. The biological CO_2 fixation and utilization project by RITE (2): screening and breeding of microalgae with high capability in fixing CO_2. Energy Conversion and Management 38 (1), S493−S497. Available from: https://doi.org/10.1016/s0196-8904(96)00316-0. Japan: Elsevier Ltd.

Naik, S.N., et al., 2010. Production of first and second generation biofuels: a comprehensive review. Renewable and Sustainable Energy Reviews 14 (2), 578−597. Available from: https://doi.org/10.1016/j.rser.2009.10.003. India.

Nyberg, M., Heidorn, T., Lindblad, P., 2015. Hydrogen production by the engineered cyanobacterial strain Nostoc PCC 7120 δhupW examined in a flat panel photobioreactor system. Journal of Biotechnology 215, 35−43. Available from: https://doi.org/10.1016/j.jbiotec.2015.08.028. Sweden: Elsevier.

Onumaegbu, C., et al., 2018. Pre-treatment methods for production of biofuel from microalgae biomass. Renewable and Sustainable Energy Reviews 93, 16−26. Available from: https://doi.org/10.1016/j.rser.2018.04.015. United Kingdom: Elsevier Ltd.

Pan, P., et al., 2010. The direct pyrolysis and catalytic pyrolysis of Nannochloropsis sp. residue for renewable bio-oils. Bioresource Technology 101 (12), 4593−4599. Available from: https://doi.org/10.1016/j.biortech.2010.01.070. China.

Passos, F., et al., 2014. Pretreatment of microalgae to improve biogas production: a review. Bioresource Technology 172, 403−412. Available from: https://doi.org/10.1016/j.biortech.2014.08.114. Spain: Elsevier Ltd.

Patil, V., Tran, K.Q., Giselrød, H.R., 2008. Towards sustainable production of biofuels from microalgae. International Journal of Molecular Sciences 9 (7), 1188−1195. Available from: https://doi.org/10.3390/ijms9071188. Norway.

Peng, W., Wu, Q., Tu, P., 2000. Effects of temperature and holding time on production of renewable fuels from pyrolysis of *Chlorella protothecoides*. Journal of Applied Phycology 12 (2), 147−152. Available from: https://doi.org/10.1023/A:1008115025002. China.

Persoone, G., et al., 1980. Air-lift pumps and the effect of mixing on algal growth. Algae Biomass Production and Use 505−522.

Pessoa, M.F., 2012. Harmful effects of UV radiation in algae and aquatic macrophytes − a review. Emirates Journal of Food and Agriculture. Portugal: United Arab Emirates University 24 (6), 510−526. Available from: https://doi.org/10.9755/ejfa.v24i6.510526.

Pragya, N., Pandey, K.K., Sahoo, P.K., 2013a. A review on harvesting, oil extraction and biofuels production technologies from microalgae. Renewable and Sustainable Energy Reviews 24, 159−171. Available from: https://doi.org/10.1016/j.rser.2013.03.034. India: Elsevier Ltd.

Prakash, J., et al., 1997. Microalgal biomass drying by a simple solar device. International Journal of Solar Energy 18 (4), 303−311. Available from: https://doi.org/10.1080/01425919708914325. India: Taylor and Francis Ltd.

Radakovits, R., et al., 2010. Genetic engineering of algae for enhanced biofuel production. Eukaryotic Cell 9 (4), 486−501. Available from: https://doi.org/10.1128/EC.00364-09. United States.

Raheem, A., et al., 2015. Thermochemical conversion of microalgal biomass for biofuel production. Renewable and Sustainable Energy Reviews 49, 990−999. Available from: https://doi.org/10.1016/j.rser.2015.04.186. Malaysia: Elsevier Ltd.

Rahman, M.M., et al., 2015. Particle emissions from microalgae biodiesel combustion and their relative oxidative potential. Environmental Sciences: Processes and Impacts 17 (9), 1601−1610. Available from: https://doi.org/10.1039/c5em00125k. Australia: Royal Society of Chemistry.

Rastogi, R.P., Incharoensakdi, A., 2013. UV radiation-induced accumulation of photoprotective compounds in the green alga Tetraspora sp. CU2551,". Plant Physiology and Biochemistry 70, 7−13. Available from: https://doi.org/10.1016/j.plaphy.2013.04.021. Thailand.

Renaud, S.M., et al., 1991. Effect of light intensity on the proximate biochemical and fatty acid composition of *Isochrysis* sp. and *Nannochloropsis oculata* for use in tropical aquaculture. Journal of Applied Phycology 3 (1), 43–53. Available from: https://doi.org/10.1007/BF00003918. Australia: Kluwer Academic Publishers.

Richardson, J.W., Johnson, M.D., Outlaw, J.L., 2012. Economic comparison of open pond raceways to photo bioreactors for profitable production of algae for transportation fuels in the Southwest. Algal Research 1 (1), 93–100. Available from: https://doi.org/10.1016/j.algal.2012.04.001. United States.

Rocha, J.M.S., Garcia, J.E.C., Henriques, M.H.F., 2003. Growth aspects of the marine microalga *Nannochloropsis gaditana*. Biomolecular Engineering. Available from: https://doi.org/10.1016/S1389-0344(03)00061-3. Portugal: Elsevier.

Roessler, P.G., 1990. Purification and characterization of acetyl-CoA carboxylase from the diatom *Cyclotella cryptica*,". Plant Physiology 92 (1), 73–78. Available from: https://doi.org/10.1104/pp.92.1.73. United States: American Society of Plant Biologists.

Salam, K.A., Velasquez-Orta, S.B., Harvey, A.P., 2016. A sustainable integrated in situ transesterification of microalgae for biodiesel production and associated co-products – a review. Renewable and Sustainable Energy Reviews 65, 1179–1198. Available from: https://doi.org/10.1016/j.rser.2016.07.068. Nigeria: Elsevier Ltd.

Saranya, D., Shanthakumar, S., 2020. An integrated approach for tannery effluent treatment with ozonation and phycoremediation: a feasibility study. Environmental Research 183, 109163. Available from: https://doi.org/10.1016/j.envres.2020.109163. Elsevier BV.

Sharma, J., et al., 2020. Upgrading of microalgal consortia with CO_2 from fermentation of wheat straw for the phycoremediation of domestic wastewater. Bioresource Technology 305. Available from: https://doi.org/10.1016/j.biortech.2020.123063. India: Elsevier Ltd.

Shin, H., et al., 2017. Harvesting of *Scenedesmus obliquus* cultivated in seawater using electro-flotation. Korean Journal of Chemical Engineering 34 (1), 62–65. Available from: https://doi.org/10.1007/s11814-016-0251-y. South Korea: Springer New York LLC.

Sialve, B., Bernet, N., Bernard, O., 2009. Anaerobic digestion of microalgae as a necessary step to make microalgal biodiesel sustainable. Biotechnology Advances. France 27 (4), 409–416. Available from: https://doi.org/10.1016/j.biotechadv.2009.03.001.

Sims, R.E.H., et al., 2010. An overview of second generation biofuel technologies. Bioresource Technology 101 (6), 1570–1580. Available from: https://doi.org/10.1016/j.biortech.2009.11.046. New Zealand: Elsevier Ltd.

Slade, R., Bauen, A., 2013. Micro-algae cultivation for biofuels: cost, energy balance, environmental impacts and future prospects. Biomass and Bioenergy 53, 29–38. Available from: https://doi.org/10.1016/j.biombioe.2012.12.019. United Kingdom.

Smith, R.E.H., et al., 1993. Growth and lipid composition of high Arctic ice algae during the spring bloom at Resolute, Northwest Territories, Canada. Marine Ecology Progress Series. Canada 97 (1), 19–29. Available from: https://doi.org/10.3354/meps097019.

Suali, E., Sarbatly, R., 2012. Conversion of microalgae to biofuel. Renewable and Sustainable Energy Reviews. Malaysia 16 (6), 4316–4342. Available from: https://doi.org/10.1016/j.rser.2012.03.047.

Suganya, T., et al., 2016. Macroalgae and microalgae as a potential source for commercial applications along with biofuels production: a biorefinery approach. Renewable and Sustainable Energy Reviews 55, 909–941. Available from: https://doi.org/10.1016/j.rser.2015.11.026. Malaysia: Elsevier Ltd.

Sukiran, M.A., et al., 2017. A review of torrefaction of oil palm solid wastes for biofuel production. Energy Conversion and Management 149, 101–120. Available from: https://doi.org/10.1016/j.enconman.2017.07.011. Malaysia: Elsevier Ltd.

Tripathi, U., Sarada, R., Ravishankar, G.A., 2002. Effect of culture conditions on growth of green alga - *Haematococcus pluvialis* and astaxanthin production. Acta Physiologiae Plantarum 24 (3), 323–329. Available from: https://doi.org/10.1007/s11738-002-0058-9. India: Polish Academy of Sciences.

Uemura, Y., et al., 2017. Torrefaction of empty fruit bunches under biomass combustion gas atmosphere. Bioresource Technology 243, 107–117. Available from: https://doi.org/10.1016/j.biortech.2017.06.057. Malaysia: Elsevier Ltd.

Vandamme, D., Foubert, I., Muylaert, K., 2013. Flocculation as a low-cost method for harvesting microalgae for bulk biomass production. Trends in Biotechnology 31 (4), 233–239. Available from: https://doi.org/10.1016/j.tibtech.2012.12.005. Belgium.

Varfolomeev, S.D., Wasserman, L.A., 2011. Microalgae as source of biofuel, food, fodder, and medicines. Applied Biochemistry and Microbiology 47 (9), 789–807. Available from: https://doi.org/10.1134/S0003683811090079. Russian Federation.

Velazquez-Lucio, J., et al., 2018. Microalgal biomass pretreatment for bioethanol production: a review. Biofuel Research Journal 5 (1), 780–791. Available from: https://doi.org/10.18331/BRJ2018.5.1.5. Mexico: Green Wave Publishing of Canada.

Wahlen, B.D., et al., 2013. Biodiesel from microalgae, yeast, and bacteria: engine performance and exhaust emissions. Energy and Fuels 27 (1), 220–228. Available from: https://doi.org/10.1021/ef3012382. United States.

Walker, J.B., 1954. Inorganic micronutrient requirements of *Chlorella*. II. Quantitative requirements for iron, manganese, and zinc. Archives of Biochemistry and Biophysics 53 (1), 1–8. Available from: https://doi.org/10.1016/0003-9861(54)90227-1. United States.

Wiley, P.E., Brenneman, K.J., Jacobson, A.E., 2009. Improved algal harvesting using suspended air flotation. Water Environment Research 81 (7), 702–708. Available from: https://doi.org/10.2175/106143009X407474. United States: Water Environment Federation.

Wiley, P.E., Campbell, J.E., McKuin, B., 2011. Production of biodiesel and biogas from algae: a review of process train options. Water Environment Research 83 (4), 326–338. Available from: https://doi.org/10.2175/106143010X12780288628615. United States: Water Environment Federation.

Wu, L.C., et al., 2005. Antioxidant and antiproliferative activities of *Spirulina* and *Chlorella* water extracts. Journal of Agricultural and Food Chemistry 53 (10), 4207–4212. Available from: https://doi.org/10.1021/jf0479517. Taiwan.

Xin, L., et al., 2010. Effects of different nitrogen and phosphorus concentrations on the growth, nutrient uptake, and lipid accumulation of a freshwater microalga *Scenedesmus* sp. Bioresource Technology 101 (14), 5494–5500. Available from: https://doi.org/10.1016/j.biortech.2010.02.016. China.

Yang, K.C., et al., 2013. Co-gasification of woody biomass and microalgae in a fluidized bed. Journal of the Taiwan Institute of Chemical Engineers 44 (6), 1027–1033. Available from: https://doi.org/10.1016/j.jtice.2013.06.026. Taiwan.

Yen, H.W., Brune, D.E., 2007. Anaerobic co-digestion of algal sludge and waste paper to produce methane. Bioresource Technology 98 (1), 130–134. Available from: https://doi.org/10.1016/j.biortech.2005.11.010. Taiwan.

Zhou, Q., Zhang, P., Zhang, G., 2014a. Biomass and carotenoid production in photosynthetic bacteria wastewater treatment: Effects of light intensity. Bioresource Technology 171, 330–335. Available from: https://doi.org/10.1016/j.biortech.2014.08.088. China: Elsevier Ltd.

Zhou, W., et al., 2014b. Environment-enhancing algal biofuel production using wastewaters. Renewable and Sustainable Energy Reviews 36, 256–269. Available from: https://doi.org/10.1016/j.rser.2014.04.073. United States: Elsevier Ltd.

Zhou, W., et al., 2013. Filamentous fungi assisted bio-flocculation: a novel alternative technique for harvesting heterotrophic and autotrophic microalgal cells. Separation and Purification Technology 107, 158–165. Available from: https://doi.org/10.1016/j.seppur.2013.01.030. United States.

Zhou, X., et al., 2015. Sustainable production of energy from microalgae: review of culturing systems, economics, and modelling. Journal of Renewable and Sustainable Energy 7 (1). Available from: https://doi.org/10.1063/1.4906919. China: American Institute of Physics Inc.

Algal cultivation in the pursuit of emerging technology for sustainable development

Achintya Das[1] and Ananya Roy Chowdhury[2]

[1]Department of Physics, Mahadevananda Mahavidyalaya, Barrackpore, India [2]Department of Botany, Chakdaha College, Nadia, India

16.1 Introduction

One of the most enticing fundamental properties of many algae species, when compared to plants, is their ability to grow vast volumes of biomass fast and affordably (Brennan and Owende, 2010). Microalgae can deliver large biomass concentrations under eutrophic conditions in nature, yet even these concentrations are insufficient for mass culture. Over the last decade, a significant amount of study has gone into establishing the ideal settings for maximizing algal growth under artificial growth conditions. Yet, building a cost-effective production method is one of the most real challenges in algal mass growing. In this regard, a variety of algae cultivation techniques can provide varying degrees of control overgrowth and product yield, as well as varying capital and running expenses. Light availability, temperature, pH, and the concentration and ratio of the key nutrients, carbon, nitrogen, and phosphorus, are all factors that can limit microalgal growth in mass culture (Sutherland et al., 2015).

The return on investment will be maximized regardless of how the biomass is produced if the harvesting and processing can be done in an integrated biorefinery that allows for the extraction of the greatest number of products and coproducts with the least amount of residual/waste. Self-adapting manufacturing processes are created by combining automation, sensors, and machine learning (ML) to create systems that can respond in real time to changes in the process (Atzori et al., 2010; Fabris et al., 2020). Not only can the algae cultivation and harvesting system be automated to save money, but a network of plug-and-play Internet of Things (IoT) sensors might also allow operators to track algal growth and productivity in real time (Whitmore et al., 2015). By combining sensor data to build a simulation of the facility and the algal culture, it is possible to predict future cellular production in real time and adjust operations to meet assumed product demand while reducing waste.

16.2 Internet of Things applied in microalgae biorefinery

The IoT is a network of physical objects. "Things" are embedded with sensors, software, and other technologies to connect and exchange data with other devices and systems over the internet. Sensing, transfer, and processing are the three essential components of the IoT, in which automation, sensors, and ML combine to produce an adaptable manufacturing process that can be altered in real time in response to changes in the process (Fabris et al., 2020). The IoT platform connects devices and objects with built-in sensors, which integrates data from numerous devices and applies analytics to share the most significant data with apps designed to address unique requirements, culminating in intelligent identification and management. Through the strong IoT platform, information may be securely and precisely discriminated against as to whether it can be used further or ignored. All this data may be used for analysis, making recommendations, and identifying potential concerns before they become a problem (Perkel, 2017).

On a microalgae biorefinery, IoT allows us to collect enough data to create better models for predicting behavior. The more data we collect, the better equipped we will be ready to deal with force majeure events. To determine the economic feasibility, a large-scale microalgal growth system must be developed. Microalgae production necessitates a lot of water and chemical fertilizers, as well as huge energy, massive monetary upfront, and it is prone to contamination. Due to the use of IoT, massive data gathering, subsequent modeling, and prediction, waste can be reduced, and water, energy, land, and other natural resources can be better conserved, especially in today's resource-scarce environment. IoT enables additional elements to be considered and more flexible industrial production tools to be developed. Another notable feature of IoT is that it may optimize the efficiency of microalgae biorefinery while also saving a large labor force and money. A hybrid semicontinuous cultivation system (a combination of airlift tubular photobioreactors and raceway ponds) appears to be the best option among the various cultivation methods in the future due to urgent requirements for cost reduction, culture contamination prevention, and high biomass productivity. Microalgae biorefinery involves both upstream and downstream processes (Tan et al., 2018). Upstream processing focuses on microalgae screening, classification, and identification, as well as microalgae cultivation and production before harvesting, whereas downstream processing (DSP) focuses on microalgae harvesting, drying, and separation, as well as the recovery and purification of high-value bioactive components from microalgae (such as unsaturated and polyunsaturated fatty acids, polysaccharides, and carotenoids) (Wang et al., 2022; Khoo et al., 2019a,b; Chew et al., 2017; Yong et al., 2021). From the standpoint of a microalgae biorefinery, an IoT sensor network not only automates the cultivation and harvesting system but also allows operators to monitor microalgal growth and production in real time, thanks to plug-and-play IoT sensor networks (Fabris et al., 2020; Hermadi et al., 2021).

16.2.1 Deployments of Internet of Things in microalgae cultivation

Due to the wide-scale potential applications in the domains of food, pharmaceuticals, cosmetics, and bioenergy production (Khoo et al., 2020a), large-scale and cost-effective microalgae cultivation is receiving interest (Khoo et al., 2020a). In a microalgal cultivation system (MCS), a high biomass output is a key to balancing cost and production. Multiple complicated parameters influence microalgae productivity, including light intensity, nutrition availability, temperature, pH, aeration rate, and fluid mechanics management (Khoo et al., 2020b). Monitoring biological variables such as cell size, morphology, and mass concentration, as well as population composition (in terms of contamination), pigment, and lipid content, all of which are direct indicators that can help to track the dynamics of an MCS. However, the tools used to measure these factors for real-time monitoring and management are time-consuming, difficult, or hazardous to the microalgae (Moi, 2010; Salas-Herrera et al., 2019; Wang et al., 2022). The lack of online sensors that can monitor microalgae appears to be the primary impediment to MCS management and online optimization (Bernard et al., 2016). In recent years, electrical technology and the Internet have combined to create different types of sensors (temperature, optical, pH, etc.) to monitor, control, and analyze complex microalgae farming systems. For measuring biomass concentration, multiwavelength-based sensors and fluorescence sensors are employed (Christian Barbosa et al., 2020; Jia et al., 2015; Shin et al., 2015). Harmful/toxic microalgae can be detected using electrochemical genosensors (Hessel and Metfies, 2014). All these sensors, which are the most important components of the IoT, were employed to continuously monitor the crucial microalgae cultivation parameters. IoT technologies can make it easier to oversee the entire factory's production. Installing, allocating, arranging, and upgrading all the sensors in the IoT architecture are also simple. The benefits of combining MCS and IoT technologies are remote control, data exchange, and continuous and real-time monitoring and updates (Wang et al., 2022). It can track and manage the growth of microalgae in real time. Microalgae culture can be automated to increase output and maintain quality. It can determine the temperature of the culture chamber, luminance (lighting level), the color of water in the photosynthesis process, and other factors that affect microalgae growth, as well as manage water circulation using an aerator. The use of IoT to research microalgae farming can aid researchers in optimizing the microalgae cultivation process by providing data from numerous sensors. Other sensors that can assist the growth of microalgae (such as CO_2 sensors, nitrate, nitrite, and other nutrition sensors) should be included in the research to make them more thorough. Similarly, research should focus on microalgae cultivation cycles so that the microalgae's specific growth behavior may be tracked in real time (Rahmat et al., 2020). Maximum output is ensured by growing microalgae under optimal culture conditions. As a result, the important process parameters (such as temperature, nutrient concentration, pH, etc., variables) must be immediately measured and regulated using safe, in situ, online, and automatic approaches. Microalgae cultivation has evolved from small-scale cultivation to large-scale microalgae biorefinery plants, including open-air or large-scale outdoor closed culture systems, because of expanding industry and social needs (Banerjee and Ramaswamy, 2019). Online monitoring could play a key role in regulating and controlling biological processes. Microalgae cultivation is typically done on a massive scale, with numerous algae species being cultivated in some circumstances, making monitoring and control difficult. The basic parameters of cell concentration in microalgal culture must be determined. Offline determination approaches in the past have been time-consuming, labor-intensive, and even influenced by external influences. As a result, researchers have created in situ microscopes to make efficiency monitoring easier (Havlik et al., 2013). For online monitoring of microalgae culture, the microscope was integrated with an image processing algorithm. The essential data such as cell size, morphology, and count may be gathered automatically throughout the culture process without the need for a sampling step when the microscope was incorporated into the photobioreactor, saving both labor and time.

16.2.2 Internet of Things applied to the downstream processing

A global monitoring program to track phytoplankton composition, notably dangerous and toxic microalgae, has been initiated (Tsaloglou, 2016). Molecular techniques for phytoplankton identification are currently available, although samples must often be transported to specialized laboratories. To monitor toxic algae, research and development of a sort of biosensor for detection (or in situ detection) of toxic algae have gotten a lot of interest. Because the phytoplankton community consists of a diverse range of species, a biosensor made up of multiprobe chips is intriguing because it can quickly assess complex samples without the need for any culture phases. However, relatively few studies have been conducted on the development of sensors for detecting harmful microalgae, also known as gene sensors. They are mostly based on electrochemical detection methods (Reverté et al., 2016), with species-specific DNA/rRNA/RNA as the probe and a sandwich hybridization assay as the detection system (Medlin and Orozco, 2017). Electrochemical biosensors have the advantages of being simple, quick to respond, accurate, and sensitive. It can directly identify nucleic acids in complicated samples without the need for target purification or amplification (Orozco and Medlin, 2013). These semiautomated or automatic DNA electrochemical biosensor-based systems could be highly useful for detecting hazardous microalgae.

The turbidimeter is commonly used to monitor the concentration of microalgae and pollutants in the culture system (Benson et al., 2009); however, the turbidimeter cannot distinguish between the two, whereas the fluorescence sensor can. The turbidimeter's main purpose is to measure the pollutants in the water because the fluorescence sensor cannot detect nonfluorescent pollutants, resulting in a complementary link when the two are coupled.

Microalgae are used to trace toxic compounds in the environment due to their high sensitivity to environmental changes and low detection limit (Boron et al., 2020; Buyong, 2019; Han, 2019; Hurtado-Gallego, 2019; O'Neill et al., 2019. Changes in photosynthetic or metabolic activity have been employed by microalgae in the biosensor business to indicate pollution-related information such as metal ions (Kashem et al., 2019; Roxby et al., 2020).

16.2.3 Robotized microalgae cells

A modern IoT is made up of a number of sensor networks as well as actuation systems. Magnetically controlled bacterial robots, nanomotors, bubble microrobots, DNA nanorobots, soft robots, physiologically activated spiral robots, and liquid metal nanorobots are examples of microrobots, which are made of nanoparticles or molecular materials and range in size from microns to millimeters (Wang et al., 2022). To run robots and complete complex tasks, it is necessary to deliver electricity to them and interface with them. Typical electrical motors, which are powered by wires or batteries, are used to power the bulk of normal-sized robots. To communicate with the robot, you can use a signal line or a wireless channel.

However, for the microrobots discussed earlier, it is surely a huge issue to provide them with minimal energy storage devices that adapt to them and include sufficient capacitance and sophisticated components for physical cables or wireless communication (Persson et al., 2013). The researchers intend to solve this challenge by developing a fully autonomous micro-robot that can automatically gather energy from its surroundings. As a result, algae robots have piqued the interest of scientists. Microalgal cells can sense and move in the microdimension, making them like microrobots in some ways. The robotized microalgae cells can be utilized to convey "goods" accurately, such as drug particles in microfluidic chips, and in biomedical processing. The microalgae microrobot (automation under IoT) can be used in DSP and have applications in post-harvesting.

16.3 Machine learning in microalgae cultivation and harvesting

Traditional data analysis frequently adopts a static data model when dealing with rapidly changing and unstructured data, which has downsides. It is not uncommon to swiftly figure out the relationship between hundreds of sensor inputs and external events that produce millions of data points. ML is the study of how computers duplicate or actualize human learning behavior to acquire new information or abilities and reorganize current knowledge structures for better performance. ML is quickly becoming a standard function of IoT devices. Fundamentally, the IoT provides enough data for ML, and IoT databases can enhance the ML algorithm. The IoT-enabled ML algorithm training addresses specific system needs such as analysis, monitoring, and prediction. Sensor data can be thought of as a one-dimensional network sampled at regular intervals, whereas picture data can be thought of as a two-dimensional network made up of pixels. Researchers have utilized ML, particularly deep learning, for the production and harvesting of microalgae in a novel way.

The most common approach for assessing the viability of microalgae cells in microalgae culture is to use machine performance. Although machines can analyze a huge number of cells in a short amount of time, effective machine operation and analysis methodologies are essential. As a result, the researchers automated the process by including ML processing (such as flow cytometry reading) into the relevant measuring procedure (Ochiai et al., 2020). Pozzobon et al. (2020) employed a clustering approach to examine the viability of microalgae using ML (Ester et al., 1996). The goal of ML processing is to automatically distinguish between living, dead, and debris cells. Semisupervised and active learning based on Gaussian mixture models can also be used to classify microalgae (Drews et al., 2013). The development of models and the implementation of advanced control mechanisms are critical.

Biofuels have been regarded as the most promising alternative to fossil fuels in relation to the change of energy instability and scarcity. Microalgae biomass energy is seen as a critical component of the world's energy structure's strategic transition in the 21st century.

The creation of microalgae biofuel is a viable solution to the energy crisis and environmental issues, with the added benefit of being environment-friendly and long-lasting. The traditional methods of analysis, on the other hand, could only identify the average lipid content of the entire population. There is yet to be created a noninvasive, noninterfering, single cell–based high-resolution analytical technique for scanning a microbial community. Guo et al. (2017) used time-stretching QPM (Quantitative Phase Microscopy) as well as image processing equipment equipped with ML and high-throughput capacity to screen lipid-producing microalgae cells to tackle this challenge. Microalgal species appropriate for biodiesel generation were examined (Guo et al., 2017).

ML has the potential to be a useful analytical tool in the development of microalgae-based biofuels. Combining optofluidics with ML algorithms to screen cells is a high-accuracy method. Researchers are encouraged to demonstrate advanced ML techniques such as deep learning to continuously enhance accuracy.

ML especially, deep learning or neural networks, has recently become more flexible for model establishment in different disciplines, as computing power has improved (Amini and Chang, 2018). These algorithms can create models only based on data, eliminating the requirement for an explicit program (Kim et al., 2017). Using this technique, Otálora et al. (2020) suggested a "black box" model for the pH of a raceway reactor (Zhang and Zaiane, 2017).

16.4 Artificial intelligence in microalgae genetic engineering

A gene is a basic physical unit of congenital features made up of DNA that provides a formula for making protein molecules to construct a microalgae cell. Current research will focus on the identified genome network and modified gene toward the accumulation of lipids, polysaccharides, and hydrocarbons in the microalgae cell (Radakovits et al., 2010; Teng et al., 2020; Ryan Georgianna and Mayfield, 2012). Genome sequencing is the key to understanding the gene's operating mechanism. Microalgae research has recently used integrated next-generation sequencing equipment that does amplification, genome sequencing, and data processing (An et al., 2018). Gene expression combined with artificial intelligence (AI) algorithms can be utilized to aid in the identification process, enhance accuracy, and automate it (Chuai et al., 2018; Teng et al., 2020; Lin and Wong, 2018). The distinction between traditional statistical methods and ML methods for genome sequencing is that ML requires less information about the sequencing data and can extract a large number of characteristics from the sequence (Bzdok et al., 2018).

AI algorithms can help with genetic modification technologies like RNAi, ZFNs, TALENs, and CRISPR-Cas9 to improve the selectivity and yield of microalgae nutritional compounds as biomolecules. In *Chlamydomonas reinhardtii*, ZFN-mediated gene editing was used to boost lipid production (Sizova et al., 2013). In *Phaeodactylum tricornutum*, TALENs-based nucleases were employed for targeted genome editing to boost nutritious components (Weyman et al., 2015). In the microalgae *P. tricornutum*, CRISPR-Cas9 was employed as an effective and rapid approach for accomplishing stable gene editing (Teng et al., 2020; Nymark et al., 2016). With recent advancements in innovative gene-editing technologies, the potential of modifying microalgae to have specific traits for specific applications is becoming a reality. Altering the genomes of microalgae can change metabolic pathways, resulting in the higher production of lipids, biomass, and other components. The potential of genetically optimized microalgae can have a "domino effect," allowing for further supply chain optimization in areas such as cultivation, processing, system design, process integration, and breakthrough products. However, because microalgae genome sequences are extensive and complex, the current degree of comprehending the functional information of diverse microalgae gene sequences is still primitive and insufficient. AI, in this perspective, has the potential to bridge the knowledge gap between microalgae genetic information and ideal bioproducts. Large and complex data from microalgae research may now be appropriately examined by combining the cutting-edge of both domains, thanks to the recent acceleration of AI research.

16.5 Conclusions

Complex industrial processes are now automated and simplified, thanks to modern technology such as AI, ML, and the IoT. The design of an integrated microalgae biorefinery system to promote the cooperative production of biofuels, high-value products, and industrial chemicals from biomass is required for the successful commercialization of the microalgae bioindustry. The use of IoT in the microalgae biorefinery business is advantageous in terms of obtaining a greener, more intelligent, automated, and a low-cost technical way for achieving people's relevant production goals. AI algorithms can extract important knowledge and predict molecular interactions in gene sequencing and editing. Recent advances in computer vision and ML algorithms have enabled accurate strain species screening and categorization, resulting in high-quality microalgae images that may be used for further investigation. Novel AI-enabled gene-editing tools enable application-specific genome modification in microalgae, resulting in higher yields in lipids, biomass, and other components.

Devices that are connected to the Internet, such as microcontrollers, sensors, wireless devices, and actuators, are frequently located in inaccessible or hard-to-reach areas of the IoT environment. A single-button battery must last years on these remote-controlled devices. Battery-powered IoT devices would have a negative impact on reducing environmental footprint (Nižetić et al., 2020; Wang et al., 2022). Finding ways to extend the tiny battery's power supply period is therefore crucial.

References

Amini, M., Chang, S., 2018. "A review of machine learning approaches for high dimensional process monitoring." In: IISE Annual Conference and Expo 2018. United States: Institute of Industrial and Systems Engineers, IISE.

An, S.M., et al., 2018. Next-generation sequencing reveals the diversity of benthic diatoms in tidal flats. Algae 33 (2), 167–180. Available from: https://doi.org/10.4490/algae.2018.33.4.3. South Korea: Korean Society of Phycology.

Atzori, L., Iera, A., Morabito, G., 2010. The Internet of Things: a survey. Computer Networks 54 (15), 2787–2805. Available from: https://doi.org/10.1016/j.comnet.2010.05.010. Italy.

Banerjee, S., Ramaswamy, S., 2019. Dynamic process model and economic analysis of microalgae cultivation in flat panel photobioreactors. Algal Research 39, 101445. Available from: https://doi.org/10.1016/j.algal.2019.101445. Elsevier BV.

Benson, B.C., Gutierrez-Wing, M.T., Rusch, K.A., 2009. Optimization of the lighting system for a Hydraulically Integrated Serial Turbidostat Algal Reactor (HISTAR): economic implications. Aquacultural Engineering 40 (1), 45–53. Available from: https://doi.org/10.1016/j.aquaeng.2008.11.001. Elsevier BV.

Bernard, O., Mairet, F., Chachuat, B., 2016. Modelling of microalgae culture systems with applications to control and optimization. Advances in Biochemical Engineering/Biotechnology 153, 59–87. Available from: https://doi.org/10.1007/10_2014_287. France: Springer Science and Business Media Deutschland GmbH.

Boron, I., Juárez, A., Battaglini, F., 2020. Portable microalgal biosensor for herbicide monitoring. ChemElectroChem 7 (7), 1623–1630. Available from: https://doi.org/10.1002/celc.202000210. Argentina: Wiley-VCH Verlag.

Brennan, L., Owende, P., 2010. Biofuels from microalgae—a review of technologies for production, processing, and extractions of biofuels and co-products. Renewable and Sustainable Energy Reviews 14 (2), 557–577. Available from: https://doi.org/10.1016/j.rser.2009.10.009. Ireland.

Buyong, M., et al., 2019. Dielectrophoresis manipulation: versatile lateral and vertical mechanisms. Biosensors 9 (1), 30. Available from: https://doi.org/10.3390/bios9010030. MDPI AG.

Bzdok, D., Altman, N., Krzywinski, M., 2018. Statistics versus machine learning. Nature Methods 15 (4), 233–234. Available from: https://doi.org/10.1038/nmeth.4642. Springer Science and Business Media LLC.

Chew, K.W., et al., 2017. Microalgae biorefinery: high value products perspectives. Bioresource Technology 229, 53–62. Available from: https://doi.org/10.1016/j.biortech.2017.01.006. Elsevier BV.

Christian Barbosa, R., Soares, J., Arêdes Martins, M., 2020. Low-cost and versatile sensor based on multi-wavelengths for real-time estimation of microalgal biomass concentration in open and closed cultivation systems. Computers and Electronics in Agriculture 176, 105641. Available from: https://doi.org/10.1016/j.compag.2020.105641. Elsevier BV.

Chuai, G., et al., 2018. DeepCRISPR: optimized CRISPR guide RNA design by deep learning. Genome Biology 19 (1). Available from: https://doi.org/10.1186/s13059-018-1459-4. China: BioMed Central Ltd.

Drews, P., et al., 2013. Microalgae classification using semi-supervised and active learning based on Gaussian mixture models. Journal of the Brazilian Computer Society 19 (4), 411–422. Available from: https://doi.org/10.1007/s13173-013-0121-y. Brazil: Springer London.

Ester, M., et al., 1996. "A Density-based algorithm for discovering clusters in large spatial databases with noise." In: Proceedings of 2nd International Conference on Knowledge Discovery and Data Mining (KDD-96).

Fabris, M., et al., 2020. Emerging technologies in algal biotechnology: toward the establishment of a sustainable, algae-based bioeconomy. Frontiers in Plant Science 11. Available from: https://doi.org/10.3389/fpls.2020.00279. Australia: Frontiers Media S.A.

Guo, B., et al., 2017. High-throughput, label-free, single-cell, microalgal lipid screening by machine-learning-equipped optofluidic time-stretch quantitative phase microscopy. Cytometry Part A 91 (5), 494–502. Available from: https://doi.org/10.1002/cyto.a.23084. Wiley.

Han, S., et al., 2019. A digital microfluidic diluter-based microalgal motion biosensor for marine pollution monitoring. Biosensors and Bioelectronics 143, 111597. Available from: https://doi.org/10.1016/j.bios.2019.111597. Elsevier BV.

Havlik, I., et al., 2013. Monitoring of microalgal cultivations with on-line, flow-through microscopy. Algal Research 2 (3), 253–257. Available from: https://doi.org/10.1016/j.algal.2013.04.001. Germany.

Hermadi, I., et al., 2021. Development of smart algae pond system for microalgae biomass production. IOP Conference Series: Earth and Environmental Science. Available from: https://doi.org/10.1088/1755-1315/749/1/012068. Indonesia: IOP Publishing Ltd.

Hessel, J., Metfies, K., 2014. "Molecular sensor-based monitoring of toxic algae." In: 5th Early Career Scientist Conference (ECC). Bremen MARUM.

Hurtado-Gallego, J., et al., 2019. Luminescent Microbial Bioassays and Microalgal Biosensors as Tools for Environmental Toxicity Evaluation. Springer Science and Business Media LLC, pp. 1–58. Available from: http://doi.org/10.1007/978-3-319-47405-2_89-1.

Jia, F., Kacira, M., Ogden, K., 2015. Multi-wavelength based optical density sensor for autonomous monitoring of microalgae. Sensors 15 (9), 22234–22248. Available from: https://doi.org/10.3390/s150922234. MDPI AG.

Kashem, M.A., et al., 2019. Development of microalgae biosensor chip by incorporating microarray oxygen sensor for pesticides sensing. Biosensors 9 (4). Available from: https://doi.org/10.3390/bios9040133. Japan: MDPI.

Khoo, C.G., Lam, M.K., Lee, K.T., 2019a. Microscale and macroscale modeling of microalgae cultivation in photo-bioreactor: a review and perspective. Advances in Feedstock Conversion Technologies for Alternative Fuels and Bioproducts: New Technologies, Challenges and Opportunities. Elsevier, Malaysia, pp. 1–19. Available from: http://doi.org/10.1016/B978-0-12-817937-6.00001-1.

Khoo, K.S., et al., 2019b. Recent advances in biorefinery of astaxanthin from *Haematococcus pluvialis*. Bioresource Technology 288. Available from: https://doi.org/10.1016/j.biortech.2019.121606. Malaysia: Elsevier Ltd.

Khoo, K.S., Chew, K.W., et al., 2020a. Recent advances in downstream processing of microalgae lipid recovery for biofuel production. Bioresource Technology 304. Available from: https://doi.org/10.1016/j.biortech.2020.122996. Malaysia: Elsevier Ltd.

Khoo, K.S., Chia, W.Y., et al., 2020b. Nanomaterials utilization in biomass for biofuel and bioenergy production. Energies 13 (4). Available from: https://doi.org/10.3390/en13040892. Malaysia: MDPI AG.

Kim, B.S., et al., 2017. Data modeling versus simulation modeling in the big data era: case study of a greenhouse control system. Simulation 93 (7), 579–594. Available from: https://doi.org/10.1177/0037549717692866. South Korea: SAGE Publications Ltd.

Lin, J., Wong, K.C., 2018. Off-target predictions in CRISPR-Cas9 gene editing using deep learning. Bioinformatics. Oxford University Press, Hong Kong. Available from: http://doi.org/10.1093/bioinformatics/bty554.

Medlin, L., Orozco, J., 2017. Molecular techniques for the detection of organisms in aquatic environments, with emphasis on harmful algal bloom species. Sensors 17 (5), 1184. Available from: https://doi.org/10.3390/s17051184. MDPI AG.

Moi, P.S., 2010. Handbook of microalgal culture: biotechnology and applied phycology. Journal of Phycology 40, 1001–1002.

Moraleda, G.O., Victoria L.R., Eduardo C.C., David H.F., Emilia M.P., 2010. Biosensors based on microalgae for the detection of environmental pollutants. U.S. Patent Application 12/667,590. https://patents.google.com/patent/US20100248286A1/en.

Nižetić, S., et al., 2020. Internet of Things (IoT): opportunities, issues and challenges towards a smart and sustainable future. Journal of Cleaner Production 274, 122877. Available from: https://doi.org/10.1016/j.jclepro.2020.122877. Elsevier BV.

Nymark, M., et al., 2016. A CRISPR/Cas9 system adapted for gene editing in marine algae. Scientific Reports 6. Available from: https://doi.org/10.1038/srep24951. Norway: Nature Publishing Group.

Ochiai, H., et al., 2020. Application of machine learning-driven label-free flow cytometry to analyze T cell products. Cytotherapy 22 (5), S132–S133. Available from: https://doi.org/10.1016/j.jcyt.2020.03.259. Elsevier BV.

O'Neill, E.A., Rowan, N.J., Fogarty, A.M., 2019. Novel use of the alga *Pseudokirchneriella subcapitata*, as an early-warning indicator to identify climate change ambiguity in aquatic environments using freshwater finfish farming as a case study. Science of the Total Environment 692, 209–218. Available from: https://doi.org/10.1016/j.scitotenv.2019.07.243. Ireland: Elsevier B.V.

Orozco, J., Medlin, L.K., 2013. Review: advances in electrochemical genosensors-based methods for monitoring blooms of toxic algae. Environmental Science and Pollution Research 20 (10), 6838–6850. Available from: https://doi.org/10.1007/s11356-012-1258-5. United States.

Otálora, P., et al., 2020. Dynamic Model for the pH in a Raceway Reactor Using Deep Learning Techniques. Springer Science and Business Media LLC, pp. 190–199. Available from: http://doi.org/10.1007/978-3-030-58653-9_18.

Perkel, J.M., 2017. The Internet of Things comes to the lab. Nature 542 (7639), 125–126. Available from: https://doi.org/10.1038/542125a. undefined: Nature Publishing Group.

Persson, M., et al., 2013. Transportation of nanoscale cargoes by myosin propelled actin filaments. PLoS One 8 (2), e55931. Available from: https://doi.org/10.1371/journal.pone.0055931. Public Library of Science (PLoS).

Pozzobon, V., et al., 2020. Machine learning processing of microalgae flow cytometry readings: illustrated with *Chlorella vulgaris* viability assays. Journal of Applied Phycology 32 (5), 2967–2976. Available from: https://doi.org/10.1007/s10811-020-02180-7. France: Springer Science and Business Media B.V.

Radakovits, R., et al., 2010. Genetic engineering of algae for enhanced biofuel production. Eukaryotic Cell 9 (4), 486–501. Available from: https://doi.org/10.1128/EC.00364-09. United States.

Rahmat, A., et al., 2020. Evaluation of system performance for microalga cultivation in photobioreactor with IOTs (Internet of Things). International Journal of Sciences: Basic and Applied Research (IJSBAR) 49, 95–107.

Reverté, L., Prieto-Simón, B., Campàs, M., 2016. New advances in electrochemical biosensors for the detection of toxins: nanomaterials, magnetic beads and microfluidics systems. A review. Analytica Chimica Acta 908, 8–21. Available from: https://doi.org/10.1016/j.aca.2015.11.050. Spain: Elsevier B.V.

Roxby, D.N., et al., 2020. Microalgae living sensor for metal ion detection with nanocavity-enhanced photoelectro-chemistry. Biosensors and Bioelectronics 165. Available from: https://doi.org/10.1016/j.bios.2020.112420. Singapore: Elsevier Ltd.

Ryan Georgianna, D., Mayfield, S.P., 2012. Exploiting diversity and synthetic biology for the production of algal biofuels. Nature 488 (7411), 329–335. Available from: https://doi.org/10.1038/nature11479. United States.

Salas-Herrera, G., et al., 2019. Impact of microalgae culture conditions over the capacity of copper nanoparticle biosynthesis. Journal of Applied Phycology 31 (4), 2437–2447. Available from: https://doi.org/10.1007/s10811-019-1747-8. Mexico: Springer Netherlands.

Shin, Y.H., et al., 2015. A portable fluorescent sensor for on-site detection of microalgae. Microelectronic Engineering 144, 6–11. Available from: https://doi.org/10.1016/j.mee.2015.01.005. United States: Elsevier.

Sizova, I., et al., 2013. Nuclear gene targeting in *Chlamydomonas* using engineered zinc-finger nucleases. The Plant Journal 73 (5), 873–882. Available from: https://doi.org/10.1111/tpj.12066. Wiley.

Sutherland, D.L., et al., 2015. Enhancing microalgal photosynthesis and productivity in wastewater treatment high rate algal ponds for biofuel production. Bioresource Technology 184, 222–229. Available from: https://doi.org/10.1016/j.biortech.2014.10.074. New Zealand: Elsevier Ltd.

Tan, X.B., et al., 2018. Cultivation of microalgae for biodiesel production: a review on upstream and downstream processing. Chinese Journal of Chemical Engineering 26 (1), 17–30. Available from: https://doi.org/10.1016/j.cjche.2017.08.010. Malaysia: Chemical Industry Press.

Teng, S.Y., et al., 2020. Microalgae with artificial intelligence: a digitalized perspective on genetics, systems and products. Biotechnology Advances 44. Available from: https://doi.org/10.1016/j.biotechadv.2020.107631. Czech Republic: Elsevier Inc.

Tsaloglou, M.-N., 2016. Microalgae: Current Research and Applications. Caister Academic Press.

Wang, K., et al., 2022. How does the Internet of Things (IoT) help in microalgae biorefinery? Biotechnology Advances 54. Available from: https://doi.org/10.1016/j.biotechadv.2021.107819. Malaysia: Elsevier Inc.

Weyman, P.D., et al., 2015. Inactivation of *Phaeodactylum tricornutum* urease gene using transcription activator-like effector nuclease-based targeted mutagenesis. Plant Biotechnology Journal 13 (4), 460–470. Available from: https://doi.org/10.1111/pbi.12254. United States.

Whitmore, A., Agarwal, A., Da Xu, L., 2015. The Internet of Things—a survey of topics and trends. Information Systems Frontiers 17 (2), 261–274. Available from: https://doi.org/10.1007/s10796-014-9489-2. United States: Kluwer Academic Publishers.

Yong, J.J.J.Y., et al., 2021. Prospects and development of algal-bacterial biotechnology in environmental management and protection. Biotechnology Advances 47. Available from: https://doi.org/10.1016/j.biotechadv.2020.107684. Malaysia: Elsevier Inc.

Zhang, S., Zaiane, O.R., 2017. "Comparing deep reinforcement learning and evolutionary methods in continuous control." arXiv. undefined: arXiv. Available at: https://arxiv.org.

Index

Note: Page numbers followed by "*f*" and "*t*" refer to figures and tables respectively.

CPI Antony Rowe
Eastbourne, UK
October 18, 2023